Practical Machinery Management for Process Plants

Volume 2: Machinery Failure
Analysis and Troubleshooting

Gulf Publishing Company
Book Division
Houston, London, Paris, Tokyo

Practical Machinery Management for Process Plants

Volume 2: Machinery Failure Analysis and Troubleshooting

Heinz P. Bloch and
Fred K. Geitner

Dedication

To Karen and Ken Bloch,
Peter and Derek Geitner

—As encouragement to go the
full distance, and then the
extra mile. You will not encounter
any traffic jams on your way.

(Freely quoted from Ecclesiastes *9:10)*

Practical Machinery Management
for Process Plants

Volume 2: Machinery Failure
Analysis and Troubleshooting

Library of Congress Cataloging in Publication Data

Bloch, Heinz P.
  Machinery failure analysis and troubleshooting.

  (Practical machinery management for process plants; v. 2)
  Bibliography: p.
  Includes index.
  1. Machinery—Maintenance and repair. 2. Plant maintenance. I. Geitner, F. K. II. Title. III. Series.
  TS191.B56 1983   670'.28'8   83-10731
  ISBN 0-87201-872-5

Acknowledgments

An experienced machinery engineer usually has a few file cabinets filled with technical reports, course notes, failure reports, and a host of other machinery-related data. But these files are rarely complete enough to illustrate *all* bearing failure modes, *all* manners of gear distress, etc. Likewise, we may have taken problem-solving courses, but cannot lay claim to recalling all the mechanics of problem-solving approaches without going back to the formal literature.

Recognizing these limitations, we went to some very knowledgeable companies and individuals and requested permission to use certain of their source materials for portions of this book. We gratefully acknowledge the help and cooperation we received from:

American Society of Lubrication Engineers, Park Ridge, Illinois (ASLE Paper 83-AM-1B-2, Bloch/Plant-Wide Turbine Lube Oil Reconditioning and Analysis).

American Society of Mechanical Engineers, New York, New York (*Proceedings of 38th ASME Petroleum Mechanical Engineering Workshop,* Bloch/Setting Up a Pump Failure Reduction Program).

American Society for Metals, Metals Park, Ohio (Analysis of Shaft Failures, etc.).

American Gear Manufacturers Association, Arlington, Virginia (Gear Failures, etc.).

Bently Nevada Corporation, Minden, Nevada (Shaft Cracking Analyzed by Proximity Probes).

Beta Machinery Analysis, Ltd. Calgary, Canada (Problem Analysis on Reciprocating Machinery).

Crane Packing Company, Morton Grove, Illinois (Mechanical Seal Distress).

Glacier Metal Company, Ltd., Alperton/Middlesex, England (Journal and Tilt-Pad Bearing Failure Analysis).

T. J. Hansen Company, Dallas, Texas (Generalized Problem-Solving Approaches).

IRD Mechanalysis, Inc., Columbus, Ohio (Vibration Analysis and Vibration Pattern Identification).

Dr. Wayne Nelson, General Electric Company, Schenectady, New York and American Society for Quality Control (Statistical Methods of Failure Analysis and Hazard Plotting).

Rome Air Development Center, Rome, New York (Sneak Analysis Methods).

SKF Industries, Inc., King of Prussia, Pennsylvania (Bearing Distress—Recognition and Problem Solving).

Brian Turner, Brega, Libya (Distinguishing Between Bad Repairs and Bad Designs).

We are certainly indebted to Exxon Chemical Americas and Esso Chemicals Canada for allowing us to publish this book. Our special thanks to Michael J. Streitman, Technical Division Manager, Baytown Olefins Plant, and Jackson Douglas, Engineering Manager (ret.), Esso Chemicals Toronto, Canada, for their commitment to support personal growth and learning. Finally, we express appreciation to our editor, Scott Becken, for patiently putting up with our technical jargon and occasional changes of mind.

Contents

fication. Six Lube-Oil Analyses Are Required. Periodic Sampling and Conditioning Routines Implemented. Lube-Oil Analysis Results. Calculated Benefit-to-Cost Ratio. Grease Failure Analysis. References.

Preface

The prevention of potential damage to machinery is necessary for safe, reliable operation of process plants. Failure prevention can be achieved by sound specification, selection, review, and design audit routines. When failures do occur, accurate definition of the root cause is an absolute prerequisite to the prevention of future failure events.

This book concerns itself with proven approaches to failure definition. It presents a liberal cross section of documented failure events and analyzes the procedures employed to define the sequence of events that led to component or systems failure. Because it is simply impossible to deal with every conceivable type of failure, this book is structured to teach failure identification and analysis methods which can be applied to virtually all problem situations that might arise. A uniform methodology of failure analysis and troubleshooting is necessary because experience shows that all too often process machinery problems are never defined sufficiently; they are merely "solved" to "get back on stream." Production pressures often override the need to analyze a situation thoroughly, and the problem and its underlying cause come back and haunt us later.

Equipment downtime and component failure risk can be reduced only if potential problems are anticipated and avoided. Often, this is not possible if we apply only traditional methods of analysis. It is thus appropriate to employ other means of precluding or reducing consequential damage to plant, equipment, and personnel. This objective includes, among others, application of redundant components or systems and application of sneak circuit analysis techniques for electrical/electronic systems.

The organizational environment and management style found in process plants often permits a "routine" level of machinery failures and breakdowns. This book shows how to arrive at a uniform method of assessing what level of failure experience should be considered acceptable and achievable. In addition, it shows how the organizational environment can be better prepared to address the task of thorough machinery failure analysis and troubleshooting, with resulting maintenance incident reduction. Finally, by way of successful examples, this book demonstrates how the progress and results of failure analysis and troubleshooting efforts can be documented and thus monitored.

H. P. Bloch
Houston, Texas

F. K. Geitner
Sarnia, Ontario

Chapter 1

The Failure Analysis and Troubleshooting System

Troubleshooting as an Extension of Failure Analysis

For years, the term "failure analysis" has had a specific meaning in connection with fracture mechanics and corrosion failure analysis activities carried out by static process equipment inspection groups. Figure 1-1 shows a basic outline of materials failure analysis steps.[1] The methods applied in our context of process machinery failure analysis are basically the same; however, they are not limited to metallurgic investigations. Here, failure analysis is the determination of failure modes of machinery components and their most probable causes. Figure 1-2 illustrates the general significance of machinery component failure mode analysis as it relates to quality, reliability, and safety efforts in the product development of a major turbine manufacturer.[2]

Very often, machinery failures reveal a reaction chain of cause and effect. The end of the chain is usually a performance deficiency commonly referred to as the symptom, trouble, or simply "the problem." Troubleshooting works backward to define the elements of the reaction chain and then proceeds to link the most probable failure cause based on failure (appearance) analysis with a root cause of an existing or potential problem. For all practical purposes, failure analysis and troubleshooting activities will quite often mesh with one another without any clear-cut transition.

However, as we will see later, there are numerous cases where troubleshooting alone will have to suffice to get to the root cause of the problem. These are the cases that present themselves as performance deficiencies with no apparent failure modes. Intermittent malfunctions and faults are typical examples and will tax even the most experienced troubleshooter. In these cases, troubleshooting will be successful only if the investigator knows the system he is dealing with. Unless he is thoroughly familiar with component interaction, operating or failure modes, and functional characteristics, his efforts may be unsuccessful.

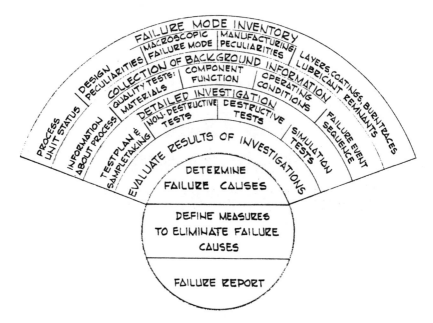

Figure 1-1. Failure analysis steps—materials technology (modified from Ref. 1).

There are certain objectives of machinery failure analysis and troubleshooting:

1. Prevention of future failure events.
2. Assurance of safety, reliability, and maintainability of machinery as it passes through its life cycles of:
   a. Process design and specification.
   b. Original equipment design and manufacture.
   c. Shipping and storage.
   d. Installation and commissioning.
   e. Operation and maintenance.
   f. Replacement.

From this it becomes very obvious that failure analysis and troubleshooting are highly co-operative processes. Because many different parties will be involved and their objectives will sometimes differ, a systematic and uniform description and understanding of process machinery failure events is important.

Causes of Machinery Failures

In its simplest form, failure can be defined as any change in a machinery part or component which causes it to be unable to perform its intended function

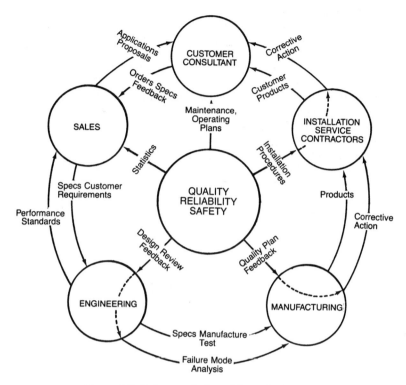

Figure 1-2. Failure analysis and the "wheel of quality."[2]

satisfactorily. Familiar stages preceding final failure are "incipient failure," "incipient damage," "distress," "deterioration," and "damage," all of which eventually make the part or component unreliable or unsafe for continued use.
  Meaningful classifications of failure causes are:

1. Design deficiencies.
2. Material defects.
3. Processing and manufacturing deficiencies.
4. Assembly errors.
5. Off-design or unintended service conditions.
6. Maintenance deficiencies (neglect, procedures).
7. Improper operation.

All statistics and references dealing with machinery failures, their sources and causes, generally use these classifications.

For practical failure analysis, an expansion of this list seems necessary. Table 1-1 shows a representative collection of process machinery failure causes. The table makes it clear that failure causes should be allocated to areas of responsibilities. If this allocation is not made, the previously listed objectives of most failure analyses will probably not be met.

Failure causes are usually determined by relating them to one or more specific failure modes. This becomes the central idea of any failure analysis activity. Failure mode (FM) is the apprearance, manner, or form in which a machinery component or unit failure manifests itself. Table 1-2 lists the basic failure modes encountered in 99 percent of all petrochemical process plant machinery failures.

In the following sections, this list will be expanded so that it can be used for day-to-day failure analysis. Failure mode should not be confused with failure cause, as the former is the *effect* and the latter is the *cause* of a failure event. Failure mode can also be the result of a long chain of causes and effects, ultimately leading to a functional failure, i.e a symptom, trouble, or operational complaint pertaining to a piece of machinery equipment as an entity.

Other terms frequently used in the preceding context are "kind of defect," "defect," or "failure mechanism." The term "failure mechanism" is often described as the metallurgical, chemical, and tribological processes leading to a particular failure mode. For instance, failure mechanisms have been developed to describe the chain of cause and effect for fretting wear (FM) in roller bearing assemblies, cavitation (FM) in pump impellers, and initial pitting (FM) on the surface of a gear tooth, to name a few. The basic agents of machinery component and part failure mechanisms are *always* force, time, temperature, and a reactive environment. Each of these can be subdivided as indicated in Table 1-3.

For our purpose, failure mechanisms thus defined will have to stay part of the failure mode definition: They will tell how and why a failure mode might have occurred in chemical or metallurgical terms, but in so doing, the root cause of the failure will remain undefined.

Root Causes of Machinery Failure

The preceding pages have shown us that there will always be a number of causes and effects in any given failure event. We need to arrive at a practical point—if not all the way to the beginning—of the cause and effect chain where removal or modification of contributing factors will solve the problem.

A good example would be scuffing (FM) as one of the major failure modes of gears. It is a severe form of adhesive wear (FM) with its own well-defined failure mechanism. Adhesive wear cannot occur if a sufficiently thick oil film separates the gear tooth surfaces. This last sentence—even though there is a long chain of cause and effect hidden in the adhesive wear failure mechanism—will give us the clue as to the root cause. What then is the root cause? We know that scuffing usually occurs quite suddenly, in contrast to the time-dependent failure mode of pitting. Thus, we cannot look for the root cause in the design of the lube oil system or in the lube oil itself—that is, if scuffing was not observed before on that particular gear set. Sudden and intermittent loss of lubrication could be the cause. Is it the root cause? No, we

Table 1-1
Causes of Failures

**Design and Specification
Responsibility**

Application
- [ ] Undercapacity
- [ ] Overcapacity
- [ ] Incorrect physical conditions
  (temperature pressure, etc.)
- [ ] Incorrect physical prop.
  (mol. wt., etc.)
- [ ] _____
- [ ] _____

Specifications
- [ ] Inadequate lubrication system
- [ ] Insufficient control instrumentation
- [ ] Improper coupling
- [ ] Improper bearing
- [ ] Improper seal
- [ ] Insufficient shutdown devices
- [ ] _____
- [ ] _____

Material of Construction
- [ ] Corrosion and/or erosion
- [ ] Rapid wear
- [ ] Fatigue
- [ ] Strength exceeded
- [ ] Galling
- [ ] Wrong hardening method
- [ ] _____
- [ ] _____

Design
- [ ] Unsatisfactory piping support
- [ ] Improper piping flexibility
- [ ] Undersized piping
- [ ] Inadequate foundation
- [ ] Unsatisfactory soil data
- [ ] Liquid ingestion
- [ ] Inadequate liquid drain
- [ ] Design error
- [ ] _____
- [ ] _____

Vendor Responsibility

Material of Construction
- [ ] Flaw or defect
- [ ] Improper material
- [ ] Improper treatment
- [ ] _____

Design
- [ ] Improper specification
- [ ] Wrong selection
- [ ] Design error
- [ ] Inadequate or wrong lubrication
- [ ] Inadequate liquid drain
- [ ] Critical speed
- [ ] Inadequate strength
- [ ] Inadequate controls and protective devices
- [ ] _____

Fabrication
- [ ] Welding error
- [ ] Improper heat treatment
- [ ] Improper hardness
- [ ] Wrong surface finish
- [ ] Imbalance
- [ ] Lub. passages not open
- [ ] _____

Assembly
- [ ] Improper fit
- [ ] Improper tolerances
- [ ] Parts omitted
- [ ] Parts in wrong
- [ ] Parts/bolts not tight
- [ ] Poor alignment
- [ ] Imbalance
- [ ] Inadequate bearing contact
- [ ] Inadequate testing
- [ ] _____

Shipping and Storage Responsibility

Preparation for Shipment
- [ ] Oil system not clean
- [ ] Inadequate drainage
- [ ] Protective coating not applied
- [ ] Wrong coating used
- [ ] Equipment not cleaned
- [ ] _____

Protection
- [ ] Insufficient protection
- [ ] Corrosion by salt
- [ ] Corrosion by rain or humidity
- [ ] Poor packaging
- [ ] Dessicant omitted
- [ ] Contamination with dirt, etc.
- [ ] _____

Physical Damage
- [ ] Loading damage
- [ ] Transport damage
- [ ] Insufficient support
- [ ] Unloading damage
- [ ] _____

Installation Responsibility

Foundations
- [ ] Settling
- [ ] Improper or insufficient
  grouting
- [ ] Cracking or separating
- [ ] _____

Piping
- [ ] Misalignment
- [ ] Inadequate cleaning
- [ ] Inadequate support
- [ ] _____

Assembly
- [ ] Misalignment
- [ ] Assembly damage (crafts)

- [ ] Defective material
- [ ] Inadequate bolting
- [ ] Connected wrong
- [ ] Foreign material left in
- [ ] General poor workmanship
- [ ] _____

**Operations and Maintenance
Responsibility**

Shock
- [ ] Thermal
- [ ] Mechanical
- [ ] Improper startup
- [ ] _____

Operating
- [ ] Slugs of liquid
- [ ] Process surging
- [ ] Control error
- [ ] Controls deactivated
  or not put in service
- [ ] Operating error
- [ ] _____

Auxiliaries
- [ ] Utility failure
- [ ] Insufficient instrumentation
- [ ] Electronic control failure
- [ ] Pneumatic control failure
- [ ] _____

Lubrication
- [ ] Dirt in oil
- [ ] Insufficient oil
- [ ] Wrong lubricant
- [ ] Water in oil
- [ ] Oil pump failure
- [ ] Low oil pressure
- [ ] Plugged lines
- [ ] Improper filtration
- [ ] Contaminated oil
- [ ] _____

Craftsmanship
- [ ] Improper tolerances
- [ ] Welding error
- [ ] Improper surface finish
- [ ] Improper fit
- [ ] General poor workmanship
- [ ] _____

Assembly
- [ ] Mechanical damage
- [ ] Parts in wrong
- [ ] Parts omitted
- [ ] Misalignment
- [ ] Improper bolting
- [ ] Imbalance
- [ ] Piping stress
- [ ] Foreign material left in
- [ ] Wrong material of construction
- [ ] _____

(table continued on next page)

Table 1-1 (cont.)

Preventive Maintenance	☐ Seal	☐ Blades
☐ Postponed	☐ Coupling	☐ Blade root
☐ Schedule too long	☐ Shaft	☐ Blade shroud
☐ _____	☐ Pinion gear	☐ Labyrinth
	☐ Bull gear	☐ Thrust bearing
	☐ Turning gear	☐ Pivoted pad bearing
	☐ Casing	☐ Roller/ball bearing
Distress, Damages, or Failed Components	☐ Rotor	☐ Cross-head piston
	☐ Impeller	☐ Cylinder
☐ Vibration	☐ Shroud	☐ Crankshaft
☐ Short circuit	☐ Piston	☐ _____
☐ Open circuit	☐ Diaphragm	☐ _____
☐ Sleeve bearing	☐ Wheel	☐ _____

Comments:_____

_____

_____

_____

_____

_____

still have to find it because we are looking for the element that, if removed or modified, will prevent recurrence or continuation of scuffing. Is it because this particular plant is periodically testing their standby lube oil pumps, causing sudden and momentary loss of lube oil pressure? Eventually, we will arrive at a point where a change in design, operation, or maintenance practices will stop the gear tooth scuffing.

Removal of the root cause of machinery failures should take place in design and operations-maintenance. Quite often the latter, in its traditional form, is given too much emphasis in failure analysis and failure prevention. In our opinion, long-term reductions in failure trends will only be accomplished by specification and design modifications. We will see again in Chapter 7 that only design changes will achieve the required results. How then does this work? After ascertaining the failure mode, we determine whether or not the failed machinery component could be made more resistant to the failure event. This is done by checking design parameters such as the ones shown in Table 1-4 for possible modification. Once a positive answer has been obtained, the root cause has also been determined and we can specify whatever is required to impart less vulnerability to the material, component, assembly, or system. As we formulate our action plan, we will test whether the mechanic's axiom holds true:

> *When in doubt*
> *Make it stout*
> *Out of something*
> *You know about.*

Table 1-2
Machinery Failure Mode Classification

Deformation—i.e. plastic, elastic, etc.
Fracture—i.e. cracks, fatigue fracture, pitting, etc.
Surface changes—i.e. hairline cracks, cavitation, wear, etc.
Material changes—i.e. contamination, corrosion, wear, etc.
Displacement—i.e. loosening, seizure, excessive clearance, etc.
Leakage
Contamination

Table 1-3
Agents of Machinery Component and Part Failure Mechanisms

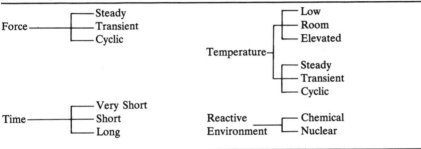

Table 1-4
Process Machinery Design Properties

Material-of-Construction Level

1. Material properties. i.e. ductility, creep resistance, heat resistance, etc.
2. Properties derived from processing, i.e. cast, rolled, forged, etc.
3. Properties resulting from heat treatment, i.e. not heat treated, hardened, stress relieved, etc.
4. Surface properties, i.e. machined, ground, lapped, etc.
5. Properties derived from corrosion and wear protection measures, i.e. overlayed, enameled, painted, etc.
6. Properties resulting from connecting method, i.e. welded, shrunk, rolled-in, etc.

Part and Component Level

7. Properties derived from shape and form, i.e. cylindrical, spherical, perforated, etc.

Part, Component, and Assembly Level

8. Suitability for service, i.e. prone to plugging, wear, vibration, etc.
9. Properties resulting from assembly type, i.e. riveted, pinned, bolted, etc.
10. Assembly quality, i.e. countersunk, flush, tight, locked, etc.

Table 1-5
Machinery Failure Modes—Process Plant

Column groups: **Deformation** (Yielding … Melting); **Fracture Separation** (Ductile Fracture … Bond Failure—Babbitt); **Surface/Material Changes** (Aging … Rubbing/seizing, including **Wear** with **Adhesive** sub-group: Fretting … Rubbing/seizing); **Displacement** (Jamming/blocking/sticking … Leakage—Contamination).

Group	Design Parameter Failure Resistance "Immediate" Cause	Yielding	Bending—Warping	Buckling	Brinelling	Creep—Bowing	Expansion/shrinking	Melting	Ductile Fracture	Brittle Fracture	Fatigue Fracture	Cracking	Bond Failure—Babbitt	Aging (bending/creep)	Rusting	Uniform Corrosion	Stress Corrosion	Pitting—Surface Fatigue	Spalling—Surface Fatigue	Cavitation Erosion	Erosion—Corrosion	Abrasive Wear	Fretting	Scuffing	Scoring	Galling	Rubbing/seizing	Jamming/blocking/sticking	Shifting	Exc. Clearance	Loosening	Leakage—Contamination
Table 1-4—Material Properties	Tensile Strength								●		●	●		●			●															
	Yield Strength	●																														
	Compressive Yield Strength			●														●	●			●										
	Shear Yield Strength	●							●																							
	Fatigue Properties										●							●	●			●										
	Ductility %RA								●	●	●	●																				
	Charpy Energy								●		●																					
	Modulus of Elasticity	●	●																													
	Creep Rate					●								●																		
	K$_{ic}$									●	●																					
	Erosion Resistance																				●											
	EMF Series															●					●	●										
	Hardness			●								●						●	●	●	●	●	●	●	●	●	●					●
	Coefficient of Thermal Expansion						●				●	●															●	●	●	●	●	●
	Melting Point							●																			●					
	Thermal Conductivity						●					●															●					
Table 1-4—Design Properties	Properties from Manufacturing Process								●	●	●	●	●				●	●	●			●	●	●	●			●	●	●		
	Properties from Corrosion Protection														●	●	●															
	Properties from Heat Treatment		●	●					●	●	●	●					●	●				●	●	●	●							
	Properties from Wear Protection																					●		●		●	●	●	●			
	Properties from Shape and Form																	●	●									●	●	●		
	Properties from Type and Quality of Assembly																										●	●	●	●	●	●
Table 1-3	Static Force	●	●	●	●				●		●	●	●				●													●	●	●
	Dynamic Force		●	●	●				●	●	●	●	●				●	●	●	●	●	●	●	●	●	●	●	●	●	●	●	●
	Coefficient of Sliding Friction						●															●	●	●	●			●	●			
	Time	●	●	●	●	●			●		●	●	●					●	●	●	●	●	●	●	●					●	●	●
	Temperature				●	●	●		●		●	●	●	●		●											●	●	●	●	●	●
	Chemical Environment											●		●	●	●	●				●				●							●

We will keep in mind our inability to influence machinery failures by simply making the part stronger in every conceivable situation. A flexibly designed component may, in some cases, survive certain severe operating conditions better than the rigid part.

Table 1-5 concludes this section by summarizing machinery failure modes as they relate to their immediate causes or design parameter deficiencies.

References

1. VDI Guidelines No. 3822., *Der Maschinenschaden*, Vol. 54, No. 4, 1981, p. 131.

2. Ludwig, G. A., "Tests Performed by the Builder on New Products to Prevent Failure", *Loss Prevention of Rotating Machinery*, The American Society of Mechanical Engineers, New York, N.Y. 10017, 1972, p. 3.

Chapter 2

Metallurgical Failure Analysis

Failure analysis of metallic components has been the preoccupation of the metallurgical community for years.[1] Petrochemical plants usually have an excellent staff of "static equipment" inspectors, whose services prove invaluable during machinery component failure analysis. The strengths of the metallurgical inspectors lie in solving service failures with the following primary failure modes and their causes:

1. Deformation and distortion
2. Fracture and separation
   a. ductile fractures
   b. brittle fractures
   c. fatigue fractures
   d. environmentally affected fractures
3. Surface and material changes
   a. corrosion
      • uniform corrosion
      • pitting corrosion
      • intergranular corrosion
4. Stress-corrosion cracking
5. Hydrogen damage
6. Corrosion fatigue
7. Elevated temperature failures
   a. creep
   b. stress rupture

The detailed analysis of these machinery component failures lies in metallurgical inspection, a highly specialized field. For an in-depth discussion of these analyses, refer to the references listed at the end of this chapter.

Metallurgical Failure Analysis Methodology

Even though the machinery failure analyst will lack the expertise to perform a detailed metallurgical analysis of failed components, he nevertheless has to stay in charge of all phases of the analysis. His job is to define the root cause of the failure incident and to come up with a corrective or preventive action. A checklist of what should be accomplished during a metallurgical failure analysis is shown in Table 2-1.

It is absolutely necessary to plan the failure analysis before tackling the investigation. A large amount of time and effort may be wasted if insufficient time is spent carefully considering the background of the failure and studying the general features before the actual investigation.[2]

In the course of the various steps listed in Table 2-1, preliminary conclusions will often be formulated. If the probable fundamental cause of the metallurgical failure has become evident early in the examination, the rest of the investigation should focus on confirming the probable cause and eliminating other possibilities. Other investigations will follow the logical sequence shown in Figure 2-1, and the results of each stage will determine the following steps. As new facts change first impressions, different failure hypotheses will surface and be retained or rejected as dictated. Where suitable laboratory facilities are available, the metallurgical failure analyst should compile the results of mechanical tests, chemical analyses, fractography, and microscopy before preliminary conclusions are formulated.

There is always the temptation to curtail work essential to an investigation. Sometimes it is indeed possible to form an opinion about a failure cause from a single aspect of the analysis procedure, such as the visual examination of a fracture surface or the inspection of a single metallographic specimen. However, before final

Table 2-1
Main Stages of a Metallurgical Failure Analysis (Modified from Ref. 1)

1. Collection of background data and selection of samples.
2. Preliminary examination of failed part (visual examination and record keeping).
3. Nondestructive testing.
4. Mechanical testing (including hardness and toughness testing).
5. Selection, identification, preservation and/or cleaning of all specimens.
6. Macroscopic examination and analysis (fracture surfaces, secondary cracks and other surface phenomena).
7. Microscopic examination and analysis.
8. Selection and preparation of metallographic sections.
9. Examination and analysis of metallographic sections.
10. Determination of failure mechanism.
11. Chemical analyses (bulk, local, surface corrosion products, deposits or coatings and microprobe analysis).
12. Analysis of fracture mechanics.
13. Testing under simulated service conditions (special tests).
14. Analysis of all evidence leading to formulation of conclusions.
15. Writing of report including recommendations.

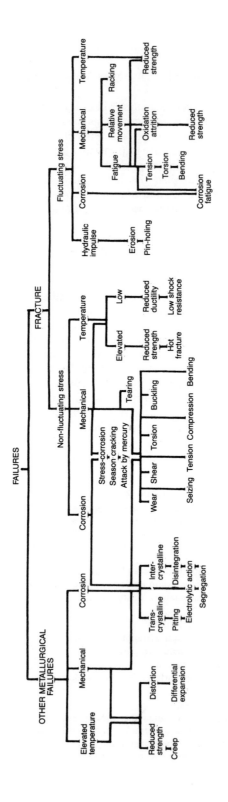

Figure 2-1. Classifications of failure causes in metals.

conclusions are reached, supplementary data confirming the original opinion should be looked for. Total dependence on the conclusions that can be drawn from a single specimen, such as from a metallographic section, may be readily challenged unless a history of similar failures can be drawn upon.[3]

Table 2-2 is a checklist that has been used as an aid in analyzing the evidence derived from metallurgical examinations and tests and in postulating conclusions.

As in other types of failure analyses, the end product of a metallurgical failure investigation should be the written failure analysis report. One experienced investigator has proposed that the report be divided into the main sections shown in Table 2-3. A detailed discussion of failure reports is given in Chapter 9.

Failure Analysis of Bolted Joints

At some time in the course of his career, the machinery failure analyst will have to deal with failures of threaded fasteners or bolted joints. This is also the time when he will find that the basic subject of "nuts and bolts" suddenly becomes complicated beyond his wildest dreams.

Anyone in the business of machinery failure analysis should be up-to-date on the design and behavior of bolted joints, for they are frequently the weakest links in engineered structures. Here is where machinery leaks, wears, slips, ruptures, loosens up, or simply fails.

Many factors contribute to failures of bolted joints. A look at available statistics reveals that problems encountered with threaded fasteners vary greatly. Consider the following: During the period 1964-1970 the research center of a large European

Table 2-2
Metallurgical Failure Examination Checklist[3]

1. Has failure sequence been established?
2. If the failure involved cracking or fracture, have the initiation sites been determined?
3. Did cracks initiate at the surface or below the surface?
4. Was cracking associated with a stress concentrator?
5. How long was the crack present?
6. What was the intensity of the load?
7. What was the type of loading: static, cyclic, or intermittent?
8. How were the stresses oriented?
9. What was the failure mechanism?
10. What was the approximate service temperature at the time of failure?
11. Did temperature contribute to failure?
12. Did wear contribute to failure?
13. Did corrosion contribute to failure? What type of corrosion?
14. Was the proper material used? Is a better material required?
15. Was the cross section adequate for class of service?
16. Was the quality of the material acceptable in accordance with specification?
17. Were the mechanical properties of the material acceptable in accordance with specification?
18. Was the component that failed properly heat treated?
19. Was the component that failed properly fabricated?
20. Was the component properly assembled or installed?

Table 2-3
Main Sections of a Metallurgical Failure Report (Modified from Ref. 2)

1. Description of the failed component.
2. Service conditions at time of failure.
3. Prior service history.
4. Manufacturing and processing history of component.
5. Mechanical and metallurgical study of failure.
6. Metallurgical evaluation of quality.
7. Summary of failure-causing mechanism(s).
8. Recommendations for prevention of similar failures.

Table 2-4
Failure Causes and Modes of Threaded Fasteners (Modified from Ref. 4)

Cause of Failure	Failure Distribution %
Product problems	50.0
Operational problems	40.0
Assembly problems	10.0
Failure Mode	
Fatigue failures	40.0
Creep failures	20.0
Sudden failures	
• brittle	10.0
• plastic	20.0
• corrosion	10.0

machinery insurance company, ATZ*, analyzed 132 cases where failures of threaded fasteners had caused damage to machinery.[4] Distribution of failure causes and failure modes are shown in Table 2-4.

Now consider a study of joint failures during live missions on the Skylab program, which produced the statistics shown in Table 2-5.[5]

From this we can see that in order to solve our problems, we have to list and document machinery failures in our own plants to obtain the necessary insight into prevailing failure causing factors specific to our environment.

Why Do Bolted Joints Fail?

It is beyond the scope of this text to give an exhaustive answer to this question. However, an overview will be provided to enable readers to ask the necessary questions when faced with a bolted-joint failure.

---

*The Allianz Center for Technology, Ismaning/Munich (Federal Republic of Germany).

Table 2-5
Summary of the Causes of Bolted Joint Failure on the Skylab Program
(All Fasteners Had Been Torqued. Modified from Ref. 5)

Failure Causes	Failure Distribution, %
Product problems	
inadequate design	24.0
parts damaged in handling	23.0
faulty parts	10.0
Assembly problems	
improper assembly	29.0
incorrect preload	14.0

Joints fail in many ways, but in all cases failure has occurred because joint members behave this way:

1. They slip in relationship to each other (displacement).
2. They simply separate (displacement).
3. Bolts and/or joint members break (fracture).

These basic failure modes are preceded in turn by the failure modes listed in Table 2-6. Table 2-6 will convey an idea of probable causes or factors that will, depending on circumstances, contribute to bolted-joint failures.

Fastener problems on machinery in the petrochemical industry will arise if the following are not considered:

1. Proper joint component selection suitable for the application.
2. Proper joint detail design parameters.
3. Importance of installation and maintenance procedures.

Some significant examples from our experience are:

Use of low-grade cap screws. If we use a cap screw with a yield strength too low for the forces being applied, it will stretch, causing "necking-out" (Figure 2-2). When the load is relaxed, the increased length will result in a loose nut which is free to vibrate off the bolt.[6] If, during preventive maintenance, the loose nut is discovered and tightened, reapplication of the load will cause the bolt to stretch at a lower load because there is less metal in the necked-out section. In many cases, the cap screw will fail completely while a mechanic is retightening the nut. Since he assumes that failure occurred because he pulled his wrench too hard, he will replace the bolt and nut with a new one of the same grade, and a vicious circle has begun.

Use of mismatched joint components. All components in a bolted-joint assembly must be matched to each other to achieve the desired holding power and service life.

Proper joint design. The kind and direction of forces to be transmitted—static or cyclic—is extremely important in the design of threaded fasteners.[7] Frequently,

Table 2-6
Failure Modes of Bolted Joints

	Primary Causes (Factors)	Fracture Under Static Load	Fatigue Failure	Vibration Loosening	Joint Leakage
Design and Manufacturing	Direction of bolt axis relative to vibration axis			•	
	Damping in Joint			•	
	Relaxation Effects		•	•	
	Radius of Thread Roots		•		
	Bolt/Joint Stiffness Ratio		•		•
	Thread Run-Out		•		
	Fillet Size and Shape		•		
	Nut Dilation	•			
	Poor Fits	•			
	Galling	•			
	Finish of Parts		•		
	Improper Heat Treatment	•			
	Tool Marks		•		
Assembly Practices	Condition of Joint Surfaces				•
	Condition of Gaskets				•
	Bolt-Up Procedure				•
	Thread Lubrication	•	•	•	
	Type of Tool Used	•			
	Improper Preload			•	•
Operating Conditions	Magnitude of Load Excursions	•	•		
	Temperature Cycling				•
	Corrosion	•	•	•	•

however, there is little known about the actual forces and loads that will be encountered in service. Consequently, the designer has to start with common assumptions regarding possible forces and moments, such as those shown in Figure 2-3.

Long-term cyclic loads as encountered in rotating/reciprocating process machinery can only be transmitted by high-tensile-strength fastener components. In order to obtain high-strength fasteners, heat treatment after fabrication is necessary. Heat treatment, however, makes steel susceptible to fatigue failure when used under variable (vibration) load conditions. The higher the grade of heat treatment, the greater the danger of fatigue if the fastener is not properly preloaded.

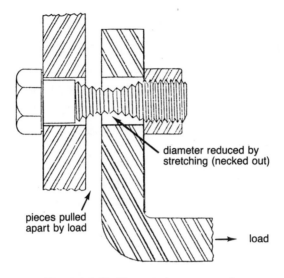

Figure 2-2. Necking-out of a cap screw.[6]

A properly designed and preloaded bolted joint is extremely reliable without additional locking devices. This is especially true for high-strength steels, provided there is sufficient bolt resilience and a minimum of joint interfaces. Design measures to increase effective bolt lengths or their resilience are shown in Figure 2-4. These measures not only have the advantage of achieving more favorable bolt load distributions but also provide greater insurance against loosening.

Failure to apply proper preload. Applying proper preload to a bolt or nut assembly is the crucial phase of many bolted joints in process machinery. J.H. Bickford[8] refers to the difficulties of bolt preload and torque control as he lists the problems associated with using a torque wrench to assemble a joint:

Friction
Operator
Geometry   FOGTAR*
Tool Accuracy
Relaxation

Most of us have wrestled with these problems, and if we do not know everything about "FOGTAR" we should get acquainted with J. H. Bickford's delightful book on the behavior of bolted joints.

---

*A Tibetan word for trouble (Ref. 8, p. 77).

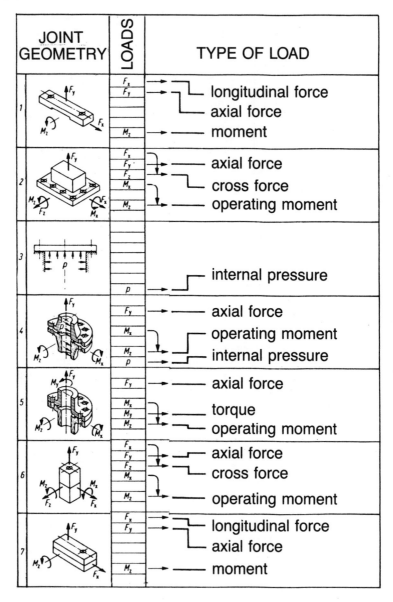

Figure 2-3. Possible operating loads encountered by bolted joints (modified from Ref. 7).

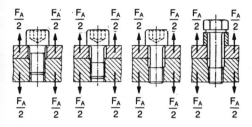

Figure 2-4. Bolted joint designs with increasing fatigue strength and resistance to loosening (modified from Ref. 7).

Carefully consider reusing fastener components in critical applications. Some critical applications are:

1. Piston-rod locknuts—reciprocating compressors
2. Crosshead-pin locknuts—reciprocating compressors
3. Impeller locknuts—centrifugal pumps
4. Thrust-disc locknuts—centrifugal compressors
5. Bolts and nuts—high-performance couplings

Typical fastener components not to be reused in any application are:

1. Prevailing torque locknuts (nylon insert)
2. Prevailing torque bolts (interference fit threads)
3. Anaerobic adhesive secured fasteners
4. Distorted thread nuts
5. Beam-type self-locking nuts
6. Castellated nut and cotter key
7. Castellated nut and spring pin
8. Flat washers
9. Tab washers
10. Vibration-resistant washers
11. Lock wire

Finally, inspect bolt and nuts for nicks, burrs, and tool marks before deciding to reuse!

Failure Analysis Steps

Failure analysis of bolted joints should consist of the following essential steps:

1. *Definition of failure mechanism.*
   a. The bolt failed under static load. Did it occur while tightening? The fracture surface will usually be at an angle other than 90° to the bolt axis. This is because the strength of the bolt has been exceeded by a combination of tension and torsional stress. A failure in pure tension will usually be at a right angle to the bolt axis.

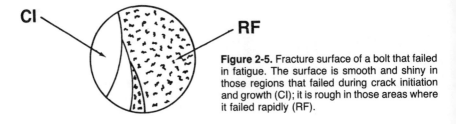

CI

RF

Figure 2-5. Fracture surface of a bolt that failed in fatigue. The surface is smooth and shiny in those regions that failed during crack initiation and growth (CI); it is rough in those areas where it failed rapidly (RF).

   b. The bolt failed in fatigue under variable and cyclic loads. High cycle fatigue will usually be indicated by "beach marks" on the fracture surface (see Figure 2-5). This might not be conclusive, as the absence of these marks will not rule out a fatigue-related failure mechanism.

   c. Static or fatigue failure from corrosion.

   d. The joint failed to perform its design function because clamping forces fell below design requirements. Possible failure modes are partial or total separation (displacement), joint slippage (displacement), fretting of the joint surfaces (corrosion), and vibration loosening (displacement) of the nut. Consequential failure mode in all these cases is "leakage."

2. *Design review.*

   a. The analyst will now estimate or calculate the operating loads and possible preloads on joint components. If failure was static, he can refer to suitable references such as the ones listed at the end of this chapter.

   b. If the failure has been caused by cyclic loads, follow up work will be much more difficult: The analyst will have to determine the endurance limit of the parts involved in the failure. This may require experiments, as published data are rare.

3. *Special-variables check.* Consider and check the factors that could contribute to the fastener failure, as shown in Table 2-6.

Shaft Failures*

Causes of Shaft Failures

Shafts in petrochemical plant machinery operate under a broad range of conditions, including corrosive environments, and under temperatures that vary from extremely low, as in cold ethylene vapor and liquid service, to extremely high, as in gas turbines.

---

* Adapted by permission from course material published by the American Society for Metals, Metals Park, Ohio 44073. For additional information on metallurgical service failures, refer to the 15-lesson course *Principles of Failure Analysis,* available from the Metals Engineering Institute of the American Society for Metals, Metals Park, Ohio 44073.

Shafts are subjected to one or more of the following loads: tension, compression, bending, or torsion. Additionally, shafts are often exposed to high vibratory stresses.

With the exception of wear as consequential damage of a bearing failure, the most common cause of shaft failures is metal fatigue. Fatigue failures start at the most vulnerable point in a dynamically stressed area—typically a stress raiser, which may be either metallurgical or mechanical in nature, and sometimes both.

Occasionally, ordinary brittle fractures are encountered, particularly in low-temperature environments. Some brittle fractures have resulted from impact or a rapidly applied overload. Surface treatments can cause hydrogen to be dissolved in high-strength steels and may cause shafts to become embrittled even at room temperature.

Ductile fracture of shafts usually is caused by accidental overload and is relatively rare under normal operating conditions. Creep, a form of distortion at elevated temperatures, can lead to stress rupture. It can also cause shafts with close tolerances to fail because of excessive changes in critical dimensions.

Fracture Origins in Shafts

Shaft fractures originate at stress-concentration points either inherent in the design or introduced during fabrication. Design features that concentrate stress include ends of keyways, edges of press-fitted members, fillets at shoulders, and edges of oil holes. Stress concentrators produced during fabrication include grinding damage, machining marks or nicks, and quench cracks resulting from heat-treating operations.

Frequently, stress concentrators are introduced during shaft forging; these include surface discontinuities such as laps, seams, pits, and cold shuts, and subsurface imperfections such as bursts. Subsurface stress concentrators can be introduced during solidification of ingots from which forged shafts are made. Generally, these stress concentrators are internal discontinuities such as pipe, segregation, porosity, shrinkage, and nonmetallic inclusions.

Fractures also result from bearing misalignment, either introduced at assembly or caused by deflection of supporting members in operation; from mismatch of mating parts; and from careless handling in which the shaft is nicked, gouged, or scratched.

To a lesser degree, shafts can fracture from misapplication of material. Such fractures result from using materials having high ductile-to-brittle transition temperatures; low resistance to hydrogen embrittlement, temper embrittlement, or caustic embrittlement; or chemical compositions or mechanical properties other than those specified.

Stress Systems in Shafts

The stress systems acting on a shaft must be understood before the cause of a fracture in that shaft can be determined. Also, both ductile and brittle behavior under static loading or single overload and the characteristic fracture surfaces produced by these types of behavior must be clearly understood for proper analysis of shaft fractures.

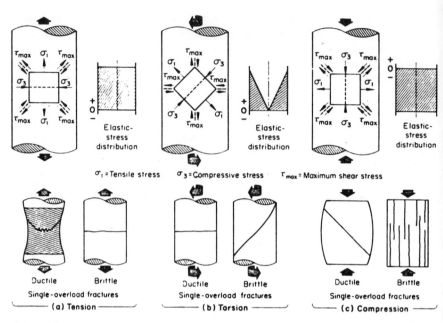

Figure 2-6. Free-body diagrams showing orientation of normal and shear stresses in a shaft under simple tension, torsion, and compression loading, and showing the single-overload fracture behavior of ductile and brittle materials.

Figure 2-6 gives simplified, two-dimensional free-body diagrams showing the orientations of the normal-stress and shear-stress systems at any internal point in a shaft loaded in pure tension, torsion, and compression. Also, the single-overload-fracture behavior of both ductile and brittle materials is illustrated with the diagram of each type of load.

A free-body stress system may be considered as a square of infinitely small dimensions. Tensile and compressive stresses act perpendicularly to each other and to the sides of the square to stretch and squeeze the sides, respectively. The shear, or sliding stresses act on the diagonals of the square, 45° to the normal stresses. The third-dimension radial stresses are ignored in this description.

The effects of the shear and normal stresses on ductile and brittle materials under the three types of loads illustrated in Figure 2-6 and those under bending load are discussed below.

Tension. Under tension loading, the tensile stresses (σ_1) are longitudinal, whereas the compressive-stress components (σ_3) are transverse to the shaft axis. The maximum-shear-stress components (τ_{max}) are at 45° to the shaft axis (see Figure 2-6, a).

In ductile material, shear stresses developed by tensile loading cause considerable deformation prior to fracture, which originates near the center of the shaft and propagates toward the surface, ending with a conical shear lip usually about 45° to the shaft axis.

In a brittle material, a fracture from a single tensile overload is roughly perpendicular to the direction of tensile stress, but involves little permanent deformation. The fracture surface usually is rough and crystalline in appearance.

The elastic stress distribution in pure tension loading, in the absence of a stress concentration, is uniform across the section. Thus, fracture can originate at any point within the highly stressed volume.

Torsion. The stress system rotates 45° counterclockwise when a shaft is loaded in torsion, as shown in Figure 2-6, b. Both the tensile and compressive stresses are 45° to the shaft axis and remain mutually perpendicular. One shear-stress component is parallel with the shaft axis; the other is perpendicular to the shaft axis.

In a ductile material loaded to failure in torsion, shear stresses cause considerable deformation prior to fracture. This deformation, however, usually is not obvious because the shape of the shaft has not been changed. If a shaft loaded in torsion is assumed to consist of a number of infinitely thin disks that slip slightly with respect to each other under torsional stress, visualization of deformation is simplified. Torsional single-overload fracture of a ductile material usually occurs on the transverse plane, perpendicularly to the axis of the shaft. In pure torsion, the final-fracture region is at the center of the shaft; the presence of slight bending will cause it to be off center.

A brittle material in pure torsion will fracture perpendicularly to the tensile-stress component, which is 45° to the shaft axis. The resulting fracture surfaces usually have the shape of a spiral.

The elastic-stress distribution in pure torsion is maximum at the surface and zero at the center of the shaft. Thus, in pure torsion, fracture normally originates at the surface, which is the region of highest stress.

Compression. When a shaft is loaded in axial compression (see Figure 2-6, c), the stress system rotates so that the compressive stress (σ_3) is axial and the tensile stress (σ_1) is transverse. The shear stresses (τ_{max}) are 45° to the shaft axis, as they are during axial tension loading.

In a ductile material overloaded in compression, shear stresses cause considerable deformation but usually do not result in fracture. The shaft is shortened and bulges under the influence of shear stress. If a brittle material loaded in pure compression does not buckle, it will fracture perpendicularly to the maximum tensile-stress component. Because the tensile stress is transverse, the direction of brittle fracture is parallel to the shaft axis.

The elastic-stress distribution in pure compression loading, in the absence of a stress concentration, is uniform across the section. If fracture occurs, it likely will be in the longitudinal direction because compression loading increases the shaft diameter and stretches the metal at the circumference.

Bending. When a shaft is stressed in bending, the convex surface is stressed in tension and has an elastic-stress distribution similar to that shown in Figure 2-6, c. Approximately midway between the convex and concave surfaces is a neutral axis, where all stresses are zero.

Fatigue in Shafts

Fatigue in shafts generally can be classified as bending fatigue, torsional fatigue, and axial fatigue. Bending fatigue can result from three types of bending loads: unidirectional (one-way), reversed (two-way), and rotating. Torsional fatigue can result from application of a fluctuating or an alternating twisting moment, or torque. Axial fatigue can result from application of alternating (tension-and-compression) or fluctuating (tension-tension) loading.

Unidirectional bending fatigue. The axial location of a fatigue crack in a stationary cylindrical bar or shaft subjected to a fluctuating unidirectional bending moment evenly distributed along the length will be determined by some minor stress raiser such as a surface discontinuity.

Beach marks of the form shown in Figure 2-7, a and b, are indicative of a fatigue crack having a single origin at the point indicated by the arrow. The crack front, which formed the beach marks, is symmetrical with respect to the origin and retains a concave form throughout. Both the single origin and the smallness of the final-fracture zone in Figure 2-7, a, suggest that the nominal stress was low. The larger final-fracture zone in Figure 2-7, b, suggests a higher nominal stress.

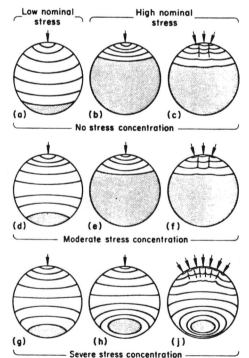

Figure 2-7. Fatigue marks, typical for a uniformly loaded shaft subjected to unidirectional loading, produced from single origins at low and high stresses and from multiple origins at high nominal stresses. Arrows indicate crack origins; final-fracture zones are shaded.

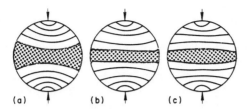

Figure 2-8. Typical fatigue marks on the fracture surface of a uniformly loaded nonrotating shaft subjected to reverse-bending stresses and having (a) no stress concentration, (b) moderate stress concentration, and (c) severe stress concentration. Arrows indicate crack origins; shaded areas are final fracture zones.

Figure 2-7, c, shows a typical fatigue crack originating as several individual cracks that ultimately merged to form a single crack front. Such multiple origins usually are indicative of high nominal stress. Note the presence of ridges, or ratchet marks, between crack origins.

Figure 2-7, d-j, describes the appearance of shaft fracture surfaces as a function of increased stress concentrations under operating conditions similar to the ones discussed previously.

Bending fatigue. When the applied bending moment is reversing (alternating), all points in the shaft are subjected alternately to tension stress and compression stress: While the points on one side of the plane of bending are in tension, the points on the opposite side are in compression. If the bending moment is of the same magnitude in both directions, two cracks of approximately equal length usually develop from origins diametrically opposite each other and often in the same transverse plane. If the bending moment is greater in one direction than in the other, the two cracks will differ in length.

Typical fatigue marks on the fracture surface of a stationary (nonrotating) shaft subjected to a reversing bending moment evenly distributed along its length are shown in Figure 2-8. The crack origins (at arrows) are shown diametrically opposite each other, but sometimes they are slightly displaced by minor stress raisers. The pattern shown in Figure 2-8, a, is typical for a single-diameter shaft with no stress concentration. The bending moment is equal in both directions.

A large-radius fillet at a change in shaft diameter imposes a moderate stress concentration. The pattern on the surface of a fracture through such a fillet is shown in Figure 2-8, b. A small-radius fillet at a change in diameter results in a severe stress concentration. The typical surface pattern of a fracture through a small-radius fillet is shown in Figure 2-8, c.

Under these loading conditions, each crack is subjected alternately to tensile and compressive stresses, with the result that the surfaces of the crack are forced into contact with one another during the compression cycle and rubbing occurs. Sometimes, rubbing is sufficient to obliterate many of the characteristic marks, and the crack surfaces may become dull or polished.

Rotational bending. The difference between a stationary shaft and a rotating shaft subjected to the same bending moment is that in a stationary shaft the tensile stress is confined to a portion of the periphery only. In a rotating shaft, every point on the periphery sustains a tensile stress, then a compressive stress, once every revolution.

Another important difference introduced by rotation is asymmetrical development of the crack front from a single origin. There is a marked tendency of the crack front to extend in a direction opposite to that of rotation. The crack front usually swings around about 15° or more, as shown in Figure 2-9, a and c. A third difference arising from rotation is in the distribution of the origins of a multiple-origin crack.

In a nonrotating shaft subjected to unidirectional bending, the origins are located in the region of the maximum-tension zone (see Figure 2-7). In a nonrotating shaft subjected to reverse bending, the origins are diametrically opposite each other (see Figure 2-8).

In rotary bending, however, every point on the shaft periphery is subjected to a tensile stress at each revolution, and therefore a crack may be initiated at any point on the periphery, as shown in Figure 2-9, b and d.

The crack surfaces are pressed together during the compressive component of the stress cycle, and mutual rubbing occurs. A common result of final fracture is that one side of the crack moves slightly in relation to the other side, frequently causing severe damage to the fracture surfaces and obliterating many marks. However, although the high spots on one surface abrade the high spots on the other, the marks in the hollows are retained. Because the hollows are mirror images of the damaged high

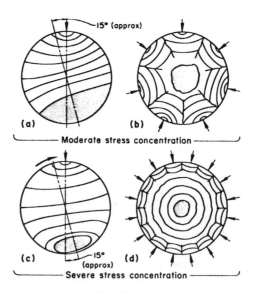

Figure 2-9. Typical fatigue marks on the fracture surface of a uniformly loaded rotating shaft produced from single and multiple origins (at arrows), having moderate and severe stress concentrations. Shaded areas are final-fracture zones; shaft rotation is clockwise.

spots on the opposing surface, they provide useful evidence; thus it is desirable to examine both parts of a cracked or fractured shaft.

The similar macroscopic appearance of fractures in shafts resulting from rotary-bending fatigue and from torsional shear frequently results in misinterpretation; therefore expert help should be obtained in these cases.

Torsional fatigue. Fatigue cracks initiated from torsional stresses show the same beach marks and ridges as those produced by bending stresses.[9] Longitudinal stress raisers are comparatively harmless under bending stresses, but are as important as circumferential stress raisers under torsional loading. This sensitivity of shafts loaded in torsion to longitudinal stress raisers is of considerable practical importance, because inclusions in the shaft material are almost always parallel to the axis of rotation. It is not unusual for a torsional-fatigue crack to originate at a longitudinal inclusion, a surface mark, or a spline or keyway corner and then branch at about 45°.

When a stress raiser such as a circumferential groove is present, different states of stresses exist around the stress raiser, and the tensile stress is increased to as much as four times the shear stress. Therefore, the tensile stress on the 45° plane will exceed the tensile strength of the steel before the shear stress reaches the shear strength of the steel. Fracture occurs normal to the 45° tensile plane, where it produces a characteristic conical or star-shape fracture surface. Shot peening of these highly stressed areas has been successfully applied.

The relative development of two torsional-fatigue cracks mutually at right angles can indicate the magnitude of the torque reversals that have been applied. Cracks of approximately the same length indicate that the torque reversals have been of equal magnitude, but only if the cracks are in a comparatively early state of development. Beyond this stage, one crack usually takes the lead, and such inferences are no longer justified. If the shaft transmits a unidirectional torque but two cracks develop mutually at right angles, it can be presumed that the torque was of a reversing character. If a bending stress is applied to a shaft that is transmitting torque, the angle at which any fatigue crack develops will be modified. Therefore, if the angle differs significantly from 45° to the shaft axis, the presence of a bending stress is indicated.

Contact fatigue. Contact fatigue occurs when components roll, or roll and slide, against each other under high contact pressure and cyclic loading. Spalling occurs after many repetitions of loading and is the result of metal fatigue from the imposed cyclic contact stresses, which are at a maximum value beneath the contact surface.

The significant stress in rolling-contact fatigue is the maximum alternating shear stress that reverses direction during rolling. In pure rolling, this stress occurs on a plane slightly below the surface and can lead to subsurface fatigue cracks. As these cracks propagate under the repeated loads, they reach the surface and produce spalling.

When sliding is imposed on rolling, the tangential forces and thermal gradient caused by friction alter the magnitude and distribution of stresses in and below the contact area. The alternating shear stress is increased in magnitude and is moved nearer to the surface by friction resulting from the sliding action.

Forged, hardened steel rolls are prone to surface spalling. Generally, spalling in a roll is attributed to an initial subsurface fatigue crack of undertermined origin or to surface-induced microscopic cracking. However, spalling cannot be attributed to a single main cause. Spalling may originate below the surface as a result of localized stress concentration, which causes plastic flow or fracture at the point of maximum reversed shear.

Ductile and Brittle Failures of Shafts

Brittle fractures are associated with the inability of certain materials to deform plastically under stress. Brittle fractures are characterized by sudden fracturing at extremely high rates of crack propagation, perhaps 6000 ft/sec (1830 m/s) or more, with little evidence of distortion in the region of fracture initiation. These fractures are characterized by marks known as herring-bone or chevron patterns on the fracture surface. The chevrons point toward the origin of the fracture.

Ductile fractures exhibit distortion (plastic flow) at the fracture surface similar to that observed in ordinary tensile-test or torsion-test specimens. When a shaft is fractured by a single application of a load greater than the strength of the shaft, there is usually considerable plastic deformation prior to fracture. This deformation often is readily apparent upon visual inspection of a shaft that fractured in tension, but often is not obvious when the shaft fractured in torsion. This ability of a material to deform plastically is a property known as ductility. The appearance of the fracture surface of a shaft that failed in a ductile manner also is a function of shaft shape, the type of stress to which the shaft was subjected, rate of loading, and, for many alloys, temperature.

Ductile fracture of shafts occurs infrequently in normal operation. However, ductile fractures may occur if operating requirements are underestimated, if the materials used are not as strong as had been assumed, or if the shaft is subjected to a single overload, such as in an accident. Fabricating errors such as using the wrong material or using material in the wrong heat-treated condition (for instance, annealed instead of quenched and tempered) can result in ductile fractures.

Shaft Failures Due to Corrosion

Most shafts are not subject to general corrosion or chemical attack. Corrosion, which takes place as general surface pitting, may uniformly remove metal from the surface or may uniformly cover the surface with scale or other corrosion products.

Corrosion pits have a relatively minor effect on the load-carrying capacity of a shaft, but they do act as points of stress concentration at which fatigue cracks can originate.

A corrosive environment will greatly accelerate metal fatigue; even exposure of a metal to air results in a shorter fatigue life than that obtained under vacuum. Steel shafts exposed to salt water may fail prematurely by fatigue even if they are thoroughly cleaned periodically. Aerated salt solutions usually attack metal surfaces at the weakest points, such as scratches, cut edges, and points of high strain. To minimize corrosion fatigue, it is necessary to select a material resistant to corrosion in the service environment, or to provide the shaft with a protective coating.

Most large shafts and piston rods are not subject to corrosion attack. Centrifugal process compressors frequently handle gases that contain moisture and small amounts of a corrosive gas or liquid. If corrosion attack occurs, a scale is often formed that may be left intact and increased by more corrosion, eroded off by entrained liquids (or solids), or slung off from the rotating shaft.

Stress-corrosion cracking occurs as a result of corrosion and stress at the tip of a growing crack. Stress-corrosion cracking often is accompanied or preceded by surface pitting, but general corrosion often is absent, just as rapid, overall corrosion does not accompany stress-corrosion cracking.

Corrosion fatigue results when corrosion and an alternating stress, neither of which is severe enough to cause failure by itself, occur simultaneously and thus can cause failure. Once such a condition exists, shaft life will probably be days or weeks rather than years. Corrosion-fatigue cracking usually is transgranular; branching of the main cracks occurs, although usually not as much as in stress-corrosion cracking. Corrosion generally is present in the cracks, both at the tips and in regions nearer the origins.

Stress Raisers in Shafts

Most service failures in shafts are attributable largely to some condition that causes stress intensification. Locally, the stress value is raised above a value at which the material is capable of withstanding the number of loading cycles that corresponds to a satisfactory service life. Only one small area needs to be repeatedly stressed above the fatigue strength of the material for a crack to be initiated. An apparently insignificant imperfection such as a small surface irregularity may severely reduce the fatigue strength of a shaft if the stress level at the imperfection is high. The most vulnerable zone in torsional and bending fatigue is the shaft surface; an abrupt change of surface configuration may have a damaging effect, depending on the orientation of the discontinuity to the direction of stress.

All but the simplest shafts contain oil holes, keyways, or changes in shaft diameter. The transition from one diameter to another, the location and finish of an oil hole, and the type and shape of a keyway exert a marked influence on the magnitude of the resulting stress-concentration and fatigue-notch factors, which often range in numerical value from 1.0 to 5.0, and which sometimes attain values of 10.0 or higher.

Most stress raisers can be classified as one of the following:

1. Nonuniformities in the shape of the shaft, such as steps at changes in diameter, broad integral collars, holes, abrupt corners, keyways, grooves, threads, splines, and press-fitted or shrink-fitted attachments.
2. Surface discontinuities arising from fabrication practices or service damage, such as seams, nicks, notches, machining marks, identification marks, forging laps and seams, pitting, and corrosion.
3. Internal discontinuities such as porosity, shrinkage, gross nonmetallic inclusions, cracks, and voids.

Most shaft failures are initiated at primary stress raisers (see number 1), but secondary stress raisers (number 2 or 3) may contribute to a failure. For instance, a change in shaft diameter can result in stress intensification at the transition zone. If there is a surface irregularity or other discontinuity in this zone, the stress is sharply increased around the discontinuity.

Changes in Shaft Diameter

A change in shaft diameter concentrates the stresses at the change in diameter and in the small-diameter portion.

The effects on stress concentration of an abrupt change and three gradual changes in diameter are shown schematically in Figure 2-10. The sharp corner at the intersection of the shoulder and shaft in Figure 2-10, a, concentrates the stresses at the corner as they pass from the large to the small diameter. The large-radius fillet shown in Figure 2-10, d, permits the stresses to flow with a minimum of restriction. However, the fillet must be tangent with the smaller-diameter section; otherwise, a sharp intersection will result, overcoming the beneficial effect of the large-radius fillet.

Press and Shrink Fitting

Components such as gears, pulleys, and impellers often are assembled onto shafts by press or shrink fitting, which can result in stress raisers under bending stress. Typical stress flow lines in a plain shaft at a press-fitted member are shown in Figure 2-11, a. Enlarging the shaft at the press-fitted component and using a large-radius fillet would produce a stress distribution such as that shown schematically in Figure 2-11, b. A small-radius fillet at the shoulder would result in a stress pattern similar to that shown in Figure 2-11, a.

Longitudinal Grooves

Longitudinal grooves, such as keyways and splines, often cause failure in shafts subjected to torsional stress. Generally, these failures result from fatigue fracture where a small crack originated in a sharp corner because of stress concentration. The crack gradually enlarges as cycles of service stress are repeated until the remaining

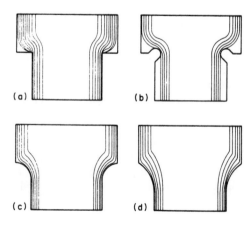

Figure 2-10. Effect of size of fillet radius on stress concentration at a change in shaft diameter.

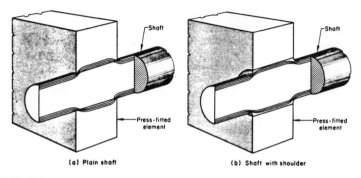

(a) Plain shaft   (b) Shaft with shoulder

Figure 2-11. Schematic diagram of stress distribution in two types of rotating shafts with press-fitted elements under a bending load.

section breaks. A sharp corner in a keyway can cause the local stress to be as much as ten times the average nominal stress.

Failures of this kind can be avoided by using a half-round keyway, which permits the use of a round key, or by using a generous fillet radius in the keyway. Good results are obtained by the use of fillets having radii equal to approximately one-half the depth of the keyway. A half-round keyway produces a local stress of only twice the average stress, thus providing greater load-carrying ability than that permitted by a square keyway. Many shafts with square keyways do not fracture in service because stresses are low or because fillets with generous radii are used.

Stress fields and corresponding torsional-fatigue cracks in shafts with keyways or splines are shown in Figure 2-12. In Figure 2-12 a, the fillet in one corner of the keyway was radiused and the fillet in the other corner was sharp, which resulted in a single crack. Note that this crack progresses approximately normal to the original stress field. In Figure 2-12, b, both fillets in the keyway corners were sharp, which resulted in two cracks that did not follow the original stress field, a condition arising from the cross effect of cracks on the stress field.

A splined shaft subjected to alternating torsion can crack along the bottom edge of the splines, as shown in Figure 2-12, c. This is another instance of a highly localized stress field strongly influencing crack development.

Failures Due to Manufacturing Processes

Surface discontinuities produced during manufacture of a shaft and during assembly of the shaft into a machine can become points of stress concentration and thus contribute to shaft failure. Operations or conditions that produce this type of stress raiser include:

1. Manufacturing operations that introduce stress raisers such as tool marks and scratches.
2. Manufacturing operations that introduce high tensile stresses in the surface, such as improper grinding, repair welding, electromachining, and arc burns.

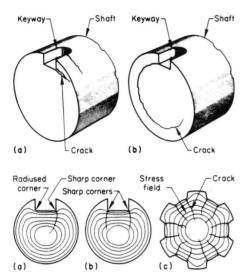

Figure 2-12. Stress fields and corresponding torsional-fatigue cracks in shafts with keyways or splines.

3. Processes that introduce metal weakening, such as forging flow lines that are not parallel with the surface; plating, causing hydrogen embrittlement; or heat treating, causing decarburization.

Fatigue strength may be increased by imparting high compressive residual stresses to the surface of the shaft. This can be accomplished by surface rolling or burnishing, shot peening, tumbling, coining, induction hardening, and sometimes case hardening.

Influence of Metallurgical Factors

Internal discontinuities such as porosity, large inclusions, laminations, forging bursts, flakes, and centerline pipe will act as stress concentrators under certain conditions and may originate fatigue fracture.

In order to understand the effect of discontinuities, it is necessary to realize that fracture can originate at any location, surface or interior, where the stress first exceeds material strength. The stress gradient must be considered in torsion and bending because the stress is maximum at the surface but zero at the center, or neutral axis. In tension, however, the stress is essentially uniform across the section.

If discontinuities such as those just noted occur in a region highly stressed in tension by bending or torsional loading, fatigue cracking may be initiated. However, if the discontinuities are in a low-stress region, such as near a neutral axis, they will be less harmful. Similarly, a shaft stressed by repeated high tensile loading must be free from serious imperfections, for there is no neutral axis. Here, any imperfection can be a stress concentrator and can be the origin of fatigue cracking if the stress is high with respect to the strength.

Nonmetallic inclusions oriented parallel to the principal stress do not usually exert as great an effect upon fatigue resistance as those 90° to the principal stress.

Surface Discontinuities

In mill operations, a variety of surface imperfections often result from hot plastic working of material when lapping, folding, or turbulent flow is experienced. The resultant surface discontinuities are called laps, seams, and cold shuts. Similar discontinuities also are produced in cold working operations, such as fillet and thread rolling. Other surface imperfections develop from foreign material embedded under high pressures during the working process. For example, oxides, slivers, or chips of the base material are occasionally rolled or forged into the surface.

Most of the discontinuities are present in the metal prior to final processing and are open to the surface. Standard nondestructive testing procedures such as liquid-penetrant and magnetic-particle inspections will readily reveal most surface discontinuities. If not detected, discontinuities may serve as sites for corrosion or crack initiation.

Because fatigue-crack initiation is the controlling factor in the life of most small shafts, freedom from surface imperfections becomes progressively more important in more severe applications. Similarly, internal imperfections, especially those near the surface, will grow under cyclic loading and result in cracking when critical size is attained. Service life can be significantly shortened when such imperfections cause premature crack initiation in shafts designed under conventional fatigue-life considerations. Surface or subsurface imperfections can cause brittle fracture of a shaft after a very short service life when the shaft is operating below the ductile-to-brittle transition temperature. When the operating temperature is above the transition temperature, or when the imperfection is small relative to the critical flaw size, especially when the cyclic-loading stress range is not large, service life may not be affected.

Examination of Failed Shafts

In general, the procedures for examining failed shafts are similar to those discussed in connection with Table 2-2. However, in summary of the foregoing, some unique characteristics of shafts require special attention:

Potential stress raisers or points of stress concentration, such as splines, keyways, cross holes, and changes in shaft material, mechanical properties, heat treatment, test locations, nondestructive examination used, and other processing requirements, also should be noted.

Special processing or finishing treatments, such as shot peening, fillet rolling, burnishing, plating, metal spraying, and painting, can have an influence on performance, and the analyst should be aware of such treatments.

Mechanical conditions. The manner in which a shaft is supported or assembled in its working mechanism and the relationship between the failed part and its associated members can be valuable information. The number and location of bearings or supports and how their alignment may be affected by deflections or distortions that can occur from mechanical loads, shock, vibrations, or thermal gradients should be considered.

The method of connecting the driving or driven member to the shaft, such as press fitting, welding, or use of a threaded connection, a set screw, or a keyway, can influence failure. Also influential is whether power is transmitted to or taken from the shaft by gears, splines, belts, chains, or torque converters.

Operating history. Checking the operating and maintenance records of an assembly should reveal when the parts were installed, overhauled, and inspected. These records also should show whether maintenance operations were conducted in accordance with the manufacturer's recommendations.

We are concluding our discussion of shaft failure analysis with an example of a typical shaft failure case and analysis.

The Case of the Boiler Fan Turbine Shaft Fracture (Figure 2-13 A and B)

After examining a failed turbine rotor shaft in a Canadian refinery, the investigating metallurgical laboratory prepared the following report for the owners:

> We have completed our examination of the F-703 turbine shaft fracture faces submitted from your refinery. The fracture faces were considerably damaged as a result of contact and relative motion subsequent to fracture, but nevertheless three distinct zones were apparent:
>
> 1. An outer ring approximately 5mm (3/16 in.) wide of relatively undamaged surface normal to the axis of the shaft. Close inspection showed that this zone had suffered some abrasion by relative motion.
> 2. A mid-radius ring extending about 20mm (3/4 in.) into the center, which displayed severe abrasion caused by contact and relative motion of the two halves of the shaft. The original fracture features were virtually obscured by the abrasion. The fracture profile in this region displayed a shallow "cup and cone" appearance.
> 3. A central region about 25mm (1 in.) in diameter in which abrasion caused by the contact and relative motion of the two halves of the shaft had resulted in continuous gouging or ploughing of the surface to produce a relatively smooth and shiny surface.
>
> The surface of the shaft was heavily scored and there was some slight surface oxidation which may have occurred as a result of frictional heating. There were no obvious corrosion products on either the shaft or the fracture surfaces.
>
> Metallographic examination showed the microstructure of the shaft to be of tempered martensite with some proeutectoid ferrite. The fracture profile was covered with what appeared to be wear transformation products formed by abrasion. Figure 2-13A shows these products at the surface and also entrained in the sublayers by severe abrasion. The figure also shows the general microstructure. Damage to the surface was

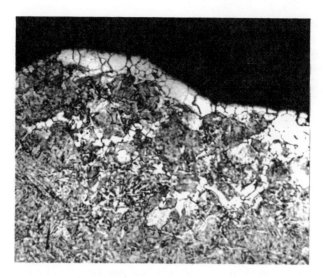

Figure 2-13A. Turbine shaft fracture profile and general microstructure. Etchant: 2% nital (200x).

too severe to tell if the original fracture was intergranular or transgranular. No corrosion at the fracture surface was apparent. Transformation products, similar to those observed on the fracture surface, were observed in patches on the scored surface of the shaft. The appearance of these products was attributed to scoring and abrasion of the shaft surface during service.

The mean hardness was 266 HV/10, corresponding by 25 HRC. There was no significant variation in hardness between the edge and the center of the shaft section.

An attempt was made to examine the fracture surface by scanning electron microscopy. Most of the surface was damaged by abrasion in some way. In the few areas in which the effects of abrasions after fracture were not apparent, the fracture surface appeared smooth and relatively featureless.

Although the shaft fracture surfaces were so badly abraded as to preclude any conclusive statement on the nature of the failure, certain characteristics do suggest that the shaft failed by fatigue. The outer zone around the periphery of the fracture face suggests that fatigue cracking was initiated at multiple points under a fairly severe stress concentration, with crack propagation occurring as the result of a bending moment applied to the rotating shaft. It appeared that stress concentration arose as a result of scoring of the shaft surface during service. Further comments are not possible because of the damage sustained by the fracture surfaces.

If you have any questions concerning this failure, I shall be pleased to answer them.

Yours truly,

James Bond,
Senior Metallurgical Engineer

Making use of this report, the owner's engineer focused his attention on root cause of failure and remedial action. He commented as follows:

The report is suggesting fatigue cracking of the shaft material as a result of a bending moment applied to the shaft while rotating. It is in agreement with our earlier assumptions as to the most probable cause of the failure.

We think that a failure like this can be avoided by careful alignment of the three sleeve bearings supporting the turbine/pinion rotor (Figure 2-13B). We are now using a mandrel technique where we assemble a dummy shaft in the bearings to obtain a blueing pattern as an indication for internal alignment. This approach is recommended. The mandrel should have oversize journal diameters to take up actual bearing clearance, and the bearing caps should be torqued down at final assembly. This would allow an indication of proper bearing alignment not only in the vertical down direction but in all other directions.

Analysis of Surface-Change Failures

Failures through material surface changes are generally caused by wear, fatigue, and corrosion. Each of these failure forms is affected by many conditions, including environment, type of loading, relative speeds of mating parts, lubrication, temperature, hardness, and surface finish.

Wear, or the undesired removal of material from rubbing surfaces, causes many surface failures. Wear can be classified as either "abrasive" or "adhesive." If corrosion is present, a form of adhesive wear occurs that is termed "corrosive wear."

Abrasive Wear

Abrasive wear occurs when a certain material scratches or gouges a softer surface. It has been estimated[10] that abrasion is responsible for 50% of all wear-related failures.

A typical example of abrasive wear is the damage of crankshaft journals in reciprocating compressors. Hard dirt particles will break through the lubricant film and cut or scratch the journal's comparatively softer surface.

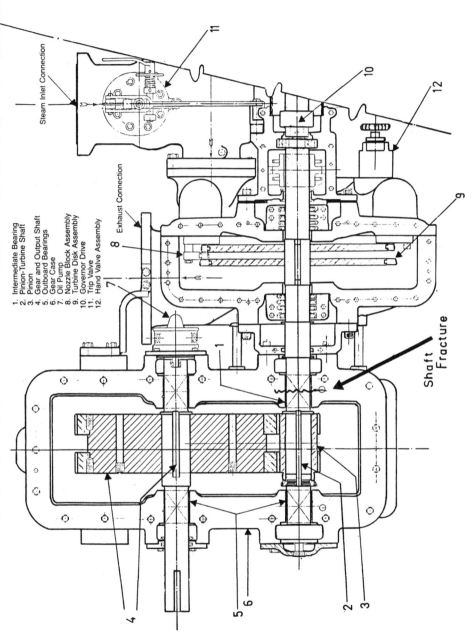

Steam Inlet Connection

1. Intermediate Bearing
2. Pinion-Turbine Shaft
3. Pinion
4. Gear and Output Shaft
5. Outboard Bearings
6. Gear Case
7. Oil Pump
8. Nozzle Block Assembly
9. Turbine Disk Assembly
10. Governor Drive
11. Trip Valve
12. Hand Valve Assembly

Exhaust Connection

Shaft Fracture

Figure 2-13B. Horizontal section of a built-in type geared turbine.

Adhesive Wear

Adhesive-wear failure modes are scoring, galling, seizing, and scuffing. They result when microscopic projections at the sliding interface between two mating parts weld together under high local pressures and temperatures. After welding together, sliding forces tear the metal from one surface. The result is a minute cavity on one surface and a projection on the other, which will cause further damage. Thus, adhesive wear initiates microscopically but progresses macroscopically.

Adhesive wear can be best eliminated by preventing metal-to-metal contact of sliding surfaces. This is accomplished either through a lubricant film or suitable coatings or through deposits such as Teflon® infusion layers, for instance. In the section on gear failure analysis in Chapter 3, adhesive wear will be discussed further.

Corrosive Wear

Corrosive wear is usually encountered on lubricated components. For example, a major source of gas-engine component wear in connection with frequent stops and starts is corrosion by water in the oil. Lube-oil management of rotating equipment is mainly directed toward eliminating the water in the oil and the corrosive wear associated with it. This topic is further developed in Chapter 3.

Effects of Corrosion

Experienced failure analysts know that corrosion often complicates machinery component failure investigations. Corrosion is the chemical or electrochemical reaction at the surface of the material which results in a change of the original surface finish or in local damage to the material.

In the case of corrosion of a metal in water, there is a reduction process taking place side by side with oxidation. Figure 2-14 shows typical types of corrosion. Porous or needle-shaped forms are referred to as "pitting corrosion."

If a corrosion-resistant material is technically or economically justified, it should be selected. If one cannot be justified, corrosion protection can be achieved by:

1. Surface coatings.
2. Inducing the formation of surface layers.
3. Corrosion preventive additives in the fluid.
4. Anodic or cathodic polarization.

Every material exhibits a given resistance to chemical or electrochemical attack. Earlier, we talked about increasing resistance to failure as a failure-fighting strategy. Naturally, in the context of corrosion, this strategy is most effective. Consequently, various materials have been evaluated for construction, and the troubleshooter may consult one of several tables published, such as those for pumps, shown in Figure 2-15.

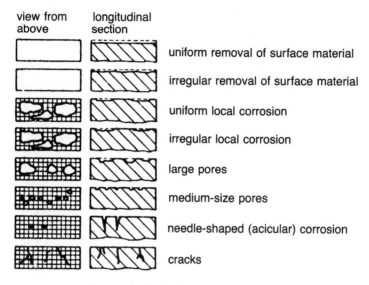

view from above	longitudinal section	

uniform removal of surface material

irregular removal of surface material

uniform local corrosion

irregular local corrosion

large pores

medium-size pores

needle-shaped (acicular) corrosion

cracks

Figure 2-14. Typical forms of corrosion.

Fretting Corrosion

Fretting wear, or fretting corrosion, is a failure mode of considerable practical significance. Fretting corrosion is also called "friction oxidation," "bleeding," and "red mud." It is defined as accelerated surface damage occurring at the interface of contacting materials subjected to small oscillatory displacements. Fretting corrosion is found in all kinds of press fits, spline connections, bearings, and riveted and bolted joints, among other places.

One important effect of fretting wear is its contribution to fatigue failures. Examinations of surface fractures have shown that fatigue cracks originate in or at the edge of a fretted area. An example of the effect of fretting on fatigue of an aged Al-Cu-Mg alloy is shown in Figure 2-16.

Detailed explanations of fretting wear have been given by Waterhouse.[12] Factors influencing fretting wear are given in the following list[13]:

1. *Slip:* No fretting can occur unless relative motion is sufficient to produce slip between the surfaces.
2. *Frequency:* Fretting wear rates increase at lower frequencies and become almost constant as frequency increases.
3. *Normal Load:* Fretting wear generally increases with applied load, provided the amplitude of slip is maintained constant.
4. *Duration:* Fretting wear increases almost linearly with the number of cycles.
5. *Temperature:* In general, fretting wear tends to increase with decreasing temperature.

MATERIALS OF CONSTRUCTION FOR PUMPING VARIOUS LIQUIDS

Column 1	Column 2	Column 3	Column 4	Column 5
Liquid	Condition of Liquid	Chemical Symbol	Specific Gravity	Material Selection
Acetaldehyde	....	C_2H_4O	0.78	C
Acetate Solvents	....	....	....	A, B, C, 8, 9, 10, 11
Acetone	....	C_3H_6O	0.79	B, C
Acetic Anhydride	....	$C_4H_6O_3$	1.08	8, 9, 10, 11, 12
Acid, Acetic	Conc. Cold	$C_2H_4O_2$	1.05	8, 9, 10, 11, 12
Acid, Acetic	Dil. Cold	....	....	A, 8, 9, 10, 11, 12
Acid, Acetic	Conc. Boiling	....	....	9, 10, 11, 12
Acid, Acetic	Dil. Boiling	....	....	9, 10, 11, 12
Acid, Arsenic, Ortho-	....	$H_3AsO_4, \tfrac{1}{2}H_2O$	2.0-2.5	8, 9, 10, 11, 12
Acid, Benzoic	....	$C_7H_6O_2$	1.27	8, 9, 10, 11
Acid, Boric	Aqueous Sol.	H_3BO_3	....	A, 8, 9, 10, 11, 12
Acid, Butyric	Conc.	$C_4H_8O_2$	0.96	8, 9, 10, 11
Acid, Carbolic	Conc. (M.P. 106 F)	C_6H_6O	1.07	C, 8, 9, 10, 11
Acid, Carbolic	(See Phenol)	....	....	B, 8, 9, 10, 11
Acid, Carbonic	Aqueous Sol.	CO_2+H_2O	....	A
Acid, Chromic	Aqueous Sol.	$Cr_2O_3 + H_2O$	....	8, 9, 10, 11, 12
Acid, Citric	Aqueous Sol.	$C_6H_8O_7 + H_2O$	....	A, 8, 9, 10, 11, 12
Acids, Fatty (Oleic, Palmitic, Ste...			....	A, 8, 9, 10, 11
Acid, Formi...			1.22	9, 10, 11
Acid, Fruit				A, 8, 9, 10, 11, 14
Acid, Hydro...			1.19 (38%)	11, 12
Acid, Hydro...				10, 11, 12, 14, 15
Acid, Hydro...				11, 12
Acid, Hydro...			0.70	C, 8, 9, 10, 11
Acid, Hydro...			....	3, 14
Acid, Hydro...			....	A, 14
Acid, Hydro...			1.30	A, 14
Acid, Lactic			1.25	A, 8, 9, 10, 11, 12
Acid, Mine...				A, 8, 9, 10, 11
Acid, Mixed				C, 3, 8, 9, 10, 11, 12
Acid, Muria...				....
Acid, Naph...				C, 5, 8, 9, 10, 11
Acid, Nitric				....
Acid, Nitric			1.50	6, 7, 10, 12
Acid, Oxalic				5, 6, 7, 8, 9, 10, 12
Acid, Oxalic			1.65	8, 9, 10, 11, 12
Acid, Ortho			....	10, 11, 12
Acid, Picric			1.87	9, 10, 11
Acid, Pyrog...			1.76	8, 9, 10, 11, 12
Acid, Pyroli...			1.45	8, 9, 10, 11
Acid, Sulfu...				A, 8, 9, 10, 11
Acid, Sulfu...			1.69-1.84	C, 10, 11, 12
Acid, Sulfu...				11, 12
Acid, Sulfu...				10, 11, 12
Acid, Sulfu...				10, 11, 12
Acid, Sulfu...				A, 10, 11, 12, 14
Acid, Sulfu...				....
Acid, Sulfu...			1.92-1.94	3, 10, 11
Acid, Sulfu...			....	A, 8, 9, 10, 11
Acid, Tanni...			....	A, 8, 9, 10, 11, 14

Summary of Material Selections and National Society Standards Designations

Material Selection	Corresponding National Society Standards Designation			Remarks
	ASTM	ACI	AISI	
1.....	A48, Classes 20, 25, 30, 35, 40 & 50	......	......	Gray Iron—Six Grades
1(a)..	A536 & A395	......	......	Ductile Cast Iron—Six Grades
2.....	B143, 1A, 1B & 2A; B144, 3A; B145, 4A	......	......	Tin Bronze & Leaded Tin Bronze—seven alloys (includes 2 alloys not covered by ASTM Specifications, as explained above under Selection #2)
3.....	A216-WCB	......	1026	Carbon Steel
4.....	A217-C5	......	501	5% Chromium Steel
5.....	A296-CA15	CA15	410	12% Chromium Steel
6.....	A296-CB30	CB30	......	20% Chromium Steel
7.....	A296-CC50	CC50	446	28% Chromium Steel
8.....	A296-CF-8	CF-8	304	19-9 Austenitic Steel
9.....	A296-CF-8M	CF-8M	316	19-10 Molybdenum Austenitic Steel
10.....	A296-CN-7M	CN-7M	......	20-29 Chromium Nickel Austenitic Steel with Copper & Molybdenum
11.....	......	......	......	A series of nickel-base alloys
12.....	A518	......	......	Corrosion Resistant High-silicon cast iron
13.....	A436	......	......	Austenitic cast iron—2 types
13(a) .	A439	......	......	Ductile Austenitic Cast Iron
14.....	......	......	......	Nickel-Copper alloy
15.....	......	......	......	Nickel

*ASTM—denotes American Society for Testing Materials
ACI—denotes Alloy Casting Institute
AISI—denotes American Iron and Steel Institute

Figure 2-15. Example for listing of materials of construction for pumps.[11]

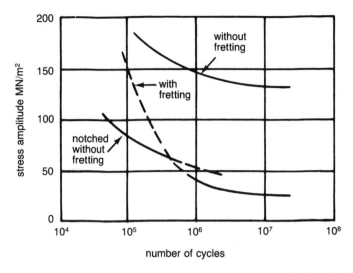

Figure 2-16. Effect of fretting on the fatigue of an aged Al-Cu-Mg alloy.[13]

6. *Atmosphere:* Fretting wear that occurs in air or oxygen atmospheres is far more severe than that produced in an inert atmosphere.
7. *Humidity:* Increased relative humidity decreases fretting wear for most metals. This is because the moisture in the air provides a lubricating film which promotes the removal of debris from contact areas.
8. *Surface Finish:* Generally, fretting wear is more serious when the surfaces are smooth, as a smooth finish will have smaller and fewer "lubricant pockets" on its surface.
9. *Lubricants:* Fretting wear is reduced by lubricants which restrict the access of oxygen. Lubricants remove debris from the fretted zone and change the coefficient of friction.
10. *Hardness:* It is generally accepted that an increase in materials hardness decreases fretting wear.
11. *Coefficient of Friction:* Fretting wear generally decreases with a decreasing coefficient of friction because of a reduced slip amplitude.

Consequently, the prevention of fretting corrosion and wear lies in the elimination of relative displacement. One way to achieve this is to decrease fit clearances. Troubleshooters encountering fretting corrosion would do well to see, for instance, that improved shaft tolerances in repair work are adopted. The tighter fits obtained by such measures will increase contact areas between shafts and bores and consequently increase frictional force.

Increasing the coefficient of friction is another preventive measure. This can be achieved by coating the contact surfaces with materials with suitable frictional properties. These coatings could be metallic or nonmetallic. Metallic coatings that have been successfully employed are cadmium, silver, gold, tin, lead, copper, and chromium. Nonmetallic coatings result from chemical treatments such as phosphatizing, anodizing, and sulfudizing, or from bonding of materials such as polymer of MoS_2 (molybdenum-disulfide) or Teflon to the contacting surfaces. Diffusion-coating techniques such as carburizing, nitriding, and cold working (i.e. cold rolling, shot-peening, and roll peening) enhance the fatigue strength of the contacting members.

Finally, looking at the factors that influence fretting wear, the analyst should always attempt to minimize their impact in order to economically prevent this failure mode.

Cavitation Corrosion

Cavitation, or the formation of cavities in a fluid, is the appearance and subsequent collapse of vapor bubbles in the flow of liquid. Vapor bubbles are formed when the static pressure in the liquid sinks so low that it attains the vapor pressure associated with the temperature of the liquid at that particular point. If the static pressure rises after this above the vapor pressure along the flow path, the vapor bubble collapses quite suddenly, followed by sudden condensation in the form of an implosion. If the implosion occurs not in the body of flowing liquid but at the wall of a component containing the flowing liquid, cavitation will result in material erosion.

Recent research in the field of cavitation has indicated that the vapor bubble inverts at first, once the implosion begins. After that, a fluid microjet is formed, directed toward the interior of the bubble, which pierces the opposite wall of the bubble. Slow-motion pictures[14] of the phenomenon indicate that where bubbles are close to a wall, liquid microjets are always directed against the wall, striking it at high speed. This causes material disintegration, which is in turn intensified by chemical action: The microjet entrains the dissolved oxygen in the liquid, which is then liberated in the vapor and forced at high pressure between the grain boundaries of the material at the wall surface. This process increases the corrosion of the wall material.

The first sign of cavitation corrosion will be pitting. In its progressed form, it will have a honey-combed, spongy appearance and structure. The amount of material removed by cavitation could be determined by:

1. Geometric measurements.
2. Weight loss.
3. Measurement of amount of built-up repair materials, i.e. metal deposited by welding or quantity of molecular metal used.
4. Mean-time-to-repair.

Figures 2-17 and 2-18 show a bronze, mixed-flow pump impeller from a cooling tower pump in a petrochemical plant after five years of operation under cavitation conditions.

Usually the troubleshooter will attempt to curtail the effects of cavitation corrosion by design or operational changes. Quite often, it will be impossible to shift the collapse of the vapor bubbles away from the wall toward the center of the flow path, just as it will be uneconomical, for instance, to change existing unfavorable

Figure 2-17. Cast-bronze mixed-flow pump impeller failed from cavitation corrosion.

Figure 2-18. Close-up of pump impeller from Figure 2-17.

Table 2-7
Cavitation-Induced Loss-of-Weight Indices
of Pump Materials[11]

Grey cast iron	1.0
Cast steel	0.8
Bronze	0.5
Cast chrome steel	0.2
Multicomponent bronze	0.1
Chrome nickel steel	0.05

submersion conditions in a vertical cooling tower pump installation. A change of materials of construction will be appropriate under these circumstances.

Materials resistant to cavitation are those with high fatigue strength and ductility, together with a high corrosion resistance. Comparative cavitation resistance indexes are listed in Table 2-7.

Analyzing Wear Failures*

An accurate analysis of a wear failure depends largely upon three sources of evidence: the worn surfaces, the operating environment, and the wear debris.

Surface damage can range from polished or burnished conditions to removal of a relatively large volume of material. Examining the worn surface can provide much information, including the amount of material removed, the type of damage or failure mode, and the existence and character of surface films. Further, it will tell whether certain constituents are being attacked preferentially, the direction of relative motion between a worn surface and abrading particles, and whether abrading particles have become embedded in the surface.

Environmental Conditions

The environment in which the failure takes place greatly affects the mechanism and rate of metal removal, and detailed knowledge of these conditions should always be sought. For instance, a coke crusher experiences erratic wear, with increased wear occurring at certain abnormal operating conditions. It would stand to reason to investigate what circumstances lead to these operating conditions and then, by eliminating them, stop the accelerated wear of the coke crusher.

Wear environments may be corrosive; may have been altered during service, such as by breakdown of a lubricant; may provide inadequate lubrication; or may differ from the assumed environment on which the original material selection was made.

---

*Adapted by permission from course material published by the American Society of Metals, Metals Park, Ohio 44073. For additional information on metallurgical service failures, refer to the 15-lesson course *Principles of Failure Analysis*, available from the Metals Engineering Institute of the American Society for Metals, Metals Park, Ohio 44073.

Wear debris, whether found between the worn surfaces, embedded in a surface, suspended in the lubricant, or beside the worn part, can provide clues to the wear mechanism. A wear particle consisting of a metallic center with an oxide covering probably was detached from the worn surface by abrasive or adhesive wear and subsequently oxidized by exposure to the environment. On the other hand, a small wear particle consisting solely of oxide may result from corrosion on the worn surface which was subsequently removed mechanically.

Analysis Procedure

Though conditions may vary greatly, the general steps in wear-failure analysis are described in Table 2-8.

Operating Conditions

Probably the first step in wear-failure analysis is to identify the type of wear or, if more than one type can be recognized, to evaluate the relative importance of each type as quantitatively as possible. This identification of the type or types of wear requires a detailed description of the operating conditions based on close observation and adequate experience. A casual and superficial description of operating conditions will be of little value.

Descriptions of operating conditions often are incomplete, thus imposing a serious handicap on the failure analyst, especially if he is working in a laboratory remote from the plant site. For instance, assume that an analyst must study the problem of a

Table 2-8
Wear-Failure Analysis

1. Identify the actual materials in the worn part, the environment, the abrasive, the wear debris, and the lubricant.
2. Identify the mechanism or combination of mechanisms of wear: adhesive, abrasive, corrosive, surface fatigue, or erosive.
3. Define the surface configuration of both the worn surface and the original surface.
4. Define the relative motions in the system, including direction and velocity.
5. Define the force or pressure between mating surfaces or between the worn surface and the wear environment, both macroscopically and microscopically.
6. Define the wear rate.
7. Define the coefficient of friction.
8. Define the effectiveness and type of lubricant: oil, grease, surface film, naturally occurring oxide layer, adsorbed film, or other.
9. Establish whether the observed wear is normal or abnormal for the particular application.
10. Devise a solution, if required.

badly seized engine cylinder. Obviously this is an instance of adhesive metal-to-metal wear or lubricated wear, because it is assumed that a suitable engine oil was used. Furthermore, assume that during an oil change the system had been flushed with a solvent to rinse out the old oil, and that inadvertently it had been left filled with the solvent instead of new oil. Also assume that a slow leak, resulting in loss of the solvent, was not detected during the operating period immediately preceding seizure. The analyst probably would receive the failed parts (cylinder block and pistons) after they had been removed from the engine, cleaned, and packed. If evidence of the substitute "lubricant" could not be established clearly, determination of the cause of failure would be extremely difficult or perhaps impossible.

Similarly, incomplete descriptions of operating conditions can be misleading in the analysis of abrasive wear. For example, in describing the source of abrasion that produces wear of a bottoms pump impeller, generalized references to the abrasive in question, such as coke particles, might be too vague. Unless the abrasives are studied both qualitatively and quantitatively, a valid assessment of wear, whether normal or abnormal, is impossible.

Solving Wear Problems

As stated earlier, wear failures differ from many other types of failures, such as fatigue, because wear takes place over a period of time. Seldom does the part suddenly cease to function properly. In most instances, wear problems are solved by two approaches: The service conditions are altered to provide a less destructive environment, or materials more wear resistant for the specific operating conditions are selected. Because the latter method is easier and less expensive, it is chosen more frequently.

Laboratory Examination of Worn Parts

Analysis of wear failures depends greatly on knowledge of the service conditions under which the wear occurred. However, many determining factors are involved that require careful examination, both macroscopically and microscopically.

Wear failures generally result from relatively long-time exposure, yet certain information obtained at the time the failure is discovered is useful in establishing the cause. For example, analyzing samples of the environment—often the lubricant— can reveal the nature and amount of wear debris or abrasive in the system.

Procedures. Examination of a worn part generally begins with visual observation and measurement of dimensions. Observations of the amount and character of surface damage often must be made on a microscopic scale. An optical comparator, a toolmaker's microscope, a recording profilometer, or other fine-scale measuring equipment may be required to adequately assess the amount of damage that has occurred.

Weighing a worn component or assembly and comparing its weight with that of an unused part can help define the amount of material lost. This material is lost in abrasive wear or transferred to an opposing surface in adhesive wear. Weight-loss estimates also help define relative wear rates for two opposing surfaces that may be made of different materials, or that have been worn by different mechanisms.

Screening the abrasives or wear debris to determine the particle sizes and weight percentage of particles of each size is often helpful. Both a determination of particle size and a chemical analysis of the various screenings can be useful when one component in an abrasive mixture primarily causes wear, or when wear debris and an abrasive coexist in the wear environment. The combination of screening with microscopy often can reveal details such as progressive alteration of the size and shape of abrasive particles with time, as might occur in a fluid coker ball mill.

Physical measurements can define the amount and location of wear damage, but they seldom provide enough information to establish either the mechanism or the cause of the damage.

Microscopy can be used to study features of the worn surface, including the configuration, distribution, and direction of scratches or gouges and indications of the preferential removal of specific constituents of the microstructure. Abrasive particles or wear debris are viewed under the microscope to study their shape and the configuration of their edges (sharp or rounded), and to determine if they have fractured during the wear process.

Metallography can determine whether or not the initial microstructure of the worn part was to specifications. It also will reveal the existence of localized phase transformation, cold-worked surface layers, and embedded abrasive particles.

Taper sectioning. For valid analyses, techniques such as taper sectioning are needed to allow metallographic observations or microhardness measurements of very thin surface layers. Almost always, special materials that support the edge of a specimen in a metallographic mount (for example, nickel plating on the specimen or powdered glass in the mounting material) must be used, and the mounted specimen must be polished with care.

Etchants, in addition to preparing a specimen for the examination of its microstructure, also reveal characteristics of the worn surface. Two features that can be revealed by etching a worn surface are phase transformations caused by localized adhesion to an opposing surface and the results of overheating caused by excessive friction, such as the "white layer" (untempered martensite) that sometimes develops on steel or cast iron under conditions of heavy sliding contact.

Macroscopic and microscopic hardness testing indicate the resistance of a material to abrasive wear. Because harder materials are likely to cut or scratch softer materials, comparative hardness of two sliding surfaces may be important. Microhardness measurements on martensitic steels may indicate for example, that frictional heat has overtempered the steel, and when used in conjunction with a tempering curve (a plot of hardness versus tempering temperature) they will give a rough estimate of surface temperature. Hardness measurements also can indicate whether a worn part was heat treated correctly.

Chemical analysis. One or more of the various techniques of chemical analysis—wet analysis, spectroscopy, calorimetry, x-ray fluorescence, atomic absorption, or electron-beam microprobe analysis—usually is needed to properly analyze wear failures. The actual compositions of the worn material, the wear debris, the abrasive, and the surface film must be known in order to devise solutions to most wear problems. An analysis of the lubricant, if a worn lubricated component is involved, should always be a matter of course.

Recording Surface Damage

Sometimes the experienced troubleshooter would like to record and preserve the appearance of surface failures. Several methods have been applied. One is the use of dental impression materials* that are cast in liquid form onto failed or damaged gear-tooth surfaces, for instance. These materials will yield highly accurate prints of surface damage.[15]

Another method, which has been successfully applied by one of the authors, is described in Ref. 9. This method is simple and can be used to record a variety of surface failures such as pitting, scoring, and general wear. With it, replicas of entire cylindrical surfaces or gear tooth flanks can be recorded permanently on a single sheet of paper. The procedure is as follows:

1. Clean the surface and leave a light oil film.
2. Brush the surface to be reproduced with graphite powder. Brush off the excess with a soft brush or tissue.
3. Apply pressure-sensitive transparent tape (matte Scotch tape), adhesive down, on the surface of interest. Rub the tape so that it adheres to the surface.
4. Strip off the tape and apply it to a sheet of white paper or a glass slide so that the surface appearance is permanently recorded.

□

In conclusion, Table 2-9 lists machinery components and their potential surface failure modes. Tables 2-10 and 2-11 will help the failure analyst identify major wear failure modes by registering their appearance and nature of debris. This is a first approach leading eventually to improved material selection as indicated in Table 2-12.

(text continued on page 52)

---

*Silicone rubbers, i.e. Reprosil or addition-cured polyvinylsiloxanes.

Table 2-9
Process Machinery Components and Wear Failure Modes

	Scuffing	Fretting wear	Galling	Corrosion	Erosion	Sticking
Trip valve stems on turbines (all)	•	•	•			•
Governor valve stems on turbines		•				•
Governor linkage pivots on turbines		•				•
Pump wear rings	•		•			
Pump impellers				•	•	
Sleeves or shafts under mechanical seals		•				•
Seal faces, high-viscosity fluid						•
Seal faces, metal on metal	•		•			
Antifriction bearings		•		•		
Centrifugal compressor impellers				•		
Centrifugal compressor labyrinths				•	•	
Centrifugal compressor shaft sleeves				•	•	
Centrifugal compressor breakdown bushings	•					
Centrifugal compressor shafts under bearings and seals		•				
Centrifugal compressor guide vanes and diaphragms		•		•		•
Pump casings				•	•	
Pump sleeves	•					
Non-lube compressor valve plates, seals, springs	•					
Compressor piston rods	•					
Compressor piston rings	•			•		
Engine pistons	•					
Engine piston rings	•					
Engine camshafts	•					
Engine valve stems	•		•			•
Coupling-hub shaft fits		•				
Gear couplings	•					
Gears	•					
Scrapers for exchangers	•					•
Gear pumps	•					
Cams or followers for turbine valve gear	•					
Reciprocating plungers	•					
Reciprocating pistons	•					
Reciprocating valves	•					
Reciprocating steam pistons	•					
Reciprocating valve gear	•					
Progressive cavity pump impellers	•					
Screw pump rotors and idlers	•					
Sliding piping supports at machinery	•	•				

Table 2-10

Surface Appearance and Failure Mode in Wear Analysis (Modified from Ref. 16)

Category	Description of Worn Surface	Melting	Exfoliation	Galling	Grooving	Scoring	Gouging	Scuffing	Scratching	Polishing	Fretting	Abrasive Wear	Erosion	Cavitation	Spalling	Pitting	Staining
Micro-Smooth	Progressive loss and reformation of surface films by fine abrasion and/or tractive stresses, mutually imposed by adhesive or viscous interaction.				•	•		•		•	•						
Micro-Smooth	Very fine abrasion, with loss of substrate in addition to loss of surface film.									•			•				
Micro-Smooth	Melting	•															
Micro-Rough	Due to tractive stresses resulting from adhesion.							•									
Micro-Rough	Micro-pitting by fatigue.													•		•	
Macro-Smooth	Abrasion by medium-coarse particles.								•			•					
Macro-Smooth	By abrasive held on or between solid backing.						•		•			•					
Macro-Rough	Abrasion by coarse particles, including carbide and other hard inclusions in the sliding materials.				•		•										
Macro-Rough	Abrasion by fine particles in turbulent fluid, producing scallops, waves, etc.												•				
Macro-Rough	Severe adhesion.			•				•									
Macro-Rough	Local fatigue failure resulting in pits or depressions by repeated rolling contact stress, and high friction sliding or impact by hard particles as in erosion.														•		
Macro-Rough	Advanced stages of micro-roughening, where little unaffected surface remains between pits.		•								•						
Shiny	Very thin—or no—surface film of oxide, hydroxide, sulfide, chloride, or other.		•		•			•		•							
Dull	Thick films resulting from aggressive environments, including high temperatures due to corrosion.										•						•

Table 2-11
Debris and Material-Loss Mechanisms
(Modified from Ref. 16)

Wear Debris	Material-Loss Mechanisms*
Long, often curly, chips or strings	Abrasion—involves particles (of some acute angular shapes but mostly obtuse) that cause wear debris, some of which forms ahead of the abrasive particle (*cutting* mechanism) but most of which is material that has been plowed aside repeatedly by passing particles and that breaks off by *low-cycle fatigue*.
	Adhesion—a strong bond that develops between two surfaces (either between coatings and/or substrate materials), which, with relative motion, produces tractive stress that may be sufficient to deform materials to fracture. The mode of fracture will depend on the property of the material, involving various amounts of energy loss or ductility to fracture, i.e.:
Solid particles, often with cleavage surfaces	low energy and ductility *brittle fracture*
	——————— or ———————
Severely deformed solids, sometimes with oxide clumps mixed in	high energy and ductility *ductile fracture*
Solid particles, often with cleavage surfaces with ripple pattern	Fatigue—due to cyclic strains, usually at stress levels below the yield of the material, also called *high-cycle fatigue*.

* In italics.

Table 2-12
Material-Loss Mechanisms and Wear Resistance (Modified from Ref. 16)

Material-Loss Mechanisms	Appropriate Material Characteristics to Resist Wear	Precautions to be Observed when Selecting a Material*
Corrosion	Reduce corrosiveness of surrounding region; increase corrosion resistance of material by alloy addition or by selecting a soft, homogeneous material.	Soft materials tend to promote galling and seizure.
Cutting	Use material of high hardness, with very hard particles or inclusions such as carbides, nitrides, etc. and/or overlaid or coated with materials that are hard or that contain very hard particles.	All methods of increasing cutting resistance cause brittleness and lower fatigue resistance.
Ductile Fracture	Achieve high strength by any method other than by cold working or by heat treatments that produce internal cracks or large and poorly bonded intermetallic compounds.	
Brittle Fracture	Minimize tensile residual stress for cold temperature; ensure low-temperature brittle transition; temper all martensites; use deoxidized metal; avoid carbides; effect good bond between fillers and matrix to deflect cracks.	Soft materials will not fail through brittleness and will not resist cutting.
Low-cycle Fatigue	Use homogeneous and high-strength materials that do not strain-soften; avoid overaged materials or two-phase systems with poor adhesion between filler and matrix.	
High-cycle Fatigue	For steel and titanium, use stresses less than half the tensile strength; for other materials to be load-cycled < 10^8 times, allow stresses less than $1/4$ the tensile strength; avoid decarburization of surfaces; avoid tensile stresses or form-compressive residual stresses by carburizing or nitriding.	Calculation of stress should include the influence of tractive stress.

* Materials of high hardness or strength usually have decreased corrosion resistance, and all materials with multiple and carefully specified properties and structure are expensive.

References

1. *Metals Handbook,* Vol. 10, "Failure Analysis and Prevention," published by the American Society for Metals, Metals Park, Ohio 44073, 8th Edition, 1975, pp. 10-26.

2. Hutchings, F. R., "The Laboratory Examinations of Service Failures", Technical Report, Vol. III, British Engine, Boiler, and Electrical Insurance Co., Ltd., London, U.K., May 1957.

3. Van der Voort, G. F., "Conducting the Failure Examination", *Met. Eng. Quart.*, May 1975.

4. *Handbook of Loss Prevention,* Editors: Allianz Versicherungs-AG, translated from the German by P. Cahn-Speyer, 1978, Berlin-Heidelburg-New York: Springer-Verlag, p. 321.

5. Investigation of Threaded Fastener Structural Integrity, Final report, Contract No. NAS9-15140, DRL-T-1190, CLIN3, DRDMA 129TA, Southwest Research Institute, Project No. 15-4665, Oct. 1977.

6. Bowman Products Division, "Fastener Facts," Cleveland, Ohio, 1974.

7. Dubbel, Taschenbuch für den Machinenbau, Editors: W. Beitz & K. H. Küttner, Berlin-Heidelberg-New York: Springer Verlag, 14th Edition, 1981, pp. 382-384.

8. Bickford, H. J., *An Introduction to the Design and Behavior of Bolted Joints,* New York and Basel: Marcel Dekker, Inc., 1981, p. 79.

9. Wulpi, D. J., "How Components Fail," American Society for Metals, Metals Park, Ohio 44073, 1966.

10. Eyre, T. S., "Wear Characteristics of Metals," *Tribology International,* October 1976, pp. 203-212.

11. Hydraulic Institute Standards, 13th Edition, Hydraulic Institute, Cleveland, Ohio, 1975, "Materials of Construction of Pumping Various Liquids," pp. 285-296.

12. Waterhouse, R. B., *Fretting Corrosion,* Pergamon Press, 1972.

13. Braunovic, M., "Tribology: Fretting Wear," *Engineering Digest,* Toronto, October 1981, pp. 48-54.

14. Klein, Schanzlin, and Becker, A. G., *Centrifugal Pump Lexicon,* Frankenthal, Germany, 1975.

15. Hennings, Cr., "Ermittlung des Schädigungszustandes an Zahnflanken," *Schmierungstechnik* 10 (1979) 10, pp. 307-310.

16. Ludema, K. E., "How to Select Material Properties for Wear Resistance," *Chemical Engineering,* July 27, 1981, pp. 89-92.

Chapter 3

Machinery Component Failure Analysis

Bearings in Distress

In the preceding pages, we looked at a variety of wear-related failures. We can readily conclude that a large portion of lubricated machinery components will experience wear damage through lubrication failure. Bearings are no exception. Lubrication-related bearing problems, according to our experience, are most frequently caused by *lack* of lubrication or lubricant *contamination.* Two examples will illustrate this.

As the first example, analysis of a pump bearing failure showed that a rolling-element bearing failed from lack of oil. However, there was a copious supply of *fresh* oil in the housing. It is not too difficult to conclude what really happened. The positive thinker will not despair but will take heart in knowing that the operating crew still owns up to their duty of keeping oil in the bearing. Perhaps somewhat late in *this* case—but better late that never as far as the rest of the many pumps in their care is concerned.

The second example: A certain engineer once described the person with the water hose as the rolling-element-bearing salesman's best friend in the paper mill.[1] No doubt, the same can sometimes be said in the chemical process industry. Nevertheless, given a proper lubrication environment and proper design service conditions, plain journal bearings will last indefinitely. However, rolling-element bearings, even under optimum conditions, will fail in wear-out mode.[2] In other words, the ultimate end of all rolling-element bearings would be fatigue if it were possible to run them long enough. In any group of rolling-element bearings, if all were run under identical conditions until all failed by fatigue, there would be a considerable spread between the longest and the shortest "lives." Figure 3-1 illustrates this fact in the form of a life dispersion curve. In short and simple terms, given proper design and application, rolling-element bearings will sooner or later fail in wear-out mode, plain bearings will last almost indefinitely, but all will fail early from abuse.

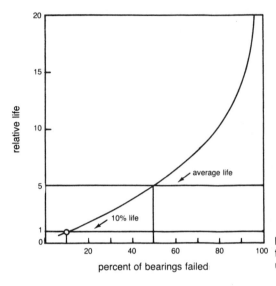

Figure 3-1. Life dispersion curve for typical group of bearings run under identical conditions.[2]

Table 3-1
Failure Causes of Rolling and Plain Bearings[3]

Failure Cause	Occurrence, %	
	Rolling Bearings	Plain Bearings
Vendor problems	30.1	23.4
Workmanship	14.4	10.7
Errors in design/applications	13.8	9.1
Wrong material of construction	1.9	3.6
User-induced problems	65.9	69.6
Operational errors, maintenance deficiencies, failure of monitoring equipment	37.4	39.1
Wear	28.5	30.5
External problems	4.0	7.0
Contaminated lubricants; intermittent failure of oil supply system	4.0	7.0

Typical bearing-failure causes have been compiled by Allianz Insurance of Munich, Germany.[3] Table 3-1 contains the results of investigations of the causes of 1400 rolling-bearing failures and 530 plain-bearing failures. It can be seen that for these components, about 30% of the functional failures are due to wear- or lubrication-related processes.

Based on our experience, we find these statistics somewhat controversial as far as the ratio of wear-related failures to other user-induced problems is concerned. Also, only a very sophisticated analysis strategy could come up with a similar separation of

Table 3-2
Antifriction Bearing Failure Modes and Their Causes

	Failure Causes	Spalling/flaking	Cracks/heat cracks	Smearing → seizing	Cage fracture	Cage deformation	Indentations	Fragment denting	Brinelling/false brinelling	Ball Path—widened	Ball Path—skew	Ball Path—uneven load zones	Fluting	Cage wear	Abrasive wear/wear	Overheating → burning → scuffing	Corrosion/etching	Fit Corrosion/fretting	Rust staining
		Fracture/Separation				**Deformation**								**Wear**			**Corrosion**		
Assembly	Excessive, uneven heat application	●																	
	Hammer blows		●				●	●											
	Improper tooling					●	●	●											
	Loose/tight fits			●						●					●			●	
	Distorted bearing housing	●							●	●									
	Rotor unbalance											●							
	Misalignment				●	●				●	●			●					
Operating Conditions	Vibration (Moving/Stationary)	●							●	●					●			●	
	Current Passage												●			●			
	Life attainment/fatigue	●																	
	Overload	●	●							●						●			
	Design error		●	●												●			
Seal	Contamination						●	●							●	●	●		
	Moisture ingress																	●	●
Lubrication	Lack of lubricant			●	●										●	●	●		●
	Excess of lubricant			●												●			
	Unsuitable lubricant			●					●						●	●	●		●

user-induced problems and external problems. Nevertheless, similar data should be compiled for your own plant, because knowledge of failure-cause distributions at the component and parts levels will eventually result in better specifications.

Tables 3-2 and 3-3 will help relate both rolling-element and plain-bearing defects or failures modes to their most probable causes.

Rolling-Element Bearing Failures and Their Causes*

Accurate and complete knowledge of the causes responsible for the breakdown of a machine is as necessary to the engineer as similar knowledge concerning a breakdown in health is to the physician. The physician cannot effect a lasting cure

---

*Copyright © SKF Industries, King of Prussia, PA. Reprinted by permission.

Table 3-3
Plain (Journal) Bearing Failure Modes and Their Causes

Failure Causes	Spalling	Cracking	Seizing	Deformation	Embedments	Uneven load patterns	Scoring → galling	Overheating → scuffing	Wear	Abrasive wear/scratching/grooving	Erosion	Cavitation	Corrosion
(Fracture/Separation)	Fracture/Separation	Fracture/Separation	Fracture/Separation	*Deformation*	*Wear*	*Wear*	*Wear*	*Wear*	*Wear*	*Wear*	*Erosion*	*Erosion*	*Erosion*
Assembly and Manufacture													
Insufficient clearance			•						•				
Misaligned journal bearing				•		•							
Rough surface finish on journal								•	•				
Pores and Cavities in bearing metal		•											
Insufficient metal bond	•												
Operating Conditions and Design													
General operating conditions											•	•	
Overload/fatigue	•	•										•	
Overload/vibration	•	•	•				•		•	•		•	
Current passage												•	•
Unsuitable bearing material				•					•				
Lubrication													
Contamination of lubricant				•		•	•		•	•			•
Insufficient or lack of lubricant				•	•	•	•	•	•	•			
Oil viscosity too low									•				
Oil viscosity too high									•				
Improper lubricant selection								•					
Lubricant deterioration									•				•

unless he knows what lies at the root of the trouble, and the future usefulness of a machine often depends on correct understanding of the causes of failure. Since the bearings of a machine are among its most vital components, the ability to draw the correct inferences from bearing failures is of utmost importance.

In designing the bearing mounting the first step is to decide which type and size of bearing shall be used. The choice is generally based on a certain desired life for the bearing. The next step is to design the application with allowance for the prevailing service conditions. Unfortunately, too many of the ball and roller bearings installed never attain their calculated life expectancy because of something done or left undone in handling, installation, and maintenance.

The calculated life expectancy of any bearing is based on the assumptions that good lubrication in proper quantity will always be available to that bearing, that the bearing will be mounted without damage, that dimensions of parts related to the bearing will be correct, and that there are no defects inherent in the bearing.

However, even when properly applied and maintained the bearing will still be subjected to one cause of failure: fatigue of the bearing material. Fatigue is the result of shear stresses cyclically applied immediately below the load-carrying surfaces and is observed as spalling away of surface metal, as seen in Figures 3-2 through 3-4. However, material fatigue is not the only cause of spalling. There are causes of premature spalling. So, although the observer can identify spalling, he must be able to discern between spalling produced at the normal end of a bearing's useful life and that triggered by causes found in the three major classifications of premature spalling: lubrication, mechanical damage, and material defects. Most bearing failures can be attributed to one or more of the following causes:

1. Defective bearing seats on shafts and in housings.
2. Misalignment.
3. Faulty mounting practice.
4. Incorrect shaft and housing fit.
5. Inadequate lubrication.

Figure 3-2. Incipient fatigue spalling.

Figure 3-3. More-advanced spalling.

Figure 3-4. Greatly advanced spalling.

6. Ineffective sealing.
7. Vibration while the bearing is not rotating.
8. Passage of electric current through the bearing.

The actual beginning of spalling is invisible because the origin is usually below the surface. The first visible sign is a small crack, and this too is usually indiscernible. The crack cannot be seen nor its effects heard while the machine operates. Figures 3-2 through 3-4 illustrate the progression of spalling. The spot on the inner ring in Figure 3-2 will gradually spread to the condition seen in the ring of Figure 3-4, where spalling extends around the ring. Figure 3-2 illustrates incipient fatigue spalling. Figure 3-3 shows more advanced spalling, with still further degrees of deterioration shown in Figure 3-4. By the time spalling reaches proportions shown in Figure 3-3, the condition should announce itself by noise. If the surrounding noise level is too great, a bearing's condition can be learned by listening to the bearing through its housing by means of a metal rod. The time between incipient and advanced spalling varies with speed and load, but it is not a sudden condition that will cause destructive failure within a matter of hours. Total destruction of the bearing and consequent damage to machine parts is usually avoided because of the noise the bearing will produce and the erratic performance of the shaft carried by the bearing.

Patterns of Load Paths
and Their Meaning in Bearing Damage

There are many ways bearings can be damaged before and during mounting and in service. The pattern or load zone produced on the internal surfaces of the bearing by the action of the applied load and by the rolling elements is a clue to the cause of failure. To benefit from a study of load zones, one must be able to differentiate between normal and abnormal patterns. Figure 3-5 illustrates how an applied load of constant direction is distributed among the rolling elements of a bearing. The large arrow indicates the applied load, and the series of small arrows show the share of this load that is supported by each ball or roller in the bearing. The rotating ring will have a continuous 360° zone, while the stationary ring will show a pattern of less than 180°. Figure 3-6 illustrates the load zone found inside a ball bearing when the inner ring rotates and the load has a constant direction. Figure 3-7 illustrates the load zone resulting if the outer zone rotates relative to a load of constant direction, or when the inner ring rotates and the load also rotates in phase with the shaft. Figure 3-8 illustrates the pattern found in a deep-groove ball bearing carrying an axial load, and Figure 3-9 shows the pattern from excessive axial load. Uniformly applied axial load and overload are the two conditions where the load paths are the full 360° of both rings. Combined thrust and radial load will produce a pattern somewhere between the two, as shown in Figure 3-10. With combined load, the loaded area of the inner ring is slightly off-center, and the length in the outer is greater than that produced by

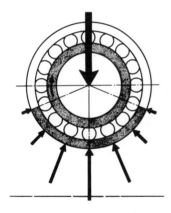

Figure 3-5. Load distribution within a bearing.

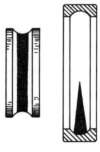

Figure 3-6. Normal-load zone, inner ring rotating relative to load.

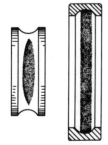

Figure 3-7. Normal-load zone, outer ring rotating relative to load, or load rotating in phase with inner ring.

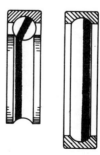

Figure 3-8. Normal-load zone, axial load.

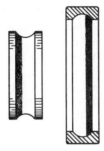
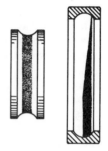

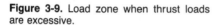

Figure 3-9. Load zone when thrust loads are excessive.

Figure 3-10. Normal-load zone, combined thrust and radial load.

radial load, but not necessarily 360°. In a two-row bearing, a combined load will produce zones of unequal length in the two rows of rolling elements. If the thrust is of sufficient magnitude, one row of rolling elements can be completely unloaded.

When an interference fit is required, it must be sufficient to prevent the inner ring from slipping on the shaft but not so great as to remove the internal clearance of the bearing. There are standards defining just what this fit should be for any application and bearing type, and a discussion of fitting practice appears later in this chapter, under "Damage Due to Improper Fits." If the fit is too tight, the bearing can be internally preloaded by squeezing the rolling elements between the two rings. In this case, the load zones observed in the bearing indicate that this is not a normal-life failure, as Figure 3-11 shows. Both rings are loaded through 360°, but the pattern will usually be wider in the stationary ring, where the applied load is superimposed on the internal preload.

Distorted or out-of-round housing bores can radially pinch an outer ring. Figure 3-12 illustrates the load zone found in a bearing where the housing bore was initially out-of-round or became out-of-round from the housing being bolted to a concave or convex surface. In this case, the outer ring will show two or more load zones, depending on the type of distortion. This is actually a form of internal preload. Figure 3-13 is a photograph of a bearing that had been mounted in an out-of-round housing that pinched the stationary outer ring. This is a mirror view and shows both sides of the outer ring raceway.

Certain types of rolling bearings can tolerate only very limited amounts of misalignment. A deep-groove ball bearing, when misaligned, will produce load zones not parallel to the ball groove on one or both rings, depending on which ring is misaligned. Figure 3-14 illustrates the load zone when the outer ring is misaligned relative to the shaft. When the inner ring is misaligned relative to the housing, the pattern is as shown in Figure 3-15. Cylindrical roller bearings, tapered roller bearings, and angular-contact ball bearings are also sensitive to misalignments, but it is more difficult to detect this condition from the load zones. Against the foregoing background of failure patterns the following failure descriptions should be meaningful.

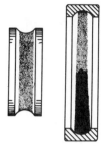

Figure 3-11. Load zone from internally pre-loaded bearing supporting radial load.

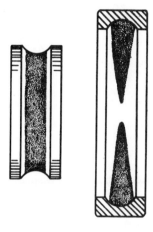

Figure 3-12. Load zones produced by an out-of-round housing pinching the bearing outer ring.

Figure 3-13. Mirror view of outer ring distorted by housing.

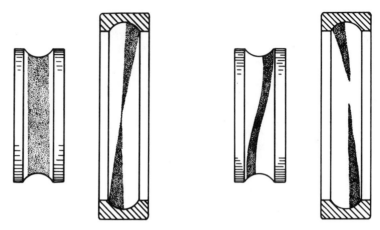

Figure 3-14. Load zone produced when outer ring is misaligned to shaft.

Figure 3-15. Load zones when inner ring is misaligned relative to housing.

Failure Due to Defective Bearing Seats on Shafts and in Housings

The calculated life expectancy of a rolling bearing presupposes that its comparatively thin rings will be fitted on shafts or in housings that are as geometrically true as modern machine-shop technique can produce. There are, unfortunately, factors that produce shaft seats and housing bores that are oversize or undersize, tapered or oval. Figure 3-16 shows another mirror view of a spherical roller bearing outer ring that has been seated in an out-of-round housing. Notice how the widest portions of the roller paths are directly opposite each other. The same condition can be produced by seating the bearing in a housing with a correctly made bore but where the housing is distorted when it is secured to the machine frame. An example would be a pillow block bolted to a pedestal that is not plane.

Figure 3-17 shows the condition resulting when a bearing outer ring is not fully supported. The impression made on the bearing OD by a turning chip left in the housing when the bearing was installed is seen in the left-hand view. This outer ring was subsequently supported by the chip alone, with the result that the entire load was borne by a small portion of the roller path. The heavy specific load imposed on that part of the ring immediately over the turning chip produced the premature spalling seen in the right-hand illustration. On either side of the spalled area is a condition called fragment denting, which occurs when fragments from the flaked surface are trapped between the rollers and the raceway.

When the contact between a bearing and its seat is not intimate, relative movement results. Small movements between the bearing and its seat produce a condition called fretting corrosion. Figure 3-18 illustrates a bearing outer ring that has been subjected to fretting corrosion. Figure 3-19 illustrates an advanced state of this condition. Fretting started the crack, which in turn triggered the spalling.

Figure 3-16. Mirror view of spherical roller-bearing outer ring pinched by housing.

Figure 3-17. Fatigue from chip in housing bore.

Fretting corrosion can also be found in applications where machining of the seats is accurate, but where on account of service conditions the seats deform under load. Railroad journal boxes are an example of this condition. Experience shows that this type of fretting corrosion on the outer ring does not as a rule detrimentally affect the life of the bearing. Shaft seats or journals as well as housing bores can yield and produce fretting corrosion. Figure 3-20 illustrates damage by movement on a shaft. The fretting corrosion covers a large portion of the surface on both the inner ring and the journal. The axial crack through the inner ring started from surface damage caused by the fretting.

Figure 3-18. Wear due to fretting corrosion.

Figure 3-19. Advanced wear and cracking due to fretting corrosion.

Figure 3-20. Fretting caused by yield in the shaft journal.

Figure 3-21. Crack caused by faulty housing fit.

Bearing damage is also caused by bearing seats that are concave, convex, or tapered. On such a seat, a bearing ring cannot make contact throughout its width. The ring therefore deflects under loads, and fatigue cracks commonly appear axially along the raceway. Cracks caused by faulty contact between a ring and its housing are seen in Figure 3-21.

Misalignment

Misalignment is a common source of premature spalling. Misalignment occurs when an inner ring is seated against a shaft shoulder that is not square with the journal, or when a housing shoulder is out-of-square with the housing bore. Misalignment arises when two housings are not on the same center line. A bearing ring can be misaligned even though it is mounted on a tight fit if it is not pressed against its shoulder and thus left cocked on its seat. Bearing outer rings in slip-fitted housings risk being cocked across their opposite corners.

Some of the foregoing misalignment faults are not cured by using self-aligning bearings. When the inner ring of a self-aligning bearing is not square with its shaft seat, the inner ring wobbles as it rotates. This results in smearing and early fatigue. If an outer ring is cocked in its housing across corners, a normally floating outer ring can become axially held as well as radially pinched in its housing. The effect of a pinched outer ring was shown in Figures 3-13 and 3-16.

Ball thrust bearings suffer early fatigue when mounted on supports not perpendicular to the shaft axis, because one short load zone of the stationary ring carries all of the load. When the rotating ring of the ball thrust bearing is mounted on an out-of-square shaft shoulder, the ring wobbles as it rotates. The wobbling ring loads only a small portion of the stationary ring and causes early fatigue.

Figure 3-22. Smearing in ball thrust bearing.

Figure 3-22 illustrates the smearing within a ball thrust bearing when either one of two conditions occurs: first, if the two rings are not parallel to each other during operation; and second, if the load is insufficient at the operating speed to hold the bearing in its designed operational position. Under the first condition, the smearing seen in Figure 3-22 occurs when the balls pass from the loaded into the unloaded zone. Under the latter condition, even if the rings are parallel to each other, centrifugal force causes each ball to spin instead of roll upon contact with the raceway, resulting in smearing. Smearing from misalignment will be localized in one zone of the stationary ring, whereas smearing from gyral forces will be general around both rings.

Where two housings supporting the same shaft do not have a common centerline, only self-aligning ball or roller bearings will be able to function without inducing bending moments. Cylindrical and tapered roller bearings, although crowned, can accommodate only very small misalignments. If misalignment is appreciable, edgeloading results, and this is a source of premature fatigue. Edgeloading from misalignment was responsible for spalling in the bearing ring seen in Figure 3-23. Advanced spalling due to the same cause can be seen on the inner ring and roller of the tapered roller bearing in Figure 3-24.

Faulty Mounting Practice

The origin of premature fatigue, or of any failure, lies too many times in abuse and neglect before and during mounting. Prominent among the causes of early fatigue is the presence of foreign matter in the bearing and in its housing during operation. We have already seen the effect of trapping a chip between the OD of the bearing and the bore of the housing, as shown in Figue 3-17. Figure 3-25 shows the inner ring of a bearing where foreign matter has been trapped between the raceway and the rollers, causing brinelled depressions. This condition is called fragment denting. Each of these small dents is the potential start of premature fatigue. Foreign matter of small particle size results in wear, and when the original internal geometry is changed, the calculated life expectancy cannot be achieved. This is illustrated later in this chapter under "Ineffective Sealing" (Figures 3-48, 3-49, and 3-50).

Figure 3-23. Fatigue caused by edge-loading.

Figure 3-24. Advanced spalling caused by edge-loading.

Figure 3-25. Fragment denting.

Figure 3-26. Fatigue caused by impact damage during handling or mounting.

Impact damage during handling or mounting results in brinelled depressions that become the start of premature fatigue. An example of this is shown in Figure 3-26, where the spacing of flaked areas corresponds to the distance between the balls. The bearing has obviously suffered impact, and if it is installed, the fault should be apparent by the noise or vibration during operation.

Cylindrical roller bearings are easily damaged in mounting, especially when the rotating part with the inner ring mounted on it is assembled into a stationary part having its outer ring and roller set assembled. Figure 3-27 shows the inner ring of a cylindrical roller bearing that has been damaged because the rollers had to slide forceably across the inner ring during assembly. Here again the spacing of the damage marks on the inner ring is the same as the distance between rollers. The smeared streak in Figure 3-27 is shown enlarged eight times in Figure 3-28.

If a bearing is subjected to loads greater than those calculated for life expectancy, premature fatigue results. Unanticipated or parasitic loads can arise from faulty mounting practice. One example of a parasitic load is found in mounting the front wheel of an automobile and not backing off the locknut after applying the specified torque to seat the bearing. Another example is a bearing that should be free in its housing, but because of pinching or cocking cannot move with thermal expansions, inducing a parasitic thrust load on the bearing. Figure 3-29 shows the effect of a

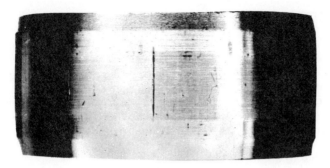

Figure 3-27. Smearing caused by excessive force in mounting.

Figure 3-28. Smearing, enlarged eight times from Figure 3-27.

Figure 3-29. Spalling from excessive thrust.

parasitic thrust load. The damaged area is not in the center of the ball groove, as it should be normally, but is high on the shoulder of the groove. The ring shown in Figure 3-30 is of a self-aligning ball bearing subjected to an abnormally heavy thrust load. Usually, in such cases, evidence of axial restraint will appear either as the imprint of a housing shoulder on the outer ring face or as areas of fretting on the OD of the bearing.

Interference between rotating and stationary parts can result in destructive cracks in the rotating bearing ring. The roller bearing inner ring in Figure 3-31 shows the effect of contact with an end cover while the bearing ring rotated.

Damage Due to Improper Fits

To decide whether a bearing ring, either inner or outer, should be mounted with an interference or a slip fit on its shaft or in its housing, determine whether the ring rotates or is stationary with reference to load direction. The degree of tightness or looseness is governed by the magnitude of the load and the speed. If a bearing ring rotates relative to the load direction, an interference fit is required. If the ring is stationary with reference to the load, it is fitted with some clearance and is called a slip fit. The degree of fit is governed by the concept that heavier loads require greater interference. The presence of shock or continuous vibration calls for a heavier interference fit of the ring that rotates relative to the load. Lightly loaded rings, or even those rings with considerable load but operating at extremely slow speeds, that rotate relative to the load, may use a lighter fit or, in some cases, a slip fit.

Figure 3-30. Spalling from parasitic thrust.

Figure 3-31. Cracks caused by contact with the end cover while the bearing ring rotated.

Figure 3-32. Scoring of inner ring caused by "creep."

Consider two examples. In an automobile front wheel, the direction of the load is constant, i.e. the pavement is always exerting an upward force on the wheel. In this case the outer rings or cups are rotating and are press-fitted into the wheel hub, while the inner rings or cones are stationary and are slip-fitted on the spindle. On the other hand, the bearings of a conventional gear drive have their outer rings stationary relative to the load and so are slip fitted, but the inner rings rotate relative to the load and are mounted with an interference fit. There are some cases where it is necessary to mount both inner and outer rings of a bearing with interference fits because of a combination of stationary and rotating loads or loads of indeterminate characteristics. Such cases require bearings that can allow axial expansion at the rollers rather than at a slip-fitted ring. Such a mounting would consist of a cylindrical roller bearing at one end of the shaft and a self-contained bearing (deep-groove ball or spherical roller bearing) at the other end.

Some examples of the effect of departure from good fitting practice follow. Figure 3-32 shows the bore surface of an inner ring that has been damaged by relative movement between it and its shaft while rotating under a constant direction load. This relative movement, called creep, can result in the scoring shown in Figure 3-32.

If lubricant can penetrate the loose fit, the bore as well as the shaft seat will appear brilliantly polished (see Figure 3-33). When a normally tight-fitted inner ring does creep, the damage is not confined to the bore surface but can have its effect on the faces of the ring. Contact with shaft shoulders or spacers can result in either wear or severe rubbing cracks (Figure 3-31), depending on the lubrication condition. Figure 3-32, also shows how a shaft shoulder wore into the face of a bearing inner ring when relative movement occurred. Wear between a press-fitted ring and its seat is an accumulative damage. The initial wear accelerates the creep, which in turn produces more wear. The ring loses adequate support and develops cracks, and the products of wear become foreign matter to fragment dent and internally wear the bearing.

Excessive fits also result in bearing damage by internally preloading the bearing, as seen in Figure 3-11, or by inducing dangerously high hoop stresses in the inner ring. Figure 3-34 illustrates an inner ring that cracked because of excessive interference fit.

Housing fits that are unnecessarily loose allow the outer ring to fret, creep, or even spin. Examples of fretting were seen in Figures 3-18 and 3-19. Lack of support to the outer ring results from excessive looseness as well as from faulty housing-bore contact. A cracked outer ring was shown in Figure 3-19.

Inadequate or Unsuitable Lubrication

Any load-carrying contact, even when rolling, requires the presence of lubricants for reliable operations. The curvature of the contact areas between rolling element and raceway in normal operation results in minute amounts of sliding motion in addition to the rolling. Also, the cage must be carried on either of the rolling elements or on some surface of the bearing rings, or on a combination of these. In most types of roller bearings, there are roller end faces which slide against a flange or a cage. For these reasons, adequate lubrication is even more important at all times. The term "lubrication failure" is too often taken to imply that there was no oil or

Figure 3-33. Wear due to "creep."

Figure 3-34. Axial cracks caused by an excessive interference fit.

grease in the bearing. While this happens occasionally, the failure analysis is normally not that simple. In many cases, a study of the bearing leaves no doubt in the examiner's mind that lubrication failed, but why the lubricant failed to prevent damage to the bearing is not obvious.

When searching for the reason why the lubricant did not perform, one must consider first its properties; second, the quantity applied to the bearing; and third, the operating conditions. These three concepts comprise the adequacy of lubrication. If any one concept does not meet requirements, the bearing can be said to have failed from inadequate lubrication.

Viscosity of the oil—either as oil itself or as the oil in grease—is the primary characteristic of adequate lubrication. The nature of the soap base of a grease and its consistency, together with viscosity of the oil, are the main quality points when considering a grease. For the bearing itself, the quantity of lubricant required at any one time is usually rather small, but sufficient quantity must constantly be available. If the lubricant is also a heat-removal medium, a larger quantity is required. An insufficient quantity of lubricant at medium to high speeds induces a temperature rise and usually a whistling sound. An excessive amount of lubricant produces a sharp temperature rise because of churning in all but exceptionally-slow-speed bearings. Conditions inducing abnormally high temperatures can render a normally adequate lubricant inadequate.

When lubrication is inadequate, surface damage will result. This damage will progress rapidly to failures that are often difficult to differentiate from primary fatigue failure. Spalling will occur and often destroy evidence of inadequate lubrication. However, if caught soon enough, one can find indications that will pinpoint the real cause of the short bearing life.

One form of surface damage is shown in stages in Figure 3-35, a, b, c, and d. The first visible indication of trouble is usually a fine roughening or waviness on the surface. Later, fine cracks develop, followed by spalling. If there is insufficient heat removal, the temperature may rise high enough to cause discoloration and softening of the hardened bearing steel. This happened to the bearing shown in Figure 3-36. In some cases, inadequate lubrication initially manifests itself as a highly glazed or glossy surface which, as damage progresses, takes on a frosty appearance and eventually spalls. The highly glazed surface is seen on the roller of Figure 3-37.

(a)

(b)

(c)

(d)

Figure 3-35. Progressive stages of spalling caused by inadequate lubrication.

Figure 3-36. Discoloration and softening of metal caused by inadequate lubrication and excessive heat.

Figure 3-37. Glazing caused by inadequate lubrication.

In the "frosty" stage, it is sometimes possible to feel the "nap" of fine slivers of metal pulled from the bearing raceway by the rolling element. The frosted area will feel smooth in one direction but have a distinct roughness in the other. As metal is pulled from the surface, pits appear, and so frosting advances to pulling. An example of pulling is seen in Figure 3-38.

Another form of surface damage is called smearing. It appears when two surfaces slide and the lubricant cannot prevent adhesion of the surfaces. Minute pieces of one surface are torn away and rewelded to either surface. Examples are shown in Figures 3-39 through 3-42. A peculiar type of smearing occurs when rolling elements slide as

Figure 3-38. Effect of rollers pulling metal from the bearing raceway.

Figure 3-39. Smearing on spherical roller end.

Figure 3-40. Smearing on cylindrical rollers caused by ineffective lubrication.

Figure 3-41. Smearing on cage pockets caused by ineffective lubrication.

Figure 3-42. Smearing on cylindrical outer raceway.

they pass from the unloaded to the loaded zone. Figure 3-43 illustrates the patches of skid smearing, one in each row. A lubricant that is too stiff also causes this type of damage. This is particularly likely to happen if the bearing is large.

Wear of the bearing as a whole also results from inadequate lubrication. The areas subject to sliding friction, such as locating flanges and the ends of rollers in a roller bearing, are the first parts to be affected. Figures 3-44 and 3-45 illustrate the damage done and the extent of wear.

Where high speeds are involved, inertial forces become important and the best lubrication is demanded. Figure 3-46 shows an advanced case of damage from high speed with inadequate lubrication. Inertia forces acting on the rolling elements at high speed and with sudden starting or stopping can result in high forces between rolling elements and the cage.

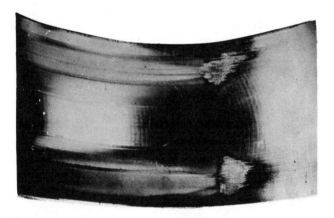

Figure 3-43. Skid smearing on spherical outer raceway.

Figure 3-44. Grooves caused by wear due to inadequate lubrication.

Figure 3-45. Grooves caused by wear due to inadequate lubrication.

Figure 3-46. Broken cage caused by ineffective lubrication.

Figure 3-47 shows a large-bore tapered-roller bearing failure due to an insufficient amount of lubricant resulting from too low a flowrate in a circulating oil system. The area between the guide flange and the large end of the roller is subject to sliding motion. This area is more difficult to lubricate than those areas of rolling motion, accounting for the discoloration starting at the flange contact area. The heat generated at the flange caused the discoloration of the bearing and resulted in some of the rollers being welded to the guide flange.

Figure 3-47. Rollers welded to rib because of ineffective lubrication.

The foregoing defines the principal lubrication-related surface failures. Importance has been assigned to lubrication adequacy and those factors attacking it. Implied, then, is the need to know how to avoid surface failures. Briefly, these are the guidelines:

1. Sufficient elastohydrodynamic film prevents surface distress (glazing, pitting).
2. Good boundary lubrication guards against smearing and sliding surface wear.
3. Clean lubricants prevent significant wear of rolling surfaces.
4. Sufficient lubricant flow keeps bearings from overheating.

As long as the surfaces of the rolling element and raceway in rolling contact can be separated by an elastohydrodynamic oil film, surface distress is avoided. The continuous presence of the film depends on the contact area, the load on the contact area, the speed, the operating temperature, the surface finish, and the oil viscosity.

SKF research has developed a procedure for determining the required oil viscosity when the bearing size, load, and speed are known and when operating temperature can be reasonably estimated. Charts enabling calculation have been published in trade journals and are available in "A Guide to Better Bearing Lubrication," from SKF, or in Volume 1 of Practical Machinery Management for Process Plants—Improving Machinery Reliability—page 233.

When the elastohydrodynamic oil film proves suitable in the rolling contacts, experience shows it is generally satisfactory for the sliding contacts at cages and guide flanges. When, in unusual applications, viscosity selection must be governed by the sliding areas, experience has proven that the viscosity chosen is capable of maintaining the necessary elastohydrodynamic film in the rolling contacts.

Ineffective Sealing

The effect of dirt and abrasive in the bearing during operation was described earlier in this chapter under "Faulty Mounting Practices." Although foreign matter can enter the bearing during mounting, its most direct and sustained area of entry can be around the housing seals. The result of gross change in bearing internal geometry has been pointed out. Bearing manufacturers realize the damaging effect of dirt and take extreme precautions to deliver clean bearings. Not only assembled bearings but also parts in process are washed and cleaned. Freedom from abrasive matter is so important that some bearings are assembled in air-conditioned white rooms. Dramatic examples of combined abrasive-particle and corrosive wear, both due to defective sealing, are shown in Figures 3-48 and 3-49. Figure 3-50 shows a deep-groove ball bearing which has operated with abrasive in it. The balls have worn to such an extent that they no longer support the cage, which has been rubbing on the lands of both rings.

In addition to abrasive matter, corrosive agents should be excluded from bearings. Water, acid, and those agents that deteriorate lubricants result in corrosion. Figure 3-51 illustrates how moisture in the lubricant can rust the end of a roller. The corroded areas on the rollers of Figure 3-52 occurred while the bearing was not rotating. Acids forming in the lubricant with water present etch the surface as shown in Figure 3-53. The lines of corrosion seen in Figure 3-54 are caused by water in the lubricant as the bearing rotates.

Figure 3-48. Advanced abrasive wear.

Figure 3-49. Advanced abrasive wear.

Figure 3-50. Advanced abrasive wear.

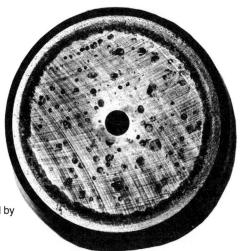

Figure 3-51. Rust on end of roller caused by moisture in lubricant.

Figure 3-52. Corrosion on roller surfaces caused by water in lubricant while bearing was not moving.

Figure 3-53. Corrosion on roller surface caused by formation of acids in lubricant with some moisture present.

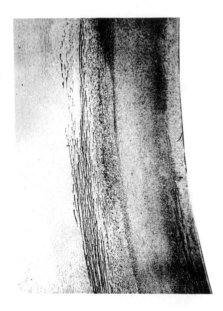

Figure 3-54. Corrosion streaks caused by water in the lubricant while the bearing rotated.

Vibration

Rolling bearings exposed to vibration while the shafts are not rotating are subject to a damage called false brinelling. The evidence can be either brightly polished depressions or the characteristic red-brown stain of fretting. Oxidation rate at the point of contact determines the appearance. Variation in the vibration load causes minute sliding in the area of contact between rolling elements and raceways. Small particles of material are set free from the contact surfaces and may or may not be oxidized immediately. The debris thus formed acts as a lapping agent and accelerates the wear. Another identification of damage of this type is the spacing of the marks on the raceway. The spacing of the false brinelling will be equal to the distance between the rolling elements, just as it is in some types of true brinelling. If the bearing has rotated slightly between periods of vibration, more than one pattern of false-brinelling damage may be seen.

A type of false brinelling with abrasive present is seen in Figure 3-55. There was no rotation between the two rings of the bearing for considerable periods of time, but while they were static they were subject to severe vibration. False brinelling developed with a production of iron oxide, which in turn acted as a lapping compound.

The effect of both vibration and abrasive in a rotating bearing is seen in the wavy pattern shown in Figure 3-56. When these waves are more closely spaced, the pattern is called fluting and appears similar to cases shown in the next section under "Passage of Electric Current Through the Bearing." Metallurgical examination is often necessary to distinguish between fluting caused solely by abrasive and vibration and that caused by vibration and passage of electric current.

Since false brinelling is a true wear condition, such damage can be observed even though the forces applied during the vibration are much smaller than those corresponding to the static carrying capacity of the bearing. However, the damage is more extensive as the contact load on the rolling element increases.

Figure 3-55. False brinelling caused by vibration with bearing stationary.

Figure 3-56. False brinelling caused by vibration in the presence of abrasive dirt while bearing was rotating.

False brinelling occurs most frequently during transportation of assembled machines. Vibration fed through a foundation can generate false brinelling in the bearings of a shaft that is not rotating. False brinelling during transportation can always be minimized and usually eliminated by temporary structural members that will prevent any rotation or axial movement of the shaft.

It is necessary to distinguish between false and true brinelling. Figures 3-57 and 3-58 are 100X photomicrographs of true and false brinelling in a raceway, respectively. In Figure 3-57 (true brinelling) there is a dent produced by plastic flow of the raceway material. The grinding marks are not noticeably disturbed and can be seen over the whole dented area. However, false brinelling (Figure 3-58), does not

Figure 3-57. Example of true brinelling—100x.

Figure 3-58. Example of false brinelling—100x.

involve flow of metal but rather a removal of surface metal by attrition. Notice that the grinding marks are removed. To further understand false brinelling, which is very similar to fretting corrosion, one should remember that a rolling element squeezes the lubricant out of its contact with the raceway, and that the angular motion from vibration is so small that the lubricant is not replenished at the contact. Metal-to-metal contact becomes inevitable, resulting in submicroscopic particles being torn from the high points. If protection by lubricant is absent, these minute particles oxidize and account for the red-brown color usually associated with fretting. If there is a slower oxidation rate, the false brinelling depression can remain bright, thereby adding to the difficulty in distinguishing true from false brinelling.

Passage of Electric Current Through the Bearing

In certain applications of bearings to electrical machinery there is the possibility that electric current will pass through a bearing. Current that seeks ground through the bearing can be generated from stray magnetic fields in the machinery or can be caused by welding on some part of the machine with the ground attached so that the circuit is required to pass through the bearing.

An electric current can be generated by static electricity, emanating from charged belts or from manufacturing processes involving leather, paper, cloth, or rubber. This current can pass through the shaft to the bearing and then to the ground. When the current is broken at the contact surfaces between rolling elements and raceways, arcing results, producing very localized high temperature and consequent damage. The overall damage to the bearing is in proportion to the number and size of individual damage points.

Figure 3-59 and the enlarged view Figure 3-60 show a series of electrical pits in a roller and in a raceway of a spherical roller bearing. The pit was formed each time the current broke in its passage between raceway and roller. The bearing from which this roller was removed was not altogether damaged to the same degree as this roller. In

Figure 3-59. Electric pitting on surface of spherical roller caused by the passage of a relatively large current.

fact, this specific bearing was returned to service and operated successfully for several additional years. Hence, moderate amounts of electrical pitting do not necessarily result in failure.

Another type of electrical damage occurs when current passes for prolonged periods and the number of individual pits accumulate astronomically. The result is the fluting shown in Figures 3-61 through 3-65. This condition can occur in ball or roller bearings. Flutes can develop considerable depth, producing noise and vibration during operation and eventual fatigue from local overstressing. The formation of flutes rather than a homogeneous dispersion of pits is not clearly explained but is possibly related to initial synchronization of shocks or vibrations and the breaking of the current. Once the fluting has started it is probably a self-perpetuating phenomenon.

Individual electric marks, pits, and fluting have been produced in bearings running in the laboratory. Both alternating and direct currents can cause the damage. Amperage rather than voltage governs the amount of damage. When a bearing is under radial load, greater internal looseness in the bearing appears to result in greater electrical damage for the same current. In a double-row bearing loaded in thrust, little if any damage results in the thrust-carrying row, although the opposite row may be damaged.

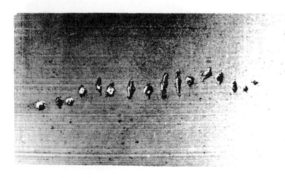

Figure 3-60. Electric pitting on surface of spherical outer raceway caused by the passage of a relatively large current.

Figure 3-61. Fluting on surface of spherical roller caused by prolonged passage of electric current.

Figure 3-62. Fluting on inner raceway of cylindrical roller bearing caused by prolonged passage of electric current.

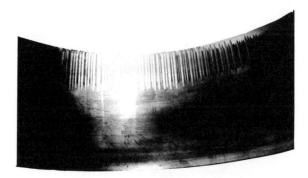

Figure 3-63. Fluting on outer raceway of spherical roller bearing.

Figure 3-64. Fluting on inner raceway of spherical roller bearing.

Figure 3-65. Fluting on outer raceway of self-aligning ball bearing caused by prolonged passage of relatively small current and vibration.

Troubleshooting Bearings

This section presents some helpful hints on bearing troubleshooting—what to look for, how to recognize the reason for the trouble, and practical solutions, wherever possible.

Observations of bearing trouble can be reduced to a few classifications, listed in the following order. For ease of relating them to conditions and solutions, they are coded A to G inclusive.

Observation or Complaint

A—Overheated bearing.
B—Noisy bearing.
C—Replacements are too frequent.
D—Vibration.
E—Unsatisfactory equipment performance.
F—Bearing is loose on shaft.
G—Hard-turning shaft.

The following table lists some typical conditions that will result in bearing failures. The first column numerically codes each typical condition (Nos. 1 to 54 inclusive). The third column is the observation or complaint code (A to G) to which the condition may apply.

TYPICAL CONDITIONS RESULTING IN BEARING FAILURES

EACH CONDITION COULD CAUSE ANY ONE OF THE COMPLAINTS LISTED OPPOSITE THE RESPECTIVE CONDITION IN COLUMN 3

CODE	CONDITION	COMPLAINT
1.	Inadequate lubrication (Wrong type of grease or oil)	A-B-C-G
2.	Insufficient lubrication (Low oil level—loss of lubricant through seals)	A-B-C-G
3.	Excessive lubrication (Housing oil level too high or housing packed with grease)	A-G
4.	Insufficient clearance in bearing (Selection of wrong fit)	A-B-C-E-G
5.	Foreign matter acting as an abrasive (Sand, carbon, etc.)	B-C-D-E-G
6.	Foreign matter acting as a corrosive (Water, acids, paints, etc.)	B-C-D-E-G
7.	Bearings pinched in the housing (Bore out of round)	A-B-C-D-E-G
8.	Bearings pinched in the housing (Housing warped)	A-B-C-D-E-G
9.	Uneven shimming of housing base (Distorted housing bore - possible cracking of base)	A-B-C-D-E-G
10.	Chips in bearing housing (Chips or dirt left in housing)	B-C-D-E-G
11.	High air velocity over bearings (Oil leakage)	C
12.	Seals too tight (Cup seals)	A-G
13.	Seals misaligned (Rubbing against stationary parts)	A-B-G
14.	Oil return holes plugged (Oil leakage)	A
15.	Preloaded bearings (Opposed mounting)	A-B-C-G
16.	Preloaded bearings (Two held bearings on one shaft)	A-B-C-E-G
17.	Bearing loose on shaft (Shaft diameter too small)	B-C-D-E-F
18.	Bearing loose on shaft (Adapter not tightened sufficiently)	B-C-D-E-F
19.	Bearing too tight internally (Adapter tightened excessively)	A-B-C-E-G
20.	Split pillow block with uneven surfaces (Oil leakage)	C
21.	Spinning of outer ring in housing (Unbalanced load)	A-C-D-E
22.	Noisy bearing (Flat on roller or ball due to skidding)	B-D-E
23.	Excessive shaft expansion (Resulting in opposed mounting)	A-B-C-E-G
24.	Excessive shaft expansion (Resulting in insufficient clearance in bearing)	A-C-E-G
25.	Tapered shaft seat (Concentration of load in bearing)	C-D-E
26.	Tapered housing bore (Concentration of load in bearing)	C-D-E
27.	Shaft shoulder too small (Inadequate shoulder support—bending of shaft)	C-D-E-G
28.	Shaft shoulder too large (Rubbing against bearing seals)	A-B-C
29.	Housing shoulder too small (Inadequate shoulder support)	C-D-E-G
30.	Housing shoulder too large (Distortion of bearing seals)	B-C-G
31.	Shaft fillet too large (Bending of shaft)	C-D-E-G
32.	Housing fillet too large (Inadequate support)	C-D-E-G
33.	Insufficient clearance in labyrinth seals (Rubbing)	A-B-C-G
34.	Oil gauge breather hole clogged (Shows incorrect oil level)	A-C
35.	Shafts out of line (Linear misalignment)	A-C-D-E-G
36.	Shafts out of line (Angular misalignment)	A-C-D-E-G
37.	Constant oil level cups (Incorrect level)	A-C
38.	Constant oil level cups (Located against rotation of bearing)	A-C
39.	Lockwasher prongs bent (Rubbing against bearing)	A-B-E-G
40.	Incorrect positioning of flingers (Rubbing against covers)	A-B-C-G
41.	Pedestal surface uneven (Bending of housing causing pinching of bearing)	A-C-D-E-G
42.	Ball or roller denting (Hammer blows on bearing)	B-C-D-E
43.	Noisy bearing (Extraneous conditions)	B
44.	Lubricant leakage and entrance of dirt into bearing (Worn out seals)	C
45.	Vibration (Excessive clearance in bearing)	D-E
46.	Vibration (Unbalanced loading)	D-E

TYPICAL CONDITIONS RESULTING IN BEARING FAILURES (Continued)

CODE	CONDITION	COMPLAINT
47.	Hard turning shaft (Shaft and housing shoulders out of square with bearing seat).......	C-E-G
48.	Bearing loose on shaft (Knurling and center punching of shaft for bearing seat).......	A-F
49.	Discoloration of bearings (Use of blow torch to remove bearing).................	B
50.	Oversized shaft (Overheating and noise).....................................	A-B-C-E-G
51.	Undersized housing bore (Overheating of bearing)..........................	A-B-C-E-G
52.	Oversized housing bore (Overheating of bearing—spinning of outer ring)...........	A-B-C-D-E
53.	Enlarged housing bore (Excessive peening of non-ferrous housings)...............	A-B-C-D-E
54.	Noisy bearing (False brinelling)...	B

The following pages offer practical solutions to the trouble conditions which were observed. Column 1 refers back to the code of the *typical condition* listed in the foregoing table. Column 2 is the *reason for that condition,* and column 3 is the *practical solution.*

TROUBLE CONDITIONS AND THEIR SOLUTION

OVERHEATED BEARING
Complaint "A"

CODE TO TYPICAL CONDITION	REASON FOR CONDITION	PRACTICAL SOLUTION	
1	Wrong type of grease or oil causing break-down of lubricant.	Consult reliable lubricant manufacturer for proper type of lubricant. Check SKF Catalog instructions to determine if oil or grease should be used.	
2	Low oil level. Loss of lubricant through seal. Insufficient grease in housing.	Oil level should be just below center of lowest ball or roller in bearing. Using grease, lower half of pillow block should be ½ to ⅔ full.	
3	Housing packed with grease, or oil level too high . . . causing excessive churning of lubricant, high operating temperature, oil leakage.	Purge bearing until only lower half of housing is ½ to ⅔ full of grease. Using oil lubrication, reduce level to just below center of lowest ball or roller.	
4	Bearings selected with inadequate internal clearance for conditions where external heat is conducted thru shaft, thereby expanding excessively the inner ring.	Replacement bearing should have identical marking as original bearing for proper internal clearance. Check with SKF if bearing markings have become indistinct.	
7-8-9 41-51	Housing bore out of round. Housing warped. Excessive distortion of housing. Undersized housing bore.	Check and scrape housing bore to relieve pinching of bearing. Be sure pedestal surface is flat, and shims cover entire area of pillow block base.	

OVERHEATED BEARING — *Complaint "A" (Continued)*

CODE TO TYPICAL CONDITION	REASON FOR CONDITION	PRACTICAL SOLUTION	
12	Leather or composition seals with excessive spring tension or dried out.	Replace leather or composition seals with ones having reduced spring tension. Lubricate seals.	
13 33-40	Rotating seals rubbing against stationary parts.	Check running clearance of rotating seal to eliminate rubbing. Correct alignment.	
14	Oil return holes blocked—pumping action of seals cause oil leakage.	Clean holes. Drain out used oil—refilling to proper oil level with fresh lubricant.	
15	Opposed mounting.	Insert gasket between housing and cover flange to relieve axial preloading of bearing.	
16 23-24	Two "held" bearings on one shaft. Excessive shaft expansion.	Back off covers in one of the housings, using shims to obtain adequate clearance of outer ring, to permit free axial bearing motion.	
19	Adapter tightened excessively.	Loosen locknut and sleeve assembly. Retighten sufficiently to clamp sleeve on shaft but be sure bearing turns freely.	
21-52	Unbalanced load. Housing bore too large.	Rebalance machine. Replace housing with one having proper bore.	
28	Rubbing of shaft shoulder against bearing seals.	Remachine shaft shoulder to clear seal.	
34	Incorrect oil level. Result: no lubricant in bearing.	Clean out clogged hole to vent oil gauge.	

OVERHEATED BEARING — *Complaint "A"* (Continued)

CODE TO TYPICAL CONDITION	REASON FOR CONDITION	PRACTICAL SOLUTION	
35-36	Incorrect linear or angular alignment of two or more coupled shafts with two or more bearings.	Correct alignment by shimming pillow blocks. Be sure shafts are coupled in straight line — especially when three or more bearings operate on one shaft.	LINEAR MISALIGNMENT / ANGULAR MISALIGNMENT
37-38	Incorrect mounting of constant oil level cup. (Too high or too low.) Cup located opposite rotation of bearing permitting excessive flow of oil, resulting in too high oil level.	The oil level at standstill must not exceed the center of the lowermost ball or roller. Sketch illustrates correct position of constant level oil cup with respect to rotation. Better replace constant level oiler with sight gage.	1-STATIC OIL LEVEL 2-OPERATING OIL LEVEL
39	Prong rubbing against bearing.	Remove lockwasher — straighten prong or replace with new washer.	RUBBING
48	Knurling and center punching of bearing seat on shaft.	Unsatisfactory because high spots are flattened when load is applied, when fit is loose, metalize shaft and regrind to proper size.	
50	Bearing seat diameter machined oversize, causing excessive expansion of bearing inner ring, thus reducing clearance in bearing.	Grind shaft to get proper fit between inner ring of bearing and shaft.	
53	"Pounding-out" of housing bore due to soft metal. Result: enlarged bore . . . causing spinning of outer ring in housing.	Rebore housing and press steel bushing in bore. Machine bore of bushing to correct size.	

NOISY BEARING

Complaint "B"

1	Wrong type of grease or oil causing break-down of lubricant.	Consult reliable lubricant manufacturer for proper type of lubricant. Check SKF Catalog instructions to determine if oil or grease should be used.
2	Low oil level. Loss of lubricant through seal. Insufficient grease in housing.	Oil level should be at center of lowest ball or roller in bearing, at standstill. Using grease, lower half of pillow block should be ½ to ⅔ full. *See Illustration—Complaint "A", Condition 2*
4	Bearings selected with inadequate internal clearance for conditions where external heat is conducted thru shaft, thereby expanding excessively the inner ring.	Replacement bearing should have identical marking as original bearing for proper internal clearance. Check with SKF if markings have become indistinct.
5	Foreign matter (dirt, sand, carbon, etc.) entering bearing housing.	Clean out bearing housing. Replace worn-out seals or improve seal design to obtain adequate protection of bearing.
6	Corrosive agents (water, acids, paints, etc.) entering the bearing housing.	Addition of a shroud and (or) flinger to throw off foreign matter.

NOISY BEARING — *Complaint "B"* (Continued)

CODE TO TYPICAL CONDITION	REASON FOR CONDITION	PRACTICAL SOLUTION
7-8 9-51	Housing bore out of round. Housing warped. Excessive distortion of housing. Undersized housing bore.	Check and scrape housing bore to relieve pinching of bearing. Be sure pedestal surface is flat, and shims cover entire area of pillow block base. *See Illustration—Complaint "A", Conditions 7-8-9-41-51*
10	Failure to remove chips, dirt, etc. from bearing housing before assembling bearing unit.	Carefully clean housing, and use fresh lubricant.
13 33-40	Rotating seals rubbing against stationary parts.	Check running clearance of rotating seal to eliminate rubbing. Correct alignment. *See Illustration—Complaint "A", Condition 13-33-40*
15	Opposed mounting.	Insert gasket between housing and cover flange to relieve axial pre-loading of bearing. *See Illustration—Complaint "A", Condition 15*
16-23	Two "held" bearings on one shaft. Excessive shaft expansion.	Back off covers in one of the housings using shims to obtain adequate clearance of outer ring to permit free axial bearing motion. *See Illustration—Complaint "A", Conditions 16-23-24*
17-18	Shaft diameter too small. Adapter not tightened sufficiently.	Metallize shaft and regrind to obtain proper fit. Retighten adapter to get firm grip on shaft.
19	Adapter tightened excessively.	Loosen locknut and sleeve assembly. Retighten sufficiently to clamp sleeve on shaft but be sure bearing turns freely. *See Illustration—Complaint "A", Condition 19*
22	Flat on ball or roller due to skidding. (Result of fast starting.)	Carefully examine balls or rollers, looking for flat spots on the surface. Replace bearing.
28	Rubbing of shaft shoulder against bearing seals.	Remachine shaft shoulder to clear seal. *See Illustration—Complaint "A", Condition 28*
30	Distortion of bearing seals.	Remachine housing shoulder to clear seal.
39	Prong rubbing against bearing.	Remove lockwasher—straighten prong or replace with new washer. *See Illustration—Complaint "A", Condition 39*
42	Incorrect method of mounting. Hammer blows on bearing.	Replace with new bearing. Don't hammer any part of bearing when mounting.
43	Interference of other movable parts of machine.	Carefully check every moving part for interference. Reset parts to provide necessary clearance.
49	Distorted shaft and other parts of bearing assembly.	Only in extreme cases should a torch be used to facilitate removal of a failed bearing. Care should be exercised to avoid high heat concentration at any one point so distortion is eliminated.

NOISY BEARING — *Complaint "B"* (Continued)

CODE TO TYPICAL CONDITION	REASON FOR CONDITION	PRACTICAL SOLUTION
50	Bearing seat diameter machined oversize causing excessive expansion of bearing inner ring, thus reducing clearance in bearing.	Grind shaft to get proper fit between inner ring of bearing and shaft.
52	Unbalanced load. Housing bore too large.	Rebalance unit. Replace housing with one having proper bore.
53	"Pounding-out" of housing bore due to soft metal. Result: enlarged bore . . . causing spinning of outer ring in housing.	Rebore housing and press steel bushing in bore. Machine bore of bushing to correct size.
54	Bearing exposed to vibration while machine is idle.	Carefully examine bearing for wear spots separated by distance equal to the spacing of the balls. Replace bearing.

REPLACEMENTS ARE TOO FREQUENT
Complaint "C"

1	Wrong type of grease or oil causing break-down of lubricant.	Consult reliable lubricant manufacturer for proper type of lubricant. Check SKF Catalog instructions to determine if oil or grease should be used.
2	Low oil level. Loss of lubricant through seal. Insufficient grease in housing.	Oil level should be at center of lowest ball or roller in bearing. Using grease, lower half of pillow block should be ½ to ⅔ full. *See Illustration—Complaint "A", Condition 2*
4	Bearings selected with inadequate internal clearance for conditions where external heat is conducted thru shaft, thereby expanding excessively the inner ring.	Replacement bearing should have identical marking as original bearing for proper internal clearance. Check with SKF if bearing markings have become indistinct.
5	Foreign matter (dirt, sand, carbon, etc.) entering into bearing housing.	Clean out bearing housing. Replace worn-out seals or improve seal design to obtain adequate protection of bearing.
6	Corrosive agents (water, acids, paints, etc.) entering the bearing housing.	Addition of a shroud and (or) flinger to throw off the foreign matter.
7-8-9 41-51	Housing bore out of round. Housing warped. Excessive distortion of housing. Undersized housing bore.	Check and scrape housing bore to relieve pinching of bearing. Be sure pedestal surface is flat, and shims cover entire area of pillow block base. *See Illustration—Complaint "A", Conditions 7-8-9-41-51*
10	Failure to remove chips, dirt, etc. from bearing housing before assembling bearing unit.	Carefully clean housing, and use fresh lubricant.

REPLACEMENTS ARE TOO FREQUENT — Complaint "C" (Continued)

CODE TO TYPICAL CONDITION	REASON FOR CONDITION	PRACTICAL SOLUTION
11	Oil leakage resulting from air flow over bearings. (Example: forced draft fan with air inlet over bearings.)	Provide proper baffles to divert direction of air flow.
15	Opposed mounting.	Insert gasket between housing and cover flange to relieve axial pre-loading of bearing. *See Illustration—Complaint "A", Condition 15*
16 23-24	Two "held" bearings on one shaft. Excessive shaft expansion.	Back off covers in one of the housings, using shims to obtain adequate clearance of outer ring, to permit free axial bearing motion. *See Illustration—Complaint "A", Conditions 16-23-24*
17-18	Shaft diameter too small. Adapter insufficiently tightened.	Metallize shaft and regrind to obtain proper fit. Retighten adapter to get firm grip on shaft. *See Illustration—Complaint "B", Conditions 17-18*
19	Adapter tightened excessively.	Loosen locknut and sleeve assembly. Retighten sufficiently to clamp sleeve on shaft but be sure bearing turns freely. *See Illustration—Complaint "A", Condition 19*
20	Oil leakage at housing split. Excessive loss of lubricant.	If not severe, use thin layer of gasket cement. Don't use shims. Replace housing if necessary.
21-52	Unbalanced load. Housing bore too large.	Rebalance machine. Replace housing with one having proper bore. *See Illustration—Complaint "A", Conditions 21-52*
25-26	Unequal load distribution on bearing.	Rework shaft, housing, or both, to obtain proper fit. May require new shaft and housing.
27	Inadequate shoulder support causing bending of shaft.	Remachine shaft fillet to relieve stress. May require shoulder collar.
29	Inadequate support in housing causing cocking of outer ring.	Remachine housing fillet to relieve stress. May require shoulder collar.
28	Rubbing of shaft shoulder against bearing seals.	Remachine shaft shoulder to clear seal. *See Illustration—Complaint "A", Condition 28*
30	Distortion of bearing seals.	Remachine housing shoulder to clear seal. *See Illustration—Complaint "B", Condition 30*
31	Distortion of shaft and inner ring. Uneven expansion of bearing inner ring.	Remachine shaft fillet to obtain proper support.
32	Distortion of housing and outer ring. Pinching of bearing.	Remachine housing fillet to obtain proper support.
33-40	Rotating seals rubbing against stationary parts.	Check running clearance of rotating seal to eliminate rubbing. Correct alignment. *See Illustration—Complaint "A", Conditions 13-33-40*

CODE TO TYPICAL CONDITION	REASON FOR CONDITION	PRACTICAL SOLUTION
34	Incorrect oil level. Result: no lubricant in bearing.	Clean out clogged hole to vent oil gauge. *See Illustration—Complaint "A", Condition 34*
35-36	Incorrect linear or angular alignment of two or more coupled shafts with two or more bearings.	Correct alignment by shimming pillow blocks. Be sure shafts are coupled in straight line—especially when three or more bearings operate on one shaft. *See Illustration—Complaint "A", Conditions 35-36*
37-38	Incorrect mounting of constant oil level cup. (Too high or low.) Cup located opposite rotation of bearing.	The oil level at standstill must not exceed the center of the lowermost ball or roller. Locate cup with rotation of bearing. Replace constant level oiler with sight gage. *See Illustration—Complaint "A", Conditions 37-38*
42	Incorrect method of mounting. Hammer blows on bearing.	Replace with new bearing. Don't hammer any part of bearing when mounting.
44	Excessively worn leather (or composition), or labyrinth seals. Result: lubricant loss; dirt getting into bearing.	Replace seals after thoroughly flushing bearing and refilling with fresh lubricant.
47	Shaft and housing shoulders and face of locknut out-of-square with bearing seat.	Remachine parts to obtain squareness.
50	Bearing seat diameter machined oversize, causing excessive expansion of bearing inner ring, thus reducing clearance in bearing.	Grind shaft to get proper fit between inner ring of bearing and shaft.
53	"Pounding-out" of housing bore due to soft metal. Result: enlarged bore . . . causing spinning of outer ring in housing.	Rebore housing and press steel bushing in bore. Machine bore of bushing to correct size.

VIBRATION

Complaint "D"

5	Foreign matter (dirt, sand, carbon, etc.) entering bearing housing.	Clean out bearing housing. Replace worn-out seals or improve seal design to obtain adequate protection of bearing.
6	Corrosive agents (water, acids, paints, etc.) entering the bearing housing.	Addition of a shroud and (or) flinger to throw off foreign matter.
7-8 9-41	Housing bore out of round. Housing warped. Excessive distortion of housing. Undersized housing bore.	Check and scrape housing bore to relieve pinching of bearing. Be sure pedestal surface is flat, and shims cover entire area of pillow block base. *See Illustration—Complaint "A", Conditions 7-8-9-41-51*
10	Failure to remove chips, dirt, etc. from bearing housing before assembling bearing unit.	Carefully clean housing, and use fresh lubricant.
17-18	Shaft diameter too small. Adapter not tightened sufficiently.	Metallize shaft and regrind to obtain proper fit. Retighten adapter to get firm grip on shaft. *See Illustration—Complaint "B", Conditions 17-18*

VIBRATION — *Complaint "D"* (Continued)

CODE TO TYPICAL CONDITION	REASON FOR CONDITION	PRACTICAL SOLUTION
21-52	Unbalanced load. Housing bore too large.	Rebalance machine. Replace housing with one having proper bore. *See Illustration—Complaint "A", Conditions 21-52*
22	Flat on ball or roller due to skidding. (Result of fast starting.)	Carefully examine balls or rollers, looking for flat spots on the surface. Replace bearing.
25-26	Unequal load distribution on bearing.	Rework shaft, housing, or both, to obtain proper fit. May require new shaft and housing. *See Illustration—Complaint "C", Conditions 25-26*
27	Inadequate shoulder support causing bending of shaft.	Remachine shaft fillet to relieve stress. May require shoulder collar. *See Illustration—Complaint "C", Condition 27*
29	Inadequate support in housing causing cocking of outer ring.	Remachine housing fillet to relieve stress. May require shoulder collar. *See Illustration—Complaint "C", Condition 29*
31	Distortion of shaft and inner ring. Uneven expansion of bearing inner ring.	Remachine shaft fillet to obtain proper support. *See Illustration—Complaint "C", Condition 31*
32	Distortion of housing and outer ring. Pinching of bearing.	Remachine housing fillet to obtain proper support. *See Illustration—Complaint "C", Condition 32*
35-36	Incorrect linear or angular alignment of two or more coupled shafts with two or more bearings.	Correct alignment by shimming pillow blocks. Be sure shafts are coupled in straight line—especially when three or more bearings operate on one shaft. *See Illustration—Complaint "A", Conditions 35-36*
42	Incorrect method of mounting. Hammer blows on bearing.	Replace with new bearing. Don't hammer any part of bearing when mounting.
45	Excessive clearance in bearing, resulting in vibration.	Use bearings with recommended internal clearances.
46	Vibration of machine.	Check balance of rotating parts. Rebalance machine.
53	"Pounding-out" of housing bore due to soft metal. Result: enlarged bore . . . causing spinning of outer ring in housing.	Rebore housing and press steel bushing in bore. Machine bore of bushing to correct size.

UNSATISFACTORY PERFORMANCE OF EQUIPMENT
Complaint "E"

CODE TO TYPICAL CONDITION	REASON FOR CONDITION	PRACTICAL SOLUTION
4	Bearings selected with inadequate internal clearance for conditions where external heat is conducted thru shaft, thereby expanding excessively the inner ring.	Replacement bearing should have identical marking as original bearing for proper internal clearance. Check with SKF if bearing markings have become indistinct.
5	Foreign matter (dirt, sand, carbon, etc.) entering bearing housing.	Clean out bearing housing. Replace worn-out seals or improve seal design to obtain adequate protection of bearing.
6	Corrosive agents (water, acids, paints, etc.) entering the bearing housing.	Addition of a shroud and (or) flinger to throw off foreign matter.
7-8-9 41-51	Housing bore out of round. Housing warped. Excessive distortion of housing. Undersized housing bore.	Check and scrape housing bore to relieve pinching of bearing. Be sure pedestal surface is flat, and shims cover entire area of pillow block base. *See Illustration—Complaint "A", Conditions 7-8-9-41-51*
10	Failure to remove chips, dirt, etc. from bearing housing before assembling bearing unit.	Carefully clean housing, and use fresh lubricant.

UNSATISFACTORY PERFORMANCE OF EQUIPMENT — *Complaint "E"* (Continued)

CODE TO TYPICAL CONDITION	REASON FOR CONDITION	PRACTICAL SOLUTION
16 23-24	Two "held" bearings on one shaft. Excessive shaft expansion.	Back off covers in one of the housings, using shims to obtain adequate clearance of outer ring, to permit free axial bearing motion. *See Illustration—Complaint "A", Conditions 16-23-24*
17-18	Shaft diameter too small. Adapter not tightened sufficiently.	Metallize shaft and regrind to obtain proper fit. Retighten adapter to get firm grip on shaft. *See Illustration—Complaint "B". Conditions 17-18*
19	Adapter tightened excessively.	Loosen locknut and sleeve assembly. Retighten sufficiently to clamp sleeve on shaft but be sure bearing turns freely. *See Illustration—Complaint "A", Condition 19*
21-52	Unbalanced load. Housing bore too large.	Rebalance machine. Replace housing with one having proper bore. *See Illustration—Complaint "A", Conditions 21-52*
22	Flat on ball or roller due to skidding. (Result of fast starting.)	Carefully examine balls or rollers, looking for flat spots on the surface. Replace bearing.
25-26	Unequal load distribution on bearing.	Rework shaft, housing, or both, to obtain proper fit. May require new shaft and housing. *See Illustration—Complaint "C", Conditions 25-26*
27	Inadequate shoulder support causing bending of shaft.	Remachine shaft fillet to relieve stress. May require shoulder collar. *See Illustration- —Complaint "C", Condition 27*
29	Inadequate support in housing causing cocking of outer ring.	Remachine housing fillet to relieve stress. May require shoulder collar. *See Illustration—Complaint "C", Condition 29*
31	Distortion of shaft and inner ring. Uneven expansion of bearing inner ring.	Remachine shaft fillet to obtain proper support. *See Illustration—Complaint "C", Condition 31*
32	Distortion of outer housing and ring. Pinching of bearing.	Remachine housing fillet to obtain proper support. *See Illustration—Complaint "C", Condition 32*
35-36	Incorrect linear or angular alignment of two or more coupled shafts with two or more bearings.	Correct alignment by shimming pillow blocks. Be sure shafts are coupled in straight line—especially when three or more bearings operate on one shaft. *See Illustration—Complaint "A", Conditions 35-36*
39	Prong rubbing against bearing.	Remove lockwasher—straighten prong or replace with new washer. *See Illustration—Complaint "A", Condition 39*
42	Incorrect method of mounting. Hammer blows on bearing.	Replace with new bearing. Don't hammer any part of bearing when mounting.
45	Excessive clearance in bearing, resulting in vibration.	Use bearings with recommended internal clearances.
46	Vibration of machine.	Check balance of rotating parts. Rebalance machine.
47	Shaft and housing shoulders, and face of locknut out of square with bearing seat.	Remachine parts to obtain squareness.
50	Bearing seat diameter machined oversize, causing excessive expansion of bearing inner ring thus reducing clearance in bearing.	Grind shaft to proper fit between inner ring of bearing and shaft.
53	"Pounding-out" of housing bore due to soft metal. Result: enlarged bore . . . causing spinning of outer ring in housing.	Rebore housing and press steel bushing in bore. Machine bore of bushing to correct size. If loads are not excessive, tighter fit in housing, without the use of the steel bushing, may correct the trouble.

CODE TO TYPICAL CONDITION	REASON FOR CONDITION	PRACTICAL SOLUTION

BEARING IS LOOSE ON SHAFT
Complaint "F"

CODE TO TYPICAL CONDITION	REASON FOR CONDITION	PRACTICAL SOLUTION
17-18	Shaft diameter too small. Adapter not tightened sufficiently.	Metallize shaft and regrind to obtain proper fit. Retighten adapter to get firm grip on shaft. *See Illustration—Complaint "B", Conditions 17-18*
48	Knurling and center punching of bearing seat on shaft.	Unsatisfactory because high spots are flattened when load is applied. When fit is loose, metallize shaft and regrind to proper size.

HARD TURNING OF SHAFT
Complaint "G"

CODE TO TYPICAL CONDITION	REASON FOR CONDITION	PRACTICAL SOLUTION
1	Wrong type of grease or oil causing break-down of lubricant.	Consult reliable lubricant manufacturer for proper type of lubricant. Check SKF Catalog instructions to determine if oil or grease should be used.
2	Low oil level. Loss of lubricant through seal. Insufficient grease in housing.	Oil level should be just below center of lowest ball or roller in bearing. Using grease, lower half of pillow block should be ½ to ⅔ full. *See Illustration—Complaint "A", Condition 2*
3	Housing packed with grease, or oil level too high . . . causing excessive churning of lubricant, high operating temperature, oil leakage.	Purge bearing until only lower half of housing is ½ to ⅔ full of grease. Using oil lubrication, reduce level to just below center of lowest ball. *See Illustration—Complaint "A", Condition 3*
4	Bearings selected with inadequate internal clearance for conditions where external heat is conducted thru shaft, thereby expanding excessively the inner ring.	Replacement bearing should have identical marking as original bearing for proper internal clearance. Check with SKF if bearing markings have become indistinct.
5	Foreign matter (dirt, sand, carbon, etc.) entering bearing housing.	Clean out bearing housing. Replace worn-out seals or improve seal design to obtain adequate protection of bearing.
6	Corrosive agents (water, acids, paints, etc.) entering the bearing housing.	Addition of a shroud and (or) flinger to throw off foreign matter.
7-8 9-41 51	Housing bore out of round. Housing warped. Excessive distortion of housing. Undersized housing bore.	Check and scrape housing bore to relieve pinching of bearing. Be sure pedestal surface is flat, and shims cover entire area of pillow block base. *See Illustration—Complaint "A", Conditions 7-8-9-41-51*
10	Failure to remove chips, dirt, etc. from bearing housing before assembling bearing unit.	Carefully clean housing, and use fresh lubricant.

HARD TURNING OF SHAFT — *Complaint "G"* (Continued)

CODE TO TYPICAL CONDITION	REASON FOR CONDITION	PRACTICAL SOLUTION
12	Leather or composition seals with excessive spring tension or dried out.	Replace leather or composition seals with ones having reduced spring tension. Lubricate seals. *See Illustration—Complaint "A", Condition 12*
13 33-40	Rotating seals rubbing against stationary parts.	Check running clearance of rotating seal to eliminate rubbing. Correct alignment. *See Illustration—Complaint "A", Condition 13-33-40*
15	Opposed mounting.	Insert gasket between housing and cover flange to relieve axial pre-loading of bearing. *See Illustration—Complaint "A", Condition 15*
16 23-24	Two "held" bearings on one shaft. Excessive shaft expansion.	Back off covers in one of the housings, using shims to obtain adequate clearance of outer ring, to permit free axial bearing motion. *See Illustration—Complaint "A", Conditions 16-23-24*
19	Adapter tightened excessively.	Loosen locknut and sleeve assembly. Retighten sufficiently to clamp sleeve on shaft but be sure bearing turns freely. *See Illustration—Complaint "A", Condition 19*
39	Prong rubbing against bearing.	Remove lockwasher. Straighten prong or replace with new washer. *See Illustration—Complaint "A", Condition 39.*
27	Inadequate shoulder support causing bending of shaft.	Remachine shaft fillet to relieve stress. May require shoulder collar. *See Illustration—Complaint "C", Condition 27*
29	Inadequate support in housing causing cocking of outer ring.	Remachine housing fillet to relieve stress. May require shoulder collar. *See Illustration—Complaint "C", Condition 29*
30	Distortion of bearing seals.	Remachine housing shoulder to clear seal. *See Illustration—Complaint "B", Condition 30*
31	Distortion of shaft and inner ring. Uneven expansion of bearing inner ring.	Remachine shaft fillet to obtain proper support. *See Illustration—Complaint "C", Condition 31*
32	Distortion of housing and outer ring. Pinching of bearing.	Remachine housing fillet to obtain proper support. *See Illustration—Complaint "C", Condition 32*
35-36	Incorrect linear or angular alignment of two or more coupled shafts with two or more bearings.	Correct alignment by shimming pillow blocks. Be sure shafts are coupled in straight line—especially when three or more bearings operate on one shaft. *See Illustration—Complaint "A", Conditions 35-36*
47	Shaft and housing shoulders, and face of locknut out of square with bearing seat.	Remachine parts to obtain squareness.
50	Bearing seat diameter machined oversize, causing excessive expansion of shaft and bearing inner ring, thus reducing clearance in bearing.	Grind shaft to get proper fit between inner ring of bearing and shaft.

Journal and Tilt-Pad Thrust Bearings*

It is not at all unusual to find unexpected bearing damage when machinery is opened for routine inspection, for turnaround maintenance, or for the purpose of troubleshooting vibration events. Very often the bearings are then blamed for faulty design or deficient manufacture when in fact an extraneous source is more likely the culprit.

This section will direct the analyzing engineer or technician on the basis of available facts to the most probable causes of bearing damage. Moreover, this section will help him decide what corrective action should be taken. This decision must be based on the nature and severity of the damage and on the significance of the type of damage.

Scoring Due to Foreign Matter or Dirt

Contamination of the lubricant includes:

1. "Built-in" dirt in crankcases, on crankshafts, in oil distribution headers, in cylinder bores, etc. present at the time of machine assembly.
2. Entrained dirt entering through breathers or air filters and particles derived from fuel combustion in internal combustion engines.
3. Metallic wear particles resulting from abrasive wear of moving parts.

"Dirt" may polish the surfaces of whitemetal-lined bearings, burnish bronze bearings, abrade overlays or other bearing linings, and score both bearing and mating surfaces, with degrees of severity depending upon the nature and size of the dirt particle or upon oil-film thickness and type of bearing material.

The effect of dirt on journal bearings is shown in Figure 3-66, which illustrates a whitemetal-lined bearing scored and pitted by "dirt." Similarly, Figure 3-67 shows a whitemetal-lined bearing with "haloes" caused by dirt particles.

Bearings in the condition shown in Figure 3-67 should be scrapped and new bearings fitted after journal, oil passages, and filters are cleaned. The oil should be changed, also.

Bearings in the condition shown in Figure 3-66 can be refitted after cleaning bearings and journal surfaces, provided any clearance increase due to wear can be tolerated.

Tilting-pad thrust bearings are similarly affected by dirt. In Figure 3-68, note concentric scoring of the thrust pad due to dirt entering the bearing at high speed. Figure 3-69 illustrates scoring by dirt entering the bearing at startup, while Figure 3-70 depicts the surface of the pad in Figure 3-69 at higher magnification. Figure 3-70 also shows irregular tracks caused by the passage of shotblast spherical steel particles.

Figure 3-66. Scoring caused by dirt.

Figure 3-67. Effects of contaminated lube oil.

Figure 3-68. Concentric scoring due to dirt entering at high speed.

Figure 3-69. Dirt-intrusion damage at low-speed operation.

Figure 3-70. Magnification of Figure 3-69. Irregular tracks were caused by spherical steel particles.

The damage shown in Figures 3-68 through 3-70 requires damaged pads to be scrapped. The lubricating system should be cleaned, the oil changed, and new pads fitted.

Wiping of Bearing Surfaces

A wiped bearing surface is where surface rubbing, melting, and smearing is evident. This is usually caused from inadequate running clearance with consequent surface overheating, from inadequate oil supply, or from both.

Damaged journal bearings are shown in Figure 3-71. This whitemetal-lined turbine bearing is wiped in both top and bottom halves because of inadequate

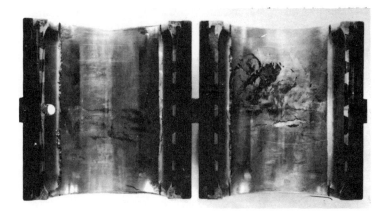

Figure 3-71. Wiping caused by tight clearance.

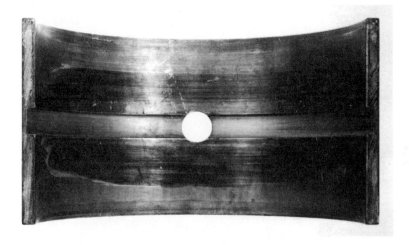

Figure 3-72. Wiping caused by barreled journal surfaces.

clearance. An overlay-plated-copper lead journal bearing wiped because of a barrel-shaped journal is illustrated in Figure 3-72.

Recommended action is as follows. If the journal bearing is lightly wiped, refit the bearing after cleaning the surface to remove any loose metal, provided the clearance can be tolerated. If it is cracked, fit a new bearing and increase the clearance when fitting. If vibration synchronous with shaft rotational frequency is present, check balance of rotor, alignment, coupling condition, etc. If vibration frequency is at half shaft speed or less, fit antiwhirl bearings.

A tilting-pad thrust-bearing segment is shown in Figure 3-73. Here, surface wiping of a whitemetal-lined pad in successive, thin layers is caused from excessive, steady load at startup. To remedy the situation, fit new pads. Improve lubrication at startup, possibly by using a prelubrication oil system. If possible, reduce loading at startup.

Fatigue Cracking

Fatigue cracking is caused by the imposition of dynamic loads in excess of the fatigue strength of the bearing material at operating temperature. The fatigue strength, especially of low-melting-point materials such as whitemetals and lead-base overlays, is greatly reduced at high temperatures, hence overheating alone may cause fatigue failure. Other causes of fatigue failure are overloading, cyclic out-of-balance loadings, high cyclic centrifugal loading due to overspeeding, and shafts that are not truly cylindrical because of manufacturing defects introduced by honing, filing, etc.

Journal bearings with fatigue defects are shown in Figures 3-74 through 3-77. Figure 3-74 illustrates fatigue cracking of a whitemetal-lined bearing due to shaft ridging caused by differential wear in the relief groove. Figure 3-75 shows fatigue cracking of 20% tin-aluminum lining due to misalignment and consequent edge loading. Fatigue cracking of a whitemetal-lined bearing due to shaft deflection and edge loading is depicted in Figure 3-76. Finally, the whitemetal-lined top-half turbine bearing in Figure 3-77 is partially fatigue cracked due to out-of-balance loading and an excessively wide cutaway section.

These journal bearings should be discarded. Investigate and, if possible, rectify the causes of misalignment, shaft deflection, and overloading. Increase the width of lands on the top half (Figure 3-77). Fit new bearings, possibly of stronger material if underdesign is indicated.

Figure 3-73. Damage caused by excessive steady load at startup.

Figure 3-74. Fatigue cracking due to shaft ridging.

Figure 3-75. Results of misalignment and high edge-loading.

Figure 3-76. Fatigue cracking due to shaft deflection and high edge-loading.

Figure 3-77. Fatigue cracking due to high rotor imbalance.

Corrosion

Corrosion of the lead in copper-lead and lead-bronze alloys, and of lead-base whitemetals, may be caused by acidic oil oxidation products formed in service, by ingress of water or coolant liquid into the lubricating oil, or by the decomposition of certain oil additives.

Removal of overlays by abrasive wear or scoring by dirt exposes the underlying lead in copper-lead or lead-bronze interlayers to attack, while in severe cases the overlays may be corroded.

The journal bearings shown in Figures 3-78 through 3-81 have been damaged by corrosion. In Figure 3-78, severe corrosion of the surface of an unplated copper-lead lined bearing is caused by the lead phase being attacked by acidic oil oxidation products. Figure 3-79 magnifies a fractured section of the bearing shown in Figure 3-78, indicating the corrosion depth of the lead layer. Corrosion of a marine turbine whitemetal bearing is illustrated in Figure 3-80. Water in the oil has caused a smooth, hard, black deposit of tin dioxide to form on the surface. Figure 3-81 captures

Figure 3-78. Corrosion caused by unserviceable lube oil.

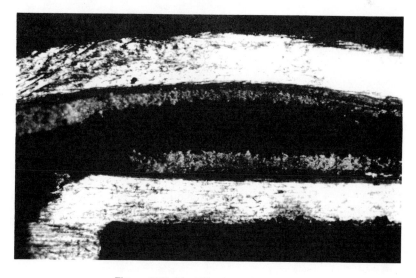

Figure 3-79. Magnification of Figure 3-78.

Figure 3-80. Tin-dioxide formation attributed to water contamination of lube oil.

Figure 3-81. Sulphur corrosion of phosphor-bronze bushing.

"sulphur corrosion" of a phosphor bronze bushing. This damage was initiated by the decomposition of a lubricating oil additive and by gross pitting and attack of the bearing surface.

In all of these damage cases it will be necessary to scrap the bearings. Investigate the condition of the oil to ascertain the cause of corrosion. Eliminate water in the oil. In case of sulphur corrosion, change the bearing material from phosphor bronze to a phosphorous-free alloy such as lead-bronze, silicon bronze, or gunmetal. It would also be appropriate to consult with oil and bearing suppliers.

Cavitation Erosion

Cavitation erosion is an impact fatigue attack caused by the formation and collapse of vapor bubbles in the oil film. It occurs under conditions of rapid pressure changes during the crank cycle in internal combustion engines. The harder the bearing material the greater its resistance to cavitation erosion.

Impact cavitation erosion occurring downstream of the joint face of an ungrooved whitemetal-lined bearing is shown in Figure 3-82. Figure 3-83 is a section through the main bearing. It illustrates the mechanism of cavitation and modifications made to the groove to limit or reduce damage. In Figure 3-84, observe the result of discharge cavitation erosion in an unloaded half-bearing caused by rapid movement of the journal in its clearance space during the crank cycle. Also of interest is Figure 3-85, a set of diesel-engine main bearings showing cavitation erosion of the soft overlay while the harder tin-aluminum is unattacked.

The journal bearings shown in Figures 3-82 through 3-85 may be put back into service if cavitation attack is not very severe or extensive. Investigate increasing the oil-supply pressure, modifying the bearing groove, blending the edges or contours of grooves to promote streamline flow, reducing the running clearance, or changing to a harder bearing material.

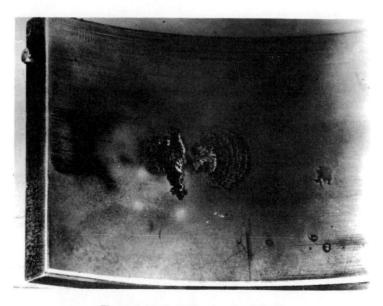

Figure 3-82. Cavitation-erosion damage.

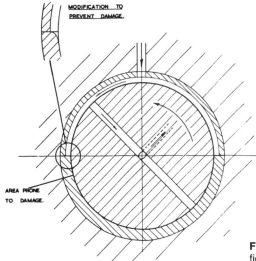

Figure 3-83. Section through modified main bearing.

Figure 3-84. Result of discharge-cavitation erosion.

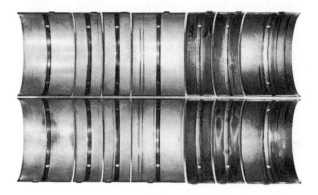

Figure 3-85. Cavitation erosion in diesel-engine main bearings.

"Black Scab" or "Wire Wool" Damage

A large dirt particle (probably not less than one mm across) carried into the clearance space by the lubricating oil and becoming embedded in the bearing may form a hard scab of material by contact with the steel journal or thrust collar. This scab will then cause very severe damage to the mating steel surface, which is literally machined away with the formation of so-called "wire wool." The action is self-propagating once it has started. The susceptibility to scab formation appears to depend upon the nature of the lubricant and the composition of the steel of the rotor shaft or collar. Steels containing chromium or manganese in excess of 1% appear to be particularly susceptible to scab formation, especially in high-speed machines with bearing rubbing speeds of over 20 meters (> 63 feet) per second.

Journal bearings exhibiting this type of damage are shown in Figures 3-86 and 3-87. In Figure 3-86, a whitemetal-lined compressor bearing shows "black scab" at its inception. In Figure 3-87, the 13% chromium-steel journal running in the bearing shown in Figure 3-86 exhibits severe machining damage.

Figure 3-86. Whitemetal-lined compressor bearing with early indication of "black scab."

Figure 3-87. Severe machining damage on 13% chrome journal.

Figure 3-88. Thrust pad with severe "black scab" formation.

Black-scab formation on a whitemetal-lined thrust pad is pictured in Figure 3-88. Black-scab or wire-wool damage renders the bearings unusable; they must be scrapped. Pay particular attention to cleanliness during assembly and avoid contaminating the bearing surface and oil passages with hard debris. Investigate changing the journal or collar surface material by sleeving with mild steel or by hard-chrome plating. Changing the bearing alloy is unlikely to be effective.

Pitting Due to Electrical Discharge

Electrical discharge through the oil film between the journal and bearing in electrical machinery or on the rotors in fans and turbines may occur from faulty insulation or grounding, or from the build-up of static electricity. This can occur at very low voltages and may cause severe pitting of bearing or journal surfaces, or both. In extreme cases damage may occur very rapidly. The cause is sometimes difficult to diagnose because pitting of the bearing surface is ultimately followed by wiping and failure, which may obscure the original pitting.

Fine hemispherical pitting and scoring of a whitemetal-lined generator bearing due to electrical discharge is evident in Figure 3-89. Figure 3-90 represents a close-up of more severe electrical discharge pitting of whitemetal.

Bearings have to be discarded if severely pitted or wiped. Examine and, if necessary, regrind the journal to eliminate pitting. Investigate grounding conditions of the machine, especially around insulation. Pay particular attention to fittings such as guards, thermocouple leads, water connections, etc. which may be bridging the insulation. Fit new bearings and run for only a short time, depending upon the period run prior to failure. Disassemble and examine.

Figure 3-89. Whitemetal-lined generator bearing with electrostatic-erosion damage.

Figure 3-90. Closeup of electrical discharge pitting of whitemetal bearing.

Damage Caused by Faulty Assembly

Under this heading are included:

1. Fretting damage due to inadequate interference fit in a weak housing.
2. Excessive interference fit, causing bearing-bore distortion.
3. Effect of joint-face offset or absence of joint-face relief in bearing. This causes overheating and damage in the region of the bearing split.
4. Fouling at crankshaft fillets due to incorrect shaft radii.
5. Misalignment or shaft deflection, causing uneven wear of bearing.
6. Entrapment of foreign matter between bearing and housing during assembly, causing bearing-bore distortion and localized overheating.
7. Effect of grinding marks on shaft.

The damage observed in Figure 3-91 represents fretting between the back of the bearing and the housing due to inadequate interference fit. Alternatively, this could also be due to a weak housing design. In Figure 3-92, observe overheating and fatigue at joint faces due to excessive interference fit, causing bearing-bore distortion. Joint face offset at assembly or a weak housing design could again cause this defect. Fretting at the joint faces of a connecting rod due to inadequate bolt tension is depicted in Figure 3-93. In Figure 3-94, observe uneven wear of an overlay-plated bearing due to misalignment. Fouling between shaft radius and the bearing-end face caused the overheating damage in Figure 3-95. Finally, Figure 3-96 illustrates severe scoring of a whitemetal-lined compressor bearing caused by shaft grinding marks.

Figure 3-91. Fretting caused by inadequate interference fit.

Figure 3-92. Overheating and fatigue damage caused by bearing-bore distortion.

Figure 3-93. Fretting caused by inadequate bolt tension.

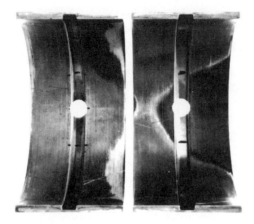

Figure 3-94. Uneven wear caused by misalignment.

Figure 3-95. Excessive heating due to inadequate oil flow caused this damage.

Figure 3-96. Scoring caused by grinding marks in shaft.

In Figure 3-97, observe the back portion of a tin-aluminum bearing showing a flat spot on the steel shell. This was caused by debris entrapped between the shell and housing. Figure 3-98 depicts the bore of the bearing shown in Figure 3-97. Entrapped debris has deformed the bearing bore, resulting in flexure and fatigue damage. In Figure 3-99, angular misalignment of shaft or housing or dirt trapped behind the carrier ring caused damage to the pads (however, on one side only).

Whenever faulty assembly is suspected or confirmed, investigate and if necessary correct the interference conditions and housing design. Correct the misalignment, discard the bearings, and fit new bearings or pads.

Figure 3-97. Flat spot on steel shell caused by careless assembly.

Figure 3-98. Bearing-bore deformation resulting from the careless assembly in Figure 3-97.

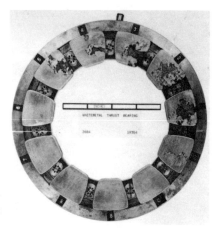

Figure 3-99. Angular misalignment or dirt trapped behind carrier ring caused damage to pads.

Inadequate Lubrication

Inadequate lubrication may be due to insufficient oil-pump capacity, inadequate header or oil passage diameters, incorrect grooving design, or accidental interruption of the oil supply.

In the case of machines provided with shaft-driven lubricating oil pumps only, the failure condition may originate with oil starvation during startup. This in turn could be caused by operating hand-priming pumps inadequately.

Depending upon the duration and severity of the lubrication failure, surface melting, wiping, or complete seizure of the bearing may occur.

Journal bearings damaged by inadequate lubrication are shown in Figures 3-100 through 3-102. Fatigue cracking caused by inadequate lubrication due to axial grooving in the loaded area is responsible for the damage in Figure 3-100. Figure 3-101 shows the effect of overheating and damage of a tin-aluminum bearing surface due to reduced oil flow. This was caused by omitting the drain hole at one end of the bearing, in effect causing the bearing cavity to be sealed and inhibiting through-flow. Finally, Figure 3-102 illustrates the seizure of a 20% tin-aluminum bearing due to lubrication failure.

Figure 3-100. Fatigue cracking caused by inadequate lubrication.

Figure 3-101. Overheating caused by insufficient oil flow.

Figure 3-102. Tin-aluminum bearing seized due to lubrication failure.

Figure 3-103. Surface damage caused by external vibration.

Recommended action is quite straightforward. Investigate and, if necessary, correct the groove design. Investigate all lubrication conditions. Scrap the bearings. Examine and, in cases of seizure, arrange for dye-penetrant checking of the journals. Fit new bearings.

Fretting Damage Due to Vibration

The damage we are talking about here is fretting damage in the bore as distinct from fretting damage elsewhere. Bearing operating surfaces may suffer fretting damage while the shaft is at rest. This could be caused by vibrations being transmitted to the machine from external sources, etc.

Figure 3-103 shows surface damage caused by external vibration while the journal was not rotating. Should this be the failure mode, eliminate the transmission of vibration from external sources, if feasible. Consider mounting the affected machine on springs or on rubber pads. If damage may occur in transit, secure the journal to prevent vibration.

Damage Due to Overheating

Lining extrusion and cracking may be due to overheating. A reduction in the strength of linings may result, so that the material yields and develops cracks because of the transmission of normal and shear forces through very thin oil films. Wiping does not necessarily occur under such conditions.

Surface deformation can be caused in an anisotropic material by thermal cycling.

Tilting-pad thrust bearings damaged by excessive temperature are shown in Figures 3-104, 3-105, and 3-106. Figure 3-104 depicts cracking of the whitemetal lining of a pad due to operation at excessively high temperatures. Note the displacement of whitemetal over the edge of the pad due to extrusion. Cracking and displacement of the whitemetal lining of a pad due to overheating under steady load conditions is shown in Figure 3-105. In Figure 3-106, observe thermal "ratcheting" of the whitemetal lining of a pad due to in-service thermal cycling through an excessive temperature range. In all three cases, it would be appropriate to investigate and, if possible, reduce maximum operating temperatures. Fit new or reconditioned pads.

Figure 3-104. Whitemetal liner damaged due to excessive temperature.

Figure 3-105. Overheating under steady load conditions caused white-metal cracking and extrusion.

Figure 3-106. Thermal "ratcheting" due to repeated cycling over an excessive temperature range.

Pivot Fatigue

Axial vibration can cause damage and fatigue of tilting-pad pivots. Both pivot and carrier ring may suffer damage by indentation or fretting. In some cases tiny hemispherical cavities may be produced. Damage may occur due to axial vibration imposed upon the journal, or may be caused by the thrust collar face running out of true.

In Figure 3-107, the pad carrier ring shows damage due to axial shaft vibrations. Similarly, Figure 3-108 depicts a pad pivot with hemispherical cavities caused by pivot fretting due to vibration.

Here, it will be necessary to recondition the carrier ring and fit new or reconditioned pads. Investigate and eliminate the cause of axial vibration.

Gear Failure Analysis

Someone once said that "Gears wear out until they wear in, and then they wear forever." The American Gear Manufacturers Association (AGMA) describes this mechanism more clearly as follows: "It is the usual experience with a set of gears on a gear unit . . . assuming proper design, manufacture, application, installation, and operation . . . that there will be an initial 'running-in' period during which, if the gears are properly lubricated and not overloaded, the combined action of rolling and sliding of the teeth will smooth out imperfections of manufacture and give the working surface a high polish. Under continued proper conditions of operation, the gear teeth will show little or no signs of wear."

Despite the truth of this observation, failures of gear teeth will occur as a result of certain extraneous influences. In many situations, early recognition may suggest a remedy before extensive damage takes place. High-speed gears on process

Figure 3-107. Pad carrier ring damaged by axial shaft vibration.

Figure 3-108. Pad pivot with hemispherical cavities caused by pivot fretting due to vibration.

machinery should therefore receive regular and thorough inspections as part of a well-managed predictive maintenance program.

The AGMA cites 25 modes of gear-tooth failure. Ku[4] thought it more logical to classify gear-tooth failure modes into two basic categories: strength-related modes and lubricant-related modes. Major modes of strength-related failure are, for instance, plastic flow and breakage. Examples of lubricant-related failure modes are rubbing wear, scuffing or scoring, and pitting.

Table 3-4 shows the principal gear failure modes and distress symptoms and relates them to their causes. Gear failure statistics are presented in Tables 3-5 and 3-6.[5]

Gear failure analysis is a complex procedure. It draws clues as to the most probable cause of failure from visual inspection, metallurgical analysis, tribological mechanisms, system dynamic analysis, design review of the casings, bearings, shafts, and seals, and, finally, the detailed analysis of the individual gears. A universally applicable approach to gear failure analysis is the subject of the following section.

Table 3-4
Gear Failure Modes and Their Causes

			Fracture/Separation					Wear				Deformation	
		Tooth		Flank									
	Failure Causes	Fracture (breakage)—overload	Fracture (breakage)—fatigue *↑	Cracking *↑	Initial pitting	Destructive pitting *↑	Spalling *↑	Wear *↑	Scuffing *↑	Scoring	Corrosion	Plastic flow *↑	Warm flow *↑
Manufacture and Design	Manufacturing problems		•	•	•	•	•	•				•	
	Overload through misalignment	•											
	Frequent load cycles		•										
	Fatigue design		•		•	•	•						
Operating Conditions	Operating Conditions (speed, loading)				•	•	•	•	•	•	•		
Lubrication	Viscosity							•	•	•	•		
	Quality							•	•	•	•		
	Quantity							•	•	•			•
	Contamination							•	•	•	•		

Legend: * In many cases leading to fatigue failure.

Table 3-5
Distribution of Stationary Gear Failure Causes[5]

Failure Cause	Distribution of Failure Cases, %
Vendor problems	36.0
Planning/design	12.0
Assembly	9.0
Manufacturing	8.0
Material of construction	7.0
Operating problems	47.0
Maintenance	24.0
Repair	4.0
Mishandling	19.0
Extraneous influences	17.0
Foreign bodies	8.0
Disturbances from the driving or driven side	7.0
Disturbances from electrical supplies	2.0

Table 3-6
Gear Failure Mode Distribution[5]

(Gear) Failure Mode	Approximate Distribution of Failure Cases, %
Forced fractures	50
Fatigue failures	16
Changes in load patterns	15
Incipient cracks	15
Distortions	4

Preliminary Considerations*

The traditional first step in gear failure analysis is to visually examine the failed parts. Gears, shafts, bearings, casings, lubricant system, and seals should be given attention. Data on the magnitude of the external loads applied by driving and driven equipment may be very helpful in differentiating between failure types. A correct evaluation of the failure type is essential in establishing the direction and depth of the remainder of the study.

---

*Based on "Introduction to Gear Failure Analysis," a paper given at the National Conference of Power Transmission, Chicago, Illinois, 1979, by P. M. Dean, MTI. By permission of the conference committee and the author.

In order to improve communication concerning gear failure problems, the AGMA published a standard in 1943 identifying some of the more frequently encountered failures. The latest revision of this standard, AGMA 110.XX, "Nomenclature of Gear Tooth Failure Modes," was published in September 1979. The photographs used in this section are taken (with permission) from this document[6] and are identified by the same captions used in the standard.

Analytical Evaluation of Gear Theoretical Capability

An important step when investigating a gear failure is to analytically evaluate the theoretical capabilities of the gears. In some cases these data will help differentiate between types or degrees of failure. More frequently, however, these data will lead the investigator toward essential bits of evidence. For example, in a case where pitting is discovered relatively early in the life of a gear set, a decision must be made: Will the gears correct themselves and the pits heal over, or will the pitting continue into destructive pitting, causing gear failure? If the investigator has made a good theoretical evaluation of the capabilities of the specific gears to withstand the known loads, he will be better guided toward the types of evidence needed. In this case, the theoretical loads may be within the capabilities of the gears, and only minor surface errors are being corrected. If the type of pits and distribution of the pitted areas support the normal-load/local-asperity thesis, the decision to run longer can be made with some confidence. On the other hand, the early discovery of pitting may have been fortuitous: The gears could be trying to say something. In one case, the pitted zones may indicate improper alignment to a degree that the internal pitting process could never correct. Even more subtle are cases wherein initial pitting cannot be explained by asperities or even by the anticipated loads. In such cases, a torsional dynamic analysis or highly instrumented measuring technique may be applied to determine if transient loads caused the observed pits.

The analytical techniques suggested here have been found to give good, realistic correlations. In cases where the calculated results agree with observed data, the investigator knows to look deeper into the problem for more subtle causes.

It is our intent to show the more generally accepted methods of establishing the surface stresses, bending fatigue stresses, flash temperature (for scoring), and lubricant film thickness (for wear). The details of applying these methods must remain beyond the scope of this text, but are covered in detail in the references noted. Key equations are shown as an aid in identification, and the terms of all equations may be found in Appendix A.

Metallurgical Evaluation

An extremely important aspect of a failure investigation is a thorough metallurgical analysis. The purpose of such an analysis is to determine if the gear did, in fact, have the physical properties anticipated in the design.

There are a number of potential defects that can only be found by sectioning the gear to obtain metallurgical specimens. Such defects include excessive grain size, nonmagnetic inclusions, seams, cracks and folds, undissolved carbides, excessive retained austenite, intermediate transformation products, grain boundary networks, excessive decarburization, banding and nonuniform transition from case to core, and excessive white layer (nitriding), depending on the type of heat-treatment process. Unfortunately, some gears are cut and finish-machined with much more attention than is given to them during heat treatment. Hence, many failures have been caused by improper metallurgy.

General Mechanical Design

Close attention must be given to evaluating the adequacy and accuracy of the mountings. A major effort in any investigation is to establish the degree, if any, to which misalignment may have contributed to failure. Misalignment can be caused either by manufacturing errors or by load-induced deflections in the shaft, bearings, or casings. In some cases, visual evidence may be found in the form of polished areas, or the lack of such areas, on contacting surfaces of the teeth, which may indicate operating alignment errors. In some cases it may be necessary to resort to deflection tests to determine the degree to which misalignment may occur under load.

In most gear failures, machine elements such as shafts, bearings, and seals will show signs of damage or will have failed. A study of these elements will help develop the clues necessary to reconstruct the failure sequence.

Lubrication

Proper lubrication is essential. The lubricant has two principal functions: to minimize rubbing friction and to carry off heat. If the lubricating film thickness is adequate and the lubricant is clean, wear will be minimized. The lubricant must sometimes provide protection against corrosion.

Some applications require lubricants having many additives so that the lubricant can adequately perform many tasks. Sometimes the lubricant itself fails. Its internal chemical compounds can break down with time, with heat, or by unforeseen reactions with chemicals from its ambience. Several types of gear failures can be traced back to lubricant failure.

Defects Induced by Other Train Components

Gearing is frequently part of a complete power train that includes a number of major components connected by shafts and couplings. In certain applications, such as in the petrochemical industry, compressor trains can be quite extensive, which in turn leads to very involved spring-mass systems. Such systems frequently have numerous potential problem areas of critical speeds and nodal points. The actual operating loads experienced by the various gears may be markedly different from the name-plate power ratings of the driving equipment.

In such applications, torsional dynamic analysis is essential during gear selection, and if failures occur, such analyses should be carefully reviewed or repeated.

Wear

Gear-tooth wear occurs as the surface material of the contacting areas is worn away during operation.[7] The type of wear that is under way can be determined by evaluating the:

1. Visual appearance.
2. Operating film thickness.
3. Particle size present in the oil.
4. Gear surface hardness.
5. Load operating profile.
6. Ambience.

It is convenient to group the more generally recognized types of wear into the following categories. These categories are based on appearance, cause, and effect.

1. Normal wear
   a. polishing
   b. moderate wear
   c. excessive wear
2. Abrasive wear
3. Corrosive wear

When evaluating wear, it is helpful to know the type of lubrication regime in which the gears are operating. At present, there are three generally recognized operating regimes:

Specific film thickness, which is the ratio of film thickness to surface asperity height, is a convenient index of the lubrication regime in effect. The risk of wear is · evaluated in Table 3-7.

Table 3-7
Risk of Wear for Specific Ratios of Film to Asperity Thickness

Film Thickness Ratio (λ)	Operating Regime	Risk of Wear
up to 1.5	boundary (wear)	high
1.51 to 4.0	mixed	moderate
4.1 and over	full elastohydrodynamic	low

A typical equation to determine film thickness in helical gearing is:

$$h = \frac{2.65 \; \alpha^{.54}}{E^{.3} \; w^{.13}} \; \left(\eta_0 \; \frac{\pi N_p}{30}\right)^{.7} \; \frac{(C \sin \phi_n)^{1.13}}{\cos^{1.56}\psi} \; \frac{m_G^{.43}}{(m_G + 1)^{1.56}}$$

An equation for specific film thickness is:

$$\lambda = \frac{2 \, h}{S_1 + S_2}$$

These equations, based on the work of Dowson and Higginson, are derived and discussed in Ref. 8. Terms are explained on page 208. A similar approach may be found in Ref. 9.

Since the film thickness is a function of the temperature of the oil within the mesh contact zone, there may be an uncertainty as to its actual value. It is helpful to plot both the film thickness and the specific film thickness for the range of temperatures that might exist within the contact zone (see Figure 3-109).

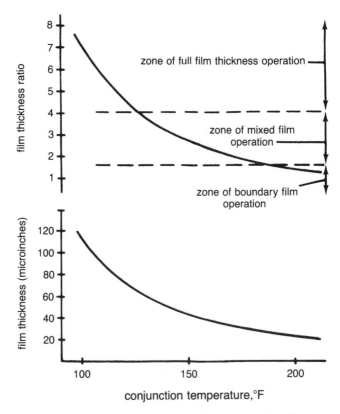

Figure 3-109. Film-thickness ratio and film thickness vs. conjunction temperature.

Normal Wear

An examination of gears operating within the "mixed" film-thickness operating regime (see Table 3-7) will usually show that they are operating with normal wear. Types of normal wear are:

Polishing. Gears that have run for a considerable period of time with good lubrication will usually show polishing. (see Figures 3-110 and 3-111). If the calculated film-thickness ratio is at the high end of the "mixed" regime, one should anticipate only polishing. A uniform polish, extending from end-to-end and from root to tip of the tooth, observed after an appreciable period of running, is an indication of a well-designed, correctly assembled and maintained gear set.

Polishing usually means a low rate of wear. Unless the gears are to have an extremely large number of mesh cycles during their anticipated life, low rates of polishing are usually of no concern. If the polishing, observed early in the life of the set, seems to indicate some wear, a heavier oil or a lower oil inlet temperature may be considered. These changes will tend to reduce wear by thickening the oil film.

Figure 3-110. Polishing. Run-in under operating conditions, this hardened hypoid pinion has developed a highly polished surface. Undoubtedly the extreme-pressure additives in the lubricant had a tendency to promote polishing on the tooth surfaces.

Figure 3-111. Polishing. This herringbone pinion, nitrided to 42 Rockwell C, shows a high degree of polish after 800 hours of operation with a mild E.P. lubricant.

Moderate wear. If the oil film is not sufficiently thick, wear may occur at a rate somewhat greater than in polishing. With this type of wear, a ridge at the operating pitch line of the teeth is noticed (see Figure 3-112). Metal is worn away in the addendum and the dedendum regions, where there is relative sliding. The operating pitch line, a zone of no sliding, then becomes evident, since it is higher than the remainder of the tooth.

Moderate wear can be controlled in some cases by an increase in film thickness, as discussed under "Polishing." This type of wear may also be a sign of dirt in the oil. In such a case the oil should be filtered to a particle size smaller than the film thickness. Moderate wear is not generally a problem unless its rate of progress causes the gear to run improperly because of vibration, breakage, or noise before the anticipated design life has expired.

Excessive wear. If the specific-film-thickness ratio is 1.0 or less, excessive wear may be anticipated, particularly for sets which will be subjected to many cycles (see Figure 3-113). If the wear rate is sufficiently high, so much material will be removed that the teeth will either break off due to thinning or run too rough to be satisfactory, long before the end of their anticipated life.

All of the modifications suggested under "Moderate Wear" may be applied to correct this problem. In addition, harder materials or a change in gear geometry may be needed.

If the gears that exhibit excessive wear are operating with a film-thickness ratio near or below 1.0, the wear can be explained on the basis of asperity contact. If, however, the specific film thickness is nearer 4, the search should concentrate on evidence of abrasive wear or the existence of high loads due to excessive vibratory or torsional activity.

Figure 3-112. Moderate Wear. Wear has taken place in the addendum and dedendum sections of the gear teeth. This causes the operating pitch line to be visible because little if any material has been removed.

Figure 3-113. Excessive Wear. Material has been uniformly worn away from the tooth surface causing a deep step. The tooth thickness has been decreased and the involute profile destroyed. Some slow-speed gears can operate on a profile of this nature until so much material has been worn away the tooth fails in beam bending fatigue.

Figure 3-114. Abrasive Wear. This herringbone pinion, nitrided to 42 Rc, shows abrasive wear from foreign particles imbedded in the mating gear.

Abrasive Wear

If dirt particles are present in the oil and their size is greater than the thickness of the lubricating film, abrasive wear will usually occur (see Figure 3-114). In most cases, abrasive wear can be reduced by attention to the cleanliness of the lubricant system. Tight enclosures, adequate seals, fine filters, and regular oil-sampling or changeout practices are required. An extreme case of abrasive wear is shown in Figure 3-115.

Figure 3-115. Abrasive Wear (Extreme Case). A large portion of the tooth thickness has been worn away due to an accumulation of abrasive particles in the lube-oil supply. Note the deep ridges and the end of the gear teeth that were not subjected to the abrasive action.

Corrosive Wear

Corrosive wear is induced on the gear-teeth surfaces as a result of a chemical action attacking the gear material (see Figures 3-116 and 3-117). The active factors attacking the materials may come from sources external to the casings of the gear set or from within the lubricant itself.

If the sources of the active factors are external to the gear casing and its lubricant system, they may enter through the breather, seals, or gaskets, in the casings, or through openings in the lubricant system. Once the method of entry is found, it is usually possible to devise means to limit further entry. In particularly hostile environments, a change in materials or special means of protecting the materials may be required.

The active factor may also result from a chemical breakdown within the lubricant itself. Lubricants containing extreme pressure additives sometime break down at high temperatures, forming very active chemical compounds. When such lubricants are used in the gear set, they should be checked at regular intervals to determine their continued suitability for service.

The original manufacturing practice itself can sometimes be the source of chemical attacks. Careful cleanup and neutralizing or passivating procedures must be undertaken after completing metal etching to remove grinding burns or chemical stripping to remove copper overlays used in case-carburizing processes.

The wear rate can sometimes be estimated by a series of measurements on specific teeth over an extended period of time. Unless wear is very pronounced, a single examination of a gear set may not be sufficient to determine if the rate of wear is so fast that the teeth will be unsatisfactory before the anticipated service life has expired.

Figure 3-116. Corrosive Wear. It is evident from this gear that considerable wear has taken place yet the surface still shows signs of being attacked chemically. Wear of this nature will continue until the gear surface is completely beyond usefulness.

Figure 3-117. Corrosive Wear. This AISI 9310 case-carburized and ground pinion shows the corrosive effects of running in a lubrication system that was contaminated with hydrogen sulfide (H_2S). The pinion has a 10 normal diametral pitch and was operating at a more than 10,000 ft/min pitch line velocity.

A gear set undergoing moderate wear (see Figure 3-112) which has been in service for some time may show the presence of a ridge at the operating pitch line near the center of the tooth. Any measurements made to establish either the amount or rate of wear should be made on the addendum or dedendum zones, away from the ridge at the operating pitch line.

Wear rate is a function of the load intensity, materials compatibility, material hardness, load-carrying characteristics of the lubricant, and operating time.

Even though all of the parameters that can be calculated—wear, scoring, and pitting—indicate a low probability of failure, some gears, such as those used in high-speed rotating machinery, accumulate so many mesh cycles during their normal life that even a low rate of wear will become appreciable in time. As a general rule, such gears should be made as hard as possible to keep to a minimum the wear from the small particulate matter in the oil and from the large number of starting cycles.

Scoring

Under certain combinations of operating load, speed, material-lubricant combinations, and operating temperatures, the oil film that is supposed to separate the active profiles of the meshing teeth may break down because of a high load temperature and allow metal-to-metal contact. This results in alternate welding and tearing actions called scoring. In some cases this may so deteriorate the active surfaces of the teeth that meshing performance is affected. The more generally recognized types of scoring may be grouped into the following categories. As in wear, these categories are based on appearance, cause, and effect:

1. Frosting
2. Moderate scoring
3. Destructive scoring

When evaluating scoring, it is helpful to analytically evaluate the degree of scoring risk inherent in the gear design. An intense effort, particularly in the field of aircraft gear design, has made possible a reasonable method of evaluating scoring risk in any given design. A typical equation for predicting flash temperature is as follows:

$$T_f = T_b + \frac{W_{te}^{.75}}{F_e} \left(\frac{50}{50-S} \right) \frac{Z_t(n_p)^{.5}}{P_d}$$

For a discussion covering the use of this equation and the evaluation of the individual terms, see Refs. 9 and 10.

Experience with this method of scoring evaluation indicates that scoring probability may be characterized as shown in Table 3-8.

For a given geometry of gears, scoring is strongly influenced by inlet temperatures, surface finish, and rubbing speed. In general, the flash temperature number at each

Table 3-8
Probability of Scoring Using AGMA Method

Calculated Flash Temperature Number	Degree of Risk
up to 275	low
275 to 350	moderate
over 350	high

5-10 points along the line of action should be determined. When tabulated, such an evaluation will appear as in Table 3-9. These data, when compared with those in Table 3-8, will give a good indication as to where, and often to what degree, scoring will likely occur.

Frosting. This is the mildest form of scoring. The affected areas of the tooth, usually the addendum, give the appearance of an etched surface (see Figures 3-118 and 3-119). When magnified, the area appears to consist of very fine pits, .001 inch or so

Table 3-9
Tabulation of Lubricant Flash Temperature
Number Along Gear Line of Action

Station	Roll Angle (Deg.)	Location	Flash Temperature Index
1	22.196	outside diameter	160.17
2	18.565		155.18
3	14.934		154.57
4	11.302	high point, single-tooth contact	162.95
5	12.474		160.88
6	13.646		158.16
7	14.818		154.92
8	15.989	low point, single-tooth contact	151.02
9	14.199		156.69
10	12.408		161.01
11	10.617	form diameter	163.82

Figure 3-118. Frosting. This nitrided helical gear shows frosting in its early stages where random patches are showing up. Close inspection shows small patches of frosting which follow along the machining marks where there are very small high and low areas. This frosting pattern shows no radial welding or tear marks.

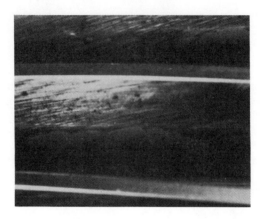

Figure 3-119. Frosting. This nitrided 4140 pinion shows early stages of frosting. Frosting appears in separate patches in the addendum and the dedendum areas of the teeth. Damage to the gear is not severe at this stage.

Figure 3-120. Moderate Scoring. This 5 diametral pitch 8620 case-carburized and ground spur gear shows scoring in the addendum and dedendum regions. Note also how scoring extends across some teeth and only part way on others.

deep. The frosted area usually marks the higher asperities or undulations that may exist on the surface of the teeth.

Frosting is frequently an indication of local errors in the heights of the tooth surfaces. If these areas can be worn away during continued running of the teeth without causing damage, the frosting will usually disappear.

Light to moderate scoring. This is a somewhat more intense form than frosting. In addition to the etched appearance, close examination shows fine scratches resulting from a sequence of welding and tearing-apart actions (see Figures 3-120 and 3-121).

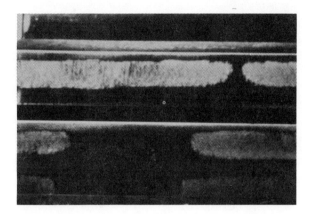

Figure 3-121. Light to Moderate Scoring. The scoring action on this gear covers a large portion of the tooth surface. It has a frosty appearance and, upon close examination, it is evident that there has been metal-to-metal contact, a welding and a tearing apart due to rotation.

Figure 3-122. Destructive Scoring. Heavy scoring has taken place above and below the pitch line leaving the material at the pitch line high or proud. As a result of this, the pitch line pits away in an effort to redistribute the load. Usually, the gear cannot correct itself and ultimately fails.

Destructive scoring. This type of scoring results in a significant amount of material removal, and it may extend well into the addendum and the dedendum zones (see Figures 3-122 and 3-123). If the load is sufficiently concentrated at the pitch line, pitting may occur at such a rate that the tooth can never recover in its effort to redistribute the load.

Scoring, unlike tooth breakage or pitting, generally occurs soon after the gears are put into service, if it is going to occur at all. Thus, it may give early indications if a new gear design is of inadequate accuracy, if the gears have not been properly aligned (Figure 3-124), or if the design is deficient in some respect.

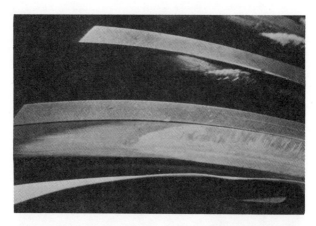

Figure 3-123. Destructive Scoring. This AISI 9310 ground aircraft spiral bevel gear shows destructive scoring.

Figure 3-124. Localized Scoring. This high-speed, high-load AISI 9310 case-carburized and ground helical gear shows localized scoring on the ends of the teeth. This failure was caused by a design oversight: when the gear casing reached operating temperature, differential thermal expansion caused a shift in alignment across the face of the gear. The resulting load distribution caused scoring.

If scoring is observed in local zones soon after a gear set has been put into service, the location and shape of the scoring patterns may indicate operating and assembly problems, which, if corrected, will eliminate further scoring (see Figure 3-125). Scoring is sometimes observed on gears that should have been given tip or profile modifications in their design (see Figure 3-126). In such cases, the addition of such modifications may correct scoring.

Since scoring is a local breakdown of the lubricating film, some control can be exercised through the lubricant itself. A change to a higher-viscosity lubricant or the use of extreme pressure additive lubricants may be considered. A lowering of the oil inlet temperature at the mesh and an increased flow of cool oil to reduce the gear-tooth temperature may also be effective.

Figure 3-125. Localized Scoring. This medium-hard (4340) marine gear which was hobbed and shaved shows scoring in a definite uniform pattern. In this case the waviness of the tooth surface contributed to the heavier scored patches superimposed on a more conventional uniform scored area.

Figure 3-126. Tip and Root Interference. This gear shows clear evidence that the tip of its mating gear has produced an interference condition in the root section. Localized scoring has taken place causing rapid removal in the root section. Generally, an interference of this nature causes considerable damage if not corrected.

Surface Fatigue

Surface fatigue results from repeated applications of compressive stress on the contacting areas of the teeth. This type of surface endurance failure is characterized by small pits in the surfaces of the teeth. These result from subsurface stresses beyond the endurance of the material.

It is convenient to group the more generally recognized types of surface fatigue into the following categories:

1. Pitting
   a. initial pitting
   b. destructive pitting
   c. spalling
2. Case crushing

The possibility of surface fatigue failure in the form of pitting or spalling can be predicted quite well by analytical techniques. The equations generally used for these predictions appear as given below and in Refs. 11 and 12.

The value of surface compressive stress number S_c is calculated as follows:

$$S_c = C_p ((W_T C_o / C_v) (C_s / d F) (C_m C_f / I))^{.5}$$

This is compared to an allowable value of surface compressive-stress number:

$$S_c' = S_{ac} (C_L C_H / C_T C_R)$$

These equations are discussed in detail in Ref. 7; the terms are defined in Appendix A. The term S_{ac} is the fatigue stress at which the material can be operated for 10^6 cycles. It is desirable to calculate a number of S_c values, one for each combination of load intensity and operating cycles at that given intensity.

When these two sets of data are plotted, the results appear as in Figure 3-127. In the particular case shown here, there is one point at which the load intensity for its number of operating cycles exceeds the capability of the material. When the gear set has run long enough to accumulate cycles at this load intensity, pitting can probably be observed. Pitting, which is a fatigue phenomenon, usually takes up to a few millions of load cycles before it can be observed, even though the load intensity is in the pitting range.

If a gear set is examined shortly before pitting has become clearly visible to the naked eye, it is usually desirable to make an estimate of the kind of pitting that is under way. Pitting tends to fall in one of two categories: "Initial pitting," a corrosive form of pitting which, when it has finished its task, will disappear; or "destructive pitting," which will continue as long as the gears are run.

If the applied stress is only slightly higher than the surface endurance limit of the material, the pits will tend to be small—just barely visible to the unaided eye. If, however, the applied stress is quite high relative to the capability of the material, the initial pits will be appreciably larger. In some cases, small pits may combine to form still larger ones if the high tooth loads are continued.

Initial pitting. This is a form of pitting that usually results in the correction of local areas, thus producing a gear set giving satisfactory operation (see Figure 3-128).

The initial contact in a new set will occur on the highest of the tooth asperities. If the load intensity on these areas is not unduly excessive, small pits will form and these areas will be corrected.

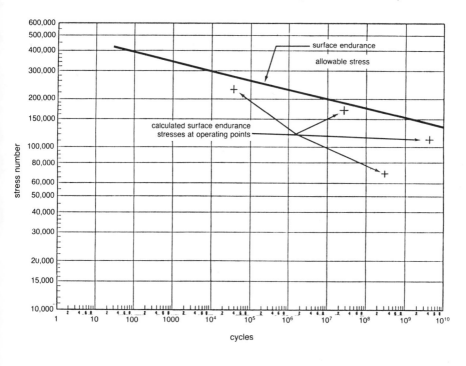

Figure 3-127. Allowable vs. calculated surface endurance stresses.

Figure 3-128. Initial Pitting. Pitting started on the outside end of the helix due to a small out-of-alignment condition. The pits were small in diameter. The pitting distress pattern progressively worked its way from the very end of the tooth toward the middle of the tooth. After a length of time, the fine pitting tapered out. Upon continued operation, the pitting stopped altogether and the pitting surface began to burnish over, indicating no new pits were being formed and that the load across the tooth had a tendency to be more evenly distributed. (This type of pitting is often called corrective pitting.)

Corrective pitting. This is a form of pitting in which local high spots are pitted and worn down to a point where a large enough area of the remainder of the active tooth surface will come into contact to reduce the load intensity to the point where the overall surface stress is less than that which will cause pitting. In this case, the initial pitted areas will usually polish over.

If a gear set is examined shortly after pitting has appeared, the pitting may be characterized as "initial pitting" if the following criteria are met:

1. The pits are very small—under .030 inch in diameter.
2. The areas showing pits are quite localized and can be logically interpreted as high spots due to manufacturing errors.
3. The analytical evaluations of the surface endurance stresses indicate that pitting should not occur.
4. The pitted or polished areas extend well along the tooth, indicating good gear alignment.

This type of pitting is usually found in gears having slight manufacturing errors, or in teeth that should have been given profile modifications.

Destructive pitting. Destructive pitting is characterized by large pits (see Figures 3-129) and is often seen first in the dedendum region of the driving gear member. In this form of pitting, the initial areas that pit have pits so deep they cannot carry load. As the pits form, the load shifts to adjacent areas which in turn pit, and which also can carry no load. The pits are at all times so large that the teeth cannot heal over with continued running. Thus, pitting proceeds over the tooth from zone to zone until the entire tooth is destroyed.

Figure 3-129. Destructive Pitting. Heavy pitting has taken place predominantly in the dedendum region. The pitted craters are larger in diameter than those denoted as corrective pitting. Some pitting has taken place in the addendum region as well.

Figure 3-130. Spalling. This hardened pinion shows an advanced stage of tooth spalling (not to be confused with case crushing). Material has been more or less progressively fatiguing away from the surface until a large irregular patch has been removed.

Figure 3-131. Spalling. This bronze worm gear shows an advanced stage of spalling in the loaded area of the tooth.

Spalling. This is a form of pitting in which large shallow pits occur (see Figures 3-130 and 3-131). Destructive pitting and spalling generally indicate that the gear set is loaded considerably in excess of the capability of the gear geometry and/or materials. Changes in material type and hardness may, in some cases, be sufficient to eliminate destructive pitting.

Failures from the Manufacturing Process

Just as some gear failures are "designed-in" to the gear through a lack of appreciation of the design requirements of the job, others may get "manufactured-in" by improper shop practices.

Grinding Cracks

Grinding cracks are an example of a manufactured-in problem. Grinding is a process that can produce an excellent gear—if it is done correctly. It does have the inherent danger that, through lack of proper attention to detail, localized areas of the teeth may suffer grinding burns. When feeds, speeds, grain size, wheel hardness, and coolant are not just right, intense localized hot spots will be developed by the wheel, and "burns" result. The localized heat will draw hardness of the steel, or in some cases quench cracks will form (see Figures 3-132 and 3-133). These cracks can become the focus of stress raisers, leading to tooth breakage.

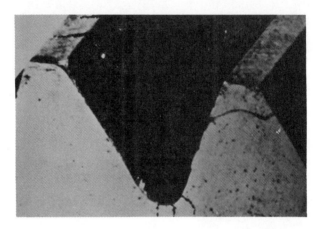

Figure 3-132. Typical quench cracks in medium hard material.

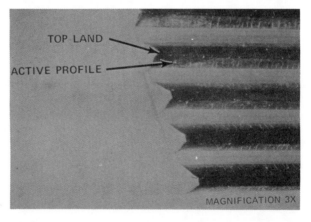

Figure 3-133. Grinding Cracks (3 × Size). Grinding cracks of the parallel type. These cracks can be seen by eye; however, they show up best when a flaw-detection procedure is utilized such as magna-flux or magna-flow. These cracks were found on a finished ground case-carburized AISI 9310 gear.

Burns can be prevented by good shop practice. They can be detected by chemical etching techniques which make the burned areas visible. The teeth may also be inspected with similar nondestructive techniques for grinding cracks.

Case Crushing

Case crushing is a subsurface fatigue failure which occurs when the endurance limit of the material is exceeded at some point beneath the surface. It is a fatigue-type failure, generally occurring in case carburized, nitrided, or induction-hardened gears.

The Hertzian stresses resulting from two teeth in contact occur along bands extending across the teeth. Both the width of the band of contact and the depth from the surface to the point of maximum shear stress can be calculated by methods given in Ref. 13. Accordingly, the maximum stress occurs at a point about .393 of the width of the band below the surface.

Figure 3-134 illustrates a section across a case-carburized and hardened gear tooth in its normal plane. Superimposed on this section is the results of a Tukon

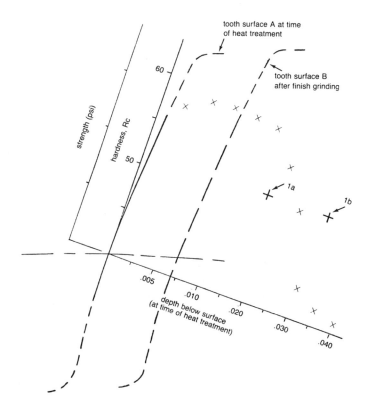

Figure 3-134. Relationship of subsurface strength to subsurface stress.

microhardness survey. Plotted perpendicularly to the surface near the pitch circle is depth to point of measurement (abscissa) versus hardness (ordinate). The X's are the values of measured hardness. In a good case-hardening job, the surface is only slightly less hard than the major part of the case because of decarburization.

The strength of the material is directly related to hardness; the maximum hardness, therefore strength, exists just below the surface and extends to a depth of about .020 inch, where it falls off to the core hardness and strength. The solid line "A" is the tooth surface at the time of the case-carburizing and hardening process.

When the tooth is run with a mating gear under load, a subsurface shear stress is developed. Its magnitude is plotted as point 1a to the same scale as that of the material strength. Since the stress is below the value that the steel can carry at this depth, the tooth will be satisfactory under this load.

Most case-carburized gears are ground to correct the almost inevitable errors that result from the distortions and changes in size that result from the heat-treating process. In some cases a highly variable amount of material must be ground from the teeth of an individual gear in order to clean it up to drawing-size tolerances. In such cases, the tooth surface may be ground down to the dashed line "B" in Figure 3-134.

When this part of the tooth is loaded up to the same value as before, the same stress will occur at the same depth below the surface, and is shown at 1b. This point now lies well inside the weaker core material, the prevailing stress is sufficiently intense to cause a subsurface shear failure. The failure can progress generally parallel to the surface and break through the surface where the stresses are right (see Figure 3-135), or progress deeper into the tooth, as in Figure 3-136.

Figure 3-135. Case Crushing. Several large longitudinal cracks, and several smaller ones, have appeared in the contact surface of this case-carburized bevel gear. The major cracks originated deep in the case-core structure and worked their way to the surface. Long chunks of material are about to break loose from the surface. The cracks should not be confused with normal fatigue cracks, which result from high root stress and normally form below the contacting face.

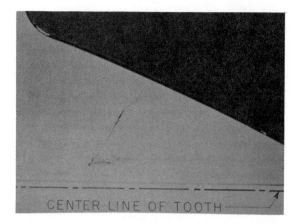

CENTER LINE OF TOOTH

Figure 3-136. Case Crushing Internal. This internal crack started just below the hardened case in the weaker core material. The material is AISI 9310; total case depth. 100 deep; origin of failure. 150 deep. The crack grew in both toward the case material in one direction and toward the opposite side root section in the other direction.

Figure 3-137. Random Fatigue Breaks. These breaks are not typical beam bending fatigue but instead originate high on the tooth flank and are somewhat uniformly dispersed across the face of the gear. Failures of this nature originate from stress raisers other than of the root fillet. In this installation, the fatigue failures were the result of grinding cracks on the tooth flanks during manufacture.

Random Failure

Sometimes only a part of a tooth will break out. In many such cases the crack did not originate in an area that is even close to the zone of the theoretical maximum (critical) bending stress within the tooth (see Figure 3-137).

Such failures may be generally attributed to the lack of proper control in the manufacturing process. The teeth may exhibit notches near the root sections, there may be cutter tears on the surface, the surfaces may have grinding burns, or the teeth may not have the proper profiles. Heat treatment may also be at fault. Improper heat treatment can produce high local stresses. The material may not be clean, being full of seams, inclusions, dirt, or other stress risers. All of these items are potential sources of local stress concentrations sufficient to cause failure (see Figure 3-138).

Local stresses due to pitting or spalling which are located away from the critical stress zones can also produce a focal point for a random failure.

A failure at the root of the tooth does not always result in an individual tooth breaking out. In some cases, the internal stress pattern will cause the fatigue failure to proceed down through the rim of the gear and into the web (see Figures 3-139 and 3-140). If this failure goes to its ultimate conclusion, an entire chunk containing several gear teeth will break out.

Breakage

Breakage is a type of failure in which part of a tooth, an entire tooth, or sometimes several teeth break free of the gear blank. It results either from a few load stresses greatly in excess of material capabilities or from a great many repeated stresses just above the endurance capability of the material.

It is convenient to group the more generally recognized breakage failure modes as follows:

1. Bending fatigue breakage.
2. Overload breakage.
3. Random fatigue breaks.

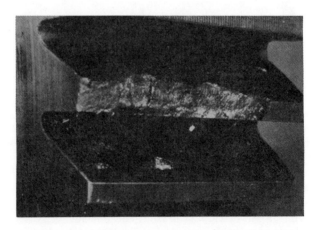

Figure 3-138. Random Fatigue Break. This AISI 4140 medium-hard gear ultimately failed from tooth fracture above the pitch line. The failure originated from large-diameter pits on the tooth surface, as can be seen on the bottom tooth. Note how a crack has propagated around the end of the gear tooth.

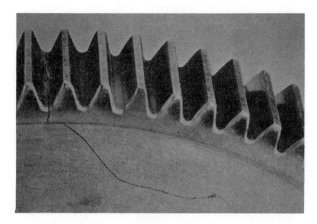

Figure 3-139. Rim & Web Failure. This hardened and ground spur gear fractured through the root section. The crack propagated from the root section down into the web of the gear. Failures of this nature are not uncommon on highly loaded thin-rim and web sections.

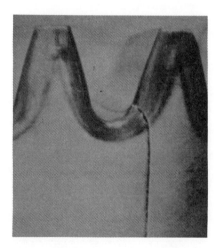

Figure 3-140. Rim Failure of a Coarse Pitch, Wide Face Width Industrial Gear. Note that the rim failure is not necessarily associated with the high root stresses but may be the result of high residual stresses in the gear wheel itself.

When evaluating a breakage failure, it is helpful to know the theoretical intensity of the bending stresses at the critical section of the tooth. It is also desirable to know the number of operating cycles at which each magnitude of bending stress will occur.

A typical equation for calculating the bending stress at the critical section of the tooth is:

$$S_t = ((W_t K_o)/K_v)\ (P_d/F)\ (K_s K_m/J)$$

A definition of each symbol is given in Appendix A. Details on the use of this equation are found in Ref. 11.

The calculated value of S_t is compared with an allowable value of S_t', determined as follows:

$$S_t' = S_{at}K_L/(K_TK_R)$$

This equation includes the effects of fatigue life and temperature, and allows for a factor-of-safety. The term S_{at} is the fatigue stress at which the material can be operated for 10^6 cycles.

In many cases, particularly where gears will accumulate a significant number of cycles at different load intensities, a plot showing the allowable stress versus fatigue cycles and the individual applied load stresses versus fatigue cycles makes it easy to determine which, if any, of the operating load points is a critical value (see Figure 3-141).

In the case of overload failures, the same equations apply, except that it is customary to use the yield stress for the value of S_{at}.

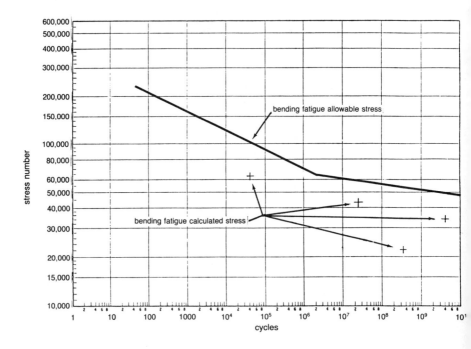

Figure 3-141. Allowable vs. calculated bending fatigue stresses.

Figure 3-142. Fatigue Failure. This tooth fracture shows the "eye" and "beach marks" that typify a subsurface fatigue failure.

Figure 3-143. Beam Bending Fatigue. This AISI 9310 case-carburized and ground aircraft power spur gear shows a fatigue crack caused by stresses at the root fillet.

Bending Fatigue Breakage

This failure mode is characterized by a fracture usually showing an "eye," beach marks, and an area of overload breakage (see Figure 3-142). The "eye" is the area where the failure started. The beach marks cover an area where the failure progresses step by step because of fatigue. These areas often show indications of fretting corrosion. After the fatigue failure has progressed well across the tooth, the remaining material will fail as a conventional fracture. Failure is the result of the continued application of load over many cycles.

The failure usually starts at the point of maximum bending stress or, if imperfections are present in the teeth, near a stress raiser (see Figure 3-143). The whole tooth or a major part may break away (see Figure 3-144).

Figure 3-144. Beam Fatigue Failure. This tooth fracture shows the type of "eye" and "beach marks" that typify a surface-oriented fatigue failure. The pinion was AISI 4340 hobbed and shaved.

This type of failure occurs because a local stress exceeds the fatigue capabilities of the material. In some gears, stress raisers are present because of manufacturing techniques. The most common of these are notches at the root fillets, cutter tears, inclusions, heat treatment and grinding cracks, and residual stresses. All of these can produce focal points from which fatigue cracks may originate.

Overload Breakage

In this type of failure, the fracture is characterized by a stringy, fibrous break, showing evidence of having been torn apart (see Figure 3-145). This type of failure is generally the result of an accidental overload far in excess of the anticipated design loads.

Lubricated Coupling Failure Analysis

There are two basic types of shaft couplings employed throughout the petrochemical industry: couplings that are mechanically flexible, and those that rely on material flexibility, such as disc and diaphragm couplings. Designers and users of mechanically flexible couplings must contend with wear, which in turn is addressed by lubrication. Figure 3-146 shows the basic styles of couplings found in the petrochemical industry. By far the most widely used lubricated coupling is the gear coupling. When the gear coupling was invented in 1918, industry was changing from steam to electric power and from belt- to direct-driven machinery. Flexible shaft couplings permitted this change to direct shaft connection, and the gear coupling played an important role in this transition.

During the last ten years, the nonlubricated, high-speed, high-torque diaphragm coupling has gained increasing popularity. In spite of this development, a large

Figure 3-145. Overload Breakage. The break in this case-carburized AISI 9310 hardened and ground spur gear has a brittle fibrous appearance and the complete absence of any of the common beach marks associated with fatigue failure. One tooth failed from shock loading, and the force was then transmitted to successive teeth. As a result, several teeth were stripped from the rotating pinion.

Table 3-10
Causes of Mechanical Coupling Failures[5]

Failure Cause	Failure Distribution, %
Vendor Problems	26.0
Planning/design	13.0
Manufacturing	7.0
Material of construction	3.0
Assembly	3.0
Operating Problems	63.0
Mishandling	38.0
Maintenance	24.0
Repair	1.0
Extraneous Influences	11.0
Foreign bodies	4.0
Overloading	7.0

portion of major process compressor trains is still equipped with gear-style shaft couplings of various sizes and configurations. Their basic purposes are to transmit rotating power to a driven machine, to accommodate known or predictable misalignment of the connected shafts, and to protect both driving and driven machines from distress caused by misalignment.

As can be expected, 75% of all gear-coupling failures are due to improper or insufficient lubrication, as is true for all other lubricated couplings. The distribution of coupling failure causes is shown in Table 3-10; the distribution of failure modes is shown in Table 3-11.

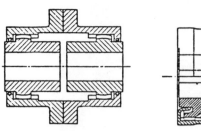

A: double-engagement gear-tooth coupling, grease (batch) lubricated; O-ring seal

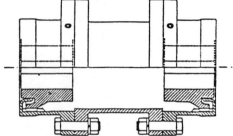

B. same as A, but spacer-type; metal lip seal

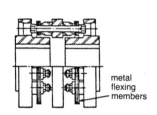

metal flexing members

C: disc-type coupling with resilient components

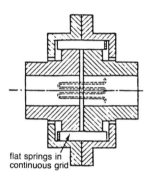

flat springs in continuous grid

D: grid coupling

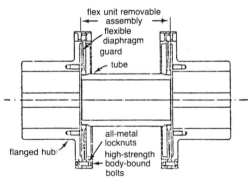

flex unit removable assembly
flexible diaphragm
guard
tube
all-metal locknuts
flanged hub
high-strength body-bound bolts

E: diaphragm-type coupling

Figure 3-146. Flexible shaft couplings used on process machinery.

Table 3-11
Distribution of Failure Modes of Mechanical Couplings[5]

Failure Mode	Distribution of Failure Cases, %
Forced fractures	60.0
Scuffing	18.0
Mechanical and corrosive surface damage	15.0
Incipient cracks	5.0
Bending, deformation	2.0

Table 3-12
Types of Typical Gear Coupling Failures[14]

Continuous Lube	*Standard or Sealed Lube*	*Excessive Misalignment*
1. Wear	1. Wear	1. Tooth breakage
2. Corrosive wear	2. Fretting	2. Scoring
3. Coupling contamination	3. Worm tracking	3. Cold flow
4. Scoring and welding	4. Cold flow	4. Wear
5. Worm tracking		5. Pitting

Gear-Coupling Failure Analysis

Few machinery components have received the amount of coverage in technical literature that has been devoted to gear couplings. No doubt, this is because lubricated couplings have frequently become the weak link in even well-managed process plants. By weak link, we are not implying that they are unreliable and prone to failure but that they need to be inspected and maintained periodically. The authors know of a well-managed petrochemical plant where during 15 years of running some 22 gear couplings on major centrifugal compressor trains, 3 forced outages were caused by gear-coupling failures. We leave it to our readers to decide whether they want to be satisfied with this record.

Machinery engineers are often fascinated by gear couplings because their "defect limit" is often difficult to determine. We believe from our experience with gear-coupling service representatives that this job is equally difficult for the expert. While it is relatively easy to judge the presence of "worm tracking," an assessment and recommendation in view of "slight wear" or even "corrosive attack" is much more difficult. Quite often the unavailability of a spare or the cost of change adds to the difficulty of making a decision after a gear-coupling inspection. K. W. Wattner of Texas Custom Builders showed clearly that wear is the predominant failure mode in gear couplings. His paper on failure analysis of high-speed couplings[14] lists the most common failure modes encountered in these couplings, as shown in Table 3-12. In short, assessing the various degrees of wear in gear couplings remains a challenge for even the most experienced machinery problem solver.

Gear-Coupling Failure Mechanisms

As gear-type couplings try to accommodate shaft misalignment, a sliding motion takes place inside the gear mesh. This motion is the main cause of coupling wear and failure. The sliding motion is composed of an oscillatory axial motion and a rocking motion that exists whenever the coupling is misaligned. The velocity of the oscillatory motion is usually called the sliding velocity. Figure 3-147 shows a gear-coupling half angularly misaligned.[15] The undulating shape of the developed gear-tooth path inside the hub is characteristic for coupling distress and failure caused by excessive misalignment.

Under proper alignment conditions and good lubrication, gear couplings can perform satisfactorily for many years. When those conditions become unfavorable, there are distress signals and forces that will adversely influence the coupled equipment before an actual coupling failure occurs. These distress signals may be subtle, but will nevertheless cause expensive downtime. Specifically there are axial forces generated by an increase in the coefficient of sliding friction and by bending moments which become excessive because of misalignment that is always present in even the most carefully assembled machinery train. The effects of misalignment are considerably more burdensome in gear couplings than in diaphragm couplings and may go so far as to weaken other major machinery components (Ref. 16).

The machinery problem solver is encouraged to get involved in regular coupling inspections. Records and photographs should be kept so that wear progress on gear couplings can be tracked. To facilitate this activity, refer to Table 3-13, which is presented in familiar matrix form.

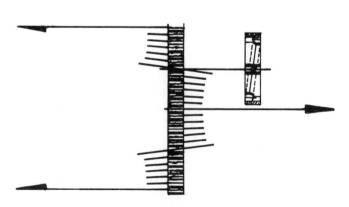

Figure 3-147. Developed plan of an angularly misaligned gear coupling half.[15]

Table 3-13
Gear Coupling Distress/Failure Analysis

Distress/failure Cause	(category)	Tooth fracture	Cracking	Pitting	Gouging △	Broken hub	Broken/sheared key	Flange fracture	Wear (adhesive)	Scuffing	Welding	Fretting/corrosive wear △△	Worm tracking	Shaft fretting	Cold flow (mesh)	Moisture/contamination	Bolt loosening	Leaking △	Overheating	Sleeve hunting	Sludging
Improper shrink fit	Design/Installation	①				①	①								②						
Random re-assembly	Design/Installation		①	①					⑤	④	③				③						
High sliding velocity	Design/Installation				①				③	⑤	④	③	③		③						
High misalignment △	Design/Installation				①			②	③	④	④	③	③								
Sealing △	Design/Installation	⑤	⑤	⑤					④	③	⑤	④	④			①		①			
Insufficient bolt torque	Design/Installation																①				
Load requirements	Operating Conditions			②				①							①						
Vibration imposed by driving/driven machinery	Operating Conditions								②		②	②	②								
High ambient temperature	Operating Conditions									②									②		
Viscosity—low	Lubrication	②	②	③					①	①	①	①	①						①	①	
Quality*/Filtration	Lubrication	⑤	③	④					①	①	①	①	①						①		
Penetration factor △	Lubrication	④	④															②			①
Quality—loss of lubrication	Lubrication	③	③	④					①	①	①	①	①			②			①		

Notes: ⊛ Distress/failure mode—cause probability ranking
*Resistance to centrifuging △
△ Low-speed/straight-tooth couplings
△ Friction oxidation/wear oxidation/chafing/bleeding/cocoa/red mud
△ Batch lubricated couplings

In all machinery component failure analysis activities, it is necessary to understand the design and functional peculiarities of the component and of the machinery system it is associated with. This is especially true when analyzing gear-coupling failures or distress signs. The following example will prove this point.

In 1972, Petronta had to shut down their 8500-hp propylene compressor after experiencing high lateral vibrations on the compressor. The compressor train configuration is similar to that in Figure 9-21. Upon inspection of the high-speed coupling, it was discovered that large portions of the hub gear teeth were broken. The coupling, a spacer type B, Figure 3-146, had been in service for 15 years and had received inspections combined with regreasing operations at two-year intervals. Because the failure was judged extremely serious, it was decided to have the failed coupling analyzed by a reputable engineering laboratory. Their failure analysis report read as follows:

1.0 Introduction (narrative deleted)

2.0 Investigation Procedures and Results
    2.1 Visual Examination
        2.1.1 The submission, as received, consisted of a male and female splined component and numerous spline segments which had separated from the male component. Some segments were jammed in the bottom of the female spline, necessitating trepanning out of a retainer insert to facilitate their removal.
        2.1.2 All surfaces were lightly rusted and dry (i.e. no trace of grease) and a dry, finely divided, red rust-like deposit was present at the bottom of the female spline. With two exceptions, the fracture faces on the male spline segments left integral with the hub were either in a circumferential plane or at an acute angle with the spline end faces. Where circumferential fracture faces were observed, most, and in some cases all, of the entire spline was absent and the surfaces were generally scalloped and heavily fretted (Figure 3-148). In all cases the scalloping had undercut the spline end-face radii at the bases of the individual teeth. The acute angled fracture faces were predominantly brittle, and characteristic "sunburst" brittle fracture patterns were seen originating at the loaded flanks on the root radii of several splines. Occasional chevron marks in the fracture surfaces indicated that most of the cracks had propagated in the direction opposite to rotation.
        2.1.3 The broken pieces of male splines had fracture surface markings which corresponded with neighboring pieces or with the fracture surfaces of male splines left integral with the hub. However, in two instances beach-like markings were observed on pieces where the fracture front was emerging at the unsupported radial/circumferential spline end-face surface at an acute angle. These pieces did not have common surfaces with the integral male splines. Many of the pieces had "sunburst" fracture patterns focusing at the loaded spline flank root radii.
        2.1.4 Examination of the female component of the coupling indicated that none of the splines were fractured or visibly cracked, but heavy fretting was

Figure 3-148. Mating fracture faces on a multispline segment showing a heavily fretted scalloped fracture and some "sunburst" fractures (magnified 1.4x).

observed on the load-bearing flanks (Figure 3-149). In the majority of cases the fretted patterns were not symmetrical across the areas in contact with the male splines; in most instances pronounced steps were visible, with up to four distinct depths of fretting present on one spline. The depths of the fretting varied from less than 0.001 in. (the highest fret marks were hardly detectable) to 0.060 in., and when viewed as a whole, the splines exhibited a definite progressive pattern from spline to spline in areas with the same depth of fretting.

2.1.5 Upon reassembly of all the pieces of the male coupling, it was apparent that the individual male splines were fretted to different degrees in much the same manner as their female counterparts. Some of the broken pieces exhibited very little fretting, while the splines still integral with the coupling exhibited the most severe fretting (Figure 3-150). Observation of the reassembled male and female portions of the coupling showed a geometrical similarity between the fretting wear patterns. The patterns on both components were mapped, the male spline pieces being in alphabetical sequence A through T in Figure 3-151.* From this exercise it was shown that the depth and shape of fretting marks on the individual male spline pieces corresponded exactly with mating areas on the female coupling, and that the depths of the fretting marks inflicted on the areas of the female coupling in contact with any particular piece of fractured male

(text continued on page 165)

---

*The Metallurgist apparently omitted the letter L.

Figure 3-149. Severe fretting on the female splines. Note the even depth and the diminishing pattern following the trapped male spline segment (magnified 0.5x).

Figure 3-150. Integral male spline and a multispline segment showing the changed dimensions resulting from fretting (magnified 2x).

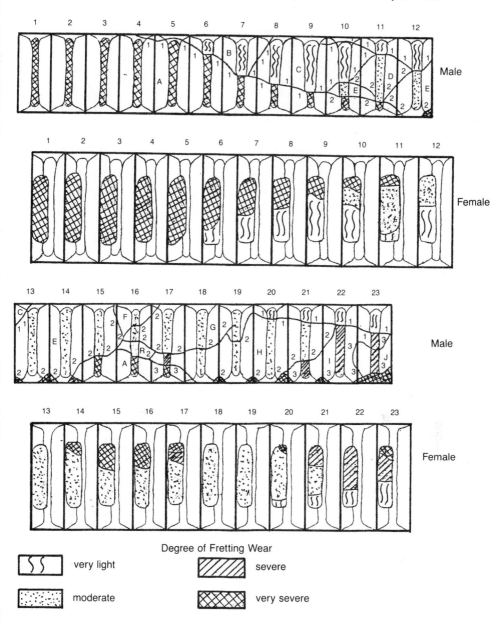

Figure 3-151. Diagram showing fracture pattern and sequence of fracture of male splines and well as the wear pattern on the corresponding female splines. The numbers along the cracks indicate the fracture phase during which the cracks occurred.

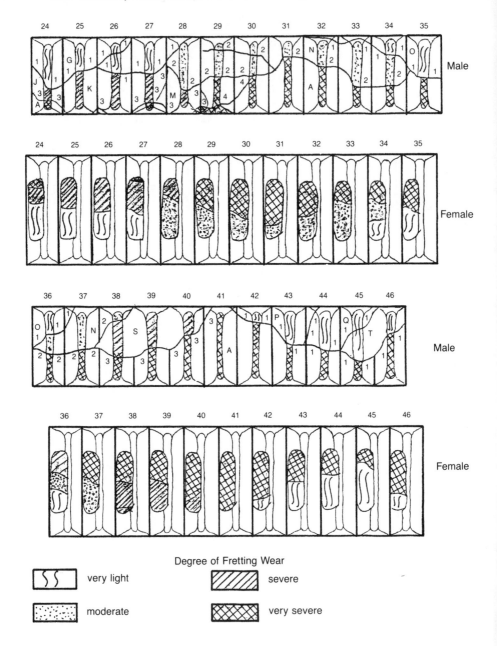

Figure 3-151. Continued.

coupling were constant from spline to spline. In general, four distinct depths of fretting were observed (Figure 3-151).

2.1.6 Three pieces of the male spline (indexed as R, S, and T on figure 3-151) were missing, but the fretting depths inflicted by them on the corresponding position of the female coupling fell into one of the four main categories.

2.2 Metallographic Examination

2.2.1 Representative sections from individual segments of spline from the failed male components of the coupling were prepared for metallographic examination.

2.2.2 The structure on all sections consisted of coarse-tempered martensite free from significant inclusions and structural defects. Grain size was predominantly #4—ASTM E112. Brittle transgranular fracture progression was observed at the "sunburst" fracture sites. No evidence of possible nucleation sites was found at the focal point of the "sunbursts."

2.2.3 Examination immediately adjacent to surfaces showing visual evidence of fretting (i.e. the load-bearing flanks on the splines and the fretted scallop-shaped fracture faces (Figure 3-148) showed no significant structural deformation. A uniform oxide film, 0.0005 in. thick, was noted on the scalloped surfaces, and fracture progression appeared to be by a brittle transgranular mechanism.

2.3 Hardness Surveys

2.3.1 A hardness traverse was made along the axis of one of the transverse male spline microspecimens, starting at the crown and finishing at the center of the hub ruling section. The readings ranged from a high of 330 HV/313 BHN (Hardness Vickers/Brinell Hardness Number) to a low of 309 HV/ 293 BHN. There was a slight tendency for the lower figures to occur at the larger ruling sections.

2.3.2 Tukon microhardness surveys normal to fretted fracture faces and normal to a "sunburst" nucleus failed to show any appreciable change in hardness at the fretted surface or at the sunburst nucleus.

2.4 Chemical Analysis

2.4.1 Drillings from the hub of the male coupling were spectrograhically analyzed. The analysis showed that the metal was a low-alloy steel of the SAE 4140 type (ref. Chemical Laboratory Report #3873).

2.5 Scanning Electron Microscope Examination

2.5.1 Examples of "sunburst" fracture faces, scalloped areas, and some coarse "beachmark"-type progression fractures were submitted to the Research Foundation for S.E.M. (Scanning Electron Microscope) examination, and analysis. The results were inconclusive, presumably due to the age of the fracture. The characteristics of brittle, ductile, or fatigue fractures normally observed at high magnifications i.e. faceted cleavage, ductile dimple and striations, were not readily identified. A small area on the heavily fretted scalloped fracture showed striation-type markings, but the validity of this observation is questionable because of the presence of an oxide layer on this surface and the visually apparent fretting, both of which would mask any true fatigue striations.

3. Discussion

3.1 Visual and metallographic examination indicated that failure of the splined coupling occurred in three or four major phases by a brittle fracture mechanism. Four distinct depths of fretting predominated, and since the shallowest fretting was less than 0.001 in. and the heaviest of the order of 0.060 in. it is surmised that breakup occurred at significant intervals of time since the last greasing operation. In general, the fretting depths were constant from spline to spline on multispline segments that had separated from the male component, suggesting rapid fracture progression. Also, several "sunburst"-type fraction patterns would occur on adjacent splines on each segment, indicating multifracture initiation points as opposed to a single crack front propagating through the segments. "Sunburst" fracture patterns are characteristic of high-energy brittle cleavage cracks radiating outward from a critical size nucleus. The majority of the male spline fracture faces exhibited this phenomenon, with the apparent nuclei being close to the root radius on the fretted load-bearing flanks of the splines.

3.2 Brittle cleavage failures of the type observed are normally associated with a combination of the following three conditions or factors.
   a) a notch "flaw" (this is essential)
   b) low operating temperatures
   c) a rapid change in strain rate
   On a gear-type coupling, condition c would arise at startup of the compressor or perhaps during a speed change if it is a variable-speed unit. Temperature effects (condition b) cannot be discussed because the time of failure and therefore the conditions prevailing at that time are not known. A "flaw" (condition a) can be mechanical (e.g. machining nicks, indentations, notches caused by fretting or wear, etc.), metallurgical (e.g. fatigue cracks, inclusions, etc.), or chemical (e.g. alloy segregation, corrosion, etc.). The nuclei of the brittle "sunburst" fractures were closely scrutinized, using visual, metallographic, S.E.M. and hardness survey techniques, for evidence of flaws. Fretting damage was the only flaw found, and some of this occurred after the initial fracture. There was no evidence to suggest a metallurgical flaw such as small fatigue cracks, but the possibility of some fatigue-initiated brittle fracture, with the fatigue portion being subsequently masked by fretting, cannot be completely discounted. The complete absence of any grease in the coupling would favor the development of fretting. The two-year regreasing cycle was presumably proven to be quite adequate, and therefore the complete absence of any grease would suggest that an abnormal condition was prevailing. Any misalignment in the coupling would favor both the fretting and a change in strain application rates.

3.3 The material was SAE-4140-type steel, heat treated to within the desired hardness range, but the relatively coarse structure noted metallographically would tend to favor brittle-fracture progression.

4. Conclusions

4.1 Failure of the splined coupling occurred by a brittle fracture mechanism in several discrete stages.

4.2 Operating conditions since its last inspection were abnormal.

4.3 Material quality was as specified.

This failure analysis report was presented because it is an excellent example of accomplished metallurgical failure analysis. What then was the root cause of this coupling failure? The answer lies in Figure 3-151: The undulating wear pattern is quite apparent. What the failure analyst described as "operating conditions" was severe misalignment accompanied by a gradual loss of the grease charge.

Determining the Cause of Mechanical Seal Distress

It has been estimated that at least 30% of the maintenance expenditures of large refineries and chemical plants are spent on pump repairs. In turn, each dollar spent on pump repairs contains a 60- or 70-cent outlay for mechanical seals.

Seal failure reduction is therefore a priority assignment for mechanical technical service personnel in the petrochemical industry. But unless troubleshooting procedures are systematic and thorough, the root cause of a failure may remain hidden, resulting in costly repeat failures. Only when the analysis includes the entire pumping circuit, its service conditions, and the physical and thermodynamic properties of the fluid in contact with the seal can the analyzing engineer or technician expect to be successful.

However, careful examination of the failed seal parts will help determine whether the problem lies in seal selection or installation, the liquid environment, or pump operation, for instance. Careful examination is possible only if the entire seal is available, in a condition as undisturbed as possible. Component inspection for wear, erosion, corrosion, binding, fretting, galling, rubbing, excessive heat, etc. will point to further clues and lead to the root causes of most mechanical seal problems. Tables 3-14 and 3-15 will allow the reader to relate predominant failure modes to the most probable failure cause.

Typical failure causes are best explained with the help of pictures from the files of a major manufacturer of mechanical seals.*

Heat Checking of a Stationary Seal Element

Figure 3-152 illustrates what occurs on the face of hard metals under abnormal operating conditions. This particular piece is a cast-stellite seal which was in a process pump handling light hydrocarbons. The rotating element of the seal operating against this piece was carbon.

It is significant that the stress cracks are confined to the surface in direct contact with the carbon rotating element, indicating a highly localized condition. The surface cracks are caused by intense heat generated on the surface of the hard metal while the balance of the part is at ambient temperature. The heat on the face of the stationary element is undoubtedly caused by loss of the lubricating film between the seal faces,

(text continued on page 170)

---

*Courtesy Crane Packing Company. Reprinted by permission.

Table 3-14
Mechanical Seal Failure Analysis

			Secondary Seal						Seal Faces									Bellows				
		Cause	Swelling	Extruding	Cracking	Corrosion	Burning/overheating	Hardening	Cracking hard face	Breakage	Corrosion	Uneven/widened wr. track	Chipping	Pitting—carbon	Grooving—hard face	Erosion—carbon	Heat checking	Breakage	Corrosion	Hardening	Coking	Clogging—I.D.
Seal Selection Strategy	Service Condition	Temperature		①			①	①									①				①	
		Pressure (PV)	①											③	③							
		Corrosives	①			①					①							①	①			
		Abrasives										④			②							
		% Solids																				①
		Coking											②	②	①						①	②
		Flashing (VP)												③	③							
		Poor lubricant					④	④										②				
		High viscosity											①	①								
	Selection	Face material combination					④	④			①		①	①	①		①			①		
		"O" ring compatibility	①		②	①	②	②														
		Flush design					③	③	①					③	②	①	②	②				
		Quench design											②	④	③						②	③
Operation and Maintenance		Misalignment—external										②										
		Misalignment—internal										①										
		Vibration—axial																		①		
		Vibration—radial																				
		Installation procedure		②					②	①		③						④				
		Run-out																				
		Loss of flush					③	③	③	④												
		Thermal shock								②												
		Frequent stop/start												⑤				③				
		Pump cavitation																				
		Secondary																⑤		①		④

Table 3-14 (continued)

Legend: Ⓝ Denotes failure mode/cause probability ranking

Breakage—springs	Corrosion—	Fretting —Drive	Fretting —Sleeve	Clogging —Springs	Clogging —Retainer	Rubbing Internal Diameter	Comments and Notes—See Table 3-15	Notes
						④	Select high temperature seal if required	(1)
								(1)
①	①						Proper material selection to resist corrosion/stress corrosion cracking	(1)
			②	①	①		⎫	
							⎬ Note	(1)
	②						⎭	
	②						Choose harder face combinations	(1)
							Elastomer compatibility should be checked	(1) (2)
							Note	(1) (3)
								(1)
							Check coupling (external) alignment	
	①	①				①	Note	(4)
	②							
						②	Note	(5)
	③					⑤		(6)
						③	Note	(7)
	④							
	⑤		②				Look for primary failure mode	(8)

Table 3-15
Mechanical Seal Failure Analysis—Notes

Note #

1. See Ref. 17 for seal selection strategy.
2. If consistent secondary sealing element deterioration is experienced, conversion to metal bellows might be indicated. If experiencing O-ring deterioration through high/low temperature, chemicals, and pressure, check standard compatibility tables.
3. Flush designs should preferably be developed in cooperation with seal *and* pump vendors.
4. Internal alignment of bearings, gland plate, and stuffing box. If experienced frequently, pump repair and installation practices may need review.
5. Identify source of vibration. Take corrective action, i.e. rotor balancing, realignment.
6. Consult one of several excellent texts on seal installation and maintenance. Develop appropriate assembly check lists.
7. Shaft run-out must be reduced.
8. Certain failure modes may occur as a consequence of, or secondary to, another seal defect, i.e. clogging of bellows caused by failure of seal faces or secondary sealing element.

Figure 3-152. Stress cracking caused by intense heat after loss of liquid film between seal faces.

which normally prevents intimate contact between the rotating and stationary elements of a seal. The moment this liquid film is lost for one reason or another, extreme heat is generated, and almost immediately the surface of the hard metal cracks.

The surface of the metal at high temperature naturally wants to expand, but the metal just below the surface, which is still at ambient temperature, prevents normal expansion of the surface metal. Inasmuch as the surface metal is not allowed to expand linearly, it is forced to buckle upward; and since it has practically no elasticity, cracking occurs. The edges of the cracks, which are slightly raised above the flat surface of the stationary element, begin to shave material from the carbon rotating element.

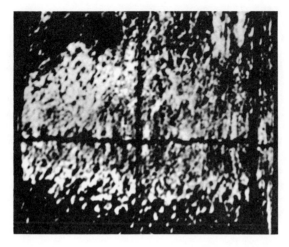

Figure 3-153. Heat-checked surface as it appears under 44:1 magnification.

Figure 3-153 shows a 44 : 1 magnification of one heat crack. The light area on either side of the crack indicates metal that was once raised above the flat surface of the stationary element, and that was subsequently worn down by the carbon rotating element. The first sign of trouble is usually the presence of large deposits of carbon dust on the outside surface of the seal plate.

Since heat checking of these hard materials is caused by heat generated at the seal faces, we must find out what causes the loss of lubricating film between the seal faces. If the pump is handling light hydrocarbons that tend to flash at low pressure, the installation should be checked to be sure proper flushing has been provided for the seal.

Good seal installation practices require all lines and openings into the seal chamber to be checked to make sure they are clear. Frequently, orifices are installed in the flushing lines to minimize fluid recirculation or to break down pressure. If an orifice is used, it should not be less than ⅛ inch in diameter to avoid clogging. Loss of pump suction will invariably cause loss of lubricating film between seal faces. Even the slightest disturbance to the full flow of liquid through the pump may result in loss of proper seal lubrication. The same condition can occur by mechanically overloading the seal faces from incorrect assembly.

Knowing what causes heat checking of these hard materials will help determine the causes of seal difficulty. If no reason can be found for continued heat checking of the hard materials, the only alternative is to use materials that have better resistance to the conditions and temperature differentials.

Unfortunately, most materials that have better resistance to this condition do not have good corrosion resistance and are limited in their use. On the other hand, there are some materials which are quite costly but which show good promise for handling some of these conditions where it is known that seal lubrication is marginal.

Worn Stationary Seal Element

Figure 3-154 is an example of extreme wear on a stellite seal which has taken place with no heat checking of the material. The indication here is that there was adequate lubrication of the seal, but that the liquid being pumped contained abrasives or tended to crystallize between the seal faces.

The carbon rotating element will usually be severely worn (but occasionally it may show no wear), while the much harder stationary element may be completely worn. This condition can occur if the abrasive particles become imbedded into the face of the carbon rotating element. It then acts as a grinding tool against the harder stationary element. The same situation can occur when handling liquids that tend to crystallize. The crystals, which are usually abrasive, completely coat the face of the carbon, thus protecting it from further wear while the stationary element is being worn down.

To correct these situations, it is apparent that abrasives must be excluded from the seal faces by one of several means for handling either abrasive liquids or those likely to deposit crystals on the seal faces.

Matched Worn Faces

Figure 3-155 illustrates how badly grooved and worn seal faces can be while still doing a fair job of sealing fluid.

Wear on these faces was gradual, so that for a considerable length of time, satisfactory service was obtained while the seal faces were no longer flat. Faces in this condition, however, can only perform satisfactorily in a unit that is very accurate and that has stable running characteristics. It can be readily seen that any runout of the shaft would disturb the mating of the surfaces and allow serious leakage. Because of the symmetrical design of the stationary element, the user was able to obtain double duty from this part, as evidenced by the same pattern of wear on the opposite side, which was used as the seal face on the original installation. Figure 3-155 thus illustrates the advantages of having smooth-running equipment for the best seal life.

Figure 3-154. Extreme wear of stellite seat due to abrasive action.

Figure 3-155. Matched worn faces of a seal which continued to operate satisfactorily.

Figure 3-156. Interrupted contact pattern due to a badly distorted hard face.

Distorted Stationary Element

It is obvious from the Figure 3-156 that the stationary element was badly distorted while in operation. This is indicated by the interrupted pattern of contact between the seal faces. There was no question about the flatness of this part before it was installed into the unit, which in this case was a multi-stage pump handling finished petroleum products at high pressure in a pipeline. When this part was returned to the manufacturer, it was found to be perfectly flat, which meant that even though it was distorted during operation, it had come back to its original flatness after removal. It was evident that distortion of the lapped stationary member, which fit into a counterbore of the seal plate that was bolted to the pump stuffing box, occurred during installation.

It was decided to inspect the replacement seal, which was leaking a small stream of fluid. The most common cause of this type of distortion is overtightening of the gland bolts. In some cases, this can actually bend the end plate, which distorts the surface supporting the lapped stationary element.

The high pressure in the seal chamber will force the lapped member to conform to the supporting surface of the end plate, which of course is no longer flat.

With this line of reasoning, the bolts on the end plate were backed off about a quarter turn, and immediately the pattern of leakage changed. This indicated that the lapped stationary element was reacting to the change in tension on the bolt. By carefully adjusting the tension on the gland bolts, leakage was reduced to almost zero. When bolt tension was again increased, the seal immediately leaked a small stream. The leakage rate could easily be controlled by the amount of tension imposed on the gland bolts, which gave definite proof of what was taking place.

Exploded Carbon

Figure 3-157 shows a very unusual condition on the face of the carbon rotating element of a seal in a large speed reducer, operating in a cold atmosphere.

The carbon supplier explained that this type of failure was caused by high surface heat. On some grades of carbon having a resin filler, high surface heat will cause immediate expansion of the resin used to fill pores in the carbon. This rapid expansion causes a small explosion below the surface of the carbon, loosening small sections of the material. This is a progressive condition wherein only a few particles of carbon are loosened on each startup of the unit. Seal failure does not actually occur until a series of these voids are connected completely across the lapped surface to form a path for leakage.

The high surface heat is caused by lack of lubrication to the seals. When these units are started in a cold atmosphere, heavy oil in the gear case clings to the gears, lowering the oil level below shaft level. Actual tests on this type of equipment show

Figure 3-157. Damage to rotating carbon element caused by high surface heat.

that the oil level immediately drops below the shaft level and remains low two to five minutes before returning to a level adequate for proper seal lubrication. This difficulty is corrected by using a resin-free carbon developed for this particular service. A combination of bronze and hardened seal faces can also be used with good results on such applications.

Heat-Damaged Rubber

The right-hand section of Figure 3-158 shows a ceramic stationary element mounted in a rubber cup which seals the ceramic to the pump body or seal plate. The rubber cup can no longer function as a resilient sealing member. Heat generated by the seal operating with insufficient lubrication has been transmitted to this part, resulting in hardening and cracking of the rubber.

The same condition will also usually result in a similar failure of the rubber rotating seal element, shown in the upper left-hand corner of Figure 3-158. The part in the lower section of the picture is the carbon rotating seal washer that turns against the assembly of ceramic and rubber. Heat generated at the seal faces is transmitted to the flanged portion of the rubber sealing element by the carbon washer. Hardening and cracking of the surface of the rubber sealing element destroys the liquid-tight fit between it and the carbon washer. The condition of the parts also helps detect incorrect seal installation or misapplication of the seal by using it at temperatures beyond the limits of the synthetic-rubber sealing elements.

Acid-Damaged Rubber

Figure 3-159 illustrates the effect of sulfuric acid traces on the synthetic-rubber seal element. On the basis of the operating information supplied to the seal manufacturer by the user, a rubber-fitted seal was perfectly satisfactory for the service. However,

Figure 3-158. Ceramic stationary element mounted in rubber cup which had become brittle in service.

Figure 3-159. Effect of trace amounts of sulfuric acid on synthetic-rubber seal element.

the user experienced seal difficulty after a short period of time and returned the damaged rubber element to the seal supplier for inspection. It was immediately apparent that the seal was handling something other than light hydrocarbons, which was the original fluid to be handled by the seal.

After reviewing actual conditions in the system, it was found that occasional traces of H_2SO_4 were present in the fluid. The problem was solved simply by supplying a seal with a Teflon sealing element.

Sealing Elements Coated with Solids

As was pointed out previously, serious damage can be caused by abrasive liquids or by liquids leaving deposits from evaporation. Figure 3-160 shows a Teflon wedge-shaped seal ring and a carbon washer that were operating in a liquid carrying some abrasives in suspension.

Both parts are coated with fine abrasives that can cause early seal failure. These same solids will usually be found on the shaft as well as directly under the seal faces, deposited out from minute seal leakage. When these deposits build up on the shaft, they will freeze the sealing elements to the shaft and eliminate all seal flexibility.

Also, on a section of a Teflon ring that has been cleaned, there is evidence that these abrasive particles have become imbedded in the Teflon seal element. This might cause wear on the shaft directly under that member.

Abrasive-Damaged Sleeve

Figure 3-161 shows the effect of abrasive particles lodged between the sealing element and the shaft. Due to slight angularity of the stationary seal member, the Teflon wedge is forced to move axially at every revolution of the shaft to compensate for the angularity of the stationary element. This motion causes the abrasive particles to wear a groove around the shaft. Either this will destroy the seal between the shaft and the Teflon wedge or the wedge may become locked in the groove and destroy the flexibility of the seals. This grooving of the shaft illustrates the importance of installing the stationary element as nearly square to the shaft as possible.

Figure 3-160. Abrasive coatings embedded in Teflon seal element caused rapid wear.

Figure 3-161. Abrasive particles lodged between sealing element and shaft caused wear.

Interference Between Shaft and Stationary Element

The cause of seal failure can sometimes be easily detected by examining the stationary seal element.

The condition shown in Figure 3-162 can result from insufficient bore clearance, runout of the shaft, or eccentricity between the shaft and pump stuffing box. This type of seal failure can be extremely dangerous when handling flammable liquids because of the heat generated from metal-to-metal contact.

In one petrochemical plant, a failure pattern as shown in Figure 3-162 was ultimately traced to a faulty piping installation which caused extreme nozzle loads on the pump casing. Localized yielding allowed extreme misalignment to result between shaft OD and stuffing-box bore.

Figure 3-162. Localized yielding of pump stuffing-box components caused the seat bore to rub on the shaft.

On critical applications handling flammable fluids, a bushing of nonsparking material should be provided. It should have an inside diameter slightly smaller than that of the stationary element. This will prevent contact between the shaft and the stationary element. Whatever the seal design, it is important that the clearance between shaft and bushing ID in the end plate or stationary element be accurately checked with feeler gauges. Clearance must be sufficient to avoid contact.

Seat Cracked by Interference

While the part shown in Figure 3-162 did not appear too badly damaged, it was damaged enough to disturb the flatness of the stationary element's lapped surface. Figure 3-162 represented a part made of a nickel iron which can withstand a considerable temperature differential. On the other hand, Figure 3-163 shows a similar part made of a cobalt-chrome alloy which actually cracked into three pieces.

Faulty Gasket

Quite frequently, leakage charged to the mechanical seal is eventually found to be from some other source. One of the most common examples is leakage between the sleeve on which the seal is mounted and the pump shaft. In some instances, the gasket has been omitted completely, which may result in several unsatisfactory seal installations until the lack of a gasket is discovered.

Improper gasketing between the sleeve and shaft results in constant leakage regardless of seal replacement. Usually, if leakage is through the seal, it will vary in degree over a period of time. In one instance, an aluminum gasket was used to seal the sleeve to the shaft. It was first assumed that the seal was leaking, but on

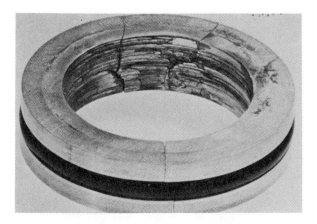

Figure 3-163. Interference fit at elevated temperature destroyed flatness of stationary part.

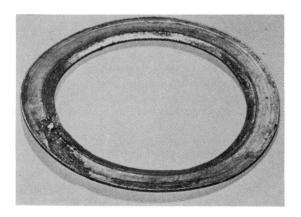

Figure 3-164. Faulty gaskets created leakage path, resulting in seal distress.

disassembly the seal appeared in perfect condition. Looking further for possible sources of leakage, it was noted that the aluminum gasket was almost eroded through in one section (Figure 3-164).

This was a high-pressure application, and the slightest imperfection on the gasket or on the surface adjacent to it would allow minute leakage that would increase because of erosion caused by high-pressure liquid. In another case, leakage in one pump out of several persisted after four or five careful seal replacements. It was finally discovered that the bronze sleeve on which the seal was mounted was porous and that the liquid was bypassing the seal.

Figure 3-165. Worn drive lugs inhibited freedom of seal movement.

Worn Drive Lugs

Figure 3-165 shows the rotating assembly of a friction-drive-type rubber-fitted seal which transmits torque generated at the seal faces through two interlocking stampings. Note that wear at the point of engagement allows the two parts to become locked together in a manner similar to a bayonet lock. This condition prevents free movement of the outer shell, which must be flexible to keep the seal faces in contact.

This type of wear is commonly found in units operating with frequent stops and starts. It has also shown up on some gear-driven units. One possible cause is angularity of the stationary seal element, which would cause some motion between the two interlocking parts, resulting in wear. This condition is corrected by furnishing a special drive band employing lugs the full length of the slot in the outside retainer shell, eliminating any possibility of a bayonet lock.

Debris in Pump

An interesting case concerns a high-pressure, chilled-water pump in air conditioning service in a skyscraper. This particular pump was experiencing an excessive amount of vibration which, it was felt, might be affecting the performance of the seals. Upon dismantling the pump, a number of various-size pieces of masonry were found lodged in the impeller. Some of these are shown in Figure 3-166 (the golf ball is for size comparison only).

Effects of Unsuitable Sealing Environment

"Unsuitable environment" can range from dry running to seal operation in excessively viscous fluids. Some typical examples are shown to illustrate the resulting damage.

Figure 3-166. Debris in pumps rarely helps the seal.

Figure 3-167. Silicon-carbide seal destroyed due to dry operation.

Figure 3-167 shows a seal removed from sewage service. The seat material is silicon-carbide converted carbon with a rubber elastomer seat. This unit failed from dry running. The ring was broken to determine the thickness of the silicon-carbide converted material.

In an agitator seal application, Figure 3-168, the primary and mating rings are made of tungsten carbide. This failure is the result of running in a low-viscosity fluid. The seal lubricant in this case was silicone oil. Seal life was 11 months. The carbon primary ring wore to a radius at the seal nose. Wear into the mating ring was approximately ⅛ inch. This condition was overcome by replacing the lubricant in the seal chamber. In this case, water was selected as the replacement fluid.

Figure 3-169 illustrates the results of running a hard-chrome plating on stainless steel at speeds to 19,000 rpm in oil mist. There was no appreciable pressure on the seal face. Seal life was just a few days. Due to the heat input on the coated surface,

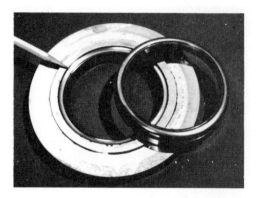

Figure 3-168. Effect of inadequate viscosity of sealing fluid.

Figure 3-169. Hard chrome-plated seat did not stand up under high p-v values.

the hard chrome plating flaked off. The seal size is 1½ inches in diameter. When applying coated seats to any application, it is necessary to watch the maximum pv (pressure velocity) values for the materials being used. For coated seats, as a rule-of-thumb, stay below 100,000 pv.

Severe damage can result also if fluid vaporization occurs in the seal cavity. Figure 3-170 shows a metal bellows seal which had been in service on a fluid operating above or near its atmospheric boiling point. Note the heavy wear on the tungsten carbide surfaces for the primary and mating rings. Also, at the inside diameter of the seal head, note the split in the first convolution of the metal bellows. When operating with fluids giving poor lubrication at higher temperatures, it is good practice to stay at least 50°F below the atmospheric boiling point.

Mechanical and Installation Problems

Although installation-related seal distress was described earlier under "Distortion" and "Interference," it is readdressed here using several additional photographs.

Figure 3-171 depicts a carbon primary ring from a high-pressure seal application. This unit had been in service on a pipeline, and as a result of gland-plate/mating-ring distortion in the presence of abrasives, wire brushing occurred at the seal face. This condition can be eliminated by identifying and correcting the cause of distortion.

Figure 3-170. Fluid vaporization in seal cavity led to failure of this metal bellows seal.

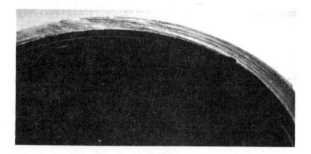

Figure 3-171. Wire brushing at seal face caused by gland plate vs. mating ring distortion.

The seal shown in Figure 3-172 had been in service handling sewage. The unit had been run only a short time before leakage occurred. Note the heavy grooving of the Ni-resist seat. This was caused by operating the seal at its solid height. Note that the back of the retainer had buckled, indicating misinstallation which developed into leakage.

Figure 3-173 shows a seal which had been in service initially for approximately 23 months. Note the heavy wear on the tungsten versus tungsten seal faces, as well as the split bellows separating the front adapter from the assembly. This condition was attributed to the mechanical condition of the pump. Upon reinstalling the same seal, it failed in a matter of days. The problem was not corrected until the pump operating parameters in terms of axial motion and alignment were corrected to operate within the manufacturer's tolerances. Note that on higher temperature applications it is always a good idea to determine whether any piping strains have been transferred to the pump casing, causing an alignment problem.

Severe starting and stopping can cause problems in marginal seal installations. In Figure 3-174, observe the wear on the carbon primary ring at the drive notch, as well as on the metal disc and springs in this assembly. Note the contact pattern on the inside surface of the retainer, as well as the contact pattern of the springs in the spring

Figure 3-172. Ni-resist seal seat damaged by faulty installation.

Figure 3-173. Pump mechanical problems caused this seal to fail.

Figure 3-174. Seal-face misalignment caused accelerated wear.

hole. As a result of the start-stop motion and the possible out of squareness, there was a buildup of material at the ID of the primary ring. As the wedge traveled along the shaft, wear of the Teflon also occurred. In this case, improving the alignment of the seal faces helped improve seal life.

Seal Design and Material Selection

Seal design peculiarities, built-in clearances, and material selection are generally under the jurisdiction of the seal manufacturer. Seal defects attributable to these parameters can be divided into two groupings: vendor errors due to misinformation furnished about use, and basic application errors or design oversights by the manufacturer.

Illustrated in Figure 3-175 are seals from an air compressor which had been exposed to deionized water. Note the corrosion products built up on the carbon, as well as on the retainer shell. The wave spring in this assembly wore into the back of the carbon primary ring. The seal failed after several months of operation because of loss of seal flexibility. This condition was overcome by improving the clearances in the seal design.

Two modes of seal distress are shown in Figure 3-176, which depicts a seal from a pipeline project. First, note the condition of the collar on the surface that faces the gland. This unit had started up with the spacer in place, causing some difficulty in heat generation at that location.

This seal also illustrates the incompatibility of a bronze washer running on a 410 stainless steel sleeve. Note the corrosion between the sleeve and the washer, caused by galvanic action between the parts. Bronze was selected to overcome an abrasion problem with the crude oil being handled. The seal failed because of a loss of flexibility from the built-up products of corrosion. The damage on the sleeve is actually severe pitting which had occurred during operation, and the bronze washer turned completely black. This problem was overcome through the use of silicon-carbide converted carbon material.

Figure 3-175. Corrosion products caused seal to lose flexibility, rendering it unserviceable.

Figure 3-176. Seal failure caused by incorrect material selection.

Figure 3-177. Cold flow of Teflon caused out-of-square condition, leading to seal failure.

Figure 3-177 represents a seal from refinery service. Cold flow of the Teflon at the back of the seat created an out-of-square condition of the mating ring, resulting in high wear at the drive dents and also wear of the springs into the retainer. This is also typical of what could be expected with an alignment or a squareness problem of the mating ring on the equipment. In this particular case, the unit was used beyond the maximum operating capabilities of a floating Teflon seat. This condition was corrected by applying a square-section tungsten-carbide seat with a Viton O-ring.

A silicon-carbide primary ring which rotated against a tungsten-carbide mating ring is shown in Figure 3-178. This seal had been in service on a fluid having extremely poor lubricating qualities. The tungsten carbide heat checked and the silicon carbide wore approximately .030 inch into the tungsten-carbide member. The

Figure 3-178. Poor lubricity of sealing fluid destroyed unsuitable components.

Figure 3-179. Cobalt-bound tungsten carbide suffered from corrosion attack.

fluid sealed was ethane at 1200 psi and 3600 rpm and the wear pattern illustrates a condition of poor lubrication. These seals have hard faces which offer no additional lubrication from the materials of construction. This problem was solved by using mechanical-grade carbons against tungsten carbide, with hydropads. These seal parts were originally mis-specified for the application.

Figure 3-179 shows the results of corrosion on a cobalt-binder tungsten carbide. This, of course, could happen even with nickel-bound tungsten carbide if it has not been properly selected for the application. Note that chemical attack has only occurred on those surfaces that have been exposed to the product being pumped. There is no corrosion on the gasketed surfaces of the clamped-in seat. Always be sure that the material selected for the application has a corrosion rate less than 0.002 inch per year.

The ultimate effect of a control-loop design deficiency is illustrated in Figure 3-180. This deficiency led to the failure of a special high-pressure seal at the inboard position of an agitator. The washer was subjected to high internal pressure that occurred when pressure was lost in the double seal chamber, causing the vessel pressure to shatter the carbon washer, thus deflecting the ears on the metal retainer. This problem was overcome by adding some additional controls to maintain the proper pressure on the OD of the seal.

The importance of proper material selection is again stressed. Figure 3-181 illustrates a metal bellows seal head which had been in chemical service. The front adapter has broken away from the bellows and material has packed into the convolutions of the bellows. This is primarily a problem of corrosion affecting the materials of construction employed for the seal head.

Figure 3-180. Carbon washer shattered by high pressure.

Figure 3-181. Metal bellows seal damaged by corrosion.

Summary

The causes of the troubles illustrated can be summed up in the following words: heat, abrasives, misalignment, faulty installation, faulty specification of conditions, errors in material selection, off-design operation of pumps, and carelessness. All of these causes of seal failure can be avoided or eliminated by careful design, installation, maintenance, and pump operation. To solve a seal problem, the trouble-shooter or analyst must examine the evidence for clues as to the root cause of the deficiency. Using a systematic approach, he is helped by observing the failure symptoms, but will direct his energies towards treating and curing only the root cause.

Lubricant Considerations

Machinery maintenance in process plants uncovers a seemingly endless variety of lubricated component failures. Frequently it appears that the lubricant or the lubrication mechanism contributed to these defects and failures. A closer look, however, often reveals that the lubricant is seldom at fault if good lubrication practices have been observed.[18]

The role of the lubricant in machinery component failures can only be determined through objective analysis. Frequent failure causes have been design- and application-related deficiencies, material and manufacturing defects, and unfavorable operating and maintenance conditions. In the last category, inadequate lubricant quality is one of the possible failure causes. Experience shows that many potential or actual machinery component defects were thus caused by user-induced problems. Therefore, failure incidents attributable to contaminated lubricant usage could at best have been deferred, but never totally prevented, even if "superior" lubricants had been used.

The following section will discuss lubrication analysis as a vital part of the root-cause assessment of lubricated machinery component failures.

Lubrication Failure Analysis

Detailed examination of the lubricating oil or grease and of the lubrication system is as important as the analysis of, say, a failed roller bearing. It would just not make sense to only look at part of the picture. Failures which, at first sight, could be attributed to application-related deficiencies or operating conditions often result from dirt and water in the oil, dirty filters, acidic oil, insufficient oil flow, the use of an improper grade of lubricant—in other words, neglect of the good lubrication practices to which we have alluded earlier (Refs. 19 and 20).

Even where the lubricant and its circulating system are not at fault (in a given bearing failure, for instance), evidence from the type of wear particles in the oil and consideration of the oil properties can often help point out the root causes of failures. However, it is not usually possible to determine "remaining bearing life" from a wear-particle analysis.

Lubricant life, if examined in the context of life expectancy, will be limited by three major factors: oxidation, thermal decomposition, and contamination. All three factors will equally influence the lubricant life characteristic, as shown in Figure 3-182, where t_D indicates the time at which the lube oil is no longer serviceable and should be discarded. In other words, linear deterioration of lube oil properties decreases remaining service life exponentially. A similar point is made in Figure 3-183, which illustrates the rapid deterioration of mineral oils exposed to high service temperatures and/or high oxygen concentrations.

Research conducted in Japan (Ref. 21) states that about ten percent of steam-turbine failures are due to problems initiated by lube-oil degradation. Besides

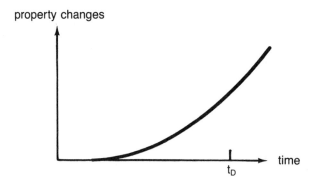

Figure 3-182. Lube-oil life curve.[18]

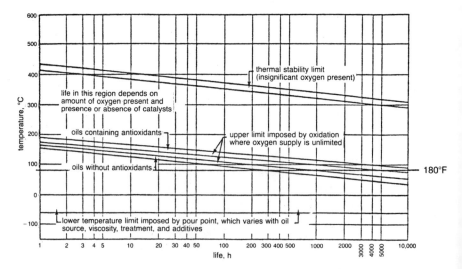

Figure 3-183. Temperature limits for mineral oils.[8]

the many factors affecting the service life of turbine oil, this research looked also at the size and operating conditions of the turbines investigated. These factors included:

1. Output of the steam turbine.
2. Steam temperature.
3. Residence time of oil in the oil tank (V/Q in minutes).
4. Heat transferred to the oil (Q•r•C• t/V).

Where:

V = Quantity of turbine oil charged, liters
Q = Main oil-pump flowrate, liters/minute
r = Turbine-oil specific gravity, kilograms/liter
C = Turbine-oil specific heat, kilocalories/kilogram, kcal/kg- °C
t = Tank to cooler-outlet oil temperature difference, °C

The findings of this research were graphically represented and show the effect of unit size on the original lube-oil life remaining (Figure 3-184) and the effect of transferred heat on the original lube oil life remaining. (Figure 3-185).

Of more practical application to machinery operation in process plants are the results of a plant-wide turbine lube-oil reconditioning and analysis program documented in Ref. 22. The point is made that reclamation, oil conditioning, or on-stream purification of turbine lube oils makes economic sense for most self-contained lube-oil systems found in typical petrochemical plants. On-stream purification of 36 lube oil reservoirs, combined with a rigorous program of periodic lube-oil analysis, has avoided machinery distress from lubrication failures at a world-scale ethylene manufacturing facility (steam cracker) since plant startup in

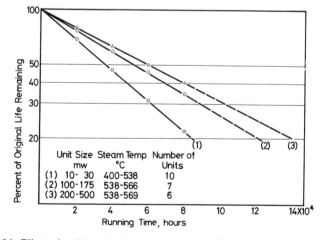

| Unit Size | Steam Temp | Number of |
mw	°C	Units
(1) 10- 30	400-538	10
(2) 100-175	538-566	7
(3) 200-500	538-569	6

Figure 3-184. Effect of turbine size on lube-oil life remaining (—measured; ---extrapolated).[21]

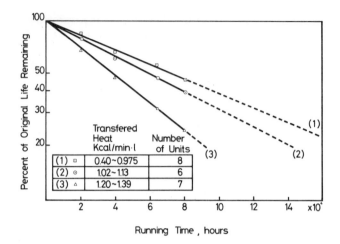

Figure 3-185. Effect of heat transferred on lube-oil life remaining (—measured; ---extrapolated).[21]

1979. The plant has made minimum downtime of major turbomachinery a priority goal, with the establishment of a lube-oil analysis program a prerequisite.

In addition to safeguarding machinery reliability, this on-stream purification and analysis program has answered the question of what to do with contaminated lube oil. Previously, waste oil had to be taken to land fills, burned in boilers and furnaces, or recycled at locations away from the user plant. Now, there is essentially no waste oil to dispose of, even during plant shutdowns for major machinery turnarounds. This fact was demonstrated during a scheduled major machinery inspection and repair downtime in 1982. Instead of discarding approximately 64,345 liters (17,000 gallons) of lube and seal oil associated with five major turbomachinery trains, two mobile lube-oil conditioners (vacuum dehydrators) were positioned near the large oil supply systems, and the reservoir contents were transferred through these units into clean holding tanks similar to the one depicted in Figure 3-186. Quite aside from solving a disposal problem, this reclamation procedure made it unnecessary to purchase new turbine lube oil, which would have cost approximately $47,000.

Vacuum Dehydration Is Used for Lube-Oil Purification

After studying such alternative methods as gravity purification, centrifuging, and coalescence, vacuum dehydration was chosen as the preferred lube-oil conditioning method for the 36 self-contained lube-oil reservoirs at this modern plant.

Vacuum dehydration is a method by which water and other undesirable constituents are removed from the oil by applying heat and a vacuum. Contaminated oil is introduced into a vacuum chamber, where it is distributed over a large surface or, in some models, sprayed at temperatures of approximately 70°C-80°C (158°F-176°F). The water is then removed in the form of vapor. The evaporated

Figure 3-186. Lube-oil conditioner (lower right) is used to purify steam-turbine lube oil drawn from the drum-type reservoir (upper left). The oil is then transferred to the 4000-gallon (15,140-liter) temporary storage drum in the center of the picture.

water is condensed before rejection from the system; noncondensible vapors are ejected through the vacuum pump exhaust.

A typical oil conditioner employing vacuum dehydration is shown in Figures 3-187 through 3-189. The oil entering the dehydrator is controlled by a solenoid valve which allows a standing reservoir in the vacuum chamber. In this dehydrator the oil is exposed to heat and a vacuum while flowing in a thin film over baffled, slanted, aluminum trays. As the oil flows over the trays, dehydration, deaeration, and degasification are accomplished. The vapors are drawn from the vacuum chamber through a condenser, and the condensed vapors settle in the distillate tanks, which are drained automatically.

Slanted aluminum trays avoid the problems sometimes encountered in dehydrators employing coalescer cartridges in the vacuum vessel. An emulsion can gradually form on the cartridge fibers as a product of the lube-oil additive package, which consists of anti-oxidants, viscosity improvers, etc. The emulsion could make the cartridge somewhat ineffective; the slanted tray design was preferred for this reason.

This oil conditioner, or vacuum dehydrator, features straightforward initial startup and simple, infrequent maintenance. The unit is trailer mounted and will purify turbine oils with an incoming temperature of 20°C (68°F) to 60°C (140°F) in a single pass at a rate of 1400 liters/hour (360 gph). It will remove solids and sludge down to 5 microns, 2% initial water content down to 10 ppm, and 10% air and dissolved gases down to 0.1% by volume, but will not remove desirable oxidation-inhibitor additives. The degasifying capability of vacuum dehydrators is important if on certain compressor trains the lube oil is used as a shaft sealing fluid. In

(text continued on page 196)

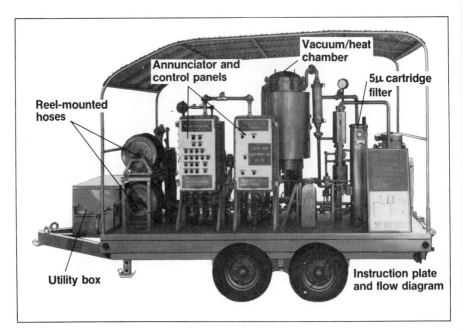

Figure 3-187. Lube-oil conditioner, front view. (Courtesy Allen Filters Inc., Springfield, Mo.)

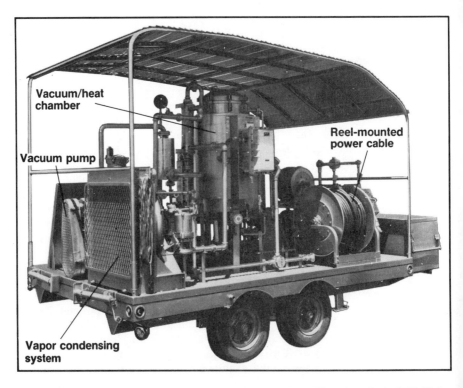

Figure 3-188. Lube-oil conditioner, rear view. (Courtesy Allen Filters Inc., Springfield, Mo.)

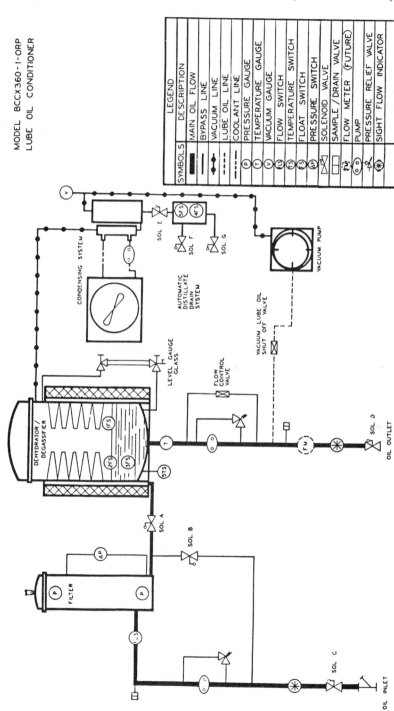

Figure 3-189. Schematic diagram of vacuum-dehydrator-type lube-oil conditioner.

this sealing mode, the oil absorbs either light hydrocarbons, hydrogen sulfide, or both. Light hydrocarbons may depress flash points and/or lube-oil viscosity. It is thus considered prudent to remove these constituents from the lube oil for safety and mechanical reliability reasons. Serious corrosion attack on the babbitt-lined seal rings of a compressor can be caused by contaminated seal oil if the process gas is sour, i.e. contains hydrogen sulfide. The compressor seal oil, after passing through the inner seal sleeve, comes into contact with the process gas and absorbs H_2S in the general region of the compressor-casing sour-oil drain cavity.[23]

Six Lube-Oil Analyses Are Required

Research efforts spearheaded years ago by the utilities industry have led to optimized lube-oil analysis methods for steam-turbine lube oils. Ref. 24 recommends testing for color, foreign solids, neutralization number, viscosity, and water content. However, more recent studies indicate that these tests alone are not sufficient to determine early on if oxidation has progressed to an undesirable degree.[25] Lube-oil oxidation can result from prolonged exposure to atmospheric oxygen, high bearing and reservoir temperatures, or possibly even excessive heating during processing in vacuum-dehydrator-type lube-oil conditioners with incorrect temperature settings.

The plant described in Ref. 22 opted for an analysis program which checks for appearance, water, flash point, viscosity, total acid number, and additive content. These tests are briefly described as follows.

Appearance Test

The test for appearance is purely visual. If free water is noted during this test, none of the remaining tests are run. If the oil is very dark, the additive test may not give sufficiently accurate results and may warrant different handling.

Testing for Dissolved Water

To determine the amount of water dissolved in the sample, automated Karl Fischer titration equipment is used and ASTM testing method D-1744 is followed. Figure 3-190 shows this testing apparatus.

Typical machinery reliability guidelines allow a maximum dissolved water content of 40 wppm in premium grade ISO-32 lube oil. This value was chosen because it generally leaves a reasonable margin for additional water contamination, and thus allows sufficient elapsed time, before saturation occurs and free water starts collecting in a given lube-oil reservoir. Free water is highly undesirable in major turbomachinery lube-oil systems because it catalyzes sludge formation.

We strongly disagree with condemnation limits as high as 2000 wppm water as advocated by some lube-oil analysis laboratories. On many occasions, the catastrophic failure of steam turbines has been attributed to the presence of free

Figure 3-190. Automated titration equipment used to test for dissolved water (Karl Fischer method).

water in the oil. As long as the oil is contaminated with free water, relay valves and pistons may stick and overspeed trip bolts may seize. This is due to certain equipment design features which allow water separation in feed lines, valves, pistons, and trip bolts. Even the occasional overspeed trip testing and exercising of moving parts cannot eliminate these risks as long as wet oil is introduced into the turbine lube and governing system.[20] While it may be true that properly designed and maintained lube-oil systems should not contain free water but should, instead, allow for gravity-settling and subsequent removal by alert operators, neither of these requirements can be depended upon in the "real world" process plant.

The guideline value of 40 wppm of dissolved water gave this plant an adequate cushion before free water started forming in the lube-oil reservoirs.

Flash-Point Testing

Flash-point testing is important whenever there is risk of light hydrocarbons dissolving in the lubricant. Using ASTM procedure D92, flash points are automatically determined by the testing equipment shown in Figure 3-191, an apparatus which employs the familiar Cleveland open cup principle. Self-imposed guidelines required the turbine lube oil in this plant to have a minimum flash point in excess of 190°C (374°F) to be considered fully satisfactory. If the oil had been diluted by a lighter hydrocarbon, the flash-point would be lowered. The test was not performed on samples containing more than one percent water.

Figure 3-191. Laboratory testing apparatus allows rapid determination of flash point (Cleveland open cup method).

Viscosity Test

The steam-turbine lube oil is measured at 38°C (100°F) on a kinematic viscosimeter, Figure 3-192, and the centistoke reading converted to Saybolt Universal Seconds (SUS) for easy reference in this non-metric plant environment. As is the case with flash-point readings, viscosity readings will fall if oil is diluted by light hydrocarbons and rise if it is contaminated by heavier components or becomes oxidized. Viscosity readings ranging from 27.1 – 37.6 cSt (140 – 194 SUS) are considered acceptable at this temperature.

Excessively low viscosity reduces oil-film strength and therefore the ability to prevent metal-to-metal contact. Other functions such as contaminant control and sealing ability may also be reduced. Excessively high viscosity impedes effective lubrication. Contaminants which cause the oil to thicken may cause accelerated wear and/or corrosion of lubricated surfaces and leave harmful deposits.

This test is performed per ASTM D-445; ASTM D-2161 is used to convert the cSt measurements to SUS units.

Figure 3-192. Kinematic viscosimeter for testing according to ASIM D-445.

Total Acid Number

Acid formation can result from high-temperature or long-term service. A titroprocessor, Figure 3-193, is used to determine the acid number according to ASTM D-664. The total acid number should not exceed 0.3 if the steam-turbine lube oil is to qualify for further long-term service.

Oil acidity can be considered a serviceability indicator. Acidity increases primarily with progressive oxidation, although on nonturbine applications products of combustion or sometimes even atmospheric contaminants can cause an increase in lube-oil acidity. The total acid number is a measure of the quantity of acidic material in the lube oil, which for test purposes must be neutralized by titration with a basic material.

Determination of Additive Content

Premium steam-turbine lube oils contain a phenolic oxidation inhibitor which may become depleted after long-term operation of lube oils contaminated by water or exposed to high operating temperatures. Oxidation inhibitors minimize the formation of sludge, resins, varnish, acid, and polymers.

New oil contains approximately 0.6 weight percent of this important additive, and depletion to less than 0.2 weight percent requires action. Additive content is, therefore, monitored with the infrared spectrophotometer shown in Figure 3-194. The presence of additives shows up in the infrared spectrum as low transmissibility,

Figure 3-193. Titroprocessor determines total acid number.

Figure 3-194. Infrared spectrophotometer measures amount of oxidation inhibitor contained in sample.

with λ around approximately 3660 cm^{-1}, where the frequency v is the reciprocal of the wave length λ.

The analysis procedure for this test is relatively straightforward and follows the instrument manufacturer's guidelines employing the test sample and a known control sample. The equipment incorporates a plotter which records the absorption points in the infrared spectrum as a function of wavelength and, therefore, frequency.

Impracticality of Wear-Particle Analysis

Lube-oil wear-particle analysis is a routine maintenance requirement for jet engines used by some major commercial airlines and military aviation services. This analysis can provide valuable information on component wear and incipient corrosion. However, when attempts were made in Europe to extend this analysis method to large steam-turbine lube-oil systems, the results proved inconclusive. Compared to aircraft jet-engine oil systems, the lube-oil quantities used with steam turbines are orders of magnitude larger. This would mandate that oil samples be concentrated or enriched in order to arrive at sample concentrations sufficient for meaningful analysis. Efforts in this regard have not given reproducible results (Ref. 25).

Periodic Sampling and Conditioning Routines Implemented

As indicated earlier, the plant described in Ref. 22 has implemented a rigorous program of periodically analyzing the condition of 36 lube-oil reservoirs. Although all of these reservoirs contain ISO Grade 32 turbine oil, only 29 are associated with steam-turbine-driven machinery. The remaining seven reservoirs perform lubrication duty on *motor*-driven pumps and compressors, which does not necessarily make them immune from the intrusion of atmospheric moisture condensation. Table 3-16 gives an overview of the various systems, the principal equipment types with which they are associated, and respective working capacities of the reservoirs.

Table 3-16
Self-Contained Lube-Oil Systems Covered by Lube-Oil
Analysis and Conditioning Program at a Large Olefins Plant

		Oil Reservoir Working Capacities				
		Capacity		No. of	Total Capacity	
Unit	Driver	Liters	Gallons	Units	Liters	Gallons
Compressor	Steam Turbine	21423	5660	1	21423	5660
Compressor	Steam Turbine	14406	3806	2	28812	7612
Compressor	Motor-Gear	10409	2750	1	10409	2750
Compressor	Motor-Gear	4542	1200	1	4542	1200
Fan	Steam Turbine	2006	530	4	8024	2120
Compressor	Turbine-Gear	1760	465	1	1760	465
Pump	Turbine-Gear	1135	300	3	3405	900
Compressor	Motor-Gear	568	150	3	1704	450
Compressor	Expander	511	135	1	511	135
Fan	Turbine-Gear	454	120	7	3178	840
Pump	Turbine-Gear	303	80	2	606	160
Pump	Turbine-Gear	227	60	5	1135	300
Pump	Turbine	189	50	3	567	150
Pump	Motor	132	35	2	264	70
				36	86340	22812

At program inception, the arbitrary assumption was made that steam-turbine-driven machinery would accumulate enough water in the oil to reach saturation after perhaps two months of operation, whereas it would take motor-driven machinery four months before free water would collect in the lube-oil reservoir. Accordingly, a schedule was prepared for the two trailer-mounted lube-oil conditioners, calling for operation at predefined locations for time periods ranging from perhaps one eight-hour shift on a small pump reservoir to as much as three or four days on a very large compressor lube-oil reservoir. These operating times were also calculated on a highly arbitrary basis, with the intention simply to reduce an assumed water contamination level of 500 wppm to as low as perhaps 20 wppm. The vacuum dehydrators would be able to remove this amount of water in one pass and send the clean oil back into the reservoir. Assuming again that during this period no significant amount of water enters the reservoir, it would take some six or eight complete passes of all the lube oil through the vacuum dehydrator before the desired water concentration level would be reached. In other words, for a 14,000-liter (approximately 3600-gallon) lube-oil reservoir, it would take the 1400-liter-(approximately 360-gallon) per-hour vacuum dehydrator 10 hours for one complete pass. The rule of thumb would thus be to leave the unit hooked up for 60-80 operating hours. A more rigorous method is available in advanced calculus texts for the more analytical souls who feel inclined to set up a differential equation to accurately calculate the whole thing.

Much was learned from these initially scheduled lube-oil reclamation procedures. Spot checking showed that the vacuum dehydrator was often hooked up to a lube-oil reservoir which at the time still had acceptable contamination levels. Both time and effort to further purify the oil were thus wasted, since achieving even lower contamination readings was academic, at best. Also, it was learned that, for organizational and logistical reasons, sending lube-oil samples to outside laboratories could be costly, time-consuming, or involve too many people. Instrumentation acquisition was justified based on the anticipated sampling frequency.

In late 1981, the lube-oil sampling and reconditioning program was modified. During the first two or three working days of each month, all 36 lube-oil reservoirs are sampled and the containers tagged as shown in Figure 3-195. The on-site laboratory immediately tests all samples for water content only and reports its findings to the mechanical planner responsible for scheduling the lube-oil conditioning tasks. The planner now singles out those reservoirs which contained excessive amounts of dissolved, or perhaps even free, water and arranges for lube-oil purification to start first on the reservoirs with the highest water concentrations. Only the reservoirs which violate the water contamination limit are purified during this particular month. The duration of hookup to one of the two vacuum dehydrators is based on experience and convenience, invoking again the rough rule-of-thumb given earlier.

After the scheduled lube-oil reservoir purification tasks are complete on the excess-water reservoirs, their contents are again sampled and taken to the laboratory. This time all six standard analyses are performed and routinely reported in tabular form. Nonroutine analyses are requested by the plant's machinery engineers as required. These would normally include analyses for all six properties described earlier.

Figure 3-195. Lube-oil samples collected during the first two or three working days of each month.

Water contamination levels are graphically represented in Figure 3-196. "Before purification" (or dehydration) vs. "after purification" water contamination levels are plotted for two compressor lube-oil reservoirs, CC-700 and CC-801 (Ref. 26). In this hypothetical installation, the guideline might be to allow water contamination to reach 250 wppm before requiring the vacuum dehydrator to be hooked up to a given lube-oil reservoir, and to purify until the residual water content is below 25 wppm. This plant may have identified CC-700 and CC-801 as requiring average purification intervals of 30 days. Using the graphical method of presenting analysis data, the technical service person, planner, or laboratory technician can rapidly spot that the month-to-month contamination of CC-700, as evidenced by the "before purification" plot, seems to remain steady, whereas the "after purification" residual contamination seems to have dropped. From this observation he may conclude that the vacuum dehydrator hookup time could be somewhat reduced while still achieving a residual contamination level of 20 wppm. Observing the "before" vs. "after" trend of CC-801, the reviewer may conclude that successive vacuum dehydrator hookup intervals can be increased in cold-weather operation, for example.

Lube-Oil Analysis Results

During the first full year of the on-stream lube-oil purification and analysis program, the plant was primarily interested in establishing the "before vs. after" results of analysis for dissolved water and analysis for oxidation inhibitor. The water check would demonstrate acceptable overall operation of the two vacuum

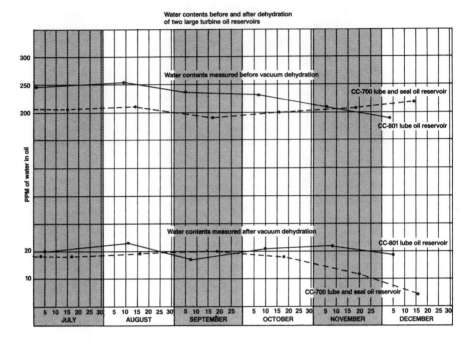

Figure 3-196. Graphical representation of "before vs. after purification" of two compressor lube-oil reservoirs.

dehydrators; checking the oxidation inhibitor content would tell if the lube oil had aged, or if perhaps the application of heat and vacuum in the unit had caused the phenolic inhibitor to be removed from the lube oil.

Four turbine-driven boiler fan lube-oil systems are typical of the reported water contamination analysis results. Each of these 2006-liter (530-gallon) reservoirs was purified during 12-hour shifts and showed "before vs. after" water contents of 317/135, 138/11, 324/47, and 351/62 wppm, respectively. The reservoir with the 135-wppm residual water content was given another 12-hour purification treatment, and all four were earmarked for 18-hour conditioner hookups in the future.

Analyses for oxidation-inhibitor depletion were made in similar "before vs. after" fashion on many lube-oil reservoirs and also by simultaneously withdrawing an oil sample from both inlet and outlet connections on the vacuum dehydrator. There were no relevant differences in the levels of inhibitor concentrations in the sample streams. Also, no relevant differences were found in the inlet vs. outlet samples when the dehydrator processing temperature was experimentally raised to 93°C (approximately 200°F). In fact, the oxidation inhibitor concentration level reportedly never dropped below the minimum acceptable level of 0.2% in any of the 36 lube-oil reservoirs since plant startup in late 1979. This leads to the conclusion that vacuum dehydration at the given processing conditions does not result in the removal of desirable additives from turbine lube oils. No effort was made to define which values of temperature and vacuum would cause concern.

Restoration of flash point was examined by comparing "before vs. after" analyses, with values expressed in degrees Fahrenheit: 405/412, 412/426, 397/419, and 415/423. As expected, vacuum dehydration removed light hydrocarbons. Very similar results were reported for seal-oil systems whose oil charge had occasionally been exposed to contact with H_2S-containing gas streams: these oils continued to show flash-point values around 210°C (410°F). In all of these cases, the accompanying change in lubricant viscosity was marginal and appeared to be within the anticipated error band for the kinematic viscosimeter described earlier and illustrated in Figure 3-192.

Finally, Ref. 22 reported observing very minor upward changes in total acid numbers for oils after conditioning. However, all of the 36 reservoirs had remained well within the specified allowable maximum total acid number value (TAN) of 0.2.

Calculated Benefit-to-Cost Ratio

The benefits of on-stream reclamation or purification of lube oil have been described in numerous technical papers and other publications. Monetary gains are almost intuitively evident from the data given earlier, and further elaboration does not appear warranted. Suffice it to say that even a gold-plated, four-wheel, canopied vacuum dehydrator would cost less than two hours of unscheduled downtime brought on by contaminated lube oil in a major turbocompressor.

Ref. 22 mentions a detailed investigation of annual costs and savings for the in-house analysis portion of their plant-wide turbine lube-oil reconditioning and analysis program. Costs consisted of laboratory technician wages and expendables such as glassware and chemicals. Higher charges for outside contract laboratory work, labor, and electric power were saved by eliminating "precautionary reclamation" practiced in the many instances where delayed reporting or logistics problems deprived the plant of timely feedback. Their net annual savings substantially exceeded the cost of acquiring supplemental laboratory instrumentation (i.e. instrumentation specifically required for in-house analysis of lube oil and not otherwise needed by their laboratory). A benefit-to-cost ratio approaching 1.8, with discounted cash flow returns exceeding 100%, was reported.

Knowledgeable sources have calculated that for most lubrication systems using more than 200 liters (approximately 50 gallons) of lubricant, oil analysis generally proves more profitable than a routine time/dump program (Ref. 27). Similar findings have been reported by large-scale users of hydraulic oil whose reconditioning efforts have proven successful and profitable (Ref. 28).

There are many reasons why thoughtful engineers should make an effort to put in place lube-oil preservation and waste-reduction programs: economy, environmental concerns, and just plain common sense, to name a few. A conscientiously implemented program of lube-oil analysis and reconditioning can rapidly pay for itself through lube-oil savings and reductions in machinery failure frequency.

Lube-oil analysis techniques are relatively easy to understand, and automated laboratory equipment makes the job more precise and efficient than it was a few decades ago. Employing these techniques in conjunction with a well-designed lube-oil conditioner is considerably better than selling, burning, or otherwise disposing of lube oil in a modern plant environment.

Grease Failure Analysis

Process machinery management should include an occasional thorough grease analysis. Most plants will have access to customer lube-oil and grease analysis laboratories maintained by the major oil suppliers.

Figure 3-197 shows the result, albeit inconclusive, of a typical grease analysis. This analysis was requested after coupling grease failure on a large turbo train had been suspected.

Major Oil Limited

82 06 09

R 235

Used Grease
C-52 Coupling

Maintenance Department

Dear Sir:

Inspections obtained on the small sample of grease from C-52 coupling from which oil had run out when shutdown, are as follows:

Ash, sulphated %	1.58
Spectrographic Analysis of Sulphated Ash	
Major Component Compounds of:	Mo, Si, Zn, Ca
Minor Component Compounds of: (High) Pb	
(Low) Na, Fe	
Water, %	Trace
Penetration (unwk)	369
(wk)	388

The spectrographic analysis of the ash shows both molybdenum and lead which are foreign to both Galena Tramo EP and Ronex EP2. The penetration of the sample is also softer than we would expect from Tramo EP or Ronex EP2. This would indicate that the grease, in the coupling, has been contaminated with another grease or is not Tramo EP or Ronex EP2. There was insufficient sample to determine the oil viscosity which would have helped to identify the grease in the coupling.

A portion of the sample was dispersed in solvent and the residue examined under a microscope. It showed little, if any, iron wear but did contain some blue-black, non-magnetic particles which may be molybdenum disulphide.

Yours very truly,

P. W. Wilson

Research Technologist

Figure 3-197. Coupling Grease Analysis.

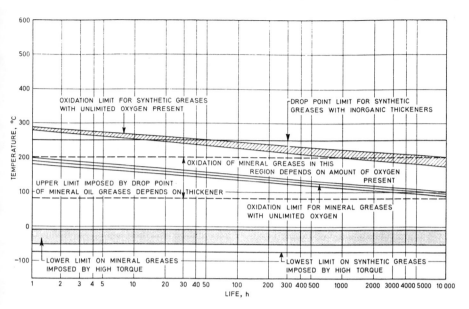

Figure 3-198. Temperature limits for greases.[18]

Temperature limits for greases are shown in Figure 3-198. In many cases grease life will be controlled by volatility or migration. This cannot be depicted simply, as it varies with pressure and the degree of ventilation. However, in general, the limits might be slightly below the oxidation limits.

Contamination of oils and greases is the most prevalent cause of premature component wear and system failures. Contaminants may be:

1. Built-in: residual sand, metal chips from machining operations, thread sealants, weld spatter, or other foreign particles that enter during fabrication.
2. Generated: pieces of metal (from wear of moving parts), elastomeric particles abraded from seals or packings, and byproducts from oil degradation.
3. Introduced: airborne dirt, oil absorbents, and water, which enter the system through seal clearances, breather caps, and other reservoir openings.

Usually, introduced contaminants have the most detrimental effect on the life of both lubricants and equipment.

Nomenclature

Symbol Definition

C	Center distance (inches).	S_t	Tensile stress number, calculated.
C_p	Influence coefficient, depending upon the elastic properties of the materials.	T_b	Temperature, gear blank.
C_H	Hardness ratio factor.	T_f	Flash temperature index (°F).
C_f	Surface condition factor.	T_i	Temperature, initial (°F).
C_L, K_L	Life factor.	w	Load per inch of face.
C_m, K_m	Load distribution factor.	W	Specific loading.
C_o, K_o	Overload factor.	W_t	Transmitted tangential load, (at operating pitch diameter).
C_R, K_R	Safety factor.	W_{tc}	Tangential load, effective (lbs).
C_S, K_S	Size factor.		
C_T, K_T	Temperature factor.	z_t	Factor, scoring geometry.
C_V, K_V	Dynamic factor.	α	Coefficient, pressure Viscosity.
d	Diameter, pitch, operating pinion (inches).	η_o	Viscosity.
E	Young's modulus.	θ	Roll angle at point of contact.
F	Face width, effective (inches).	λ	Specific film thickness.
h	Film thickness, minimum (inches).	ρ_p	Radius of curvature pinion.
		ρ_G	Radius of curvature, gear.
J	Geometry factor (bending fatigue).	ϕ_n	Pressure angle, normal (degrees).
m_N	Ratio, loading sharing.		
N_G	Number of teeth, gear	ϕ_t	Pressure angle, transverse (degrees).
N_P	Number of teeth, pinion	ψ	Helix angle.
n_p	Revolutions per minute, pinion number.	$C_c = \dfrac{\cos\phi_t \, \sin\phi_t}{2}\left(\dfrac{m_G}{m_G + 1}\right)$	
P_d	Diametral pitch transverse, operating.		
S	Surface finish, RMS (after running in).	$I = C_c/m_N$	
S_{ac}	Contact stress number.	$m_G = N_G/N_P$	
S_{at}	Bending stress number.		
S_c	Contact stress number, calculated.	$Z = \dfrac{.0175\left[(\rho_p^{.5} - (N_P/N_G^{.5}\,\rho_G)\right]P_d^{.25}}{\cos\phi_t^{.75}\left[(\rho_P\,\rho_G)/(\rho_P + \rho_G)\right]^{.25}}$	

References

1. Good, W. R., "A Few Observations About Paper Machine Lubrication," SKF *Ball Bearing Journal* No. 202, p. 8.

2. Walp, H. O., "Recognition of the Causes of Bearing Damage as an Aid in Prevention," SKF Industries, 1958.

3. *Handbook of Loss Prevention.* Edtors: Allianz Versicherungs-AG. Translated from the German by P. Cahn-Speyer. Berlin-Heidelberg-New York: Springer-Verlag, 1978, p. 33.

4. Ku, P. M., "Gear Failure Modes—Importance of Lubrication and Mechanics," ASLE *Transactions,* Vol. 19, 1976, p. 239.

5. *Handbook of Loss Prevention,* Editors: Allianz Versicherungs-AG. Translated from the German by Cahn-Speyer. Springer Verlag, Berlin, 1978.

6. AGMA 110.xx, The American Gear Manufacturers Association, Nomenclature of Gear-Tooth Failure Modes (1979).

7. AGMA 215.01, The American Gear Manufacturers Association, (R 1974) Information Sheet for Surface Durability (Pitting) of Spur, Helical, Herring-bone and Bevel Gear Teeth (1966).

8. EHD Lubricant Film Thickness Formulas for Power Transmission Gears (L. S. Akin), Journal of Lubrication Technology—Transactions of the ASME, July 1974, pgs. 426-436.

9. AGMA 217.01, The American Gear Manufacturers Association, Information Sheet—Gear Scoring Design Guide for Aerospace Spur and Helical Power Gears (1965).

10. AGMA 411.xx, The American Gear Manufacturers Association, Gear Scoring Design Guide for Aerospace Spur and Helical Power Gears (undated).

11. AGMA 225.01 The American Gear Manufacturers Association, Information Sheet for Strength of Spur, Helical, Herringbone and Bevel Gear Teeth (1967).

12. AGMA 226.01, The American Gear Manufacturers Association, Information Sheet, Geometry Factors for Determining the Strength of Spur, Helical, Herringbone and Bevel Gear Teeth (1970).

13. Dudley, D. W., "Practical Gear Design," McGraw-Hill Book Co., 1954.

14. Wattner, K. W., "High Speed Coupling Failure Analysis", Proceedings of 4th Turbomachinery Symposium, Texas A & M University, College Station, TX, 1975, pp. 143-148.

15. Calistrat, M. M., "Some Aspects of Wear and Lubrication of Gear Couplings," ASME Paper No. 74-DET-94, Presented at the Design Engineering Technical Conference, New York, N.Y., Oct. 5-9, 1974.

16. Bloch, H. P., "How to Uprate Turboequipment by Optimized Coupling Selection," *Hydrocarbon Processing,* Jan. 1976, pp. 87-90.

17. Bloch, H. P., "Selection Strategy for Mechanical Shaft Seals In Petrochemical Plants", Proceedings of 38th ASME Petroleum Mechanical Engineering Workshop and Conference, Philadelphia, Pa., Sept. 12-14, 1982, pp. 17-24. Also reprinted in *Hydrocarbon Processing,* January, 1983.

18. Neale, M. J., *Tribology Handbook,* Halsted Press, New York-Toronto, 1973.

19. Bloch, H. P., "Optimized Lubrication of Antifriction Bearings for Centrifugal Pumps," ASLE Paper No. 78-AM-1D-2, Presented at the 33rd Annual Meeting, Dearborn, Michigan, April 17-20, 1978.

20. Bloch, H. P., "Criteria for Water Removal from Mechanical-Drive Steam-Turbine Lube Oils," ASLE Paper No. 80-A-1E-1, Presented at the 35th Annual Meeting, Anaheim, California, May 5-8, 1980.

21. Watanabe, H., and C. Kobayashi, "Degradation of Lube Oils—Japanese Turbine Lubrication Practices and Problems," *Lubrication Engineering,* Vol. 34, August 1978, pp. 421-428.

22. Bloch, H. P., "Results of a Plant-Wide Turbine Lube-Oil Reconditioning and Analysis Program," Presented at the 38th Annual Meeting of the American Society of Lubrication Engineers, Houston, Texas, April 25-28, 1983.

23. Bloch, Heinz P., "Reclaim Compressor Seal Oil," *Hydrocarbon Processing,* October 1974, pp. 115-118.

24. "Oelbuch," Anweisung fuer Pruefung, Ueberwachung und Pflege der im Elektrischen Betrieb verwendeten Oele und synthetischen Fluessigkeiten mit Isolier- und Schmiereigenschaften/Herausgeber: VDEW. 4. Auflage, 1963, Frankfurt: Verlags- und Wirtschaftsgesellschaft der Elektrizitaetswerke (VWEW).

25. Grupp, H., "Moderne Untersuchungsverfahren Fuer Schmieroele und Hydraulikfluessigkeiten von Dampfturbinen," *Der Maschinenschaden,* 55(1982) Heft 2, pp. 84-87.

26. Bloch, Heinz P., "Vacuum Oil Conditioner Removes Contaminants from Lubricating Oil," *Chemical Processing,* March 1982, pp. 84-85.

27. Jacobson, D. W., "How to Handle Waste Oil," *Plant Engineering,* June 10, 1982, pp. 107-109.

28. Sullivan, J. R., "In-Plant Oil Reclamation—A Case Study," *Lubrication Engineering,* Volume 38, July 1982, pp. 409-411.

Chapter 4

Machinery Troubleshooting

Machines should work;
people should think
　　　　IBM Pollyanna Principle[1]

Proof is often no more than lack
of imagination in providing an
alternative explanation
　　　　De Bono's 2nd Law[2]

If failure analysis represents the post mortem of a failed machinery component, then troubleshooting—as an extension of failure analysis—encompasses all root cause determination activities. Often, the troubleshooting process will not stop there but will lead to problem elimination. Two basic cases will illustrate the principle: One, a machinery manufacturer experiences consistent fatigue failures of a drive shaft during prototype tests of his newly designed equipment. Two, a machinery owner experiences similar failures after years of trouble-free operation. In the first case, failure analysis will uncover the cause of the failure, and remedial action through a change of design parameters can be achieved relatively fast. In the second case, failure analysis will be part of a process generally called troubleshooting. Here, once failure analysis is completed and, say, "fatigue fracture," due to low-cycle torsional vibration has been determined as the immediate cause-and-effect relationship, troubleshooting begins. What caused torsional vibration suddenly to appear? The experienced engineer will say: "But it does not usually work this way! How about diagnostics? Is that not a part of troubleshooting?" A justifiable question, because quite often machinery engineers are called upon to solve problems associated with machinery performance deficiencies as well as to assess machinery health. Typical process machinery performance deficiencies, or simply problem symptoms, are listed in Table 4-1.

Table 4-1
Typical Process Machinery Performance Deficiencies

SYMPTOMS Abnormal Performance Parameters	Process Plant Machinery								
	Compressors	Blowers	Fans	Pumps	Turbines	Electric Motors	Engines	Gears	Lubrication Systems
Change in Efficiency	•			•	•	·			
No Delivery				•					•
Insufficient Capacity	•	•	•	•	•		•		•
Abnormal Pressure	•	•	•	•					•
Excessive Power Requirement	•	•	•	•					
Excessive Leakage	•	•		•	•			•	•
Excessive Noise	•	•	•	•	•	•	•	•	
Overheating	•	•		•	•	•	•	•	
Knocking	•			•			•		
Excessive Discharge Temperature	•	•			•		•		
Fails to Start	•	•	•	•	•	•	•		•
Fails to Trip					•	•	•		
Lack of Power					•	•	•		
Excessive Steam/Fuel Consumption					•		•		
Failure of Automatic, Control, or Safety Devices	•	•	•	•	•	•	•	•	•
Freezing	•			•					
Abnormal or Fluctuating Speed					•	•	•		
Stoppage	•	•	•	•	•	•	•	•	•
Vibration—Excessive	•	•	•	•	•	•	•	•	

The appearance of one or more of these symptoms will require the machinery engineer to answer the following questions:

1. If the machine is operating, must it be shut down to reduce consequential damage? How serious is the problem? How fast must we react to it? Is the problem increasing, is it holding, or has it decreased?
2. If the machine is available for maintenance, must it be opened for inspection and repair?
3. What components and their failure mode and effect chains might be causing the symptoms to appear, and what parts and facilities must be on hand to reduce downtime?

How does the experienced troubleshooter go about answering these vital questions? In the following pages, we will explore several competing approaches and demonstrate methods that have been applied successfully.

Competing Approaches

Organizational approaches. Quite often the organizational approach to trouble-shooting does not help. The structuring of organizations in process plants to provide "fixes" for symptoms and failures will not lead to conclusive identification and elimination of the true cause. Instead of doing something about the *cause* of a machinery problem, the "maintenance" approach of just correcting symptoms or effects is employed. Later in this chapter, in a section about organizing for successful failure analysis and troubleshooting, the shortcomings of the purely organizational, very conventional approach will be addressed in detail.

Learning to live with the problem. "The problem is no problem any more because we have learned to live with it," one of the authors overheard somebody say in connection with a machinery problem. We are all familiar with the thought behind it. If it is an ongoing problem but only occurs, say, every three months or so, we will have adapted to it with all kinds of contingent actions: "When it appears, we do this—and that is quite normal!" It goes without saying that lack of proper motivation is the reason for this approach, but another reason is a lack of understanding of machinery failure modes.

Well-meaning approach. With this approach, the troubleshooting effort is well intended but ends up creating new problems in the course of curing the old ones. It represents failure to compare the advantages and disadvantages of a solution.

Another characteristic of this approach is the failure to follow up on recommendations made once the failure cause has been determined. We often fail to translate the correction of a problem into preventive action in areas where the same problem might occur. Parochial attitudes sometimes will deter us from doing so.

Mr. Machinery approach. This approach actually belongs in the categories of organizational approaches and the well-meaning approach. It is the approach taken when sole responsibility of a plant's machinery fate rests on one single person—Mr.

Machinery. His batting average is usually good. How does one know whether or not it can be improved? Quite often he is in charge because he is an efficient firefighter. He can make the trouble go away. He is used to calling the shots. He is the machinery troubleshooter by definition. How successful might he be?

Throughout this text we will show how failure analysis and troubleshooting must be a cooperative effort, with very little room for Mr. Machinery going it alone.

Approaches caused by wrong thinking patterns. Machinery failure analysis and troubleshooting is really the process of *understanding* what causes machinery to fail or malfunction. "Understanding is thinking"[2], and as one successful troubleshooter once said, troubleshooting is knowing when to think and when to act. If failure analysis and troubleshooting at times are not successful, it may be that our thinking patterns do not allow us to be efficient and successful.

Our thinking process is impeded by two factors: one is an extraneous factor; the other is a factor determined by the thinking process itself. The first factor is the recognition that machinery component life can be either predictable or not (we alluded to this fact before, and we have addressed the need to recognize this when undertaking failure analysis and troubleshooting of process machinery). The second factor is the subject of the following discussion.

We should begin by saying that the same processes that make the mind so effective a thinking device are also responsible for its mistakes.[2] Consequently, there is this first mistake:

The monorail mistake. Because machinery vibration signature analysis has worked so well in the inspection of jet engines, it is easy to assume the same would be true for major unspared process machinery such as centrifugal compressors and turbines. However, where in one case there are literally hundreds of identical machines, in the other case, among process machines, really no two are the same. Whereas, in one case, preliminary diagnostics of one jet engine could be validated in a number of others, with heavy process machinery this is not possible. So consequently, miracles, in terms of troubleshooting results, are expected from the process plant machinery engineer equipped with all the paraphernalia of vibration analysis. The monorail mistake simply involves going directly from one idea to another in an inevitable manner, ignoring all qualifying factors.[2] Another example: "This is a vibration severity chart published by experts, and according to it, this machine runs with excessive vibration!" The qualifying factor being neglected is the fact that all machines are not created equal. If one cannot think of any qualifying factors in machinery troubleshooting, it is easy to make the monorail mistake.

The magnitude mistake. This is the second major thinking mistake that arises directly from the way the mind works. The mind moves from one idea to the other in what seems a valid way if only the "names" of the ideas are looked at. Repair reports on machinery are full of magnitude mistakes. We see "normal wear," "failed from excessive misalignment," etc. Magnitude mistakes are unlikely to occur in situations where the engineer has had direct experience. The machinery engineer has to ask the questions: "How much has it worn—in what time?" "What was the amount of misalignment?" etc.

The misfit mistake. You are walking down the street and recognize the general appearance of someone you know very well. You quicken your pace to meet the person, only to find a total stranger. This is the misfit mistake, because the idea of what something is does not fit reality. One recognizes certain features but does not wait until all the features are registered before jumping to a conclusion. The more familiar the situation, the quicker the jump to recognition. "Heard this noise before and it turned out to be. . . ." Misfit mistakes are easy to make in troubleshooting; in fact, the troubleshooting matrices presented in this text provide a good idea as to how many misfit mistakes can be made if this danger is not recognized.

The must-be mistake. This mistake is also called the "arrogance mistake." There may be nothing wrong with the way information has been put together to reach a conclusion, but arrogance fixes this conclusion so that no change or improvement is possible. When new evidence in the process of machinery troubleshooting emerges, the must-be mistake prevents us from changing our conclusions. This mistake goes so far as shutting out the possibilities of a completely new viewpoint which has not even been generated. "Don't confuse me with facts—I have made up my mind!"

The miss-out mistake. This mistake arises when someone considers only part of a situation and yet reaches conclusions that are applied to the whole situation. Quite often this will happen to the troubleshooter relying on second-hand information. For instance, in the hunt for input into the background of a thrust bearing failure of a large process compressor, an investigator could be misled by the fact that there is no record of a recent process upset which caused the machine to be liquid slugged. We have all been in the situation where we were left with the vague impression that there must be a fuller picture somewhere—fuller than the one we have. When we sense this in a machinery troubleshooting situation, we have several choices:

1. To reject the conclusion that is put forward because we are somehow convinced that it is based on only part of the whole picture.
2. To reject the conclusion that has been offered because we dislike it and, consequently, claim that it must be based on only part of the picture.
3. To accept the conclusion with reservations, but still look for the whole picture.
4. To accept the conclusion because we like it, and elect that the picture is, after all, complete.
5. To conclude that the remainder of the picture is not really there, since we cannot find it and that we have the whole picture.

The Professional Problem Solver's (PPS) Approach

We have been moving toward the PPS approach to failure analysis and troubleshooting by discussing the impediments to the thinking process. The last "mistake" seems to be the most critical one, as selection plays an important role in all information processing. There are three elements of selection involved in problem solving and troubleshooting[3]:

1. *Selective encoding.* One must choose elements to encode from the often numerous and irrelevant bits of information. The trick is to select the right elements and eliminate the wrong ones.
2. *Selective combination.* There may be many possible ways for the encoded elements to be combined or otherwise integrated. The trick is to select the right way of combining them. This process is usually based on a thorough knowledge of the interrelated functions of the various machinery parts and the effect of adverse conditions. An important insight in this context is that most machinery troubles have several causes which combine to give the observed result. A single-cause failure is a very rare occurrence.
3. *Selective comparison.* New information must be related to older pieces of information. There are any number of analogies or relations that might be drawn. The trick is to make the right comparisons.

These three ingredients of our mental process are commonly called insight. They have one factor in common: *selection.* However, there are other ingredients of a professional problem solver's approach to failure analysis and troubleshooting.

Prior knowledge. Even apparently simple problems often require prior knowledge. Consider the performance deficiencies listed in Table 4-1. It just does not make sense to try to diagnose these symptoms without at least some knowledge of what their causes could be.

Prior knowledge is based on experience, and if our experience has been the right kind, we will have built up certain basic troubleshooting axioms. Good examples are:

On lubrication:

1. Lubrication- or tribology-related problems on machinery often arise because deflections increase with size, while lubricant film thicknesses generally do not.
2. Temperatures have a very significant effect on lubricated components directly and indirectly because of differential expansions and thermal distortions. Therefore, we must always check temperatures, steady temperature gradients, and temperature transients when troubleshooting lubricated components.

On "trouble spots":

1. Trouble is likely to occur at interfaces, i.e. tight clearances, so we should expect fit corrosion and rubs. Cavitation will occur where an interface exists between vapor bubbles and a liquid, and between the liquid and an adjacent wall.[4] Only when both interfaces are present at the same time (see Figure 4-1), does cavitation occur.
2. Components that are designed to move relative to each other, but that are not in fact moving, will seize up.
3. Experienced maintenance people know to match-mark an assembly before taking it apart, to make sure that they can put it back together. But handling similar jobs a number of times can lead to overconfidence. Disregarding a simple precaution such as match-marking can then cause improper reassembly and serious operating problems after startup.

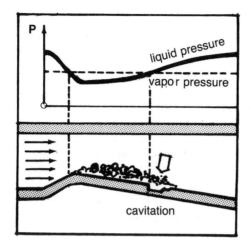

Figure 4-1. Cavitation event (modified from Ref. 5).

On problem solving:

1. Problems are best solved one at a time.
2. Problems have causes. Identify the cause before trying to solve the problem.
3. Before doing something, know what you want to accomplish.
4. Problems occur. It's better to plan for them than to be surprised.

And, on the lighter side: By definition, when you are investigating the unknown, you do not know what you will find.[1]

Executive processes. These are the processes involving the planning, monitoring, and performance evaluation in failure analysis and troubleshooting. To start with, one must first study the problem carefully, define it with regard to identity, location, time, and extent, and modify the perception of the problem as new information becomes available. Another important executive process involves monitoring the solution process—keeping track of what has been done, is being done, and still needs to be done—and then switching strategies if progress is too slow.

Motivation. Really challenging problems often require a great deal of motivation on the part of the problem solver. Successful problem solvers are often those who are simply willing to put in the necessary effort.

Style. People approach problems with different styles of thinking and understanding. Some tend to be more impulsive, and others more reflective. The most successful problem solvers are the ones who manage to combine both styles, for certain points in the troubleshooting process require following impulses and other points require reflection or rigorous calculation.

In summary, the professional problem solver's approach to failure analysis and troubleshooting involves the coordinated application of good selection process, prior knowledge, executive processes, motivation, and style.

The Matrix Approach to Machinery Troubleshooting

In the preceding chapters it became evident that process machinery trouble-shooting must contain certain diagnostic steps. We will encounter machinery trouble situations where the cause is known (no diagnostics required), where the cause is suspected, and where the cause is unknown. A responsible troubleshooter will not walk away from any of these situations—including the one where the cause is known. The following will cover diagnostics and address the approach to the situations where the cause is unknown or suspected.

Diagnostics is the method of reaching a diagnosis. Diagnosis, in turn, is simply another word for "recognition," as, for instance, the recognition of a disease.[5] In preceding sections we have alluded to certain aspects of machinery diagnostics. We should be able to differentiate between two basic forms: one, where diagnostics or failure analysis is performed after the failure has occurred; and two, where diagnosis is part of "machinery health monitoring" efforts, as shown in Table 4-2, i.e. where we attempt to recognize a machine's health or condition by assembling all pertinent performance and condition data. This approach is also referred to as condition monitoring, which plays an important role in predictive maintenance.

A good application of this form of diagnostics is in connection with health monitoring of large diesel engines and gas-engine compressors. Here, exhaust-gas temperature-monitoring programs have been used as an inexpensive and general machinery health indicator, but seldom have these programs provided adequate early warning of impending failure. Consequently, sophisticated diagnostic techniques were developed in the form of the engine analyzer. Typical diagnostic displays obtained with an engine analyzer are shown in Figures 4-2 and 4-3. Few other process machines allow the extent of failure diagnosis that can be achieved with this type of monitoring on these particular machines. Figure 4-4 and the inset text in Figure 4-5 give a good idea of what is possible in other fields with respect to failure and malfunction diagnosis.[6]

Closer to home, automatic fault diagnosis is available for motor shutdown trip conditions on large pump drives, as shown in Table 4-3.[7]

Frankly, as much as we would like to subscribe to the more predictive, on-line type diagnosis, its undeniable disadvantage is a frequently difficult-to-define cost/benefit ratio.

It has been said that 99% of all machinery failures are preceded by some nonspecific malfunction sign. The corollary is that with the exception of reciprocating compressors, internal combustion engines, and integrated electrical systems, most process machinery systems will give little warning of *specific* individual component distress prior to total failure. It therefore stands to reason that we have to be prepared to apply both post-mortem and on-line diagnostics in our failure-fighting strategy.

Table 4-2
Process Machinery Diagnostics

We thought it a good idea to look to the medical world for suitable analogies pertaining to our topic; after all, we borrowed the terms diagnosis and diagnostics from there. Three analogies come to mind:

1. *Symptoms.* Like the human body, a working process machine has certain "vital signs" that reveal to the experienced troubleshooter its internal state of health. Like the physician, the machinery analyst must rely primarily on his senses to obtain the symptomatic data he needs. Unlike the physician however, he very seldom has had those ten years or so of university training on the same model. The machinery troubleshooter must above all use his general experience and

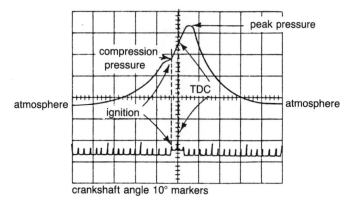

Figure 4-2. Display of pressure versus crank angle showing typical waveform characteristics of an engine analyzer display.[7]

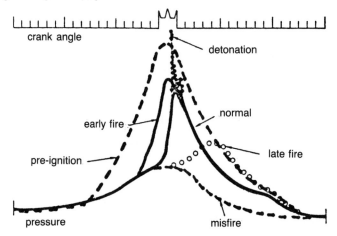

Figure 4-3. Two-cycle engine cylinder pressure display illustrating various malfunction symptoms.[8]

. . . the "Forward Facing Crew Cockpit" (FFCC-Boeing 767) affords total on-screen information and control: The flight monitoring system projects visible to the pilot onto the screen at any time engine operating data and flight systems conditions. Additionally, it indicates "Maintenance work due" for the ground maintenance crew. Computers, able to store some 200 malfunction conditions, warn the flight crew of approaching problems and indicate what measures are required for failure elimination.

The onboard computer is able to search within seconds for the root cause along the "trouble chain", and—above all—it does this quietly: Three quarters of all usual audible signals are now silenced in the computer cockpit . . . (From *Aviation Week Space Technology*)

Figure 4-4. The computer cockpit.

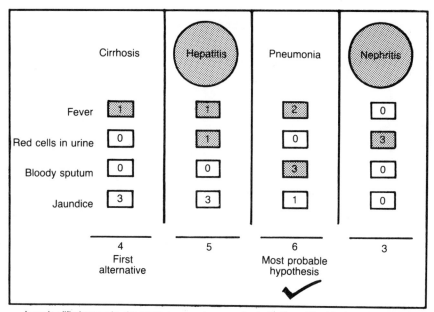

In a simplified example, the computer examines a patient with fever, blood in the urine, bloody sputum from the lungs, and jaundice. The program adds together numbers that show how much each symptom is related to four possible diagnoses—cirrhosis of the liver, hepatitis, pneumonia, and nephritis (kidney disease)—and picks pneumonia as top contender. The runner-up in score is hepatitis. But because hepatitis has one symptom not shared with pneumonia (blood in the urine), the computer chooses cirrhosis as first alternative. This process, called partitioning, focuses the computer's attention on groups of related diseases.

Figure 4-5. Computerized medical diagnosis (modified from Ref. 6).

his five senses. These, unaided by suitable instrumentation, will often tell him that "something" is wrong, without being specific. Consequently, in machinery troubleshooting as in medical diagnostics one has to wait for significant changes in a disease or distress situation to obtain from a given quantity of nonspecific symptoms specific ones that allow a valid diagnosis. Examples of nonspecific symptoms are shown in Table 4-4. Quite often the machinery troubleshooter must work with nonspecific symptoms and act timely or the specific symptom will show itself in a forced shutdown or even a wreck.

The following typical case will serve as an example of nonspecific machinery trouble symptoms. The patient: An 800-hp, 320-rpm induction motor driving a balanced-opposed reciprocating two-stage process gas compressor. The symptoms: Operating technicians have noticed the following gradual change in the motor's operating behavior:

- Unusual axial vibrations at 0.1 IPS (usual level: 0.05 IPS (inches/second peak-to-peak velocity).

Table 4-3
Motor Fault Diagnosis Display[7]

U/B	O/L	TRIPPED RELAY INDICATION START	RTD	G/F	TRIP	U/C	ON	PROBABLE CAUSE OF TRIP
	●	●			●		●	START TIMER EXCEEDED • locked rotor • too many starts • start timer set too low
	●				●		●	OVERLOAD • excessive overloads during running • mechanical jam, rapid trip • short circuit
			●		●		●	HIGH TEMPERATURE (RTD OPTION) • blocked ventilation • cooling fins blocked • high ambient temperature
●	●	●			●		●	BLOWN FUSE START • check fuses
●					●		●	EXCESSIVE UNBALANCE • blown fuses • single phase supply • loose wiring connection • short between motor windings
				●	●		●	GROUND FAULT • motor winding to case short • wiring touching metal ground
●	●			●	●		●	GROUND FAULT • motor winding to case short • wiring touching metal ground
					●		●	MEMORY LOCKOUT • memory still locked out after power is reapplied
						●	●	UNDERCURRENT ALARM • pump loss of suction • pump closed discharge valve • blower loss of airflow • conveyor overload warning

Table 4-4
Nonspecific Symptoms in Diagnostics

Machinery Trouble Diagnosis	Medical Diagnosis
Increased Temperature	Fever
Noise	Looking Ill
Vibration	Skin Rash
High Bearing Temperature	Coated Tongue

- Discrete rubbing sound from inside the motor.
- Rubbing sound decreasing, disappearing with decreasing load.
- Ampere reading fluctuating ± 15 around 86% of full electrical load.

What is the troubleshooter to do who gets called upon the scene? Will he wait until further, more *specific* symptoms appear? Will he have the motor uncoupled from its driven machine in order to perform an idle test run and perhaps some electrical tests for rotor bar looseness? Is there perhaps a bearing problem in connection with water-contaminated oil that should be investigated? And, finally, are there even enough distress symptoms to warrant any action at all?

These and other thoughts will go through the troubleshooter's head as he looks at the symptoms before him.

2. *Diagnosis* in machinery troubleshooting and in the medical profession is far from an exact science. Respected specialists in medicine will examine the same set of symptoms and arrive at different conclusions,[9] most likely because there are often too many nonspecific distress symptoms.

3. *Computerized Diagnosis.* Computer diagnosis in medicine is already functioning well.[6] In a way this is not surprising, as it is applied, again, to only one model and not to a diverse population of process machinery. The approach, however, is interesting in our context. It functions as described in Figure 4-5.

We are now ready to look at the information needed to perform machinery troubleshooting and diagnostics. Here are the requirements:

1. List distress or *trouble symptoms.* Start with the ones that are most easily perceptible:
   a. *Taste/odor:* gas, acid leakage, etc.
   b. *Touch:* overheating, vibration, etc.
   c. *Sound:* excessive noise, knocking, piston slap, rubbing, detonation, annunciator alarms, etc.
   d. *Sight:*
      1. *Direct sight or observation:* vapor, fume, and fluid leakage, vibration, loosening, smoke, fire, etc.
      2. *Indirect sight or observation:*
         a. Changes (up/down) in indicator readings:
            - Pressure
            - Temperature
            - Flow
            - Position
            - Speed
            - Vibration
         b. Changes (up/down) in performance:
            - Pressure ratios
            - Temperature ratios
            - Power demand
            - Product loss
            - Efficiencies

e. Internal inspection results.

f. Failure analysis (post-mortem) results.

2. List *possible cause or hypothesis:* Begin by listing the machine's major components, systems, and auxiliaries. It is here where the troubleshooter's intimate knowledge of the machinery in his care finds it application.

3. Indicate *tie-in or relationship* of symptom to cause. Make use of experience by asking: What can and usually does go wrong with this component? What are the symptoms associated with this distress or malfunction? A typical aid for this effort will be a fault chart such as shown in Figure 4-6.

4. Indicate symptom/cause *probability ranking.* For instance, if port plugging, piston-ring sticking, excessive ash, and varnish and carbon deposits are

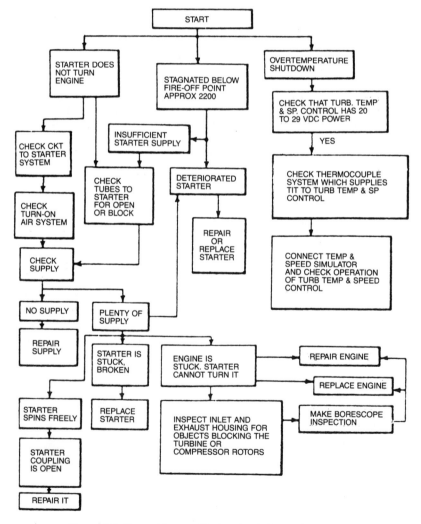

Figure 4-6. Gas-turbine troubleshooting (adapted from ref. 10).

Table 4-5
Typical Yearly Repair Summary,
Centrifugal Pumps

Centrifugal pumps installed: 2560
Centrifugal pumps operating at any given time: 1252 (average)
  Total pumps repaired in 1979: 768
  Pumps repaired at site location: 382
  Pumps repaired at own shops: 267
  Pumps repaired at outside shops: 119

Failure Causes	Distribution (%)
Mechanical seals	34.5
Bearing distress	20.2
Vibration events	2.7
Packing leakage	16.3
Shaft problems/couplings	10.5
Case failure/auxiliary lines	4.8
Stuck	4.3
Bad performance	2.5
Other causes	4.2
Total	100.0

symptomatic of overlubrication in internal combustion engines, which symptom is most indicative of overlubrication and which one less so? It is a good idea to rank relative probabilities of symptom/cause relationships in machinery troubleshooting, as shown in Figure 4-5.

5. Indicate remedial action—*remedy*. Quite often remedial action is obvious and commonplace. Typical actions taken after the most probable cause of a trouble has been identified are checking, inspection, adjustment, lubrication, cleaning, balancing, replacing, and, of course, analysis and follow-up.

The following sections show troubleshooting guides in matrix form accompanied by appropriate comments. Test these against your own troubleshooting efforts, and keep in mind that in some cases the ultimate development could well be computerized machinery fault diagnosis.* Where possible, relevant failure statistics are presented for various types of process machinery. This will help in problem symptom and cause probability ranking or partitioning.

Troubleshooting Pumps

The centrifugal pump is the workhorse of petrochemical-plant pumping applications. As far as process pumps are concerned, Table 4-5 gives the reader, by way of a statistic, an indication of the most frequently encountered troubles.

Table 4-6 is an appropriate troubleshooting guide for the most common on-line distress situations. In Reference 11 Igor Karassik refers to these situations as "field

(text continued on page 229)

---

* Current machinery operating and maintenance manuals have troubleshooting information of varying quality. We believe suitably formatted troubleshooting information should be made a specified requirement for new process machinery purchases.

Table 4-6
Troubleshooting Guide—Centrifugal Process Pumps

Symptoms		Symptoms	
D Insufficient Disch. Pressure		**Short Bearing Life** **E**	
C Intermittent Operation		**Short Mech. Seal Life** **F**	
B Insufficient Capacity		**Vibration and Noise** **G**	
A No Liquid Delivery		**Power Demand Excessive** **H**	

	Possible Causes	#	A	B	C	D	E	F	G	H	#	Possible Remedies
Suction Problems	Pump Is Cavitating (Symptom For Liquid Vaporizing In Suction System)	1	2	1	1			9	1		1	* Check NPSHa/NPSHr Margin * If Pump Is Above Liquid Level, Raise Liquid Level Closer To Pump * If Liquid Is Above Pump, Increase Liquid Level Elevation
	Insufficient Immersion Of Suction Pipe Or Bell (Vert. Turbine P.)	2	1	1	1				1		2	* Lower Suction Pipe Or Raise Sump Level * Increase System Resistance
	Pump Not Primed	3	1		2						3	* Fill Pump & Suction Piping Complete With Liquid * Eliminate High Points In Suction * Remove All Non-Condensibles (Air From Pump, Piping And Valves) * Eliminate High Points In Suction Piping * Check For Faulty Foot Valve Or Check Valve
Hydraulic System	Non-Condensibles In Liquid	4		2	3	1					4	* Check For Gas/Air Ingress Through Suction System/Piping * Install Gas Separation Chamber
	Supply Tank Empty	5	3								5	* Refill Supply Tank
	Obstructions In Lines Or Pump Housing	6		9		7			7		6	* Inspect And Clear
	Possible Causes	**#**	**A**	**B**	**C**	**D**	**E**	**F**	**G**	**H**	**#**	**Possible Remedies**

Table 4-6 (cont.)

	Symptoms					Symptoms				
D Insufficient Disch. Pressure					Short Bearing Life **E**					
C Intermittent Operation					Short Mech. Seal Life **F**					
B Insufficient Capacity					Vibration and Noise **G**					
A No Liquid Delivery					Power Demand Excessive **H**					
Possible Causes	**#**	**A**	**B**	**C**	**D**	**E**	**F**	**G**	**H**	**#**	**Possible Remedies**
Strainer Partially Clogged	7		3							7	* Inspect and Clean
Pump Impeller Clogged	8	8	8					5		8	*Check For Damage And Clean
Suction or/& Dischrg. Valve(s) Closed	9	9								9	* Shut Down & Open Valves
Viscosity Too High	10		7		5				4	10	*Heat Up Liquid To Reduce Viscosity * Increase Size Of Discharge Piping To Reduce Pressure Loss * Use Larger Driver Or Change Type Of Pump * Slow Pump Down
Specific Gravity Too High	11							2		11	* Check Design Specific Gravity
Total System Head Lower Than Design Head Of Pump	12				4		11	3		12	* Increase System Resistance To Obtain Design Flow * Check Design Parameters Such As Impeller Size etc.
Total System Head Higher Than Design Head Of Pump	13	6	5	4			10	2		13	* Decrease System Resistance To Obtain Design Flow * Check Design Parameters Such As Impeller Size, etc.
Unsuitable Pumps In Parallel Operation	14	7	6		6					14	* Check Design Parameters
Improper Mechanical Seal	15						1			15	* Check Mechanical Seal Selection Strategy
Possible Causes	**#**	**A**	**B**	**C**	**D**	**E**	**F**	**G**	**H**	**#**	**Possible Remedies**

Hydraulic System (rows: Strainer Partially Clogged through Total System Head Higher Than Design Head Of Pump)

Mechanical System (rows: Unsuitable Pumps In Parallel Operation, Improper Mechanical Seal)

(table cont. on next page)

Table 4-6 (cont.)

	Symptoms						Symptoms				
D	**Insufficient Disch. Pressure**					**Short Bearing Life**			**E**		
C	**Intermittent Operation**					**Short Mech. Seal Life**			**F**		
B	**Insufficient Capacity**					**Vibration and Noise**			**G**		
A	**No Liquid Delivery**					**Power Demand Excessive**			**H**		

Possible Causes	#	A	B	C	D	E	F	G	H	#	Possible Remedies
Speed Too High	16								1	16	* Check Motor Voltage—Slow Down Driver
Speed Too Low	17	4	4		2					17	* Consult Driver Troubleshooting Guide
Wrong Direction Of Rotation	18	5			3				6	18	* Check Rotation With Arrow On Casing—Reverse Polarity On Motor
Impeller Installed Backward (Double Suction Imp.)	19		10						12	19	* Inspect
Misalignment	20					1	2	4	7	20	* Check Angular And Parallel Alignment Between Pump & Driver
Casing Distorted From Excessive Pipe Strain	21					2	3	5		21	* Check For Misalignment * Check Pump For Wear Between Casing And Rotating Elements * Analyze Piping Loads
Inadequate Grouting Of Base	22							6		22	* Check Grouting & Regrout If Required
Bent Shaft	23					3	4	7	8	23	* Check Deflection (Should Not Exceed 0.002″). Replace Shaft & Bearings If Necessary
Internal Wear	24				8				9	24	* Check Impeller Clearances
Possible Causes	#	A	B	C	D	E	F	G	H	#	Possible Remedies

Mechanical System

Table 4-6 (cont.)

	Symptoms					Symptoms				

D	Insufficient Disch. Pressure		Short Bearing Life	E
C	Intermittent Operation		Short Mech. Seal Life	F
B	Insufficient Capacity		Vibration and Noise	G
A	No Liquid Delivery		Power Demand Excessive	H

	Possible Causes	#	A	B	C	D	E	F	G	H	#	Possible Remedies
Mechanical System	Mechanical Defects Worn, Rusted, Defective Bearings	25					5	8	10		25	* Inspect Parts For Defects—Repair Or Replace. Use Bearing Failure Analysis Guide * Check Lubrication Procedures
	Unbalance—Driver	26					5	7	9		26	* Run Driver Disconnected From Pump Unit—Perform Vibration Analysis
	Unbalance—Pump	27					4	6	3		27	* Investigate Natural Frequency
	Motor Troubles	28					6	8	10	11	28	* Consult Motor Troubleshooting Guide
	Possible Causes	#	A	B	C	D	E	F	G	H	#	**Possible Remedies**

Note: for vibration & short bearing life related causes, refer to special sections on bearing failure analysis and vibration diagnostics.

troubles," and no doubt his check chart for centrifugal pump problems is one of the more comprehensive troubleshooting guides for centrifugal pumps. The numbers listed in columns A through H indicate the probability: the lower the number, the higher the probability.

If we look at Table 4-6 to determine the most probable cause for insufficient pressure generation (Symptom "D"), we find that we should look at possible causes in this sequence:

1. Noncondensibles (air) in liquid.
2. Pump speed too low.
3. Wrong direction of rotation.
4. Total system head lower than design head of pump—pump is "running out."
5. Viscosity too high.
6. Two or more pumps in parallel operation and unsuitable.
7. Internal obstructions in lines ahead of pressure gauges.
8. Internal wear, i.e. wear rings worn.

If more than one symptom is observed, the checking sequence may be established by obtaining the average ranking number by cause. If the pump also works intermittently, alternately losing and regaining its prime after startup, the sequence of probable cause would be that shown in Table 4-7. Tables 4-8 through 4-12 have been compiled with the same partitioning approach in mind.

(text continued on page 240)

Table 4-7
Establishment of Average Ranking Numbers by Combining Probable Causes

Possible Cause	Symptom (C)	Symptom (D)	Sequence (#)
1. Pump is cavitating	1		1
2. Pump not primed	2		2
3. Non-condensibles in liquid	3	1	2*
4. Speed too low		2	2
5. Wrong direction of rotation		3	3
. . . and so forth.			

* Average = (3 + 1)/2

Table 4-8
Troubleshooting Guide—Vertical Turbine Pumps

	Symptoms						Symptoms				
	C Insufficient Disch. Pressure						Vibration D				
	B Insufficient Capacity						Abnormal Noise E				
	A Liquid Delivery						Power Demand Excessive F				
	Possible Causes	**#**	**A**	**B**	**C**		**D**	**E**	**F**	**#**	**Possible Remedies**
Hydraulic System	Pump Suction Interrupted (Water Level Below Bell Inlet)	1	1							1	* Check sump level
Hydraulic System	Low Water Level	2		8						2	* Check water level
Hydraulic System	Cavitation Due to Low Submergence	3						7		3	* Check submergence
	Possible Causes	**#**	**A**	**B**	**C**		**D**	**E**	**F**	**#**	**Possible Remedies**

Table 4-8 (cont.)

Symptoms					Symptoms				

		C Insufficient Disch. Pressure			Vibration D				
		B Insufficient Capacity			Abnormal Noise E				
		A Liquid Delivery			Power Demand Excessive F				

Possible Causes	#	A	B	C		D	E	F	#	Possible Remedies
Vortex Problems	4					8			4	* Install vortex breaker shroud
Suction or Discharge Recirculation	5					7			5	* Establish design flows
Operation Beyond Max. Capacity Rating	6					6	6		6	* Establish proper flow rate
Entrained Air	7			6					7	* Install separation chamber
Suction Valve Closed	8	2							8	* Shut down or open valve
Suction Valve Throttled	9		7						9	* Open valve
Strainer Clogged	10	4	6						10	* Inspect & clean
Impeller Plugged	11	3		4					11	* Pull pump & clean
Impeller or Bowl Partially Plugged	12		4				8		12	*Pull pump & clean
Impellers Trimmed Incorrectly	13		2	2				3	13	* Check for proper impeller size
Improper Impeller Adjustment	14					9		2	14	* Check installation/repair records
Impeller Loose	15	7	3	3					15	* Pull pump & analyze
Impeller Rubbing On Bowl Case	16						5		16	* Check lift
Wear Rings Worn	17			5					17	* Inspect during overhaul
Shaft Bent	18					5		5	18	* Pull pump & analyze
Shaft Broken, or Unscrewed	19	6							19	* Pull pump & analyze
Enclosing Tube Broken	20						4		20	* Pull pump & analyze
Possible Causes	#	A	B	C		D	E	F	#	Possible Remedies

Hydraulic System (rows: Vortex Problems through Impeller Plugged)
Mechanical System (rows: Impeller or Bowl Partially Plugged through Enclosing Tube Broken)

(table continued on next page)

Table 4-8 (cont.)

	Symptoms					**Symptoms**				
	C Insufficient Disch. Pressure					Vibration D				
	B Insufficient Capacity					Abnormal Noise E				
	A No Liquid Delivery					Power Demand Excessive F				
	Possible Causes	**#**	**A**	**B**	**C**	**D**	**E**	**F**	**#**	**Possible Remedies**
Mechanical System	Bearings Running Dry	21					2		21	* Provide lubrication
	Worn Bearings	22				6			22	* Pull pump & repair
	Column Bearing Retainers Broken	23					3		23	* Pull pump & analyze
	Wrong Rotation	24	5	9	8				24	* Check rotation
	Speed Too Slow	25		1	1				25	* Check rpm
	Speed Too High	26						1	26	* Check rpm
	Misalignment Of Pump Assembly Through Pipe Strain	27				4		4	27	* Inspect
	Leaking Joints	28		4	5				28	* Inspect
	Pumping Sand, Silt Or Foreign Material	29						6	29	* Check liquid pumped
	Motor Noise	30					1		30	* Check sound level
	Motor Electrical Imbalance	31				1			31	* Perform phase check
	Motor Bearing Problems	32				2			32	* Consult motor/bearing troubleshooting guide
	Motor Drive Coupling Out of Balance	33				3			33	* Inspect
	Resonance: System Natural Frequency Near Pump Speed	34				10			34	* Perform vibration analysis
	Possible Causes	**#**	**A**	**B**	**C**	**D**	**E**	**F**	**#**	**Possible Remedies**

Table 4-9

Fire Pump Troubleshooting Guide[12]

Possible Causes	#	Disch. Press. Too Low For gpm. Discharge	Insufficient Water Discharge	Pump Loses Suction After Starting	Disch. Press. Not Constant For Same gpm	Too Much Power Required	Pump Is Noisy or Vibrates	No Water Discharge	Pump Unit Will Not Start	Pump or Driver Overheats	Excessive Leakage at Stuffing Box	
Symptoms →												
Suction System												
Suction lift too high.	1	•	•	•	•		•	•				1
Foot valve too small, partially obstructed or of inferior design causing excessive suction head loss.	2	•	•	•								2
Air drawn into suction connection.	3	•	•	•	•		•					3
Air drawn into suction connection through leak.	4	•	•	•	•		•	•				4
Suction connection obstructed.	5	•	•	•				•				5
Air pocket in suction pipe.	6	•	•	•				•				6
Hydraulic cavitation from excessive suction lift.	7	•	•				•					7
Well collapsed or serious misalignment.	8					•	•		•	•		8

• Check this cause (no probability ranking assigned).

(table continued on next page)

Table 4-9 (cont.)

	Possible Causes	#	Symptoms: Disch. Press. Too Low For gpm. Discharge	Insufficient Water Discharge	Pump Loses Suction After Starting	Disch. Press. Not Constant For Same gpm	Too Much Power Required	Pump Is Noisy or Vibrates	No Water Discharge	Pump Unit Will Not Start	Pump or Driver Overheats	Excessive Leakage at Stuffing Box	
Pump	Stuffing box too tight or packing improperly installed, worn, defective, too tight, or incorrect type.	9	●			●	●	●		●	●	●	9
	Water-seal or pipe to seal obstructed.	10	●	●	●	●					●		10
	Air leak into pump through stuffing box.	11	●	●	●	●							11
	Impeller obstructed.	12	●	●			●	●	●		●		12
	Wear rings worn.	13	●	●			●						13
	Impeller damaged.	14	●	●				●					14
	Wrong diameter impeller.	15	●	●			●				●		15
	Actual net head lower than rated.	16	●	●									16
	Casing gasket defective, permitting internal leakage. (Multi-stage pumps only.)	17	●	●			●						17
	Bad pressure gage on top of pump casing.	18	●										18
	Incorrect impeller adjustment. (Vertical pumps only.)	19		●			●			●	●		19
	Impellers locked.	20								●			20
	Pump is frozen.	21								●			21

Group	Cause	No.									
Pump	Pump shaft or shaft sleeve scored, bent, or worn.	22				•	•				•
Pump	Pump not primed.	23						•			•
Pump	Seal ring improperly located in stuffing box, preventing water from entering space to form seal.	24	•						•		•
Driver/Pump	Excess bearing friction due to wear, dirt, rusting, failure, or improper installation.	25				•	•			•	•
Driver/Pump	Rotating element binds against stationary element.	26				•	•				•
Driver/Pump	Pump and driver misaligned. Shaft running off center because of worn bearings or misalignment.	27		•		•	•				•
Driver/Pump	Foundation not rigid.	28				•	•				•
Driver/Pump	Engine cooling system obstructed. Heat exchanger or cooling water system too small. Cooling pump faulty.	29								•	•
Driver	Faulty driver.	30				•	•			•	•
Driver	Lack of lubrication.	31								•	•
Driver	Speed too low.	32			•						
Driver	Wrong direction of rotation.	33			•	•	•				•
Driver	Speed too high.	34				•	•				•
Driver	Rated motor voltage different from line voltage, i.e., 220- or 440-volt motor on 208- or 416-volt line.	35			•	•					•
Driver	Faulty electric circuit, obstructed fuel system or obstructed steam pipe, or dead battery.	36								•	

Table 4-10
Troubleshooting Guide—Rotary Pumps

Possible Causes	#	A	B	C	D	E	F	G	#	Possible Remedies
		A No Liquid Delivery	**B** Insufficient Capacity	**C** Starts, But Loses Prime	**D** Excessive Wear	**E** Excessive Heat	**F** Vibration and Noise	**G** Excessive Power Demand		
Suction Filter Or Strainer Clogged	1	1	1				1		1	* Clean strainer or filter.
Pump Running Dry	2	2			2	1	4		2	* Reprime
Insufficient Liquid Supply	3		2	4	3		3		3	* Look for suction restriction or low suction level.
Suction Piping Not Immersed in Liquid	4	4		3					4	* Lengthen suction pipe or raise liquid level
Liquid Vaporizing In Suction Line	5		4	2			5		5	* Check NPSH. Check for restriction in suction line.
Air Leakage Into Suction Piping Or Shaft Seal	6		3	1			6		6	* Tighten & seal all joints. * Adjust packing or repair mechanical seal.
Solids Or Dirt In Liquid	7				1				7	* Clean system. * Install filtration.
Liquid More Viscous Than Designed For	8							1	8	* Reduce pumped medium viscosity. * Reduce pump speed. * Increase driver HP
Excessive Discharge Pressure	9		5		7		9	2	9	* Check relief valve or by-pass setting. * Check for obstruction in discharge line.
Pipe Strain On Pump Casing	10				5	4	8	5	10	* Disconnect piping and check flange alignment.
Coupling, Belt Drive, Chain Drive Out Of Alignment	11				6	3	2	4	11	* Realign
Rotating Elements Binding	12				4	2	7	3	12	* Disassemble & Inspect.
Possible Causes	#	A	B	C	D	E	F	G	#	Possible Remedies

Hydraulic System (rows 1–9) and *Mechanical Problems* (rows 10–12)

Table 4-10 (cont.)

		Symptoms							Symptoms	
	D		**Excessive Wear**							
	C	**Starts, But Loses Prime**				**Excessive Heat**		**E**		
	B	**Insufficient Capacity**				**Vibration and Noise**		**F**		
	A	**No Liquid Delivery**				**Excessive Power Demand**		**G**		
	Possible Causes	**#**	**A**	**B**	**C**	**D**	**E**	**F**	**G**	**#**	**Possible Remedies**

Mechanical Problems	Possible Causes	#	A	B	C	D	E	F	G	#	Possible Remedies
	Internal Parts Wear	13		7						13	* Inspect & replace worn parts
	Speed Too Low	14		6						14	* Check driver speed
	Wrong Direction Of Rotation	15	3							15	* Check & reverse if required
	Possible Causes	**#**	**A**	**B**	**C**	**D**	**E**	**F**	**G**	**#**	**Possible Remedies**

Table 4-11
Troubleshooting Guide—Reciprocating (Power) Pumps

		Symptoms									Symptoms	
	D	**Excessive Wear-Liquid End**				**Excessive Wear—Power End**			**E**			
	C	**Short Packing Life**				**Excessive Heat—Power End**			**F**			
	B	**Insufficient Capacity**				**Vibration and Noise**			**G**			
	A	**No Liquid Delivery**				**Persistent Knocking**			**H**			
	Possible Causes	**#**	**A**	**B**	**C**	**D**	**E**	**F**	**G**	**H**	**#**	**Possible Remedies**

Suction System	Possible Causes	#	A	B	C	D	E	F	G	H	#	Possible Remedies
	Excessive Suction Lift	1	1								1	* Decrease suction lift—Consider charge or booster pump
	Not Enough Suction Pressure Above Vapor Pressure	2	2								2	* Remove all non-condensibles (air) from pump, piping, & valves * Eliminate high points in suction piping * Check for faulty foot valve or check valve
	Non-Condensibles (Air) In Liquid	3	3	1		5			3		3	* Install gas separation chamber
	Possible Causes	**#**	**A**	**B**	**C**	**D**	**E**	**F**	**G**	**H**	**#**	**Possible Remedies**

Table 4-11 (cont.)

	Symptoms				Symptoms				
D	**Excessive Wear-Liquid End**				**Excessive Wear—Power End E**				
C	**Short Packing Life**				**Excessive Heat—Power End F**				
B	**Insufficient Capacity**				**Vibration and Noise G**				
A	**No Liquid Delivery**				**Persistent Knocking H**				

	Possible Causes	#	A	B	C	D	E	F	G	H	#	Possible Remedies
Suction System	Cylinders Not Filling	4	2		2				4		4	* Attempt to prime pump. * Install foot valve at bottom of suction pipe if pump has to lift * Vortex in supply tank. Consider charge or booster pump
	Low Volumetric Efficiency	5	3								5	* Liquid with low specific gravity or high discharge pressure compressing & expanding in cylinder
	Supply Tank Empty	6	6								6	* Refill supply tank
Operating Conditions	Obstruction In Lines	7	4						5		7	* Inspect & clean
	Abrasives or Corrosives In Liquid	8			4	1					8	* Establish proper maintenance intervals * Consider design change
	Relief or Bypass Valve(s) Leaking	9	4								9	* Inspect & repair
Valves/Cylinders/Pistons/P. Rods	Pump Speed Incorrect	10	7					5			10	* Check for slipping belts. Consult belt drive & driver TSG * Check driver speed
	One or More Cylinder(s) Not Operating	11	5								11	* Stop & prime all cylinders
	Pump Valve(s) Stuck Open	12	6								12	* Check for damage & clean
	Broken Valve Springs	13	8		3				6		13	* Inspect & replace
	Worn Valves, Seats, Liners, Rods, &/or Plungers	14	5	9		4					14	* Inspect & repair
	Possible Causes	**#**	**A**	**B**	**C**	**D**	**E**	**F**	**G**	**H**	**#**	**Possible Remedies**

Table 4-11 (cont.)

Symptoms

- **D** Excessive Wear-Liquid End
- **C** Short Packing Life
- **B** Insufficient Capacity
- **A** No Liquid Delivery

Symptoms

- Excessive Wear—Power End **E**
- Excessive Heat—Power End **F**
- Vibration and Noise **G**
- Persistent Knocking **H**

	Possible Causes	#	A	B	C	D	E	F	G	H	#	Possible Remedies
Valves/Cylinders/ Pistons/P. Rods	Improper Packing Selection	15			1						15	* Develop selection guide-lines
	Scored Rod or Plunger	16		10							16	* Regrind or replace
	Misalignment of Rod or Plunger	17			2						17	* Check & realign. Tolerance: Eccentricity in box within 0.003″ max.
	Loose Piston or Rod	18								1	18	* Inspect & repair
Crosshead Bearings	Loose Cross-Head Pin or Crank Pin	19								2	19	* Adjust or replace * Check for proper clearances
	Worn Cross-Heads or Guides	20			3			4			20	* Adjust or replace
	Other Mechanical Problems: Worn, Rusted, Defective Bearings	21					1	3	1	3	21	* Inspect part for defect limits & repair or replace * Consult bearing failure analysis guide * Check lubrication procedures
	Inadequate Lubrication Conditions	22					2	1	2		22	* Establish lubrication schedule * Review design assumptions for crankcase lubrication, etc.
Loading Drive	Liquid Entry Into Power End	23					5	2			23	* Replace packing & parts affected
	Gear Problems	24					4		7	4	24	* Consult gear TSG†
	Driver Troubles	25							8		25	* Consult appropriate TSG†
	Overloading	26					3				26	* Check design assumptions.
	Possible Causes	**#**	**A**	**B**	**C**	**D**	**E**	**F**	**G**	**H**	**#**	**Possible Remedies**

Table 4-12
Troubleshooting Guide—Liquid Ring Pumps

Possible Causes	Symptoms				
	Noise Excessive	Vibration Up	Overheating Temperature Up	Capacity Down	Power Consumption Up
Speed too low				③	
Suction line leakage				①	
Service liquid temperature too high*			①	②	
Service liquid level too high	①	②		④	①
Service liquid insufficient			②	⑤	
Coupling misalignment	②	①	③		②

*Insufficient heat rejection (cooler make-up liquid).
Note: The numbers indicate what to check first, or symptom-cause probability rank.

Troubleshooting Centrifugal Compressors, Blowers, and Fans

Centrifugal compressors, blowers, and fans form the basic make-up of unspared machinery equipment in petrochemical process plants. Single, unspared centrifugal compressor trains support the entire operation of steam crackers producing as much as 800,000 metric tons (approximately 1.76 billion lbs) of ethylene per year. When plants in this size range experience emergency shutdowns of a few hours' duration, flare losses alone can amount to $400,000 or more. Timely recognition of centrifugal compressor troubles is therefore extremely important. Table 4-13 shows what is most likely to happen to a centrifugal compressor or blower.

Table 4-14 presents a troubleshooting guide for centrifugal compressors and their lubrication systems. Similiar guides may be constructed for other types of centrifugal machinery such as fans and blowers. Since Table 4-14 cannot possibly be all-encompassing, refer to special sections in this book addressing such related topics as vibration-causing syndromes and bearing, gear, and coupling failure analysis and troubleshooting.

Table 4-13
Typical Distribution of Unscheduled Downtime Events for
Major Turbocompressors in Process Plants

Cause of Problem	Estimated Frequency	Estimated Average Hrs/Event	Downtime Events/Yr.	Hrs/Yr
	Approximate number of shutdowns per train per year: 2			
Rotor/shaft	22%	122	44	54
Instrumentation	21%	4	42	2
Radial bearings	13%	28	26	7
Blades/impellers	8%	110	16	18
Thrust bearings	6%	22	12	3
Compressor seals	6%	48	12	6
Motor windings	3%	200	06	12
Diaphragms	1%	350	02	7
Miscellaneous causes	20%	70	40	28
All causes	100%		2.00	137 hours

Table 4-14
Troubleshooting Guide—Centrifugal Compressor and Lube System

Symptoms					Symptoms				
D	Low Lube Oil Pressure								
C	Loss Of Disch. Pressure				E	Excessive Brg. Oil Drain Temp.			
B	Compressor Surges				F	Units Do Not Stay In Alignment			
A	Excessive Vibration				G	Water In Lube Oil			

Possible Causes	#	A	B	C	D	E	F	G	#	Possible Remedies
Excessive Bearing Clearance	1	13							1	* Replace bearings
Wiped Bearings	2					7			2	* Replace bearings * Determine & correct cause
Rough Rotor Shaft Journal Surface	3					9			3	* Stone or restore journals * Replace shaft
Bent Rotor (Caused By Uneven Heating or Cooling)	4	8							4	* Turn rotor at low speed until vibration stops, then gradually increase speed to operating speed. * If vibration continues, shut down, determine & correct the cause
Possible Causes	#	A	B	C	D	E	F	G	#	Possible Remedies

Table 4-14 (cont.)

Symptoms: A = Excessive Vibration; B = Compressor Surges; C = Loss Of Disch. Pressure; D = Low Lube Oil Pressure; E = Excessive Brg. Oil Drain Temp.; F = Units Do Not Stay In Alignment; G = Water In Lube Oil

	Possible Causes	#	A	B	C	D	E	F	G	#	Possible Remedies
Rotor/Bearing System	Operating In Critical Speed Range	5	9							5	* Operate at other than critical speed
	Build-Up of Deposits On Rotor	6	10	4						6	* Clean deposits from rotor * Check balance
	Build-Up Of Deposits in Diffuser	7		3						7	* Mechanically clean diffusers
	Unbalanced Rotor	8	11							8	* Inspect rotor for signs of rubbing * Check rotor for concentricity, cleanliness, loose parts * Rebalance
	Damaged Rotor	9	12							9	* Replace or repair rotor * Rebalance rotor
	Loose Rotor Parts	10	15							10	* Repair or replace loose parts
Coupling	Shaft Misalignment	11	5							11	* Check shaft alignment at operating temperatures * Correct any misalignment
	Dry Gear Coupling	12	6							12	* Lubricate coupling
	Worn or Damaged Coupling	13	7							13	* Replace coupling * Perform failure analysis
Operating Conditions	Liquid "Slugging"	14	14							14	* Locate & remove the source of liquid * Drain compressor casing of any accumulated liquids
	Operating In Surge Region	15	16							15	* Reduce or increase speed until vibration stops * Consult vibration analysis guide
	Insufficient Flow	16		1						16	* Increase recycle flow through machine
	Possible Causes	#	A	B	C	D	E	F	G	#	Possible Remedies

Table 4-14 (cont.)

	Symptoms							Symptoms	
	D	**Low Lube Oil Pressure**							
	C	**Loss Of Disch. Pressure**					**Excessive Brg. Oil Drain Temp. E**		
	B	**Compressor Surges**					**Units Do Not Stay In Alignment F**		
	A	**Excessive Vibration**					**Water In Lube Oil G**		

	Possible Causes	#	A	B	C	D	E	F	G	#	Possible Remedies
Operating Conditions	Change In System Resistance Due to Obstructions or Improper Inlet or Disch. Valve Positions	17		2						17	* Check position of inlet/discharge valves * Remove obstructions
	Compressor Not Up To Speed	18			1					18	* Increase to required operating speed
	Excessive Inlet Temperature	19			2					19	* Correct cause of high inlet temperature
	Leak In Discharge Piping	20			3					20	* Repair leak
	Vibration	21					7			21	* Refer to "A" in symptom column
	Sympathetic Vibration	22	4							22	* Adjacent machinery can cause vibration even when the unit is shut down, or at certain speeds due to foundation or piping resonance. A detailed investigation is required in order to take corrective measures
Assembly	Improperly Assembled Parts	23	1							23	* Shut down, dismantle, inspect, correct.
	Loose or Broken Bolting	24	2							24	* Check bolting at support assemblies. * Check bed plate bolting * Tighten or replace * Analyze
Support System	Piping Strain	25	3					1		25	* Inspect piping arrangements and proper installation of pipe hangers, springs, or expansion joints
	Possible Causes	**#**	**A**	**B**	**C**	**D**	**E**	**F**	**G**	**#**	**Possible Remedies**

(table continued on next page)

Table 4-14 (cont.)

Symptoms										Symptoms
D Low Lube Oil Pressure										
C Loss Of Disch. Pressure						Excessive Brg. Oil Drain Temp. E				
B Compressor Surges						Units Do Not Stay In Alignment F				
A Excessive Vibration						Water In Lube Oil G				
Possible Causes	#	A	B	C	D	E	F	G	#	Possible Remedies
Support System — Warped Foundation or Bedplate	26						2		26	* Check for possible settling of the foundation support * Correct footing as required * Check for uneven temperatures surrounding the foundation causing
Faulty Lube Oil Pressure Gauge or Switch	27				1				27	*Calibrate or replace
Faulty Temperature Gauge or Switch	28					2			28	* Calibrate or replace
Oil Reservoir Low Level	29				2				29	* Add oil
Clogged Oil Strainer/Filter	30				5				30	* Clean or replace oil strainer or filter cartridges
Relief Valve Improperly Set or Stuck Open	31				8				31	* Adjust relief valve * Recondition or replace
Incorrect Pressure Control Valve Setting on Operation	32				9				32	* Check control valve for correct setting and operation
Poor Oil Condition/Gummy Deposits On Bearings	33					4			33	* Change oil * Inspect and clean lube oil strainer or filter * Check and inspect bearings * Check with oil supplier to ascertain correct oil species being used
Possible Causes	#	A	B	C	D	E	F	G	#	Possible Remedies

Table 4-14 (cont.)

Symptoms						Symptoms				
D Low Lube Oil Pressure										
C Loss Of Disch. Pressure					**Excessive Brg. Oil Drain Temp. E**					
B Compressor Surges					**Units Do Not Stay In Alignment F**					
A Excessive Vibration					**Water In Lube Oil G**					
Possible Causes	**#**	**A**	**B**	**C**	**D**	**E**	**F**	**G**	**#**	**Possible Remedies**
Inadequate Cooling Water Supply	34					5			34	* Increase cooling water supply to lube oil cooler * Check for above design cooling water inlet temperature
Fouled Lube Oil Cooler	35					6			35	* Clean or replace lube oil cooler
Operation at a Very Low Speed Without The Auxiliary Oil Pump Running (If main L.O. pump is shaft driven)	36				7				36	* Increase speed or operate aux. lube oil pump to increase oil pressure
Bearing Lube Oil Orifices Missing or Plugged	37				11				37	* Check to see that lube oil orifices are installed and are not obstructed * Refer to lube oil system schematic diagram for orifice locations
Oil Pump Suction Plugged	38				3				38	* Clear pump suction
Leak In Oil Pump Suction Piping	39				4				39	* Tighten leaking connections * Replace gaskets
Failure Of Both Main & Auxiliary Oil Pumps	40				6				40	* Repair or replace pumps
Oil Leakage	41				10				41	* Tighten flanged or threaded connections * Replace defective gaskets or parts
Clogged Or Restricted Oil Cooler Oil Side	42					1			42	* Clean or replace cooler
Possible Causes	**#**	**A**	**B**	**C**	**D**	**E**	**F**	**G**	**#**	**Possible Remedies**

Lube Oil System

(table continued on next page)

Table 4-14 (cont.)

	Symptoms						Symptoms			
D	**Low Lube Oil Pressure**									
C	**Loss Of Disch. Pressure**					**Excessive Brg. Oil Drain Temp. E**			
B	**Compressor Surges**					**Units Do Not Stay In Alignment F**			
A	**Excessive Vibration**					**Water In Lube Oil G**			
	Possible Causes	**#**	**A**	**B**	**C**	**D**	**E**	**F**	**G**	**#**	**Possible Remedies**

	Possible Causes	#	A	B	C	D	E	F	G	#	Possible Remedies
	Inadequate Flow Of Lube Oil	43					3			43	* Refer to "D" in symptom column * If pressure is satisfactory, check for restricted flow of lube oil to the affected bearings
	Water in Lube Oil	44					8			44	* Refer to "G" in symptom column
	Leak In Lube Oil Cooler Tube(s) or Tube Sheet	45							1	45	* Hydrostatically test the tubes and repair as required * Replace zinc protector rods (if installed) more frequently if leaks are due to electrolytic action of cooling water
Lube Oil System	Condensation in Oil Reservoir	46							2	46	* During operation maintain a minimum lube oil reservoir temperature of 120 deg. F to permit separation of entrained water * When shutting down, stop cooling water flow to oil cooler * Commission lube oil conditioning unit * Refer to lube oil management guide
											NOTE: Vibration may be transmitted from the coupled machine. To localize vibration, disconnect coupling and operate driver alone. This should help to indicate whether driver or driven machine is causing vibration.
	Possible Causes	**#**	**A**	**B**	**C**	**D**	**E**	**F**	**G**	**#**	**Possible Remedies**

Table 4-15
Distribution of Basic Failure Causes—Reciprocating Compressors

Basic Failure Cause	Distribution of Incidents (%)
Valves	41.0
Piston Rings	14.0
Cylinders	1.0
Pistons	3.0
Piston Rods	10.0
Packings	10.0
Lubricator/System	18.0
Crossheads	1.0
Crankshafts	1.0
Bearings	1.0
Controls	1.0

Troubleshooting Reciprocating Compressors

Valves and piston rings, for obvious reasons, are frequently the weak link of reciprocating compressor operation. A typical statistic taken from a five-year operating history of three Canadian refineries with comparable compressor populations is shown in Table 4-15.

Valves.* A successful compressor valve application requires the careful analysis of the nature, properties, and quality of the gas that is compressed, and dimensions and specifications of the compressor, and the compressor's operating conditions. Then the proper-size valve is selected, with an acceptable compromise between efficiency and durability.

There are many different types of compressor valves manufactured by different manufacturers. The following is a list of the more representative valve designs.

1. Single- and multi-ring valves.
2. Plate valves.
3. Ring-plate valves.
4. Channel valves.
5. Feather-strip valves.

1. The *single- and multi-ring valve* is the valve most commonly used in reciprocating process gas compressors. The valve has one or more concentric valve rings that are generally spring loaded separately, as shown in Figure 4-7.

---

*Courtesy IIC, 364 Supertest Road, Downsview (Toronto), Ontario, M3J 2M2, Canada.

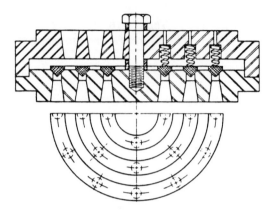

Figure 4-7. Pyramid ring valve.

A ring-type valve using rings with a contoured cross section, such as the IIC pyramid ring valve, is designed to lower the resistance to gas flow and improve efficiency.[13]

Attention to the distribution of the spring loads on the different valve rings is important in order to achieve proper timing in the opening and closing of the multiple valve rings.

2. *The plate valve* illustrated in Figure 4-8 uses a slotted disc for the valve element and springs for damping. This Hoerbiger design generally has an evenly distributed spring load over the entire valve area. Achieving proper damping without adversely affecting the flow efficiency is generally difficult with this design, as strong spring load is usually required for proper damping.

The successful use of relatively thick damping plates, spring loaded separately from the Hoerbiger valve plate, is an attempt to achieve a fast opening of the valve plate (lightly spring loaded), with sufficient damping to avoid valve slamming.

3. The *ring-plate valve* is a hybrid between the multi-ring valve and the plate valve. Generally it uses two concentric, slotted ring plates as the valve elements. It is a well-designed valve, but it is somewhat limited in cushioning or damping. The major manufacturer of this type of valve is Cooper-Bessemer.

4. The *channel valve* in Figure 4-9 is designed by Ingersoll-Rand. It consists of a valve seat, a guard, channels with leaf springs, and guides to hold the channels in place. Sometimes a lapped, stainless-steel seat plate is used to improve wear and to facilitate repair.

Each channel, complete with its leaf spring, operates individually by closing over a corresponding slot-shaped port. The distribution of spring load on the different channels inside the same valve is very important to achieve proper timing in the opening and closing of the channels.

A small volume of gas is trapped between the channel and its leaf spring. This gives a pneumatic cushioning effect so that the channel can ease to a stop when opening.

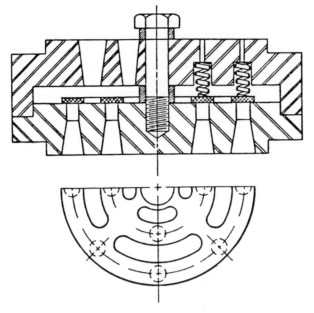

Figure 4-8. Plate valve.

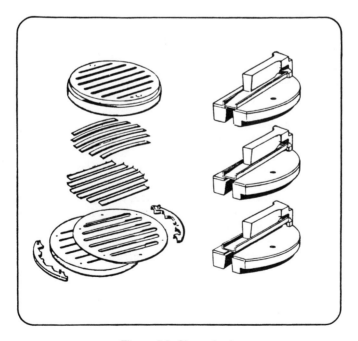

Figure 4-9. Channel valve.

5. The *feather-strip valve* in Figure 4-10 is very simple in design. Manufactured by Worthington, this valve consists of a seat, a guard, and several feather strips. These feather strips are made of light strips of steel whose flexing action permits the passage of gas. The valve seats merely by increasing contact from each end of the strip to the center, not by impact. It is generally an efficient valve, with a large seat-port area, lift area, and guard-port area. Since the feather strips are not cushioned, they are subjected to comparatively high mechanical wear.

Valve failures causes and their remedies.* Valve failures can generally be classified into three categories:

1. Mechanical wear and fatigue.
2. Foreign matters in the gas stream.
3. Abnormal action of valve elements and compressor-valve equipment.

Mechanical wear and fatigue. Mechanical wear in a compressor valve normally occurs where the valve components contact each other. Prolonged wearing of the moving valve element guides can cause sloppy valve action, cocking, and poor seating of the valve elements, leading to leakage, poor performance, and eventual failure. Wear on components should be checked and repaired regularly.

The moving valve elements are also subjected to cyclical, mechanical, and thermal stresses in their normal operation. The strength of the valve components must be considered in each application. Wear on valve parts can be minimized by lubricating the valve parts affected in a particular application. Synthetic lubricants will often outperform mineral oils in this service.

Mechanical fatigue load can be minimized by proper damping or cushioning of the moving valve parts. Proper cylinder cooling and interstage cooling will minimize heat stresses on the valve components.

Wear and fatigue on compressor valves can be sharply aggravated by the presence of foreign matter in the gas stream and by abnormal mechanical valve action.

Foreign matter in the gas stream. Foreign matter may include dirty gas, liquid carryover, coking or carbon formation, and corrosive chemicals.

1. *Dirty gas.* These are generally materials left inside the gas stream by an improper filtration system. There may be rust; fine, sandy, or abrasive particles; loose particles from the gas passages leading to the compressor; and sometimes even remnants from previous valve failures left inside the compressor. These materials accelerate wear on valve components such as the rings, the plates, and the guides. Foreign matter between the coils of the springs can cause rapid spring failures. This has a direct effect on the performance and service life of other valve parts.
2. *Liquid carryover.* Liquid may be carried into the gas stream in chemical processes, or it can be formed from condensation inside the interstage cooler of

---

* Modified from Reference 13.

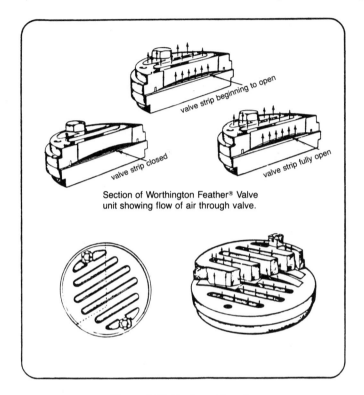

Section of Worthington Feather® Valve
unit showing flow of air through valve.

Figure 4-10. Feather-strip valve.

the compressor system. In many instances, liquid is present in the first compression stage because excessively cold jacket cooling water promotes condensation or "sweating" of the suction gas stream. Liquid slugs are particularly hard on valves. Because of their high bulk modulus slugs of liquid can break the valve seats. The presence of liquid in the gas stream can dilute the lubricants used inside the compressor, causing accelerated wear on mechanical components.

Moisture traps should be used in the piping to avoid liquid damage. They should be periodically drained, together with the interstage separators. Piping should be designed such that there are no low spots where liquid may collect, and a sight glass or similiar devices can be used for visual inspection.

3. *Coking and carbon formation.* Excessive heat can cause coking of the lubricating oil and polymerization of gases containing unsaturates. These contaminants can reduce compressor efficiency and cause serious valve problems. A spray of steam or a suitable solvent in the form of a mist may be injected periodically into the compressor to clean the cylinders and the valves. Again, consideration should be given to dibasic ester-type synthetic lubricants which greatly resist coking and have very low coefficients of friction.

4. *Corrosive chemicals.* Corrosion can cause leakage at valve seating surfaces and can weaken the valve body to cause valve failures. It can be minimized or eliminated by using different valve materials or by using chemical scrubbers to remove or reduce the corrosive agents before they enter the compressor. Also, removing the moisture or heating up the gas before it enters the compressor can reduce or eliminate the formation of corrosive agents such as acids.

Corrosion takes place gradually, usually over several months. Upon failure of a valve, all the valves in the machine should be checked and repaired. Generally, it is good practice to keep a complete set of spare valves in stock in order to minimize downtime.

Abnormal actions of valve elements. These include slamming, fluttering of the valve elements, disturbances caused by the gas flow pattern, and the effects of resonance and pressure pulsations in suction and discharge pipes.

During normal compressor operation, the discharge valves are more susceptible to *slamming* during opening when the piston is pushing the gas out of the cylinder. This can be reduced by using either damping springs and/or pneumatic cushioning to slow down the valve elements as they move off the seat toward the guard. In other situations, slamming can also occur to compressor valves upon closing.

When opening and closing a valve, the inertia of the valve element has to be overcome. If the inertia of the valve element is high, the finite time required for it to open or close is comparatively longer than required for one with a lower inertia. With high inertia, the valve will close late and the backflow of gas, instead of the valve springs, will push the valve closed, causing the valve elements to be slammed against the seat. Slamming can usually be identified by a mottled appearance where the valve element contacts the seat. It can frequently be heard outside the compressor as clattering.

This problem can be solved by reducing the lift and/or increasing the spring rate to regulate the valve timing; or, where conditions allow, by changing the valve element to a lighter-weight material. A phenolic or fiber-reinforced plastic valve element is generally much lighter than a similar steel unit and, by its nature, more shock absorbent. However, these materials are limited by temperature and strength in their applications.

Valve fluttering occurs when the flow of gas passing through the valve is insufficient to lift the valve element fully to the guard containing the valve springs. The valve element will then oscillate between the seat and the guard. This can accelerate spring wear, and if the valve element strikes the seat and guard several times, the wear and fatigue of it is accelerated. Also, valve slamming is likely to occur when the valve is closing. Fluttering is generally identifiable when no definite pattern or impact trace of the valve springs exists on the valve elements, or when the valve elements appear to have spun. Fluttering may be eliminated by lowering the lift and/or by using lighter springs.

The flow pattern into and out of the compressor cylinder can be affected by obstructions in the compressor gas passages, and sometimes by the orientation of the valves inside the compressor. These problems are rare, but they are also less apparent and more difficult to troubleshoot.

A problem related to the flow pattern is the vibration of valve unloaders on the suction valves, activated by the drafting effect of the gas flowing into the cylinder. This is usually caused by unloader parts protruding across the direction of the gas stream, causing it to be pulled into vibration. The vibration may eventually cause some of the unloader fingers to break off, and leakage can occur in the suction valves due to unseated valve elements.

The solution is to use, where possible, a stronger retention spring on the unloader and/or to redesign the parts of the unloader protruding into the gas flow.

Resonance and pressure pulsations in the compressor gas passages can affect the timing in opening and closing the compressor valves. These problems can be corrected by changing the piping of the compressor and by using settling chambers, surge bottles, and pulse dampers. The application of surge bottles and pulse dampers is a science in itself and is beyond the scope of this text; however, their purpose is simply to prevent pressure pulses from traveling back into the compressor cylinder, thus unseating the valves and affecting the compressor. Sometimes a properly sized orifice installed in the compressor piping is all that is needed to cure the problem. The following section will describe suitable approaches to troubleshooting piping-system-related resonance and pressure pulsations.

General troubleshooting. Very few pieces of petrochemical process machinery lend themselves to early symptom-cause identification as do reciprocating compressors. This becomes evident when considering how the neglect of subtle performance changes can result in costly wrecks. For instance, high discharge temperatures can result from the simple primary cause of insufficient cooling water. Not responding to this cause will lead to an overheated cylinder and ultimately to such final events as piston seizure, ring breakage, and piston cracking. Similarly, not responding to the audible symptom of "knocking in cylinder," where the most probable cause is incorrect piston-to-head clearance will ultimately result in piston failure or rod breakage with attendant damage to the crosshead. A detailed analysis of the symptom-cause-failure chain would be in order. Table 4-16 will serve this purpose.

When troubleshooting reciprocating compressors, the most important symptoms to watch for are unusual sounds and changes in pressures, temperatures, and flowrates. Consequently, the primary troubleshooting tools are our five senses, two pressure gauges, two temperature indicators, and a flowmeter. Generally, flowmeters are not available for each individual stage of compression, but considering that what goes in the front end comes out the back end, flow measurement at one stage along the way is sufficient in cases where no interstage inlets or knockouts are involved.

Compressor temperatures and pressures are basic to design calculations and help determine compressor health. The difference between the observed and calculated temperatures, ΔT, should be more or less constant from day to day. The actual observed or calculated temperature may vary; when suction temperature increases, so does discharge temperature. If the compression ratio across a cylinder increases, its discharge temperature also increases. Comparing calculated and actual discharge temperatures provides a yardstick with which to determine operating deviations.

(text continued on page 257)

Table 4-16
Troubleshooting Guide—Reciprocating Compressor

Group	Possible Causes	Knocking	Vibration	Discharge Pressure Up	Discharge Pressure Down	Inter-Cooler Press. Up	Inter-Cooler Press. Down	Discharge Temp. Up	Outlet Cooling Water Temp. Up	Overheating Valves	Overheating Cylinder	Overheating Frame	Capacity Down	Carbon Deposits Abnormal	Piston Rings Cyl. Wear Up	Valve Wear/Breakage Up
Valves	L.P. Valves Wear/Breakage				②		①	①	③	③	②		①	④		⑤
Valves	H.P. Valves Wear/Breakage				①	①										
Valves	L.P. Unloading System Defective		③				②	②	④	④	⑧		②	⑤		⑦
Valves	H.P. Unloading System Defective					②										
Pistons/Cylinders	L.P. Piston Rings Worn				④		⑤						⑦	⑨	⑥	
Pistons/Cylinders	H.P. Piston Rings Worn					③										
Pistons/Cylinders	Piston Rod Nut Loose	④														
Pistons/Cylinders	Piston Loose	⑥														
Pistons/Cylinders	Head Clearance Too Small	②														

Column groupings (under "Symptoms"): Noise Vibration (Knocking, Vibration); Pressure (Discharge Pressure Up/Down, Inter-Cooler Press. Up/Down); Temperature (Discharge Temp. Up, Outlet Cooling Water Temp. Up, Overheating: Valves/Cylinder/Frame); Flow (Capacity Down); Int. Inspection Result (Carbon Deposits Abnormal, Piston Rings Cyl. Wear Up, Valve Wear/Breakage Up).

Category	Cause													
Frame	Bearing Clearance Too High													⑤
	Flywheel Or Pulley Loose												②	⑦
	Crosshead Clearance Too High													③
Support/Cooling/Lubrication	Cooling Water Quty. Too Low						④	①	④					④
	Cylinder Lubrication Inadequate	①	①				⑥		⑦				⑥	⑨
	Frame Lubrication Inadequate					①								①
	Cylinder Lubrication Excessive	⑧		②									②	⑧
	Lubricating Oil Incorrect Spec.	②	②	①			⑦		⑧					⑩
	Foundation/Grouting Inadequate												④	⑧
	Piping Support Inadequate												①	
Piping System	Resonant Pulsations (Suction or Discharge)	⑨			③						③			⑨
	Suction Filter Dirty/Defective	⑤	⑤		③					③				
	Suction Line Restricted				④					④				

(table continued on next page)

Note: The numbers indicate what to check first, or the probability ranking.

Table 4-16 (cont.)

Possible Causes	Noise Vibration: Knocking	Vibration	Pressure: Discharge Pressure Up	Discharge Pressure Down	Inter-Cooler Press. Up	Inter-Cooler Press. Down	Discharge Temp. Up	Temperature: Outlet Cooling Water Temp. Up	Overheating: Valves	Cylinder	Frame	Flow: Capacity Down	Int. Inspection Result: Carbon Deposits Abnormal	Piston Rings Cyl. Wear Up	Valve Wear/Breakage Up
Piping System															
System Leakage Excessive				③								⑤			
System Demand Exceeds Compressor Capacity				⑤											
Operating Conditions															
Discharge Pressure Too High	⑪	⑦	②				③	⑤	①	①	③		⑥		
Discharge Temperature Too High													⑦		
Intercooler Fouled					④		⑥	⑥					⑪		
Liquid Carry-Over														③	③
Dirt/Corrosion Products Into Cyl.														④	④
Cylinder Cooling Jackets Fouled							⑤	②		⑤			⑩		
Running Unloaded Too Long									②						
Speed Incorrect		⑤		⑥						③	②	⑥	⑧		

NOTE: The numbers indicate what to check first, or the probability ranking.

Calculations help. The procedure for analyzing reciprocating compressors is to measure the temperatures and pressures across each cylinder when the unit is known to be in good operating condition. The discharge temperature is calculated from the compression ratio, suction temperature, and k-value of the gas. The k-value, a physical property of the gas, is the ratio of the specific heats. Figure 4-11 shows the necessary equations and k-values for some of the more common process gases.

The observed and calculated discharge temperatures are seldom exactly the same. There might be variations in pressure and temperature indicator readings as well as in gas properties. However, results should be comparable and differences between actual and calculated discharge temperatures reasonably constant from day to day if the compressor is operating properly. When the temperature difference increases, a problem with the compressor valves or piston rings can be suspected (see Table 4-16).

Discharge temperature is the most powerful tool in analyzing the unit for potential problems. The troubleshooter should be able to rely on a record of observed discharge temperatures and pressures when the valves and piston rings are known to be in good, but not necessarily new, condition.

Readings must be made with reliable gauges. With the information recorded, it is not necessary to calculate discharge temperature every day, only to have the raw data available when a problem is suspected. One should then go back and analyze the actual and calculated conditions.

Diagnostics. Other tools for diagnosing reciprocating compressor troubles are lube-oil analyses for metal content, vibration readings, and Beta analyzer-type diagnostics. In lube-oil analyses, particular emphasis is on finding the metals that are

---

$$T_2 = T_1 \cdot \left(\frac{P_2}{P_1}\right)^{\frac{K-1}{K}}$$

where for a single cylinder:

T_1 = Inlet Temperature, °Rankine (°F + 460)
T_2 = Outlet Temperature, °R
P_2 = Outlet Pressure, psia
P_1 = Inlet Pressure, psia

K = Ratio of Specific Heats, $MC_p/MC_v = \dfrac{MC_v - 1.986}{MC_v}$

K = 1.4 for a perfect diatomic gas

K-Values of Common Gases
Air, N_2, O_2, H_2 ~1.4
CO_2 1.28
Propane   1.15
Natural Gas   Ranges from 1.1 to 1.25

Figure 4-11. Discharge temperature calculations for reciprocating compressors.

in the bearings and running gear. Increases in metals contents from the previous analysis indicate the beginning of wear of these parts; we can frequently predict the point when wear becomes critical.

Vibration analysis is usually applied to rotating equipment such as centrifugal compressors. It can also be useful for reciprocating compressors. Vibration analysis can alert us to coupling misalignment, even with relatively low-speed reciprocating compressors. Some troubleshooters have used vibration measurements in the analysis of piping fatigue failures and fractures caused by inherent system resonance and pulsations. Some of these problems are caused by first or second orders of running speed, exciting resonant frequencies in pipe runs. Usually, if the vibration frequency is higher than twice the running speed of the machine, acoustic pulsations are suspected.

Beta analyzer indicator card readings allow us to directly recognize what is happening inside a compressor cylinder. This method of analysis, further explained in Chapter 5, can be extremely helpful in solving problems related to valve losses and piston-ring leakages. Such analytical equipment can be expensive, but there are companies supplying this service on either a contract or a one-time basis. However, we must alway remember that from a diagnostic point of view, a one-time reading is no substitute for recorded historical operating data. After all, the most decisive feature of a trouble symptom is its upside/downside *change*.

This concludes our discussion of driven process machinery troubleshooting. As an introduction to our discussion of drivers we include Table 4-17, a troubleshooting guide for V-belt drives. Coupling failure analysis and troubleshooting was discussed earlier in Chapter 3.

Table 4-17
Troubleshooting Guide—V-Belt Drive

	Symptoms										Symptoms
	C Repeated Belt Fracture						Rapid Belt Wear D				
	B Bearings Overheating							Checked Or Cracked Belts E			
	A Belt Squeal								Belts Turn Over in Sheave F		
	Possible Causes	**#**	**A**	**B**	**C**		**D**	**E**	**F**	**#**	**Possible Remedies**
Wear and Tear	Rubbing Belt Guard	1					1			1	* Check clearance
	Sheave Diameter too Small	2					3			2	* Check sheave side walls
	Mismatched Belts	3					5			3	* Replace with matched setor banded belts
	Slippage	4	3	2			6	1		4	* Check tension
	Replacing One Belt Only in Multibelt Drive	5					8			5	* Replace as complete set
	Misalignment	6					10			6	* Realign
	Possible Causes	**#**	**A**	**B**	**C**		**D**	**E**	**F**	**#**	**Possible Remedies**

Table 4-17 (cont.)

Symptoms						Symptoms					
C Repeated Belt Fracture						Rapid Belt Wear D					
B Bearings Overheating						Checked Or Cracked Belts E					
A Belt Squeal						Belts Turn Over in Sheave F					
	Possible Causes	**#**	**A**	**B**	**C**	**D**	**E**	**F**	**#**	**Possible Remedies**	
Wear and Tear	Insufficient Contact Arc	7	2						7	* Increase center distance	
Wear and Tear	Belt Bottoming In Grooves	8	4						8	* Replace sheaves or belt	
Wear and Tear	Small Sheaves	9		4			3		9	* Redesign or use cog belt	
Design/Installation	Backward Bending	10					4		10	* Use cog belt	
Design/Installation	Overtensioned Drive	11		1					11	* Recheck tension	
Design/Installation	Sheaves Too Far From Bearing	12		3					12	* Move sheaves closer to bearing block	
Design/Installation	Sheaves Too Small	13		4			3		13	* Redesign drive	
Design/Installation	Improper Installation	14			2		11	5	14	* Check if belt pried over sheave	
Design/Installation	Misplaced Slack	15			3				15	* Keep slack on one side	
Design/Installation	Extended Or Improper Storage	16				7			16	* Replace	
Operating Conditions	Overloads	17	1			4		2	17	* Redesign drive	
Operating Conditions	Overheating	18				9			18	* Improve operating environment * Use cog belt	
Operating Conditions	Shock Loads Or Vibration	19			1			3	19	* Use banded belts	
Operating Conditions	Foreign Materials In Grooves	20			4			4	20	* Improve belt shielding	
Operating Conditions	High Ambient Temperatures	21					2		21	* Provide adequate ventilation * Use cog belts	
Operating Conditions	Worn Sheave Grooves	22				2			22	* Check Sheave side walls	
Operating Conditions	Broken Tensile Cords	23						1	23	* Check if belt was pried over sheave	
	Possible Causes	**#**	**A**	**B**	**C**	**D**	**E**	**F**	**#**	**Possible Remedies**	

Note: The numbers indicate what to check first, or the probability ranking.

Troubleshooting Engines

Internal combustion engines, such as large four- or two-cycle integral gas engine compressors, gas and diesel engine drivers, also belong to the diverse population of petrochemical process plant machinery. These services range as high as 16,000 BHP with brake-mean-effective-pressure ratings of up to 250 psi for 20-cylinder engines. Usually, this equipment consists of high-performance, finely tuned machines that must be installed with extreme care and maintained equally thoroughly. A typical duty for an integral gas engine compressor would be hydrogen gas compression service using refinery off-gas as fuel. Other typical applications are diesel engines as drivers for cooling tower pumps and emergency generators.

During the years 1971-1974, the Hartford Steam Boiler Inspection and Insurance Company compiled the failure statistics shown in Table 4-18. The equipment covered consisted of a population of 180 engines, including diesel and dual-fuel engines, spark-ignited gas and gasoline engines, and internal combustion engines driving reciprocating compressors. Table 4-19 shows a breakdown of failures by cause and of parts initially failed.

Earlier we alluded to the fact that large internal combustion engines, along with other reciprocating machinery, usually respond well to monitoring and on-line diagnostics. Here, as in the case of reciprocating compressors, final dramatic failures will result because subtle initial distress symptoms were either not recognized or not acted upon. Consider bearing failures, which head the list in both the number of engine failures and cost. Again, just the "middle" link in a chain of symptoms and causes often beginning with lubrication deficiency (see Chapter 3) and ending with a crankshaft failure. Consequently, paying attention to bearing wear and failure symptoms will eliminate a major portion of crankshaft failure causes.

Engine bearing wear or damage, although directly related to crankshaft alignment, is also the result of many other factors. Table 4-20 represents an engine troubleshooting chart that shows the following causes for bearing wear and failure symptoms as itemized on page 266.

(text continued on page 266)

Table 4-18
Distribution of Large Internal Combustion Engine Failures and their Cost by Failure Modes[14]

Failure Mode	(%)	Distribution $-Value of Total (%)
Breaking/tearing	48.3	46.7
Cracking	26.1	19.4
Deforming	1.1	0.4
Scoring	5.6	8.2
All others	18.9	25.3
Totals	100.0	100.0

Table 4-19
Distribution of Large Internal Combustion Engine Failures and their
Cost by Cause and Parts Failed First[14]

Failure Cause	Distribution Incidents (%)	$—Value of Total (%)
Cause undetermined	28.9	17.2
Lubrication problems	12.8	17.1
Cooling problems (loss of, insufficient, inadequate)	10.6	6.1
Fatigue	10.0	5.2
Overheating (except from overload)	5.6	3.5
Material, defective	5.6	3.5
Misalignment	2.2	13.9
Overspeeding	2.2	8.4
Accumulation of liquid in cyl.	2.2	2.4
Cracking, progressive	2.2	1.1
Not otherwise classified	2.2	1.0
Vibration	1.7	6.3
Maintenance, inadequate	1.7	1.3
Properties, change in metallurgical	1.7	0.6
Wear (age or service)	1.7	0.4
Repair, improper	1.1	1.4
Device, loosening of locking	1.1	0.6
Workmanship, poor (owner)	1.1	0.6
Freezing, low ambient	1.1	0.1
All others	4.3	9.3
	100.0	100.0

Initial Part to Fail		
Bearings, all	24.4	33.5
Piston, piston rings	19.4	17.3
Cylinder, head, block, liner	16.7	8.3
Crankshaft	6.1	24.2
Valves	5.6	3.2
Rod, connecting, piston	4.4	3.3
Manifold	4.4	2.2
Lube-oil system	2.2	0.8
Gears	2.2	0.6
Cam shaft	1.7	0.6
Frame, cast	1.7	0.4
Couplings	1.7	0.2
Turbocharger rotor	1.1	1.1
Control, pressure, temperature	1.1	0.3
All others	7.3	4.1
	100.0	100.0

Table 4-20
Engine Troubleshooting Guide

Symptoms		Valve Angle—Incorrect	Adjustment—Incorrect	Metallurgy—Incorrect	Ring Size—Incorrect	Ring Groove—Worn	Oil Ctrl. Ring—Worn	Oil Ctrl. Slot—Plugged	Side Clearance—Excessive	Piston/Ring Design Inadequate
	Possible Cause	*Valves*			*Piston/Rings*					
Monitoring/Surveillance Results	Hammering/Knocking									
	Pre-Ignition					8				
	Detonation		7							
	Misfiring									
	Overheating	14	15		16					
	Soot Formation (Exhaust)		3							
	Valve Leaking	2	5	3						
	Port Plugging (*)									
	Piston Blow-By				7	9		10		
	Bearing Wear									
	High Oil Consumption						1			
	Short Oil Life					12	13			14
	Short Filter Life					10				
	High Maintenance									
	Power Loss				2					
Internal Inspection Results	Varnish/Sludge									7
	Ash Deposits	11			2					3
	Carbon Deposits									
	Spark Plug Fouling		5		6			11		7
	Piston Ring Sticking				7	9		10	5	
	Cylinder/Liner Wear						17			6
	Valve Failure	2	5	3						
	Bearing Failure									
	Piston Seizure				3					4
	Cylinder Hd. Cracking									3
	Turbocharger Failure				2					

	Symptom																		
Cylinder	Liner Distortion			3			17		8							8	3		8
	Surface Finish Inadequate																2		
	Metallurgy—Incorrect																4		
	Break-In Procedure—Incorrect																1	1	
Ignition System	Timing—Incorrect		2	3	1	11	4	4		3		4		2	8	4	1	2
	Ignition Elements—Faulty				2	9					5		3			6		
	Spark Plug Gap—Incorrect				3						6		4					
Air/Fuel System	Compression Ratio—Too High	5	6		4	5	6							11		6		
	Air/Fuel Ratio—Incorrect	9	10		6	7	1	9	4	1	8		9	12				
	Intake Air Temp.—Too High	3	4			4	5	5			3			9		5		
	Filtration—Inadequate			3				9	15	2	12			15	8	9		
	Intake Air Restriction					12						2						
	Scavenging—Inadequate	8	9	3	5	6		3	5	6	7	2	8					
	Fuel Quality—Inadequate				7	3		4	4	9				7				
	Fuel—Wet	1	2		7	8		2	13	4						2		
	Fuel—Unstable HC		1					3								3		

(*) Two-Cycle Engine

(table continued on next page)

Table 4-20 (cont.)

Symptoms — Internal Inspection Results / Monitoring/Surveillance Results

System	Possible Cause	Turbocharger Failure	Cylinder Hd. Cracking	Piston Seizure	Bearing Failure	Valve Failure	Cylinder/Liner Wear	Piston Ring Sticking	Spark Plug Fouling	Carbon Deposits	Ash Deposits	Varnish/Sludge	Power Loss	High Maintenance	Short Filter Life	Short Oil Life	High Oil Consumption	Bearing Wear	Piston Blow-By	Port Plugging (*)	Valve Leaking	Soot Formation (Exhaust)	Overheating	Misfiring	Detonation	Pre-Ignition	Hammering/Knocking
Cooling System	Cylinder Temp.—Too Low		1	2		1	10	6			10				9	3	4		6	4	1	2	2		4	5	
Cooling System	Cooling—Inadequate																										
Cooling System	Cooling Water Temp. Too Low														11	7											
Cooling System	Coolant Leakage Into Lube Oil				14							5				6		14									
Lube Oil System	Oil Quality—Inadequate				15	6		5	10		9	2		1	1	1	5		5	6	6						
Lube Oil System	Oil Viscosity—Incorrect				13	7	14											13					13				
Lube Oil System	Oil Filtration—Inadequate				8		16	2				8		2	8	16		8	2	1	7						
Lube Oil System	Overlubrication	1						1	1	1	1			3	6		2		1						6	7	

Table 4-20 (cont.)

Possible Cause	Hammering/Knocking	Pre-Ignition	Detonation	Misfiring	Overheating	Soot Formation (Exhaust)	Valve Leaking	Port Plugging (*)	Piston Blow-By	Bearing Wear	High Oil Consumption	Short Oil Life	Short Filter Life	High Maintenance	Power Loss	Varnish/Sludge	Ash Deposits	Carbon Deposits	Spark Plug Fouling	Piston Ring Sticking	Cylinder/Liner Wear	Valve Failure	Bearing Failure	Piston Seizure	Cylinder Hd. Cracking	Turbocharger Failure
Lube Oil System																										
Lack of Lubrication							4			12											18	4	12			
Lubricator Failure					1			5				2				1										3
Oil Drain Interval Excessive									4				7			6				4						
Contamination												10	3			4										
Oil Temp. Too Low												8														
Misc.																										
Metallurgy—Incorrect										10													10			
Crankshaft Alignment/Brg. Fit.										1													1			
Loose Parts	1																									
Overload					10			2		7			12		5						13		7			

Symptoms	
Monitoring/Surveillance Results	Internal Inspection Results

(*) Two-Cycle Engine

☐1☐ Improper crankshaft alignment and bearing fit. Some remedial actions are:
  a. Inspect and adjust alignment and bearing clearances, i.e. "crush."
  b. Take crankshaft web deflection readings periodically for early detection of misalignment problems.
  c. Inspect foundation for any movement between engine base and grout.
  d. Correct a loose-grout problem as soon as possible. Use an oil-resistant polymeric grout.
  e. Ascertain that foundation bolt torque values are adequate.

☐2-7☐ ☐9☐ ☐11☐   Fuel quality: overload/unsteady, rough-running-related causes.

☐8☐ Inadequate oil filtration. Some remedial actions are:
  a. Obtain periodic lab analysis of crankcase oil to check for dilution and oxidation.
    • Lube-oil dilution on diesel engines, for instance, results from:
    • High-pressure tube fittings leaking between fuel-injection pumps and spray nozzles.
    • Clogged camshaft trough drains.
    • Wear of fuel-injection pump pushrod bushings.
    • Loose pushrod in the fuel injector.
    • Incorrectly tightened fittings on injection pumps and fuel return piping.
  b. Review procedure for keeping duplex oil strainers clean and available for changeover.

☐10☐ Incorrect metallurgy. Refer to Chapter 3.

☐12☐ Lack of lubrication. Remedial actions could be:
  a. Inspect oil pump drive.
  b. Monitor oil supply pressure.

☐13☐ Incorrect oil viscosity. Remedial actions could be:
  a. Review manufacturer's recommendations.
  b. Obtain periodic lab analysis.

☐14☐ Coolant leakage into lube oil. Remedial action is shown in 8a.

The foregoing will have given the reader an impression of the complexity of an engine troubleshooting system. We recommend, therefore, that Table 4-19 be referred to as a first guide only.

Troubleshooting Steam Turbines

Failure statistics. Steam turbines in the petrochemical industry range from smaller than 10 hp, stand-by pump drives, to larger than 60,000 hp, unspared process gas compressor drives. Table 4-21 shows a representative statistic of failure cases and their major causes and affected components.

Similarly, Table 4-22 shows failure-mode distributions from the same source. Accordingly, rubbing through axial and radial contact can be caused by:

1. Thermal distortion of rotor and casing.
2. Alignment changes—external and internal.
3. Unbalance.

Table 4-21
Distribution of Steam Turbine Failure
Incidents by Cause and Affected Components[15]

Failure Causes	Distribution of Incidents (%)	Components	Distribution of Incidents (%)
Vendor Problems	64.1	Rotor blades	29.0
Planning, design, and calculation	16.5	Bearings Radial bearings (12.5) Thrust bearings (4.2)	16.7
Assembly	16.0		
Technology	10.6	Shaft seals, balance pistons	15.6
Manufacturing	8.7	Rotors with discs	10.3
Material	8.0	Casings with baseplates, screws	9.8
Repair	4.3	Strainers, fittings	4.0
Operational Problems	15.3	Control	4.0
Surveillance	10.6	Guide blades and diaphragms	3.4
Maintenance	4.7	Gears, transmissions	2.4
External Influences	20.6	Pipelines	0.8
Foreign bodies	7.2	Other parts	4.0
From the electrical grid	4.1		
Others	9.3		
	100		100

Table 4-22
Steam Turbine Failure Mode Distribution[15]

Failure Mode	(%) of Incidents
Rubbing	23.0
Fatigue and creep failure	18.5
Damage to bearings	14.6
Thermal stress cracking	11.7
Sudden failure	9.3
Incipient cracks	8.0
Mechanical surface damage	5.4
Corrosion and erosion	3.3
Shaft bending	2.4
Wear	2.3
Abrasion	1.5
	100

4. Damage to axial or radial bearings.
5. Off-design condition leading to thermally-induced dimensional changes.

Rotor blade failures can be caused by:

1. Dynamic overloading.
2. Loosening of blade seating.
3. Seating design deficiencies.

Bearing failures are discussed in Chapter 3. Thermal stress cracking occurs primarily in cast-steel components in the region of highest temperatures and at points of changes in wall thickness. Thermal stress cracking of turbine rotors is a rare occurrence. Other components affected by thermal stress cracking are trip and throttle valve casings, steam chests, and the wheel-chamber sections of high-pressure casings.

Table 4-23 represents a troubleshooting guide for general-purpose steam turbine drivers. Table 4-24 is a troubleshooting guide for a commonly used hydraulic governor, the TG-10 governor made by Woodward Governor Co. Finally, Figure 4-12 is a stair step-type diagnostic guide for governor-system troubles on larger steam turbines. *(text continued on page 275)*

Table 4-23
Troubleshooting Guide—Steam Turbines

	Symptoms						Symptoms			
D	**Speed Increases as Load Decreases**				**Governor Not Operating / Excessive Speed Variation**			**E**		
C	**Insufficient Power**				**O.S.T.† on Load Changes**			**F**		
B	**Slow Start-Up**				**O.S.T. at Normal Speed**		**G**			
A	**Turbine Fails to Start**				**Leaking Glands**			**H**		

	Possible Causes	#	A	B	C	D	E	F	G	H	#	**Possible Remedies**
Casing/Rotor System	Too Many Hand Valves Closed.	1			2		9	5			1	* Open additional hand valves.
	Nozzles Plugged Or Eroded	2		6	7						2	* Remove nozzle pipe plugs and hand valves. Inspect nozzle holes. Clean as required.
	Dirt Under Carbon Rings	3								1	3	* Steam leaking under carbon rings may carry scale or dirt which will foul rings. Remove, inspect, replace
	Worn or Broken Carbon Rings	4								2	4	* Replace with new rings.
	Possible Causes	#	A	B	C	D	E	F	G	H	#	**Possible Remedies**

†Overspeed trip.

Table 4-23 (cont.)

	Possible Causes	#	A	B	C	D	E	F	G	H	#	Possible Remedies
			D	Speed Increases as Load Decreases				Governor Not Operating / Excessive Speed Variation		**E**		
			C	Insufficient Power				O.S.T.† on Load Changes		**F**		
			B	Slow Start-Up				O.S.T. at Normal Speed		**G**		
			A	Turbine Fails to Start				Leaking Glands		**H**		
		#	A	B	C	D	E	F	G	H	#	
Trip and Throttle Valve	Shaft Scored	5								3	5	* Shaft surface under carbon rings must be smooth to prevent leakage. Observe proper run-in procedures.
	Leak-Off Pipe Plugged	6								4	6	* Make sure all condensate is draining.
	Throttle Valve Travel Restricted	7	1	1	1	3	2				7	* Close the main admission valve and disconnect throttle linkage. Valve lever should move freely from full open to full closed.
	Throttle Assembly Friction	8	4		2	3	3				8	* Disassemble throttle valve. Inspect for freedom & smoothness of movement of all parts. * Inspect valve stem for straightness and for build-up of foreign material.
	Valve Packing Friction	9			3	4	4				9	* Inspect valve packing for excessive compression. If it is compressed to the point where "drag" on valve stem exists, replace.
	Throttle Valve Loosening	10				5					10	* Some "floating" valve assemblies have critical end play. Replace or repair valve & stem as necessary.
	Possible Causes	#	A	B	C	D	E	F	G	H	#	**Possible Remedies**

†Overspeed trip.

(table continued on next page)

Table 4-23 (cont.)

	Symptoms				Symptoms					
D	**Speed Increases as Load Decreases**				**Governor Not Operating / Excessive Speed Variation**					**E**
C	**Insufficient Power**				**O.S.T.† on Load Changes**					**F**
B	**Slow Start-Up**					**O.S.T. at Normal Speed**				**G**
A	**Turbine Fails to Start**						**Leaking Glands**			**H**
Possible Causes	**#**	**A**	**B**	**C**	**D**	**E**	**F**	**G**	**H**	**#**	**Possible Remedies**
---	---	---	---	---	---	---	---	---	---	---	---
Throttle Valve & Seats Cut or Worn	11				4	5				11	* Remove valve assembly & check valve & seats for wear & steam cutting.
Trip Valve Set Too Close to Operating Speed	12						1	3		12	* Consult operating manual for proper trip-speed setting.
Trip Valve Does Not Open Properly	13	3		6						13	* Assure trip levers are properly latched.
Dirty Trip Valve	14							2		14	* Inspect & clean.
Steam Strainer Plugging	15	2	2	8						15	* Inspect & clean from all foreign matter.
Oil Relay Governor Set Too Low	16			3						16	* Refer to operating manual for speed adjustment & speed range limits.
Governor Droop Adjustment Needed	17					1				17	* An increase in internal droop setting will reduce variation or hunting.
No Governor Control At Start-Up	18		5			8				18	* Check for proper direction of rotation. * Check for proper governor speed range. * Consult governor trouble-shooting guide
Governor Lubrication Problem	19					2				19	* Low governor oil level, dirty, contaminated, or foamy oil, will cause poor governor response. * Drain, flush, and refill with proper spec. oil.
Possible Causes	**#**	**A**	**B**	**C**	**D**	**E**	**F**	**G**	**H**	**#**	**Possible Remedies**

†Overspeed trip.

Table 4-23 (cont.)

Symptoms					Symptoms				
D Speed Increases as Load Decreases					**Governor Not Operating / Excessive Speed Variation** **E**				
C Insufficient Power					**O.S.T.†** on Load Changes **F**				
B Slow Start-Up					**O.S.T. at Normal Speed** **G**				
A Turbine Fails to Start					**Leaking Glands** **H**				
Possible Causes	**#**	**A**	**B**	**C**	**D**	**E**	**F**	**G**	**H**	**#**	**Possible Remedies**

Possible Causes	#	A	B	C	D	E	F	G	H	#	Possible Remedies
Inlet Steam Pressure Too Low or Exhaust Pressure Too High	20	1	4	5						20	* Use accurate gauges to check steam pressure at turbine inlet and the exhaust pressure close to the exhaust flange. * Low inlet pressure may be the result of auxiliary control equipment being too small, improper pipe sizing, excessive piping lengths, etc.
Light Load & High Inlet Steam Pressure	21					6	6			21	* The tendency for excessive speed variation is quite high when reserve capacity is available and steam pressure is high. * Open hand valves or install smaller throttle valve & body.
Load Higher Than Turbine Rating	22		3	4						22	* Ascertain the actual load requirements of the driven equipment.
Rapidly Changing Load	23					7				23	* Rapidly changing load can cause governor hunting. * Check turbine application.
High Starting Torque Of Driven Equipment	24									24	*Check the required starting torque. * Ascertain turbine is not overloaded.
Excessive vibration	25							1		25	* See vibration diagnostics and note below.
Possible Causes	**#**	**A**	**B**	**C**	**D**	**E**	**F**	**G**	**H**	**#**	**Possible Remedies**

Note: For vibration & short bearing life-related causes consult special sections on bearing failure analysis and vibration diagnostics.

†Overspeed trip.

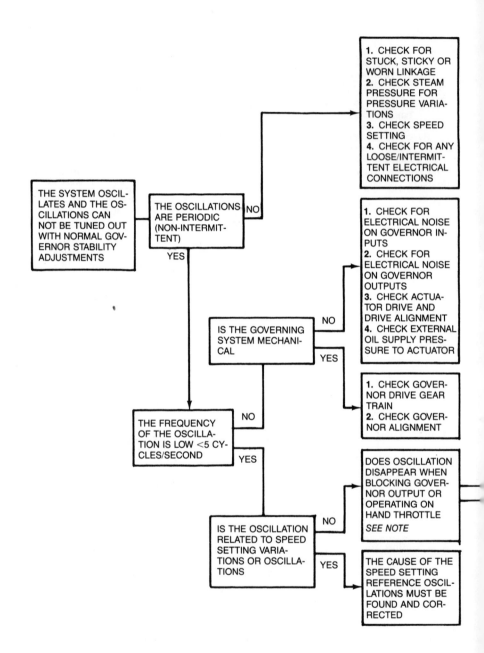

Figure 4-12. Steam-turbine governor troubleshooting guide.

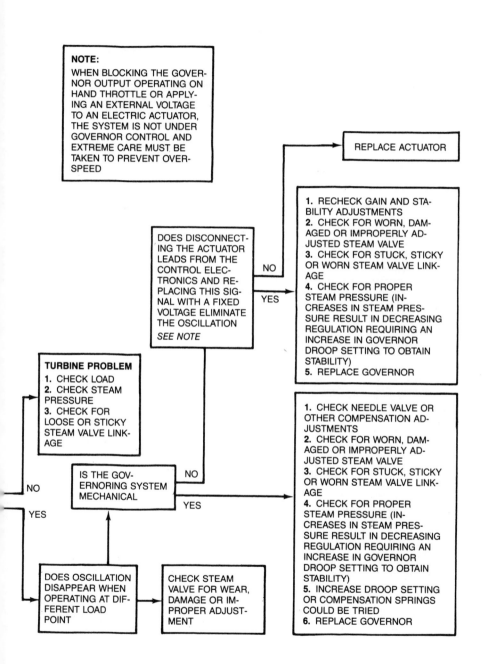

NOTE:
WHEN BLOCKING THE GOVERNOR OUTPUT OPERATING ON HAND THROTTLE OR APPLYING AN EXTERNAL VOLTAGE TO AN ELECTRIC ACTUATOR, THE SYSTEM IS NOT UNDER GOVERNOR CONTROL AND EXTREME CARE MUST BE TAKEN TO PREVENT OVERSPEED

REPLACE ACTUATOR

DOES DISCONNECTING THE ACTUATOR LEADS FROM THE CONTROL ELECTRONICS AND REPLACING THIS SIGNAL WITH A FIXED VOLTAGE ELIMINATE THE OSCILLATION

SEE NOTE

NO

YES

1. RECHECK GAIN AND STABILITY ADJUSTMENTS
2. CHECK FOR WORN, DAMAGED OR IMPROPERLY ADJUSTED STEAM VALVE
3. CHECK FOR STUCK, STICKY OR WORN STEAM VALVE LINKAGE
4. CHECK FOR PROPER STEAM PRESSURE (INCREASES IN STEAM PRESSURE RESULT IN DECREASING REGULATION REQUIRING AN INCREASE IN GOVERNOR DROOP SETTING TO OBTAIN STABILITY)
5. REPLACE GOVERNOR

TURBINE PROBLEM
1. CHECK LOAD
2. CHECK STEAM PRESSURE
3. CHECK FOR LOOSE OR STICKY STEAM VALVE LINKAGE

IS THE GOVERNORING SYSTEM MECHANICAL

NO

YES

NO

YES

1. CHECK NEEDLE VALVE OR OTHER COMPENSATION ADJUSTMENTS
2. CHECK FOR WORN, DAMAGED OR IMPROPERLY ADJUSTED STEAM VALVE
3. CHECK FOR STUCK, STICKY OR WORN STEAM VALVE LINKAGE
4. CHECK FOR PROPER STEAM PRESSURE (INCREASES IN STEAM PRESSURE RESULT IN DECREASING REGULATION REQUIRING AN INCREASE IN GOVERNOR DROOP SETTING TO OBTAIN STABILITY)
5. INCREASE DROOP SETTING OR COMPENSATION SPRINGS COULD BE TRIED
6. REPLACE GOVERNOR

DOES OSCILLATION DISAPPEAR WHEN OPERATING AT DIFFERENT LOAD POINT

CHECK STEAM VALVE FOR WEAR, DAMAGE OR IMPROPER ADJUSTMENT

Figure 4-12. Continued.

Table 4-24
Troubleshooting Guide—TG-10 Governor

	Symptoms				Symptoms				
	C Governor Unstable/Oscillations				**Engine/Turbine Cannot Obtain D**				
	B Governor Output Jiggles				**Rated Speed**				
	A Engine/Turbine Hunts				**Gov'nr. Does Not Start/Control E**				
	and Surges				**Gov'nr. Starts But Remains F**				
					at Maximum				
	Possible Causes	**#**	**A**	**B**	**C**	**D**	**E**	**F**	**#**	**Possible Remedies**
---	---	---	---	---	---	---	---	---	---	---
Governor Adjustment	Speed Setting Too Low	1				1			1	* Increase governor speed setting.
	Speed Setting Too High	2						1	2	* Reduce speed setting until governor controls, then adjust for desired speed.
	Insufficient Droop Adjustment	3	4						3	* Reposition droop adjusting lever to increase droop
	Too Much Droop	4			2				4	* Re-position the droop adjusting lever to increase droop.
	Governor Speed Range Is Incorrect For This Application	5				3			5	* Check speed range of governor.
Design/Installation/Maintenance	Wrong Governor Drive Rotation	6					1		6	* Check governor drive. * Reverse pump parts for different rotation.
	Worn Flyweight Pins	7	2						7	* Check flyweights and pins for smooth movement. * Replace both parts if one is defective.
	Pump Drive Pin Broken	8					3		8	* Disassemble pump housing to check pin & replace if required.
	Dirt/Contamination In Governor Oil	9	2		4				9	* Drain, flush, and refill with fresh oil of proper specification.
	Low Oil Level	10	1						10	* Add oil to level visible in the oil level indicator.
	Possible Causes	**#**	**A**	**B**	**C**	**D**	**E**	**F**	**#**	**Possible Remedies**

Table 4-24 (cont.)

	Symptoms				Symptoms				
C Governor Unstable/Oscillations **B** **Governor Output Jiggles** **A** **Engine/Turbine Hunts** **and Surges**					**Engine/Turbine Cannot Obtain** **D** **Rated Speed** **Gov'nr. Does Not Start/Control** **E** **Gov'nr. Starts But Remains** **F** **at Maximum**				
Possible Causes	**#**	**A**	**B**	**C**	**D**	**E**	**F**	**#**	**Possible Remedies**
Improper Alignment Of The Governor Coupling	11	1						11	* Check and realign as required
Drive Key Missing Or Improperly Installed	12				2			12	* Check drive installation
Binding Terminal Shaft Linkage	13	3						13	* Realign linkage as necessary.
Insufficient/Incorrect Terminal Shaft Travel	14			1	2			14	* Check linkage. Review recommended travel from no load to full load.*
High Steam Valve Gain	15			3				15	* Ascertain steam valve is not too large for this application.
Possible Causes	**#**	**A**	**B**	**C**	**D**	**E**	**F**	**#**	**Possible Remedies**

(Left margin label: Design/Installation/Maintenance)

Troubleshooting Gas Turbines

Although gas turbines are not as wide-spread in petrochemical plants as steam turbines, they have unique features that make them ideally suited as drivers for a number of refinery and chemical plant services. Gas turbines are able to operate on a wide variety of fuels, including low-grade fuels, and under severe environment conditions. Gas turbines have been found to experience more frequent failures than steam turbines. Figure 4-13 gives the distribution of primary failure causes for industrial gas turbines from 1970 to 1979. Figure 4-14 shows the component damage distribution for the same population of machines.

A typical gas turbine consists of three major systems: the compressor section, the combustor, and the turbine. For effective troubleshooting, a good understanding of these systems and their interaction is necessary. The following gas-turbine distress or failure modes must be understood for successful trouble diagnosis:

1. Air inlet system and filtration.
   a. Combustion air problems.
      1. Compressor fouling.

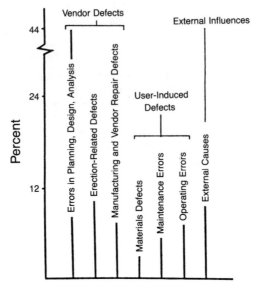

Figure 4-13. Distribution of primary causes of failure for industrial gas turbines, 1970–1979 (Source: *Der Maschinenschaden,* No. 53, 1980).

    2. Compressor erosion.
    3. Cooling air blockage.
    4. Rotor unbalance.
    5. Locking of turbine blade roots.
    6. Hot corrosion or sulfidation.
 2. Fuel system and treatment.
    a. Start-up and lightup failures.
    b. Tripping due to hydrate formation and freezing.
    c. Hydrocarbon liquid carryover.
    d. Combustor fouling.
 3. Rotor system.
    a. Rubbing in hot-gas path.
    b. Bearing and seal failures.
 4. Turbine and exhaust.
    a. Governor malfunction.
    b. Vibration problems.
 5. Control system.
    a. Instrument air problems.

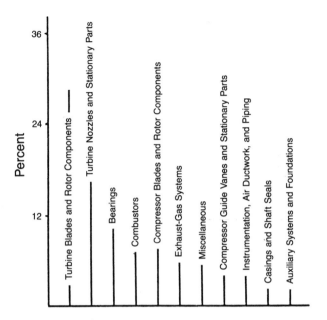

Figure 4-14. Distribution of gas-turbine component damage, 1970–1979 (Source: *Der Maschinenschaden,* No. 53, 1980).

    b. Loss of electrical continuity.
    c. Failure of air intake hoar-frost monitoring and deicing system.
6. Lube-oil system.
    a. Filter problems.
    b. Leakage.
    c. Pressure control problems.
7. Coupling and load system.
    a. Misalignment.
    b. Coupling distress.
    c. Gearbox and ancillary problems.
8. Environment.
    a. Power outage.
    b. Fuel supply.
    c. Operator error.
    d. Changes in load demand.

Finally, Table 4-25 is an appropriate gas-turbine troubleshooting guide.

**Table 4-25
Troubleshooting Guide—Gas Turbine**

Symptoms

Possible Cause	Combustion Noise	Vibration	Bearing Pressure		Cooling Air Pressure	Bleed Chamber Pressure	Fuel Pressure		P₂/P₁	P₃/P₄	Bearing Temperature	Wheel Space Temperature	Exhaust Temperature		Exhaust Temperature Variance	T₂/T₁		T₃/T₄	Mass Flow	Comp. Eff.	Therm. Eff.
	Erratic	Up	Up	Down	Down	Erratic	Up	Down	Down	Down	Up	Up	Up	Down	Up	Up	Down	Down	Down	Down	Down
Compressor — Blade Damage		②				②			③							②			③	③	
Compressor — Bearing Failure		①		①							①										
Compressor — Filter Clogging									②										②		
Compressor — Surging		③	①			①					②								④	②	
Compressor — Fouling		③							①							①			①	①	
Combustor — X-Over Tube Failure							③OR③								③						
Combustor — Liner Cracking/Loosening	③						④OR④								④						
Combustor — Fouling	①						②OR②						①		①						
Combustor — Clogging	②						①					①			②						

Possible Cause	Combustion Noise Erratic	Vibration Up	Bearing Pressure Up	Bearing Pressure Down	Cooling Air Pressure Down	Bleed Chamber Pressure Erratic	Fuel Pressure Up	Fuel Pressure Down	P_2/P_1 Down	P_3/P_4 Down	Bearing Temperature Up	Wheel Space Temperature Up	Exhaust Temperature Up	Exhaust Temperature Down	Exhaust Temperature Variance Up	T_2/T_1 Up	T_3/T_4 Down	Mass Flow Down	Comp. Eff. Down	Therm. Eff. Down
Turbine — Blade Damage		②															②			②
Nozzle Distortion		④								①		③					③			③
Bearing Failure		①		①							①									
Cooling Air Failure					①						②	①								
Fouling		③										②					①			①

Troubleshooting Electric Motors

Electric motors are the most commonly encountered prime movers in petrochemical process services. In the typically found two- or four-pole induction design, they range in size from fractional horsepower to more than 20,000 BHP in, for instance, refinery cat plant air blower service. Electric motor troubles appear in two forms: mechanical rotor/bearing difficulties and electrical, winding and supply-system problems. However, it has been estimated that more than 90% of electrical problems, mainly winding failures, can again be traced back to pre-occurring mechanical distress. Typical mechanical failure modes are overloading, flooding, and blockage of cooling air, single phasing, and bearing failures. Some of these failures, in turn, happen because of lack of maintenance or for reasons external to the motor itself. As an example, most single-phasing failures happen because of lack of maintenance on exactly those protective devices installed to guard against this danger.

Consider the case of a 3500-hp, four-pole, pipe-ventilated, squirrel-cage motor driving a centrifugal process compressor. The motor suffered a winding failure and caught fire. The fire was successfully extinguished and the motor disconnected and removed to a motor repair shop. The following failure diagnosis was made:

1. *Results of examination:*
   a. The fire in the motor was caused by a winding failure at that point where the windings emerge from the slots in the stator.
   b. There was a buildup of oil-impregnated dust, coke and fibrous material on the windings that fueled the fire.
   c. That portion of the windings normally encased in the slots of the stator showed widespread signs of overheating, evidenced by the fact that the insulating enamel powdered off at the slightest touch. That this same enamel was relatively intact in the fire-damaged area leads us to conclude that the deterioration was due to internal overheating.
2. *Most probable cause identification:*
   a. Multiple starts, i.e. starting the machine or subjecting it to inrush current levels more frequently than designed for.
   b. Operating the machine in a sustained overload condition. This is most unlikely here, since the machine has accurate relay protection schemes applied to prevent this, and since the compressor is assumed to be operating within design.
   c. Inadequate cooling air: fouled filters, fan failure, or operating the machine without fans.
   d. Fouled windings, i.e. the cooling air is unable to dissipate the normal heat buildup, which is caused by inadequate preventive maintenance (P.M.) on the motor to ensure clean windings and filter system.
3. *Conclusions:*
   a. It is suspected that this machine failure was caused by a combination of 2a, 2c, and 2d.

b. *Remedial actions:*

1. All large motors will be scheduled out of service at least every five years for P.M. Where the service conditions warrant, this frequency will be higher. The mechanical department is responsible for scheduling this work.
2. Filter and fan operation should be monitored closely and any failure given high priority for repair.
3. Multiple-start criteria as laid down in the operating manual must be strictly adhered to. This varies with the design of the machine, but generally the machine must be allowed sufficient time between successive starts to return to normal operating temperature.

This failure could perhaps have been prevented by appropriate troubleshooting routines. Table 4-26 represents a typical troubleshooting guide for electrical motors.

Troubleshooting the Process

From time to time, the machinery troubleshooter will be called upon to solve problems in cases where machinery is a more integral part of the total process system. A refrigeration system utilizing compression machinery with different types of drivers is a good example. Figure 4-15 shows a simplified schematic of this system; Table 4-27 shows an appropriate troubleshooting guide.

Usually, the following problems my be encountered in a refrigeration system:

1. Compressor problems:
   a. Centrifugal compressor—See the troubleshooting guide in Table 4-14.
   b. Reciprocating compressor:
      1. Open or faulty clearance pockets.
      2. Leaking valves.
      3. See the troubleshooting guide in Table 4-16.
   c. Steam-turbine drivers:
      1. High exhaust-steam pressure.
      2. Governor not opening fully.
      3. Turbine hand valves not open.
      4. Fouling of turbine wheels.
      5. Overspeed trip set too low.
      6. See the troubleshooting guide in Table 4-23.
   d. Electric-motor drivers:
      1. Motor running too hot.
      2. Amperage trip set too low.
      3. See the troubleshooting guide in Table 4-26.

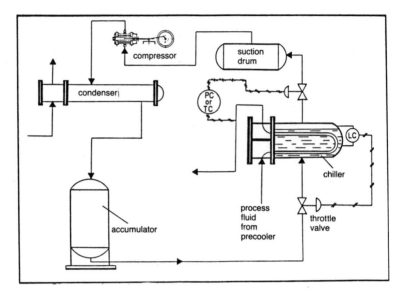

Figure 4-15. Simplified process refrigeration system.

2. Impurities in the refrigerant/composition:
   a. Formation of ice and hydrates.
   b. Insufficient circulation with speed-limited compressors.
   c. Violation of compressor minimum suction pressure.
3. Malfunction of throttle valve and accumulator:
   a. Throttle valve excessively open.
   b. Throttle valve excessively closed.
   c. Accumulator wrong liquid level.
4. Chiller:
   a. Low refrigerant level.
   b. Inadequate refrigerant feed.
   c. Low circulation on refrigerant side.
   d. Fouling on the process side.
   e. Seal-oil accumulation in refrigerant.
5. Condenser:
   a. Fouling due to low water velocities.
   b. Not enough air flow through air-cooled condensers.
   c. Recirculation of warm air through air fin coils.
   d. Poor refrigerant drainage from condenser.
   e. Excessive subcooling of refrigerant.

The foregoing concludes our discussion on the use of troubleshooting matrices.

Table 4-26
Electric Motor Troubleshooting Guide

Symptoms												
A — Will Not Start												
B — Runs Backward												
C — Noise—Excessive												
D — Rapid Knocking Noise												
E — Fails to Pull in—Synchronous Motor												
F — Will Not Come Up To Speed—Synchronous Motor												
G — Overheating												
H — Hot Sleeve Bearing												
I — Hot Antifriction Bearing												
J — Hot Bearings (General)												
K — Vibration												
L												
Probable Cause	#	A	B	C	D	E	F	G	H	I	J	K	L	#	Remedy
Line Trouble—Overload Trip—Loose Connections	1	•					•							1	Check Out Power Supply
Reverse Phase Sequence	2		•											2	Interchange Two Line Connections At Motor
Open Circuit, Short Circuit, Or Winding Ground	3	•					•							3	Check For Open Circuits and Ground
Motor Single Phased (3-Phase Machine)	4			•										4	Stop—Then Start Again—Motor Will Not Start
Unbalanced Load Between Phases	5			•				•						5	Check For Single Phasing—All Three Leads
Low Voltage/Improper Line Voltage	6	•						•						6	Check Motor Name Plate—Check Wire Size
Short Circuit In Stator Winding	7	•						•						7	Check Insulation Resistance
Ground In Stator Winding	8	•						•						8	Check For Ground

(table continued on next page)

Table 4-26 (cont.)

Symptoms

- A — Will Not Start
- B — Runs Backward
- C — Noise—Excessive
- D — Rapid Knocking Noise
- E — Fails to Pull In—Synchronous Motor
- F — Will Not Come Up To Speed—Synchronous Motor
- G — Overheating
- H — Hot Sleeve Bearing
- I — Hot Antifriction Bearing
- J — Hot Bearings (General)
- K — Vibration
- L —

	Probable Cause	#	A	B	C	D	E	F	G	H	I	J	K	L	#	Remedy
Elec System	Field Excited	9													9	Check Field Control
	No Field Excitation	10					•								10	Check Field Contactor & Conn./Exc. Output
Mechanical System	Uneven Air Gap	11			•										11	Center Rotor. Replace Bearings If Needed
	Unbalanced Rotor	12			•								•		12	Check Mechanical & El. Balance
	Foreign Matter In Air Gap	13							•						13	Clean Motor
	Loose Parts Or Loose Rotor On Shaft	14			•								•		14	Repair
	Rotor Rubbing On Stator—Bent Shaft	15				•			•			•			15	Check Shaft Run-Out/Center Rotor/Repl. Bearing
	Ventilation Restricted	16							•				'		16	Inspect Motor Interior For Dirt
	Motor Tilted (Shaft Bumping)	17				•					•	•			17	Check Level & Realign If Needed
	Misalignment Or Excessive Thrust	18				•					•	•			18	Realign & Check Limited End Float Cplg.

	Cause	A	B	C	D	E	F	G	H	I	J	K	L	Remedy
Mechanical System	19 Jammed	•							•					19 Disconnect Motor From Load, Check Solo Run
	20 Overloaded	•					•	•	•					20 Reduce Load Or Install Larger Motor
Bearings	21 Loose Coupling			•										21 Check Alignment & Tighten Coupling
	22 Motor Loose On Foundation Or Base Plate			•	•		•							22 Check Hold-down Bolts & Grout
	23 Noisy Ball Bearings			•	•									23 Check Lubrication & Replace Bearings
	24 Ball Bearings Installed Incorrectly			•							•	•		24 Replace
	25 Ball Bearings Damaged—Worn			•							•	•	•	25 Replace
Lubrication	26 Insufficient Oil						•				•	•	•	26 Add Proper Amount Of Oil
	27 Wrong Grade Of Oil										•	•	•	27 Use Recommended Lubricant
	28 Oil Contaminated									•	•	•	•	28 Drain Oil & Relubricate
	29 Oil Rings Not Rotating										•		•	29 Oil May Be Too Heavy, Check For Burrs
	30 Too Much Grease											•	•	30 Remove Excess Grease
	31 Insufficient Grease											•	•	31 Add Proper Amount Of Grease
	32 Wrong Grade Of Grease											•	•	32 Use Recommended Grease
	33 Grease Contaminated										•	•	•	33 Relubricate, Assure Grease Supply Is Clean

Table 4-27
Troubleshooting Guide—Refrigeration System

Possible Causes	#	A	B	C		D	E	#	Possible Remedies
Symptoms						**Symptoms**			
C — Low Suction Pressure / B — High Discharge Pressure / A — Low Discharge Pressure						High Suction Pressure — D / Process Side of Chiller — E / Temperature Too High			
Turbine Driver Problem (HP Limiting)	1	1				4	1	1	* Consult steam turbine TSG
Centrifugal Compressor Problem (Overload)	2	1				4	2	2	* See centrifugal compressor TSG
Recip. Compressor Problem (Valves etc.)	3	1				4	3	3	* See TSG reciprocating compressor
Wrong Composition For Speed Limited Compressors	4	2		3		5	4	4	* Adjust composition
Wrong Composition For HP-Limited Compressors	5	2					5	5	* Adjust composition
Wrong Composition	6	2	3	4			4	6	* Change composition
Shortage Of Refrigerant	7	3					10	7	* Find & repair leak. * Recharge System
Wrong Accumulator Level	8	4	4	6			9	8	* Adjust
Throttle Valve Plugged	9						8	9	* Change or repair on line
Throttle Valve Out Of Adjustment	10	5	5	5		1	7	10	* Adjust superheat setting
Insufficient Process Fluid In Chiller	11			1				11	* Check superheat at chiller outlet * Check for obstructions etc.
Low Refrigerant Level	12					2	6	12	* Check LIC
Inadequate Refrigerant Feed	13					3	5	13	* Check for restrictions in refrigerant flow
Low Circulation Ratio On Refrigerant Side	14			2		2		14	* Investigate Cause
Possible Causes	**#**	**A**	**B**	**C**		**D**	**E**	**#**	**Possible Remedies**

Compressor & Drivers — Refrigerant — EV — Chiller

†EV = Throttle/Expansion valve

Table 4-27 (cont.)

	Symptoms						Symptoms			
	C Low Suction Pressure						High Suction Pressure D			
	B High Discharge Pressure						Process Side of Chiller E			
	A Low Discharge Pressure						Temperature Too High			
	Possible Causes	**#**	**A**	**B**	**C**		**D**	**E**	**#**	**Possible Remedies**
Chiller	Oil Accumulation	15						3	15	* Wash out with low temperature solvents
Chiller	Fouling On Process Side	16						1	16	* Clean
Condenser	Fouling On Water Side	17		6					17	* Check cooling water velocities * Backflush
Condenser	Vapors Leaking Out	18	6						18	* Find leak & recharge
Condenser	Not Enough Air Flow In Air Cooled Cond./Fouling	19		1					19	* Check for obstructions on condenser coils * Check for faulty fan operation * Ambient conditions too hot
Condenser	Improper Refrigerant Drainage From Cond.	20		2					20	* Check for obstructions
Condenser	Excessive Subcooling Of Refrigerant	21	7						21	* Check cooling water temperature
	Possible Causes	**#**	**A**	**B**	**C**		**D**	**E**	**#**	**Possible Remedies**

References

1. Bloch, A., *Murphy's Law and Other Reasons Why Things Go Wrong,* Price, Stern, Sloan, Los Angeles 1979, "Machinemanship" pp. 40-45, "Metalaws," p. 87.
2. De Bono, E., *Practical Thinking—4 ways to be right. 5 ways to be wrong. 5 ways to understand,* Penguin Books, 1981.
3. Sternberg, R.J. and Davidson, J.E., "The Mind of the Puzzler," *Psychology Today,* June 1982 pp. 37-44.
4. Hänny, J., "Interfaces—The Achievement of Engineering Excellence," *Sulzer Technical Review,* 3/1981, pp, 77-81.

5. Selye, H., *The Stress of Life,* McGraw-Hill Book Company, 1956, p. 308.

6. McKean, K., "Diagnosis by Computer—Caduceus is learning fast, has a prodigious memory and, some believe, a brilliant future as a diagnostician," *Discover,* September 1982, pp. 62-65.

7. Multilin Inc., "139 Series Motor Protection Relays," Multilin Inc., Markham, Ontario, Canada.

8. Vesser, W.C., "Predictive Maintenance of Heavy-Duty Engines Through Premalfunction Waveform Analysis," ASME Publication 74-Pet-34, p. 6.

9. Stengel, R., "Calling Dr. Sumex—The Diagnostician's New Colleague Needs No Coffee Breaks," *Time,* May 17, 1982, p. 73.

10. *Sawyer's Turbomachinery Maintenance Handbook,* First Edition, Turbomachinery International Publications, Norwalk, Conn., 1980, Volume I, p. 6/29.

11. Karassik, I.J., *Centrifugal Pump Clinic,* Marcel Dekker, Inc., New York, 1981, pp. 365-369.

12. National Fire Protection Association, "Centrifugal Fire Pumps 1976," ANSI Z277.1—1976, pp. 20/108-109.

13. Sum, C., *A Practical Guide to Compressor Valves,* The IIC Group of Companies, 364 Supertest Rd., Downsview, (Toronto), Ontario, Canada, 1982, p. 8.

14. Blue, A.B., "Internal Combustion Engine Failures: A Statistical Analysis," ASME Publication 75-DGP-16, 1975, pp. 2-4.

15. *Handbook of Loss Prevention,* Allianz Versicherungs—AG, Springer Publishing, Berlin, 1978, pp. 137-140.

Chapter 5

Vibration Analysis*

Machinery distress very often manifests itself in vibration or a change in vibration pattern. Vibration analysis is, therefore, a powerful diagnostic tool, and troubleshooting of major process machinery would be unthinkable without modern vibration analysis.

There are many ways that vibration data can be obtained and displayed for detecting and identifying specific problems in rotating machinery. Some of the more common techniques include:

1. Amplitude versus frequency.
2. Amplitude versus time.
3. Amplitude versus frequency versus time.
4. Time waveform.
5. Lissajous patterns (orbits).
6. Amplitude and phase versus rpm.
7. Phase (relative motion) analysis.
8. Mode shape determination.

Each of these analysis techniques is discussed in detail in the following sections, with emphasis on practical applications of diagnosing specific machinery malfunctions. However, before an analysis is undertaken, it is helpful and advisable to consider the machine and the circumstances which lead to a requirement for vibration analysis.

Machine history. When the complaint is an excessive increase in vibration, much can be learned by reviewing the history of the machine. For example, if components such as couplings, sheaves, or bearings have been replaced, there is a good possibility

---

*Copyright © 1980 IRD Mechanalysis, Inc., Columbus, Ohio. Reprinted by permission.

that balance or alignment may be affected. Additional machines added to a structure can easily change the natural frequency of the structure, causing resonance. Changes in normal operating conditions of speed, load, temperature, or pressure may produce significant changes in machine vibration from unbalance, cavitation, aerodynamic/hydraulic forces, etc. Improper grounding when arc welding on machines can cause extensive damage to bearings and couplings. In any case, when the vibration of a machine increases, it is generally caused by wear or deterioration in the machine's mechanical condition or by changes which have been made to the machine or structure. In many cases, considerable analysis time and effort can be saved by reviewing the history of the machine to see if such changes might be associated with the vibration increase.

Machine characteristics. A review of machine characteristics such as rpm, type of bearings, gear frequencies, aerodynamic/hydraulic frequencies, etc. can help establish the expected vibration frequencies. These will help determine the type of instrumentation and vibration transducers needed for the analysis.

The operating characteristics of the machine may also dictate the type of analysis equipment required. For example, production machine tools with very short duty cycles may require a real-time spectrum analyzer because of the limited analysis time available. Machines which operate with continuously variable speed or load conditions may require real-time analysis or special instrumentation with tracking-filter capability. Machines with extremely complex, random, or transient vibration may require the spectrum averaging or "transient-capture" features of a real-time analyzer.

To summarize, a thorough evaluation of the machine and its history leading to the requirement for analysis can greatly simplify the analysis procedure by specifying the appropriate instrumentation and analysis data. Listing the faults typical of the machine, combined with past experience on this or similar installations, can suggest likely trouble areas. Discussion with the machinery manufacturer might also provide clues to common problems which have been encountered.

Data Acquisition

Vibration analysis is nearly always a two-step proposition. The first step is to obtain the necessary vibration data in a systematic way; the second step is to evaluate the data to identify the specific problem(s) with the machine. In this section, the data-acquisition, display, and analysis techniques listed on page 289 are outlined in detail, along with examples of practical applications. A more detailed coverage of the identification of specific machinery problems, together with a discussion of problems associated with specific machines, is presented later in this chapter under "Data Interpretation."

Vibration Amplitude Versus Frequency Analysis

The procedure of obtaining and displaying the amplitudes of vibration for all the frequencies present is perhaps the most useful of all analysis techniques. It is estimated that over 85% of the mechanical problems occurring on rotating machinery can be identified by displaying the vibration amplitude versus frequency data. However, successful identification of the problem(s) in a machine utilizing this or any other analysis technique requires that complete data be obtained in a systematic way that will simplify interpretation.

Importance of Tri-axial Readings

It can be noted from the sample data in Figures 5-1 and 5-2 that it is common practice to record the amplitude-versus-frequency data measured in the horizontal, vertical, and axial pickup directions at each bearing of the machine being analyzed. Obtaining measurements in all three directions is extremely important for distinguishing between various mechanical problems. For example, unbalance, misalignment, and bent shafts will generally cause vibration at a frequency of 1 × rpm. However, unbalance (except for overhung rotors) will almost always produce high amplitudes in the radial directions (horizontal and/or vertical) while revealing

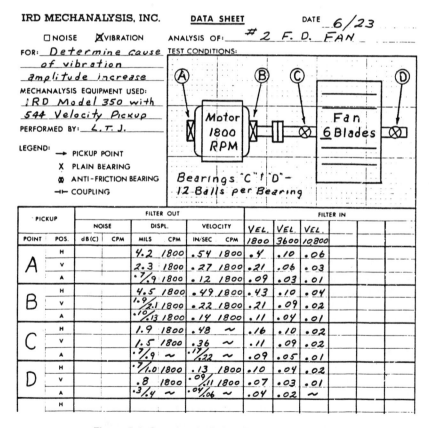

Figure 5-1. Completed tabular vibration analysis data.

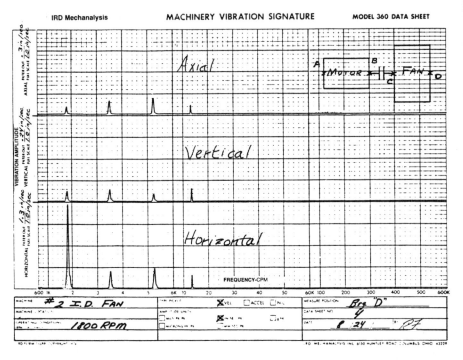

Figure 5-2. The high amplitude ratio between the horizontal and vertical vibrations at 1800 cpm is somewhat abnormal for simple unbalance and may indicate other problems such as resonance or looseness.

considerably lower amplitudes in the axial direction. By comparison, misalignment of couplings and bearings or a bent shaft will generally show a relatively high amplitude of vibration in the axial direction, along with high radial amplitudes. In general, whenever the amplitude of axial vibration exceeds 50% of the radial amplitude, misalignment or a bent shaft should be suspected.

In addition to comparing the radial and axial readings, it is also important to compare the radial horizontal and radial vertical readings as well. Much can be learned about the machine from such comparisons. For example, from the tabular analysis data in Figure 5-1, it can be seen that the horizontal amplitudes are typically higher than the vertical amplitudes. This is generally considered normal for rigid-mounted machines where vertical stiffness is typically higher than horizontal stiffness.

Horizontal-to-vertical amplitude ratios of 2:1, 3:1, 4:1, or perhaps even 5:1 can usually be considered normal. In the case of machines mounted on elevated structures or on resilient vibration isolators such as coil springs or rubber pads, the vertical amplitudes may be slightly higher. In cases where the amplitude ratio is unusually high, as in Figure 5-2 where the horizontal amplitude at 1800 CPM is nearly eight times higher than the vertical amplitude at the same frequency, additional readings should be taken to determine whether any unusual problems exist. For example, the analysis data in Figure 5-2 strongly suggest that the machine or supporting structure is resonant in the horizontal direction at 1800 CPM. This is indicated by the fact that the horizontal amplitude is nearly eight times the vertical amplitude at 1800 CPM, while for the other vibration frequencies present, the amplitude ratios are considerably less.

The tabular analysis data in Figure 5-3 illustrate another unusual comparison of radial readings. Note in Figure 5-3 that the horizontal amplitudes at 1200 CPM are slightly higher than the vertical readings at bearings A, B, and C. However, at bearing D, it was found that the vertical amplitude was more than *twice* the horizontal amplitude. While this may be quite normal for this particular machine on this structure, it does appear strange and, as such, deserves additional inspection. Such comparative readings have been found to be caused by wiped bearings, excessive bearing clearance, or other sources of looseness.

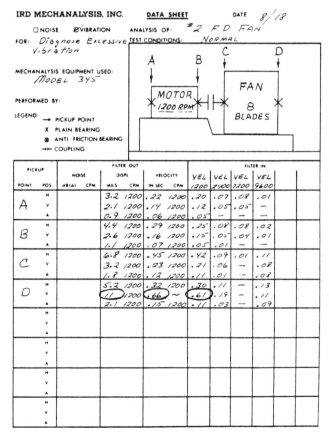

Figure 5-3. The unusually high vertical amplitude at bearing "D" may be due to looseness, a wiped bearing or other problem aside from simple unbalance.

The Machine Sketch

A very important part of analysis data which is often overlooked is the machine sketch. In order to identify the points at which vibration and noise readings are taken, both for immediate use as well as for future reference, it is necessary to make a complete sketch of the machine. A block diagram such as that shown in Figure 5-1 is normally all that is required. Although such a sketch is not very elaborate or artistic, it covers all the basic elements necessary. The sketch must show all of the essential components of the machine, including the driver and driven units as well as major accessories such as exciters or gear drives. It is important that the sketch incorporate a basic layout readily recognizable by others.

The sketch should show the rotating speed (rpm) of each component. Machine size is indicated by noting the horsepower of the driver or the output of the driven unit.

Since vibration readings must be taken at each bearing of the machine, the location of each bearing point should be clearly identified. Note in Figure 5-1 that each bearing point is identified by a letter of the alphabet. Note also that standard symbols are used to identify plain bearings, antifriction (ball or roller) bearings, and couplings.

In addition to the information needed to clearly identify the machine and the pickup locations, it is recommended that any additional information be included which may help identify the sources of vibration. This may include information such as gear-meshing frequencies (number of gear teeth × gear rpm), the number of blades on a fan, the number of vanes on a pump impeller, or the presence of large, nearby machines which may contribute "background" vibration.

Sometimes a more elaborate sketch is required to show important details. Perhaps the machine has a history of structural or piping problems. Thus, it may be necessary to show the relationship of piping, pipe hangers, duct work, or structural details, particularly if vibration readings will be taken at points other than the bearings of the machine.

Supporting Information

You will note on the sample form, Figure 5-1, that space is provided for entering important supporting information. This information should be filled in completely, particularly if the analysis data will be retained and filed for future reference.

The specific machine being analyzed should be clearly identified by noting the manufacturer, machine type, serial number, location, or other details which will distinguish the machine from other, perhaps identical, machines.

Under *Test conditions,* enter information defining the operating conditions of the machine at the time the analysis is taken. This may include conditions such as speed, load, temperature, flow, etc. Normally, analysis data should be taken with the machine operating as it normally does. However, a change in operating condition may make a noticeable change in the vibration characteristics. If the machine operates under varied conditions, a few sample readings should be taken to detect any significant variations in vibration.

Under *date,* enter the day, month, and year the analysis is taken. In some instances, the time of day should be entered also, because some machines will change vibration characteristics from one time of the day to the next. This change may correspond with other plant operations, or it may correspond to a temperature change from the warm of mid-day to the cool of the evening, which could affect piping strains or machine alignment.

Obtaining Amplitude-Versus-Frequency Data

Vibration amplitude-versus-frequency data can be obtained and recorded in several ways. When using a basic analyzer with a manually tuned filter, such as the instrument shown in Figure 5-4, the filter is manually tuned over the analyzer frequency ranges and the significant vibration amplitudes and corresponding frequencies identified by carefully observing the instrument amplitude and frequency meters. Data are manually tabulated on a form such as that illustrated in Figure 5-1. When the analysis data are being manually recorded, the first step is to measure and record the overall or "filter-out" amplitude and predominant frequency readings at each measurement point in the horizontal, vertical, and axial directions.

Vibration velocity measurements are generally preferred for most analyses. However, displacement measurements may also be taken where it is likely that vibration frequencies below 600 CPM will be encountered. If very high vibration frequencies are anticipated (above 100,000 CPM), vibration acceleration measurements may also be recorded.

The filter-out amplitude and frequency readings are valuable, for a number of reasons. First, the amplitude readings reveal the extent of the problem and provide a basis for comparison with filter-in amplitudes to check for completeness of the

Figure 5-4. Vibration Analyzer/Dynamic Balancer (IRD 345).

analysis data. In addition, the predominant frequency readings may quickly direct attention to the problem source. However, conclusions about the nature of the problem should not be based on filter-out readings alone; a thorough frequency analysis of the vibration should be made before final decisions are made.

To obtain the necessary "filter-in" measurements, each frequency range of the analyzer is carefully scanned with the tunable filter. By scanning each frequency range, all vibration frequencies of significance can be found without trying to anticipate which frequencies will be present. This means that even those persons who are not totally familiar with the machine can still obtain good, complete analysis data. Begin by slowly tuning the filter through the frequency range while observing the frequency meter. Of course, make sure that the amplitude meter is reading up-scale and on-scale at all times. Continue to tune slowly until the frequency meter settles down and "locks on," indicating that the filter is approaching a particular frequency. Once the frequency meter has locked on, note the reading on the frequency meter and slowly turn the filter tuning dial until the reading on the dial is approximately the same as the reading on the frequency meter. Now, make fine adjustments to the tuning dial to obtain the peak reading on the amplitude meter. The filter is now tuned to the first significant vibration. Without making any further adjustments to the filter, measure and record the amplitudes of vibration for this frequency in the horizontal, vertical, and axial directions at each bearing of the machine.

After the amplitude readings have been taken and recorded for the first vibration frequency, continue scanning with the tunable filter until the next frequency is found. Again, fine-tune for the peak amplitude and record the amplitude for this frequency at each measurement point and pickup direction. Continue the filter scanning procedure until all significant vibration frequencies have been discovered and their amplitudes properly recorded. Figure 5-1 shows an example of tabular analysis data obtained by this scanning technique.

After manual analysis data have been obtained, a quick check can be made to determine whether or not all the important data have been obtained for each measurement point. This is done by comparing the filter-in amplitudes with the recorded filter-out amplitudes. As a general rule, the sum of the filter-in amplitudes should equal or exceed the filter-out amplitude measurement. For example, in Figure 5-1, the filter-out amplitude in the horizontal direction at bearing A is 0.54 in/sec. Adding the filter-in amplitudes (0.4 @ 1800 + 0.1 @ 3600 + 0.06 @ 10,800) results in a sum of 0.56 in/sec, which only slightly exceeds the filter-out amplitude. This indicates that all the important vibration data have been found at that particular point. On the other hand, at point C, the filter-out reading is 0.48 in/sec, while the sum of the filter-in readings is only 0.28 in/sec. This means that there are some additional vibration data to be found, and the frequency ranges should be scanned again with the pickup at bearing C.

Plotting Amplitude-Versus-Frequency (X-Y) Signatures

Some analyzers can be connected to a standard X-Y recorder to obtain graphic (X-Y) plots, or "signatures," of machinery vibration amplitude versus frequency, as illustrated in Figure 5-2. Such plots are obtainable with the stand-alone equipment shown in Figure 5-5 (right side), and Figure 5-6 (left side). On older analyzers it was generally necessary for the operator to manipulate the filter. Thus, as the operator would manually adjust the filter over the frequency range, the recorder automatically plotted the amplitude-versus-frequency data. The equipment shown in Figures 5-5 through 5-8 is intended to be used with the X-Y recorder to obtain graphic vibration signatures; however, with these instruments, the filter is automatically swept over the various frequency ranges, thus eliminating the need for manual tuning by the operator.

Figure 5-5. Vibration Spectrum Analyzer with X-Y recorder. (Ono Sokki/Shigma, Inc., Elk Grove Village, IL 60007.)

Figure 5-6. Vibration Analyzer/Dynamic Balancer with X-Y recorder (IRD Model 840).

Figure 5-7. Automatic Spectrum Analyzer/Balancer (IRD Model 360).

Figure 5-8. On-Line Real Time Spectrum Analyzer.

Plotting vibration amplitude-versus-frequency signatures has many advantages over manual (tabular) analysis. First, many sources of human error in observing and recording the data are eliminated. There is less chance of missing significant vibration frequencies, and analysis time is greatly reduced by eliminating the need to fine-tune to each frequency found. However, as with tabular analysis, for a complete analysis of the vibration, plots must be made in the horizontal, vertical, and axial directions at each bearing of the machine. A recommended procedure is to record all three signatures for a particular bearing on the same sheet as illustrated in Figure 5-2. This greatly simplifies data interpretation by concentrating all the vibration data for a particular bearing on one sheet. In almost all cases the amplitude resolution provided with data recorded this way is quite adequate for making necessary comparisons.

Another recommendation to simplify data interpretation is to use the same amplitude range for all vibration signatures wherever possible. It should be apparent that if the horizontal signature was plotted with the analyzer amplifier range selector on the 1.0 in/sec range, the vertical signature obtained on the 0.3 in/sec range, and the axial signature obtained on the 0.1 in/sec range, it would be possible to obtain three signatures which visually appeared quite similar. However, these signatures could differ quite noticeably in terms of true significance. Therefore, it is advisable to note and record the overall or filter-out amplitudes for each bearing before carrying out the analysis. Based on filter-out readings, one amplitude range should be selected which is appropriate for all measurement points. With all signatures plotted on the same range, data evaluation and interpretation will be greatly simplified and far less confusing than if different amplitude ranges had been used. For example, the three signatures in Figure 5-2 were all obtained on the 1.0 in/sec range. As a result, by just glancing at the comparative signatures it is obvious that the most significant vibration occurs in the horizontal direction at 1800 CPM.

Linear Versus Logarithmic (dB) Display

Linear amplitude scales (mils peak-peak, in/sec peak, microns, G's, etc.) are by far the most common scales and are familiar to most industrial personnel. This, plus the fact that there are many industrial standards based on linear amplitude scaling, is a strong argument for their continued use. There are occasions, however, when a dB (log) amplitude scale may be preferred.

Noise measurement is an example where the dB (log) amplitude scale is traditionally used. The dB (log) amplitude scale may also be used to advantage under certain conditions for vibration measurements. The on-line real-time spectrum analyzer can be used to display the data on either the conventional linear amplitude scale or on a dB (log) scale.

A dB (log) amplitude scale has the advantage of compressing widely varying vibration or noise amplitudes into a smaller physical range, making it easier to plot them graphically. Essentially what this type of scaling does is to deemphasize the high amplitudes and instead emphasize the low amplitudes. This concept is illustrated in Figure 5-9, which shows the same vibration data plotted on both linear and log scales. Note that the high-amplitude signals appear essentially the same on

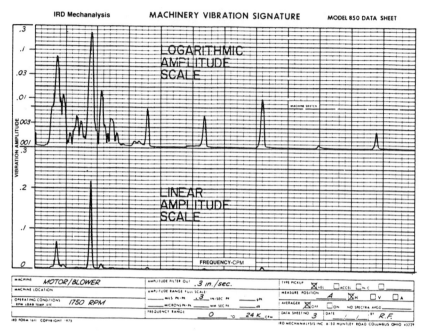

Figure 5-9. Plotting vibration data on a logarithmic (dB) scale provides additional emphasis to low amplitude signals.

both plots; however, the low-amplitude signals, which were barely noticeable on the linear scale, take on added significance when plotted on a dB (log) scale.

Although analysis data are somewhat more difficult to interpret when expressed in dB (log) format, for critical machines where it may be important to detect changes in low-amplitude vibration signals, the recording of baseline data as well as the future periodic recording of signatures on a dB (log) scale should be seriously considered. For example, on some machines the first signs of mechanical deterioration occur with small increases in the vibrations generated by the bearings. These are often low amplitude signals that are initially overshadowed by vibrations caused by unbalance, misalignment, etc. They are, however, an early warning signal. By using a dB (log) amplitude scale these low-amplitude vibrations are emphasized so that they are not overlooked in the vibration analysis.

Real-Time Spectrum Analysis

Standard frequency analysis procedures are carried out with the assumption that the vibration being analyzed is reasonably steady-state and will be present long enough to allow the frequency ranges to be scanned with the tunable filter and the data recorded manually or with an X-Y recorder. Unfortunately, not all analysis situations meet these ideal requirements. Some machines may operate under continually varying speeds, loads, or temperatures, resulting in wide variations in the machine's vibration characteristics. Other machines have such random vibration resulting from combustion, boiling, flow turbulence, or rolling contact that a single signature of the vibration cannot accurately reflect the true nature of the vibration. In still other cases, the vibration may be present for such a short period of time that frequency analysis by standard procedures is impossible. Production machine tools, punch presses, and forging hammers are typical examples of short-lived or "transient," vibrations. For situations where frequency analysis data must be obtained very quickly, a real-time spectrum analyzer such as the unit illustrated in Figure 5-8 can be used.

The real-time spectrum analyzer can analyze complex vibration and noise hundreds of times faster than conventional analyzers. An oscilloscope built into the front panel of the analyzer provides an instantaneous and continually updated display of the vibration amplitude-versus-frequency signature so that the analysis is essentially displayed as it occurs. The unit illustrated in Figure 5-8 also incorporates a built-in strip-chart X-Y recorder to provide hard-copy signatures in just a few seconds to minimize the time required for data recording and comparison.

Without question, real-time analysis offers many advantages over conventional analysis techniques, particularly when dealing with highly complex, random, or transient vibration. Observing the actual vibration or noise spectra while a machine undergoes changes in various operating parameters provides information that would be difficult, if not impossible, to obtain in any other way.

Spectrum Averaging with the Real-Time Analyzer

If the vibration or noise spectrum of a machine does not fluctuate or change significantly as it is observed on the analyzer scope display, then a single, instantaneous spectrum is entirely adequate to describe the vibration or noise characteristics of the machine. However, when conducting an analysis, it is not unusual to find that while some frequencies have relatively constant amplitudes, others have amplitudes that vary considerably about some average value. In other cases, the spectrum may vary so much that it is almost impossible to accomplish the analysis with any confidence in the results. When the vibration and noise being measured is random, it is very difficult to draw conclusions as to the amplitudes and frequencies when only a single spectrum is available. Such vibrations and noise are not unusual, since they are caused by flow of steam, gas, and liquids through piping, fittings, pumps, etc. Random spectra are also caused by combustion, boiling, and rolling contact of metal on metal, such as occurs in ball bearings, roller bearings, steel wheels, etc. Spectrum averaging with the on-line real-time spectrum analyzer is an extremely valuable tool as it provides stable and repeatable spectra of these random vibrations and noise. Figures 5-10 illustrates the difference between an average and a single spectrum of random vibration.

Averaging, in its simplest form, is the process of taking a number of samples, adding them together, and then dividing the sum by the number of samples taken. The result will approach the typical value to be expected if an infinitely large number

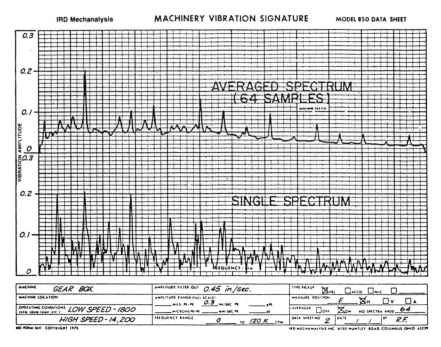

Figure 5-10. When vibration or noise data is random, spectrum averaging can be used to clearly identify significant components.

of samples were taken. The spectrum averager incorporated in the on-line real-time spectrum analyzer performs linear averaging of the amplitude-versus-frequency spectra over a preselected number of samples, ranging from 1 to 512. For example, if it is desired to average, say, 16 samples, the averager, once activated, will automatically proceed to sample the data 16 times, each time accepting one-sixteenth of the data present. After the sixteenth sample has been completed, the averaged spectrum, retained in memory, is displayed on the scope and can be recorded for permanent record on the X-Y recorder.

As previously mentioned, there are certain types of machinery vibration and noise spectra where the information content can be clarified and/or improved by averaging. Spectrum averaging can be extremely important where:

1. Maximum repeatability of a machine's spectrum is required.
2. The steady-state spectrum of a machine is changing rapidly and unpredictably.
3. Important discrete frequencies are masked by random vibration or noise.

If a machine's periodic spectrum is to be compared with a reference or "baseline" signature, as would be the case in an advanced preventive maintenance program, the average spectrum may be of more value than a single spectrum. The reason for this is that when an average spectrum of a machine in good condition is compared with the average spectrum of the same machine at a later date, a more reliable indication of any changes in the vibration or noise levels of the machine will be obtained. That is, more confidence can be placed in the fact that small changes are real, not just momentary variations in the vibration or noise signature. While such additional effort may not be justified in all cases, for large, critical machines, this improved capability to detect defects in their earliest stage may more than justify the effort.

An increase in random vibrations from a bearing can be detected from the average spectrum, which will provide an indication of bearing deterioration or, perhaps, erosion of material in the fluid flow path.

Still another application in which averaging can be used to advantage occurs when low-amplitude, discrete frequencies are hidden by random vibration or noise. These low-amplitude, discrete signals can be enhanced relative to the random signals occuring at the same frequencies by spectrum averaging. This is illustrated in Figure 5-10. The reason for this is that the discrete frequency amplitude remains fairly constant, while the random signal amplitude will fluctuate between zero and some peak value. As a result, the average amplitude of the random signals will be significantly less than their peak values.

Of course, obtaining averaged spectra will require slightly more data-acquisition time than is required for obtaining single, instantaneous signatures. Therefore, it is important to evaluate the vibration and noise and select the minimum number of averaged samples required for each situation. Averages of 1, 2, 4, 8, 16, 32, 64, 128, 256, and 512 spectra can be selected. No firm rule can be given for selecting the appropriate number of samples; experience is truly the best guide. Observing the continuously updated spectrum on the scope will give the operator a feel for the variability of the signals. If the spectrum pattern is well defined, relatively few samples will be required. However, if the frequencies and amplitudes are rapidly changing unpredictably, more samples may be necessary. A good way to determine the required number of samples is to begin with a low number, such as four. When the averaging is complete for four samples, note the spectrum on the plotter or recorder. Next, switch the averager to eight samples and resume data collection. The averager will then add an additional four samples to the average of four already in memory, bringing the average up to a total of eight samples. This average spectrum (eight samples) is then compared with the previous average spectrum (four samples) to determine whether there is any significant difference between the two. If there is little difference, then the lower average (four samples) is probably adequate. However, if the difference is significant, select the next higher number of spectra and again obtain additional samples. This procedure can be repeated until no significant difference between succeeding averages is found. The number of spectra averaged can often be as few as eight and rarely more than 64.

To summarize, spectrum averaging can play an extremely valuable part in establishing the vibration and noise characteristics for machines. Of course, whenever the data include averaged spectra, it is most important that the number of averaged spectra be recorded along with frequency range, amplitude range, amplitude units (displacement, velocity, or acceleration), linear vs. dB (log) readout, etc. This will ensure that all signatures are obtained in the same way for direct comparisons.

Transient Capture with the Real-Time Analyzer

One of the most valuable features of real-time analyzers is their capability to analyze transient vibrations and noise. In fact, it is real-time analyzers that made analysis of transients a practical procedure.

Transients are generally considered vibrations or noise having a rapid onset and which are single events rather than steady-state, repetitive signals. In a broader sense, the changing vibrations and noise of a machine as it is slowly brought from standstill to operating speed are also transients. For the purposes of this discussion, however, only those short transient vibrations and noise which result in a momentary change from some steady-state condition will be considered. The reason for this is that long types of transients are often analyzed differently from the short types.

In order to analyze a transient using the real-time analyzer, the vibration or noise signals containing the transient information must be held in the analyzer's memory so that they can be studied by the operator. One method of doing this with short transients is to watch the oscilloscope, using either the SPECTRUM or TIME analysis mode. The mode selected will depend upon which is more important, the time signature or the frequency spectrum analysis.

When the transient occurs, the frequency spectrum peaks will increase in amplitude, reach a peak, and then decrease in amplitude. To hold the transient in the analyzer memory for further study, the operator depresses the HOLD button just as the transient frequency spectrum reaches its maximum amplitude. On analyzers equipped with automatic PEAK HOLD feature, the maximum value will be automatically captured. Once held in memory, the operator can perform a number of operations using the analyzer to obtain a better description of the transient.

Manually capturing the transient at its maximum amplitude is entirely satisfactory, however, as long as two conditions exist:

1. The operator knows approximately when the transient will occur.
2. The transient occurs slowly, so that the operator has time to depress the HOLD button before the transient has passed. Generally, transients lasting .5 to 1 second, or longer, can be captured by manually depressing the HOLD button.

Obviously, while there are many transients that meet these conditions, there are also many which do not. It is for these latter transients that the transient-capture or PEAK HOLD mode was built into the real-time analyzer.

Basically, the transient-capture mode replaces the operator by automatically triggering the HOLD button after the transient signals are in the analyzer memory.

The first step in setting up the transient-capture mode is to select the level at which the capturing process will be triggered. The level control on the real-time analyzer, Figure 5-8, has numbers from 0 to 10. These represent the percentages of full scale at which the transient capture will be triggered. For example, if the level control is set at "5", the capture process will be triggered when the signal level reaches 50% of full scale. Thus, the level control keeps the operator from triggering the capture process on the inevitable low-level background signals or on small transients which are not of interest.

Once the desired trigger level has been set, the PRESS TO ARM button is depressed, which arms the transient-capture circuit. When the transient vibration (or noise) occurs with an amplitude exceeding the preset trigger level, the data are automatically captured and held in the real-time analyzer memory. Once captured, the data can be displayed in the time function or frequency spectrum mode on the analyzer oscilloscope, or the data can be recorded on the X-Y plotter.

The transient-capture feature of the real-time analyzer is extremely useful for obtaining vibration or noise data for conditions such as compressor surges, startup transients, load changes, and similar problems where the data are present for only a brief period of time. The transient-capture feature is also useful for performing "bump" tests for determining the natural frequencies of machines and structures—especially where the test must be performed by one person.

Amplitude-Versus-Frequency-Versus-Time Analysis

Amplitude-versus-frequency data are extremely useful for identifying most machinery problems; however, a single signature can only reveal the characteristics of vibration at a single instant in time with the machine operating at a particular speed and under a specific load condition. There are many occasions where it is useful to be able to observe and record the vibration amplitude-versus-frequency data repeatedly and quickly. For example, during machine startup, it may be important to know where resonant conditions or critical speeds are being excited by the various forcing frequencies generated by machine components. Or, it may be necessary to evaluate the vibration amplitude and frequency characteristics of a machine during a transition in load, temperature, or other operating variables.

The high-speed analysis capability of the real-time spectrum analyzer is ideally suited for these requirements for amplitude-versus-frequency-versus-time data. For example, the analyzer pictured in Figure 5-8 incorporates a built-in strip-chart X-Y recorder which, when placed in the RECUR mode of operation, produces a new vibration amplitude-versus-frequency signature every four seconds. A sample of recurring signatures obtained with this recorder during the startup of a machine is illustrated in Figure 5-11, where the resonant frequency of the machine is clearly revealed by noting the frequency at which the vibration reached an amplitude peak.

Where it is necessary to obtain vibration or noise signatures more rapidly, the analyzer in Figure 5-8 can be used with an optional high-speed X-Y recorder to provide chronological signatures at a rate of one every half second. Figure 5-12 illustrates the type of data provided by the real-time analyzer and high speed recorder during the startup of a machine.

Plots such as this are sometimes referred to as "waterfall" diagrams. The occurrence of an oil whirl during startup is clearly shown in Figure 5-12, as are the resonant/critical speeds excited by rotor unbalance. Plots such as this are also useful for evaluating the effects of load changes or other changes in machinery operation, since the effect of the change on *each* frequency can be detected. The occurrence of new or additional vibration frequencies or the disappearance of existing frequencies can also be detected.

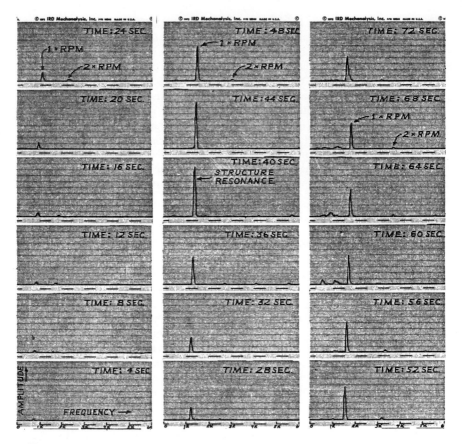

Figure 5-11. The X-Y recorder built into the real time analyzer can be placed in the recur mode to provide updated signatures every 4–5 seconds.

Amplitude/Phase Versus Machine rpm

All objects, including machines and their supporting structures, have resonant frequencies, where very high amplitudes of vibration can result from a relatively small exciting force. And, since machines and their structures are generally complex systems consisting of many spring masses with various natural frequencies, resonance is a relatively common problem.

Much can be learned about the response of a machine to the forces that cause vibration by obtaining plots of vibration amplitude and phase as a function of machine rpm. Figure 5-13 illustrates a typical recording of vibration amplitude and phase obtained during machine startup or coastdown. Such recordings, sometimes referred to as Bodé plots, clearly identify the resonant frequencies by the

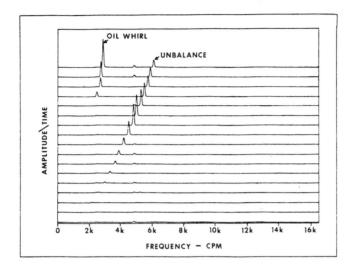

Figure 5-12. Cascade (waterfall) diagram of amplitude vs. time.

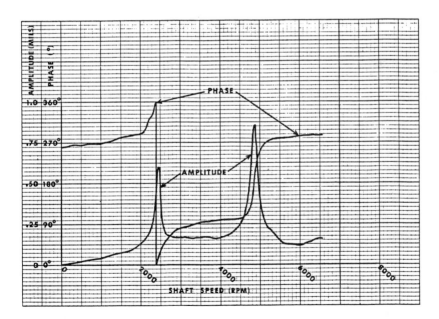

Figure 5-13. Plots of amplitude and phase versus machine rpm clearly identify critical speeds and resonant conditions.

characteristics peak amplitudes and corresponding 180° phase shift. From the sample plot in Figure 5-13, it is apparent that the machine in question has two significant resonant frequencies, one at approximately 2450 rpm and another at 4850 rpm. Of course, if at the normal operating speed(s) of the machine there is an exciting force which corresponds to either of these frequencies, and which arises from normal unbalance, misalignment, aerodynamic or hydraulic forces, torque pulses, reciprocating forces, looseness, oil whirl, etc., then an undesirable and probably destructive vibration will likely result. For this reason, it is helpful to know:

1. The exciting frequencies of vibration inherent to the machine at its normal operating speed(s). These can be identified through normal amplitude-versus-frequency analysis techniques already discussed.
2. The natural or resonant frequencies of the machine or structure. As illustrated in Figure 5-13, a plot of amplitude and phase versus machine rpm will reveal these.

Obtaining Amplitude/Phase Versus rpm Data

A general indication of the resonant frequencies of a machine can sometimes be obtained using a basic vibration analyzer by simply switching the analyzer filter to the OUT position and carefully observing the amplitude and frequency meters as the machine coasts down or is brought up in speed. Of course, this does not provide a permanent record like the recorded plot in Figure 5-13, and it requires careful observation of the instrument meters to note when the vibration amplitude has peaked and the frequency (rpm) at which the peak occurs. In addition, the phase-change information is often an important factor when analyzing system response. However, using a basic analyzer, it is usually difficult to observe the changing phase of vibration with the stroboscopic light while trying to monitor the amplitude and frequency meters at the same time. For truly accurate and complete amplitude/phase-versus-rpm data, such as the recorded data in Figure 5-13, a specialized instrument incorporating a tracking filter is generally used.

The analyzer shown in Figure 5-7 includes a tracking filter for obtaining graphic recordings of vibration amplitude and phase versus machine rpm. This instrument utilizes a reference pickup at the shaft of the machine to provide a voltage pulse on each shaft revolution. Reference pickups normally used include photocells and electromagnetic pickups; however, noncontact pickups may also be used where the shaft target area has been properly prepared, to provide a 1 × rpm pulse or "spike-like" signal. For example, an existing key or keyway is an excellent target for either an electromagnetic or noncontact reference pickup.

The 1 × rpm signal from the reference pickup essentially serves three functions:

1. The reference signal automatically tunes the analyzer filter to the rpm of the shaft. If shaft rpm changes, the frequency of the reference signal changes accordingly. In this way, the instrument filter is locked or synchronized to shaft rpm.
2. It controls the reference signal. This reference signal is actually a D.C. voltage which is proportional to shaft rpm. This voltage is utilized to drive the X axis of

an X-Y or XY_1Y_2 recorder for obtaining graphic plots of amplitude and/or phase versus rpm.

3. The reference signal provides a fixed reference for comparison with the signal from a vibration pickup, resulting in a D.C. voltage proportional to the relative phase between the two signals. This D.C. voltage is indicated on the phase meter as a relative phase measurement (0°-360°). This D.C. voltage is also available to drive a Y axis of an X-Y or XY_1Y_2 recorder for plotting phase versus rpm (or phase versus amplitude).

Interpreting Amplitude/Phase-versus-rpm Plots

As mentioned previously, when a machine passes through a resonance frequency during coastdown or startup, the usual indications will be a peak in amplitude at the resonance speed, accompanied by the characteristic phase shift of approximately 180° (see Figure 5-13). Occasionally, however, the recorded data may reveal some unusual conditions of system response, as illustrated by the following examples:

Example 5-1. Referring to the data in Figure 5-14, two distinct amplitude peaks are noted at 500 rpm and 1200 rpm, suggesting two resonant speeds. However, the recorded phase data show the characteristic 180° phase shift for only the lower speed (500 rpm) amplitude peak. There is virtually no phase shift associated with the higher 1200-rpm peak. Based on the recorded data it can be concluded that the lower 500-rpm peak is, in fact, a true resonance. However, the 1200-rpm amplitude peak is *not* resonance due to the lack of a corresponding phase shift. The question now is "what would cause a peak in 1 × rpm vibration which is not due to resonance?"

One possible cause of the 1200-rpm amplitude peak in Figure 5-14 might be the presence of significant background vibration at the indicated frequency. For

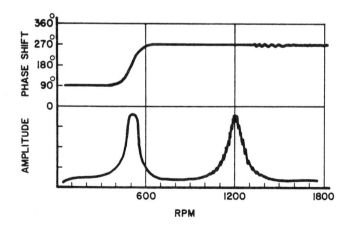

Figure 5-14. True resonance at 500 rpm only.

example, suppose a plot was being taken during the coastdown of an 1800-rpm motor, and at the same time, a 1200-rpm fan operating nearby was contributing substantial background vibration to 1200 CPM. Obviously, as the 1800-rpm motor coasts down, its speed must eventually coast through 1200 rpm. When this happens, the motor vibration will momentarily add to the background vibration from the fan to produce an amplitude peak at 1200 rpm.

Example 5-2. Referring to the amplitude and phase plots in Figure 5-15, an amplitude peak is noted at approximately 600 rpm. A 180° phase shaft corresponding to the amplitude peak confirms this as being a resonance. In addition, it is also noted that a 180° phase shift occurred at approximately 1400 rpm, but there is no amplitude peak at this speed to suggest resonance. The 180° phase shift, however, verifies that there is, in fact, a resonance at 1400 rpm. How can there be a resonance without a peak in amplitude?

The absence of an amplitude peak at resonance might be the result of one of the following:

1. If the exciting force corresponding to the resonant frequency is very low, or if the system is heavily damped, little amplification in amplitude may occur at resonance (see Figure 5-16 for the effect of damping on system response).
2. If the pickup is located at a nodal point of the resonant system, there may be little or no noticeable increase in amplitude at the resonance frequency. When a rotor shaft or structure is excited to vibrate at resonance, it will likey assume one of the vibratory modes illustrated in Figure 5-17. Each mode has one or more nodal points which are points of minimal amplitude. Of course, if the vibration pickup was located at a nodal point, there may be little noticeable amplification of the vibration amplitude when passing through this resonance. However, the phase plot will reveal the characteristic 180° shift. When this occurs, the vibration pickup should be repositioned to a new location and a second plot obtained to confirm the resonance.

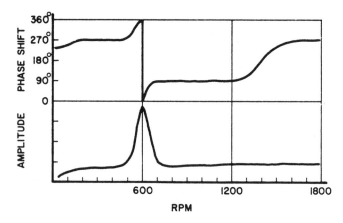

Figure 5-15. Phase shift without amplitude peak at 1400 rpm.

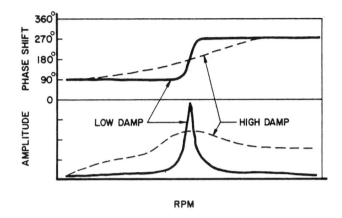

Figure 5-16. Effect of damping on system response.

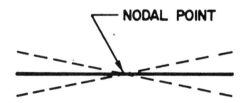

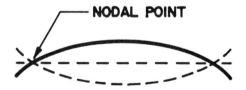

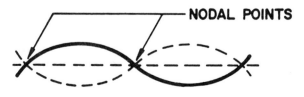

Figure 5-17. Nodal points.

Example 5-3. Referring to the amplitude and phase-versus-rpm data in Figure 5-18, it can be noted that an amplitude peak occurred at approximately 900 rpm. However, checking the phase information it is noted that the amplitude peak is accompanied by a phase shift of approximately 360° instead of the usual 180°. In this case, the phase data suggest that there are actually *two* systems in resonance at or near the same frequency, and each of the two resonances is contributing a 180° phase shift for a total phase shift of 360°. In this case, identifying and correcting only one of the resonant systems may not solve the problem.

Example 5-4. Referring to the sample data in Figure 5-19, the somewhat unusual portion of the data which deserves explanation is the dip in vibration amplitude which occurs at approximately 1000 rpm, along with the accompanying 360° shift in phase. As in example 3, the 360° phase shift at 1000 rpm in Figure 5-19 suggests two spring-mass systems in resonance at or very near the same frequency. The dip or reduction in vibration amplitude at this frequency, sometimes referred to as an "anti-node," can be explained by referring to the spring-mass systems in Figure 5-20.

First, assume that the initial exciting force is the unbalance, U. If spring-mass system A is resonant with the unbalance frequency, then we know from vibration theory that the *actual vibration* of mass A lags the unbalance force by 90°. Now, if spring-mass system B is resonant at the same frequency, then its exciting force will be derived from the motion of mass A. Therefore, the actual vibration of mass B must lag the motion of A by 90°. As a result, the actual vibration of mass B will lag the initial unbalance force by a total of 180° (90° + 90° = 180°). This produces two opposite forces acting on mass A, one being the initial unbalance force and the second being the force of vibration of mass B. Since the two forces acting on mass A are opposite, mass A will exhibit a minimal vibration amplitude, as illustrated by the dip in Figure 5-19. Mass B, on the other hand, may reveal an extremely high amplitude of vibration. For this reason, when the data suggest a possible anti-node, it

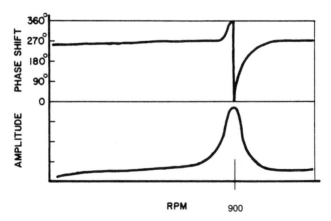

Figure 5-18. Two systems at or near resonance at approximately same frequency.

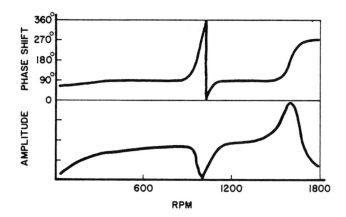

Figure 5-19. Anti-node near 1000 rpm.

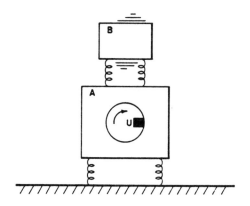

Figure 5-20. Typical spring-mass system.

would be advisable to identify the second system in resonance and plot its response as well, to ensure satisfactory performance.

The principles of the anti-node are sometimes used beneficially to control a resonance. For example, if a machine or structure was found vibrating at resonance, the addition of a second resonant spring-mass system may control or minimize the vibration of the machine or structure. For example, if the bearing pedestal in Figure 5-21 was vibrating at resonance, the addition might be used to minimize bearing vibration in a situation where changing the exciting frequency (rpm) or changing the bearing natural frequency was not possible or practical. The resonant system added to the machine or structure is often called a "dynamic absorber."

The position of the weight on the bar in Figure 5-21 can be adjusted to accurately fine-tune the resonant frequency of the bar to the known exciting frequency.

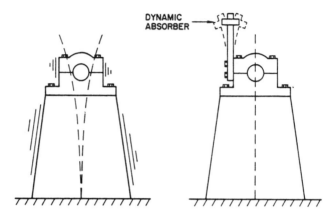

Figure 5-21. A dynamic absorber applied to control bearing resonance.

Example 5-5. Plots similar to the one in Figure 5-22 are sometimes obtained when noncontact pickups are used to identify rotor critical speeds. The plot does display a peak amplitude and the normal 180° phase shift; however, it should be further noted that the amplitude plot includes a dip at approximately 5000 rpm—in this case, just slightly below the shaft critical speed. The amplitude dip in Figure 5-22 does not have a substantial corresponding phase shift and, thus, is not likely the result of an anti-node, which is plotted in Figure 5-19. The cause of the indicated amplitude reduction (dip) in the plot in Figure 5-22 can usually be traced to excessive electrical and/or mechanical run-out of the shaft at the noncontact pickup target area. The noncontact or proximity pickup cannot distinguish between actual shaft vibration and any run-out or eccentricity of the shaft journal. As a result, the noncontact pickup provides a signal proportional to the *vector sum* of run-out and actual shaft vibration.

Operating well below its critical speed, a shaft is considered to be rigid, and thus shaft vibration will essentially be in-phase with unbalance. However, as the rpm of the shaft approaches its critical speed, shaft deflection will progressively increase. In addition, the deflection in the shaft (actual shaft vibration) begins to lag the heavy spot of unbalance. In fact, when operating at critical speed, shaft deflection lags the unbalance heavy spot by 90°; and at operating speeds above shaft critical, deflection will lag the unbalance by 180°. The important point to note here is that the amplitude *and phase* of shaft vibration are undergoing a change as the shaft passes through critical speed. As a result, the *vector sum* of shaft vibration and run-out is also undergoing a change. The dip, or amplitude reduction, plotted in Figure 5-22 results when the shaft vibration and run-out vectors become momentarily out of phase and, thus, tend to cancel one another. Of course, instead of cancelling one another, the run-out and shaft vibration vectors could momentarily add to reveal a plot such as the one in Figure 5-23. In either case, it is important to note that when using noncontact pickups, the presence of excessive run-out may give a very distorted picture of rotor response. The recorded data may suggest a condition which is far better or far worse than the true response of the rotor. Therefore, it is advisable to carefully measure

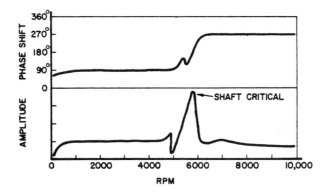

Figure 5-22. Amplitude reduction due to out-of-phase relationship of shaft vibration and run-out vectors near 5000 rpm.

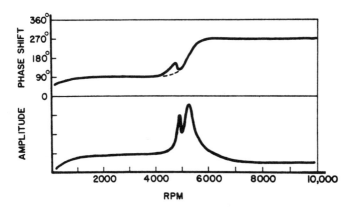

Figure 5-23. Amplitude increase due to in-phase relationship of shaft vibration and runout vectors near 5000 rpm.

run-out amplitude before response measurements are taken. Of course, excessive run-out should be physically or electronically eliminated so that plots of true rotor response can be obtained.

Example 5-6. The recorded plot in Figure 5-24 illustrates a situation where the recorded data can be somewhat misleading. The suspicious data in the plot, Figure 5-24, are the somewhat unusual phase shifts that occur at 6000 and 7000 rpm. While these apparent phase shifts may suggest resonant frequencies not indicated on the amplitude plot, it is possible that these are not phase shifts at all. From our previous discussion, recall that *two* signals are required to obtain a D.C. voltage proportional to phase for plotting purposes. These signals are a voltage pulse from a reference pickup at the shaft of the machine and a vibration signal from a vibration pickup. If either one of these signals is lost or lacks sufficient amplitude, the phase voltage will

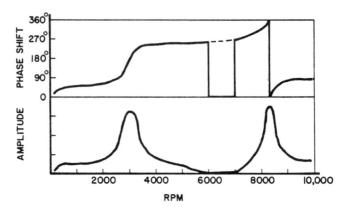

Figure 5-24. Apparent phase shift attributed to absence of sufficient vibration signal to provide a true phase reading.

drop to zero. Referring to the amplitude plot in Figure 5-24, note that the vibration amplitude is at a relatively low value between 6000 and 7000 rpm. Note also that the phase plot drops to 0° between 6000 and 7000 rpm. Therefore, it is more likely that the apparent phase shifts are nothing more than the absence of a sufficient vibration signal to provide a true-phase reading. Below 6000 rpm and above 7000 rpm the vibration amplitude has increased sufficiently to provide these indications.

Example 5-7. The phase plot in Figure 5-25 illustrates another situation where the data can be somewhat misleading. From the phase plot, it appears that there were three abrupt phase shifts of 360°, each within the speed range of 4500-6000 rpm. However, this type of phase plot is not unusual where the phase indication tends to vary around 0° (360°), and it does not suggest anything unusual or abnormal as far as system response is concerned. More likely, the unusual phase shifts between 4500 and 6000 rpm in Figure 5-25 result from a phase indication which may have been varying only slightly, say between 1° and 359°, which is only a 2° variation. Of course, if the phase changes from 1° to 359°, the plotter pen must travel all the way from the bottom to the top of the phase chart. And, if the phase should shift slightly from 359° back to 1°, then the pen must again travel the full distance from the top of the phase chart to the bottom. This tends to produce a cluttered and somewhat confusing phase picture; however, as long as the cause is understood, the data should not be misinterpreted.

Filter-In and Filter-Out Amplitude Versus rpm

In the preceding paragraphs, a number of plots of vibration amplitude versus rpm were illustrated and discussed. In each example, the recorded vibration amplitude occurred at the rotating speed frequency (1 × rpm) of the machine and/or structure. However, these plots alone may not reveal the system response to other vibration frequencies which can affect overall machine performance.

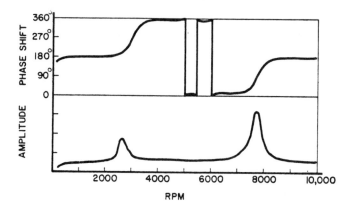

Figure 5-25. Insignificant phase shift from, say 359 deg. to 1 deg., can cause instrument pen to travel and give erroneous impression relative to phase reversals.

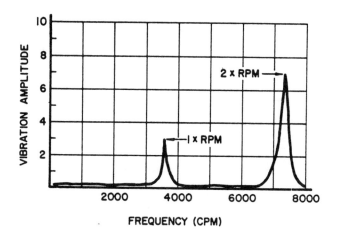

Figure 5-26. Once and twice rpm indications often appear in amplitude vs. frequency signatures.

To illustrate, assume that a 3600-rpm pump has a vibration amplitude-versus-frequency signature as illustrated in Figure 5-26, obtained with the pump operating at 3600 rpm. In this instance, the pump has two significant exciting frequencies of vibration—one occurring at 1 × rpm, due perhaps to unbalance; and another vibration at 2 × rpm (7200 CPM), possibly the result of misalignment. Further assume that the pump in question has a resonant frequency at 2000 CPM. Obviously, as the pump coasts down in speed, the unbalance at 1 × rpm will excite the resonance when the speed of the pump reaches 2000 RPM. In addition, as the pump continues to coast down, resonance will again be excited when the 2 × rpm vibration reaches the resonant frequency of 2000 CPM. In other words, when pump speed equals 1000 rpm, the 2 × rpm (2000 CPM) misalignment vibration will excite the 2000 CPM system resonance. This excited resonance will not appear on the filter-in amplitude-versus-rpm plot because in this mode, the analyzer filter is automatically tuned to 1 × rpm and rejects all other frequencies.

Because a filter-in amplitude-versus-rpm plot does not always give a complete picture of total system response, a common practice is to obtain two plots of amplitude versus rpm—one filter-in, synchronized to 1 × rpm, and a second filter-out amplitude-versus-rpm plot which gives a more complete picture of total system response. The two plots in Figure 5-27 are the comparative filter-in and filter-out plots one might expect from the 3600-rpm pump described above. Note in the filter-in plot that the amplitude peaks only at 2000 rpm—the resonant frequency. However, in the filter-out plot, an additional, even more severe, amplitude peak occurred at 1000 rpm, when the 2 × rpm misalignment vibration excited the 2000 CPM resonant frequency. This is not unusual, since the 2 × rpm vibration in Figure 5-26 had a higher initial amplitude to begin with. The value of filter-out amplitude-versus-rpm plots should be obvious from this example.

When obtaining comparative filter-out and filter-in versus rpm plots during machinery startup or coastdown, the objective is to detect significant differences in the two plots—in particular, any amplitude peaks which appeared in the filter-out plot but not in the filter-in plot. When this occurs, as it did in Figure 5-27, the cause can usually be traced to one of the following:

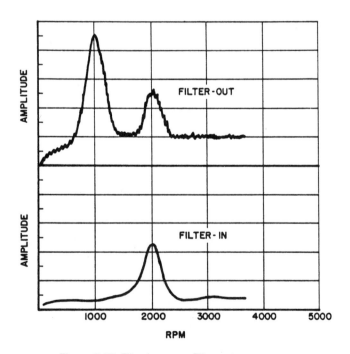

Figure 5-27. Filter-in versus Filter-out, amp-vs.-rpm.

1. *A harmonic (or sub-harmonic) of rotating speed exciting resonant frequencies.* These can normally be identified because the additional peak(s) which appear on the filter-out amplitude plot will generally occur at exact submultiple frequencies of the resonant frequencies identified by the filter-in amplitude plot. This is illustrated in Figure 5-27, where the additional peak occurs at 1000 rpm, which is one-half the known resonant frequency of 2000 CPM.

   Of course, harmonic vibration frequencies may also excite resonant frequencies which are *above* the maximum rpm of the machine. For example, suppose that the machine analyzed in Figure 5-26 has a resonant frequency at 6200 CPM. Since this is above the maximum rpm (3600 rpm) of the machine, the 6200 CPM resonance will not be excited by 1 × rpm vibration during coastdown. However, the 6200 CPM resonance will be excited when the 2 × rpm vibration reaches 6200 CPM. This will occur at 3100 rpm. Therefore, the filter-out amplitude plot would show an additional amplitude peak at 3100 rpm.

2. *Another rotating component operating at a different rpm exciting resonant frequencies.* To illustrate how this generates additional peaks in the filter-out amplitude plot, consider a system consisting of a 3600-rpm motor driving a fan at 1100 rpm through a 3.27:1 gear reducer. If the system has a resonance at, say, 800 CPM, a filter-in amplitude-versus-fan rpm plot would reveal an amplitude peak at 800 rpm. However, the motor might also excite the 800 CPM resonance when the motor reaches 800 rpm. And, when the motor is operating at 800 rpm, the fan speed is approximately 245 rpm (based on the gear reduction ratio of 3.27:1). Therefore, the filter-out amplitude-versus-fan rpm plot would indicate an additional amplitude peak at 245 rpm.

3. *An additional (new) frequency of vibration which arises at a particular speed range during startup or coastdown.* One example of this condition has been observed on centrifugal compressors and other high-speed machines which, on occasion, have excited an oil whirl or other instability at a particular rpm during startup or coastdown. Since the frequency of oil whirl vibration is slightly less than one-half rpm, the increase in vibration due to momentary oil whirl does not appear on the filter-in amplitude plot. However, the oil whirl would produce an additional amplitude peak on the filter-out amplitude-versus-rpm plot.

Stroboscopic Observation

It was mentioned previously that most basic vibration analysis instruments are typically furnished with a high-intensity stroboscopic (strobe) light. The strobe light, actually triggered by the vibration, is an extremely valuable tool for phase analysis and in-place dynamic balancing; but quite often the strobe light must be used to simply locate the source of vibration.

When an analysis of vibration amplitude versus frequency is performed as illustrated in Figures 5-1 or 5-2, it is often assumed that certain frequencies found are those occurring at one, two, three, or some other multiple of rotating speed, based on relating machine rpm with the frequency reading obtained from the X-Y chart or the frequency meter of the analyzer. However, the accuracy of the X-Y chart is only as good as the initial set-up and calibration, and frequency-meter accuracy is generally limited to two or, at most, three significant figures. As a result, it is not realistic to assume that the measured frequency of a vibration is exactly equal to some multiple of the rotating speed of a machine component; instead, this should be positively confirmed by observing the various rotating components with the strobe light. Since the strobe light and analyzer frequency meter are, for all practical purposes, "triggered" by the same vibration signal, if the measured frequency of vibration and machine rpm (or multiple) are the same, the strobe light will make the rotor appear to stand still, with one or more reference marks visible.

However, if the frequency being measured is not exactly the same as shaft speed or some multiple of shaft speed, the rotor will not appear to stand still when observed with the strobe light. If the vibration is actually coming from another part of the machine or perhaps from a nearby machine, the strobe image may appear erratic or the reference mark may appear to rotate slowly.

Some machines will have mechanical problems or rotating components which can produce vibration frequencies very close together. In such cases, it is essential that the strobe light be used to confirm the source of vibration. For example, an induction motor with a rotating speed of 1750 rpm might have an unbalance in the rotor which will generate a vibration frequency of 1750 CPM. However, the motor might also have an electrical problem such as an open or short circuit in the stator windings, unbalanced phases, or unequal air gap, which will result in vibration at a frequency equal to the rotating speed of the motor's magnetic field, which in this case will be 1800 CPM. Since these two sources of motor vibration have frequencies very close together, separated by only 50 CPM, it would be difficult to distinguish one from the other based on the frequency resolution provided by the frequency meter or plotted signature. However, observing the motor rotor with the strobe light will quickly confirm the source of vibration. If the rotor appears to stand still when viewed with the strobe light, the vibration is likely due to unbalance; if the rotor appears to rotate slowly when viewed with the light, the problem is not unbalance but is likely due to electrical problems.

The importance of utilizing the strobe light to confirm the source of vibration or noise can be best illustrated with two case histories.

Case History 5-1. This example involved the analysis of excessive vibration of an 1100-rpm fan, belt driven by a 1725-rpm motor. The vibration signatures obtained on the unit disclosed a predominant vibration at approximately 1100 CPM, with high amplitudes in both the horizontal and vertical directions. The evaluation of the vibration signatures clearly suggested that fan unbalance was the cause of vibration. However, when the fan rotor was observed with the strobe light, it appeared to rotate very slowly, indicating that the vibration was not occurring exactly at fan rpm. Thus, the problem was *not* fan unbalance. Observing the other machine components with the strobe light, it was discovered that the drive belts appeared perfectly stationary, identifying the drive belts as the actual source of trouble. The vibration being generated was occurring at a multiple of belt rpm.

Case History 5-2. A major paper manufacturer decided to increase the speed of an older paper machine in order to increase production. However, it was soon discovered that the machine had to pass through a severe critical (resonant) speed before the higher operating speed could be achieved. As machine speed approached the critical, the vibration amplitudes became so severe that operators refused to increase speed further for fear of causing extensive damange to the machine and structure. A vibration frequency analysis was performed with the machine operating slightly below critical in an attempt to discover the exciting force. It was believed that roll unbalance was the cause, and that if the roll(s) at fault could be identified and balanced, the machine could be run safely through critical to the desired operating speed. The frequency analysis confirmed that the predominant vibration occurred at roll rpm; however, there were several rolls in the section, all approximately the same diameter and rotating at approximately the same rpm. The problem at this point was to determine which roll was causing the predominant vibration. The strobe light, triggered by the vibration, solved the problem. By carefully observing each roll with the strobe light, it was noted that all but one of the rolls appeared to rotate very slowly. Since unbalance vibration occurs at exactly 1 × rpm of the unbalanced roll, that one roll which did not rotate when viewed with the strobe light was identified as the problem roll. Once identified, the problem roll was replaced with one that had been carefully balanced. This resulted in a substantial reduction in vibration and made it possible to increase speed safely through the critical.

In addition to triggering the stroboscopic light with the vibration signal, the light can also be flashed at any desired rate by adjusting an internal oscillator. This provision makes it possible to perform "slow-motion" studies with the strobe light by simply adjusting the flash rate slightly faster or slightly slower than the rpm of the part. To illustrate the value of the strobe light for slow-motion studies, a belt-driven dryer fan was producing excessive vibration, and a frequency analysis revealed that almost all of the vibration was occurring at 1 × rpm of the fan. This, combined with low axial amplitudes, suggested that unbalance was the major problem. However, several attempts to balance the fan produced little improvement. Then, while adjusting the analyzer filter with the internal oscillator and strobe light, the operator observed the fan pulley in slow motion and noted that the drive belt appeared to ride up and down in the pulley groove, revealing uneven wear of the pulley, resulting in eccentricity of its pitch diameter. After a new pulley was installed, the vibration levels were sufficiently reduced to make trim balancing unnecessary.

Amplitude Versus Phase (Nyquist Plots)

In the preceding section, several examples were presented where amplitude and phase-versus-rpm plots were obtained to identify resonances and other unusual machinery problems. These plots, sometimes called Bodé plots, are typically recorded on a Cartesian (rectangular) coordinate graph. Another technique for presenting these data is to plot the unbalance amplitude-versus-phase vector on polar-coordinate graph. Such plots are often referred to as Nyquist plots. The plots in Figure 5-28 illustrate how the amplitude-versus-phase data recorded on a polar-coordinate graph compare to the familiar amplitude and phase-versus-rpm data plotted on a Cartesian-coordinate graph.

Obtaining polar amplitude-versus-phase plots requires an instrument such as the IRD Model 360, Figure 5-7, with tracking filter and polar amplitude and phase output signals for driving a standard X-Y recorder. The X-Y recorder is adjusted so

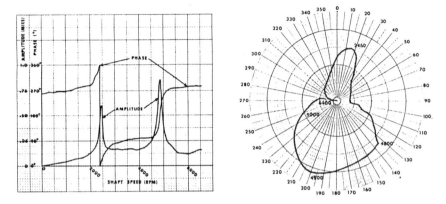

Figure 5-28. Comparing Bode (left) and Nyquist plots (right).

that a zero-amplitude signal positions the pen at the origin or center of the polar graph paper. The amplitude FULL SCALE control on the recorder is adjusted so that any increase in vibration amplitude will cause the pen to move radially outward from the center a proportionate amount. The angular direction in which the recorder pen moves when amplitude increases will, of course, be governed by the phase of the vibration.

Polar amplitude-versus-phase plots offer many advantages over Bodé plots. A Nyquist plot:

1. Provides an immediate indication of the unbalance vector without the need to compare separate amplitude and phase plots.
2. Eliminates confusing phase discontinuities that appear on Bodé plots as the phase changes from 0° to 360° (see Figure 5-25).
3. Eliminates phase discontinuities that appear on Bodé plots when the vibration amplitude has reduced to a low level where there is not sufficient vibration signal to maintain a phase indication (see Figure 5-24).
4. Where only a single pen recorder is available, the Nyquist plot allows both amplitude and phase data to be obtained during a single run-up or coast-down of the machine.
5. Where noncontact pickups are being used, the vector resulting from combined mechanical and electrical run-out can be eliminated by recalibrating the recorder amplitude ZERO control to the origin of the polar graph while the run-out vector is being displayed with the machine on slow roll. Figure 5-29 shows comparative Nyquist plots obtained with and without recalibrating for shaft run out. As illustrated, run-out can result in a substantial unbalance vector at any one operating speed.
6. Nyquist plots obtained for each bearing of a machine can be easily compared to provide information on the "mode shape" of each resonance detected. To illustrate, assume that the two plots in Figure 5-30 were obtained at the two bearings of a large turbine during startup. Note that both plots reveal amplitude

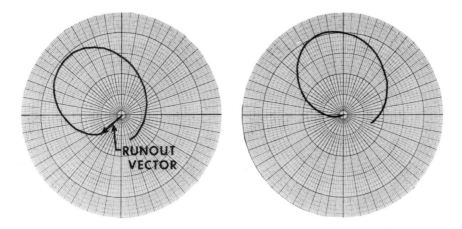

Figure 5-29. Nyquist plots with and without runout compensation.

peaks at speeds of 1200, 2200, and 3000 rpm. The amplitude peaks, together with the corresponding shifts in phase, confirm that all three are resonant frequencies. In addition, both bearings show the same phase indication for the resonant amplitude peak at 1200 rpm. This suggests that the resonance at 1200 rpm is the first rigid mode of the bearings (structure) diagrammed in Figure 5-31. Further, the plots show that the amplitude peaks at 2200 rpm have opposite (180°) phase. This indicates that the resonance at 2200 rpm is the second rigid mode of the bearings, as illustrated in Figure 5-31. Finally, since the amplitude peaks at 3000 rpm again show identical phase indications on the two plots, it can be concluded that the resonance at 3000 rpm is the first bending critical of the turbine rotor, as illustrated in Figure 5-31. Although this same information and interpretation could be obtained from Bodé plots, the comparison is made considerably easier with polar amplitude and phase (Nyquist) plots.

Despite their advantages and many useful applications, Nyquist plots nevertheless have certain limitations when compared to amplitude and phase-versus-rpm plots. Major limitations include:

1. While the Nyquist plot does reveal the presence and significance of resonant conditions during run-up or coast-down, it does not provide a speed reference to indicate the rpm at which resonance occurs. As a result, it is necessary to monitor machine speed and manually record rpm values as noted on the sample plots in Figures 5-28 and 5-30.
2. The Nyquist plot is a plot of the unbalance (synchronous) vibration and, thus, is restricted to examining only the vibration occurring at 1 × rpm. As indicated previously, severe resonant conditions can be excited by vibration frequencies other than 1 × rpm unbalance. Or, during startup or coastdown, additional

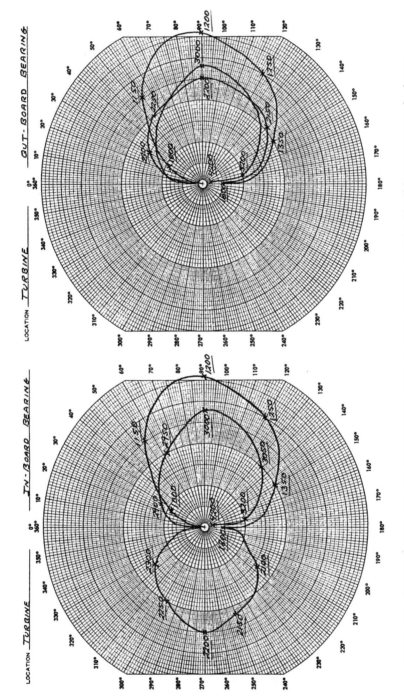

Figure 5-30. Comparative Nyquist plots obtained for each bearing of the machine can aid in identifying resonant mode shapes.

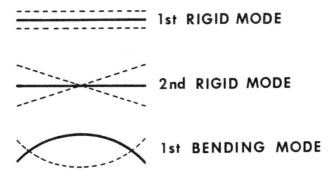

Figure 5-31. Resonant mode shapes.

problems such as oil whirl, resonance whirl, or rubs may occur, resulting in nonsynchronous vibration frequencies. Evidence of these and other problems can be seen when the overall or unfiltered vibration is plotted versus rpm, Figure 5-27. Such problems are likely to go undetected if only Nyquist plots are obtained.

Time Waveform Analysis

Although vibration analysis using the amplitude-versus-frequency characteristics is generally adequate for identifying most machinery problems, sometimes additional information is needed to diagnose a particular defect or to study the dynamic behavior of a machine under specific operating conditions. One additional technique which is often very useful is the observation of the time waveform of a vibration signal on an oscilloscope, as illustrated in Figure 5-32. The vibration signal is applied to the scope vertical input, and the vertical axis on the CRT (cathode-ray tube) is scaled in amplitude. The horizontal axis of the scope is scaled in time, such as seconds or milliseconds.

Significance of Time Waveform Analysis

An amplitude-versus-frequency analysis does not always provide complete information about a vibration. Some mechanical problems may have identical frequencies yet differ considerably in terms of dynamic behavior. To illustrate, consider the vibration generated by a gear having one chipped, broken, or deformed tooth. The defective tooth will impact the mating gear each revolution, resulting in a vibration at a frequency of 1 × rpm of the bad gear. This is the same frequency generated by unbalance. Thus, if only an amplitude-versus-frequency analysis was made, it is quite possible that the defective gear tooth could be mistaken for an unbalance condition. However, the time waveform generated by the defective gear

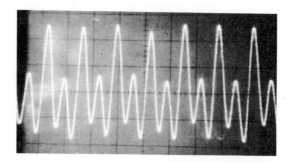

Figure 5-32. Typical time-waveform display.

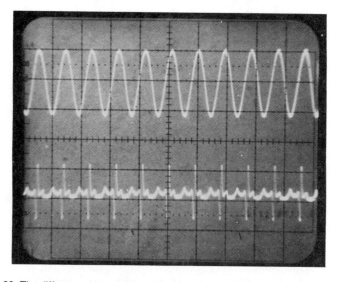

Figure 5-33. The difference between unbalance vibration and that generated by a defective gear tooth would be clearly indicated by observing the vibration time wave-form.

tooth would differ noticeably from that caused by unbalance, as illustrated in Figure 5-33. The unbalance would produce a sinusoidal waveform, whereas the defective tooth, being in contact for only a brief instant in the 1 × rpm cycle, would show a distinctive spike-like appearance. In addition, due to the short duration of the tooth-excited vibration, it is doubtful that the filtered amplitude-versus-frequency analysis would reveal the true amplitude of the vibration. When a spike-like signal such as that in Figure 5-33 is filtered, the amplitude may be significantly attenuated. The oscilloscope allows the true peak amplitude to be observed.

Another benefit of viewing the vibration time waveform results from the instantaneous, undamped response provided by the oscilloscope. This makes the scope an ideal tool for evaluating short-lived, transient types of vibration.

For example, the waveform in Figure 5-34 shows the transient vibration resulting from the startup of an electric motor. This vibration is present only for a very short period of time, yet the true peak amplitude is easily noted. By comparison, the amplitude meters incorporated in most vibration meters and analyzers will damp the meter movement in order to minimize erratic movement of the pointer. However, this damping tends to slow the response of the meter, so that the meter does not register the true peak amplitude. Even digital meters used on some instruments do not provide an indication of the true peak amplitudes of transient vibrations, since these meters must "count" the data for a selected period of time before the reading is displayed.

From this discussion, it can be seen that there are definite benefits to viewing the time waveform of the vibration on an oscilloscope. Additional applications are suggested by numerous examples included later in this chapter.

Obtaining Waveform Data

All high-quality portable vibration analyzers provide a receptacle for connecting an oscilloscope to display the time waveform of a vibration signal. Practically any standard or memory oscilloscope can be used. The real-time analyzer, Figure 5-8, can also be used to display the time waveform from data locked in its digital memory. The real-time analyzer does not require the controls to be constantly readjusted in an attempt to obtain a stationary display of the signals of interest. This is true for steady-state signals, but is even more important for nonsteady state and short transient data. For this reason, most of the time waveform examples presented were obtained with the real-time analyzer.

When a standard, high-quality vibration analyzer and oscilloscope are being used, simply connect the oscilloscope (SCOPE) output of the analyzer to the vertical input of the oscilloscope. Before applying the vibration signal, the scope horizontal and vertical position controls should be adjusted to center the trace on the CRT.

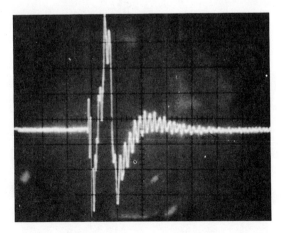

Figure 5-34. Waveform analysis of startup transient of an electric motor.

In order to obtain reasonably accurate amplitude readings from the time waveform, the oscilloscope vertical gain control must be properly adjusted. The simplest way to accomplish this is to tune the analyzer filter to a steady-state vibration such as 1 × rpm of the machine. Then simply note the amplitude of vibration from the analyzer amplitude meter and adjust the scope vertical gain control until the peak or peak-to-peak value of the waveform displayed on the CRT corresponds to the reading on the meter. This is illustrated in Figure 5-35, where the vertical gain control has been adjusted so that each division on the scope graticule represents 1 mil displacement.

If the oscilloscope is not being used with an analyzer instrument and the transducer signals are being applied directly to the scope, it will be necessary to know the sensitivity of the input signal (i.e. mv/mil, mv/in/sec, mv/g., etc.) so that the vertical gain can be set to an appropriate range.

To illustrate, suppose that the signal from a noncontact signal sensor is being applied to the oscilloscope, and that the signal sensor has a sensitivity of 200 mv/mil. If the vertical gain control has been set to a range of 100 millivolts/division to achieve a good display, the amplitude of the vibration can be found by noting the height of the waveform in divisions. For example, if the waveform is 7.5 divisions high, the signal voltage is 750 millivolts (7.5 divisions × 100/mv/div = 750 mv). Dividing the signal level by the input sensitivity provides the amplitude of the vibration being measured. Thus, 750 mv ÷ 200 mv/mil = 3.75 mils peak-to-peak.

To obtain the most benefit from time waveform analysis, the vibration signal should be examined in several different ways. Just as filtering is important to amplitude-versus-frequency analysis, filtering can prove very useful, and even essential, in time analysis. This is particularly true when the signal is complex, with many high-frequency components. In such a case, the time waveform often appears

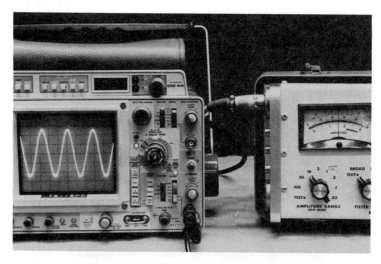

Figure 5-35. Here the oscilloscope has been calibrated such that each vertical division represents 1 mil. The amplitude indicated is 5 mils peak-to-peak displacement.

as nothing more than random vibrations having few distinctive characteristics. This will contribute little additional understanding of the machine's mechanical condition. With the use of variable high-pass, low-pass, and band-pass filters, however, the signal can be broken down into simpler vibration components, with resulting clarification of the vibration. Where filtering the signal is not possible, some frequency discrimination can be achieved by selecting a more appropriate parameter of amplitude displacement, velocity, or acceleration. For example, if a vibration signal consists of both low and high frequencies, the time waveform of vibration velocity may appear quite complex and confusing.

If there is only interest in studying the low-frequency vibrations, the displacement mode could be selected to provide additional emphasis on the low frequencies. Where the interest is in studying high frequencies, using the acceleration (g's) mode of measurement will provide additional emphasis to the high frequencies and will tend to reject the low-frequency signals.

Along with the use of filtering, it is often helpful to analyze the time waveform from the viewpoint of what might be called the "long term" and "short term." In the long-term time analysis the time display may cover a period of 5-10 seconds; or putting it differently, it would cover a period of about 50-100 cycles of the dominant frequency of interest. The waveform shown in Figure 5-36 is typical of a long-term display of the time waveform and covers approximately five seconds. Such long-term displays are particularly useful for studying low-frequency "beats," low-frequency amplitude modulations, and the behavior of a mechanical system as it passes through resonance (see Figure 5-37).

The short-term time analysis typically covers 50-100 milliseconds. Figure 5-32 illustrates a short-term display where only a few cycles of the vibration are presented. Generally, the time display is adjusted so that the details of the vibrations of interest can be examined.

Obtaining Frequency Data from the Waveform

Calibrating the vertical scope display for amplitude readout has already been discussed. Frequency data can also be obtained from the time waveform by simply counting the number of cycles of a vibration occurring in a known time interval. For

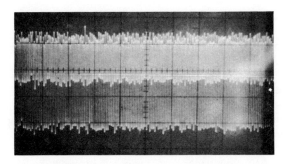

Figure 5-36. Typical long-term waveform.

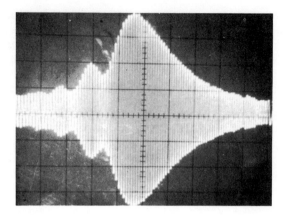

Figure 5-37. Long-term waveform obtained during machine start-up or coast-down identifies resonances.

example, the long-term display in Figure 5-36 represents a time period of five seconds. Since the scope graticule is divided into 10 divisions, each division in this case represents one-half second; and by counting, it is found that 15 cycles of the vibration occur within one division, or a half-second. Thus, the frequency of this vibration is 30 cycles per second, or 30 Hz. The frequency in cycles per minute is found by multiplying the frequency in cycles per second by 60 (30 Hz × 60 = 1800 CPM). Therefore, the fundamental frequency of vibration in Figure 5-36 is approximately 1800 CPM.

For the short-term display in Figure 5-38, where less than two complete cycles are displayed, the approximate frequency can be found by first determining the period for one complete cycle. The total time displayed on the scope in Figure 5-38 is 0.1 second, and it can be noted that one complete cycle of the vibration required approximately 7 of the 10 graticule divisions, or 70% of the time displayed. Therefore, the one cycle requires a period of 0.7 × 0.1 second, or 0.07 second. The frequency of this vibration is the reciprocal of its period, or 1.0/0.07. From this, the frequency is found to be approximately 14.3 Hz, or 858 cycles per minute.

Examples of Time Waveform Analysis

The benefits which can be derived from observing the time waveform of vibration can best be illustrated by some typical examples.

Startup transient analysis. Machines such as electric motors used in applications requiring frequent starts and stops may experience premature bearing, coupling, or even shaft failures because of high vibration forces which may be generated each time the machine is started or stopped. The time waveform in Figure 5-34 shows the high amplitude of vibration generated during startup of an electric motor. As can be seen, the overall background vibration is very low in amplitude, as is the vibration level under normal operating conditions. As a result, a normal amplitude-versus-fre-

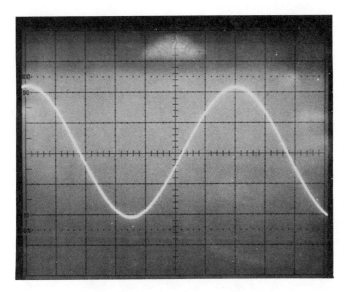

Figure 5-38. Typical short-term display.

quency analysis of the background vibration under normal operating conditions would not reveal the problem source. Further, due to its short duration, the true significance of the startup transient would not likely be indicated by observing the amplitude meter of the analyzer or vibration monitor, since meter damping would prevent the instrument from responding to the true peak amplitude of vibration. However, observing the time waveform during startup, Figure 5-34, reveals the true dynamic response of the machine and the severity of the startup transient.

Unbalance versus other problems. As illustrated in Figure 5-33, a gear having a single broken, chipped, or deformed tooth could generate vibration at a frequency equal to 1 × rpm and appear quite similar to unbalance if only the amplitude-versus-frequency data were obtained. The time waveform data in Figure 5-33 clearly show how these problems can be distinguished. There are, of course, other mechanical problems which cause vibration at a frequency of 1 × rpm, including reciprocating forces, eccentricity (reaction forces), rubbing, and misalignment. All of these may be distinguished from unbalance by analyzing the time waveform data.

Measuring electrical vibration. Electric motors can develop problems such as open or shorted windings, broken rotor bars, unequal rotor-to-stator air gap, unbalanced phases, and other conditions which result in unbalanced magnetic forces acting on the rotor and stator. These problems are generally termed "electrical," since the forces and resultant vibrations are only generated when electrical power is being applied. As a result, the usual check to determine whether or not a vibration is due to electrical problems is to observe the change in overall vibration amplitude the instant electrical power is removed from the machine. If the vibration amplitude decreases rapidly as soon as power is disconnected, the problem is very likely electrical. If the amplitude does not decrease instantly but decreases only gradually after electrical power has been disconnected, then the vibration is more likely due to mechanical problems such as unbalance or misalignment. In any event, this check is adequate for many applications, but may not be adequate in all cases. For example, some machines may come to rest very quickly once power is disconnected, and damping of the amplitude meter may prevent a true indication of how quickly the amplitude declined. Or, the vibration may be complex, consisting of two or more significant frequencies. For example, the waveform in Figure 5-39 shows the instantaneous change in the vibration of an electric motor the instant power was disconnected. The fundamental frequency which is present both before and after power was removed is a 1 × rpm vibration of approximately 1800 CPM, caused by unbalance. The vibration component which disappeared the instant power was removed was a vibration at a frequency of 7200 CPM due to electrical "torque pulses." The time waveform in Figure 5-39 clearly shows the extent of the electrically excited vibration on this motor. It is doubtful that observing the overall vibration amplitude on the analyzer amplitude meter would have provided this same information. As can be seen from the time waveform, the overall vibration amplitude after power was disconnected is not much less than the initial overall amplitude. In fact, it appears from the waveform that the overall amplitude actually begins to *increase* noticeably approximately four cycles (.14 second) after power was cut, indicating that the motor may be coasting down to a resonant frequency.

Confirming harmonic vibration frequencies. When performing a vibration analysis, confirming that a particular vibration frequency is an *exact* multiple or submultiple of another frequency can be extremely important when making the final diagnosis. Normally, such confirmation can be made by observing the rotating components with a stroboscopic light, since vibration frequencies which are exact multiples or submultiples of rotating speed will make the rotor appear to stand still under the strobe light. However, if the machine rotor is totally enclosed, use of the strobe light may not be possible. In such cases, the harmonic relationship between

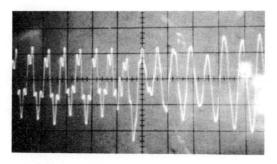

Figure 5-39. Time waveform showing vibration just before and after disconnecting power to motor clearly shows significance of electrically excited vibration.

two or more vibration frequencies can be determined by observing the time waveform of the vibration. To illustrate, Figure 5-40 shows the amplitude-versus-frequency data obtained on a reciprocating chiller compressor direct-coupled to a drive motor.

The vibration frequencies noted "appear" to be those at one and two times rotating speed. However, due to limited frequency resolution, it is not possible to determine from this signature whether the higher frequency is in fact the second harmonic or merely a frequency close to that of the second harmonic. By observing the time waveform of the vibration, Figure 5-41, it can be seen that both frequencies are quite steady and that the phase between them is essentially constant, which confirms that the two frequencies are harmonically related.

In comparison with the previous example, Figure 5-42 shows the vibration signature which was obtained on the motor driving the chiller compressor. Again, the signature reveals two vibration frequencies which appear to be one and two times rpm. However, observing the time waveform of these vibrations, Figure 5-43, it can be seen that the phase relationship is continually changing with time, indicating that

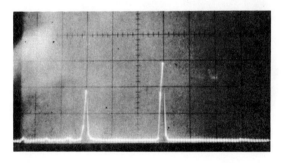

Figure 5-40. Amplitude vs. frequency analysis reveals apparent 1 × rpm and 2 × rpm vibration.

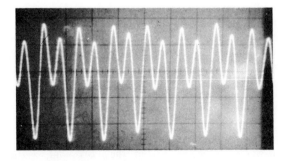

Figure 5-41. Observing the vibration waveform confirms that the two frequencies (Figure 40) are, in fact, harmonically related.

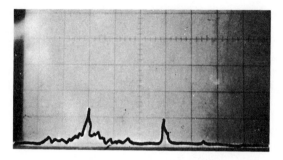

Figure 5-42. Frequency analysis of drive motor reveals apparent 1 × rpm and 2 × rpm vibration.

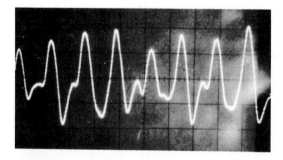

Figure 5-43. Observing the vibration waveform reveals that the two frequencies (Figure 42) are *not* harmonically related.

these two frequencies are *not* harmonically related. Further checks disclosed that the lower frequency was, in fact, at 1 × rpm, or approximately 1750 CPM. However, the higher frequency of vibration was caused by electrical sources at a frequency of 60 Hz, or 3600 CPM. The 2 × rpm vibration, if present, would have a frequency of 2 × 1750, or 3500 CPM.

The preceding examples represent only a very few instances where time waveform data can be beneficial for vibration analysis. In the section "Data Interpretation," later in this chapter, additional examples are included to illustrate the application of time waveform presentations for bearing analysis, reciprocating machines, machine tools, electric motors, and others.

Lissajous Pattern (Orbit) Analysis

Most machinery problems can be readily identified by obtaining comparative amplitude-versus-frequency signatures in the horizontal, vertical, and axial directions at each bearing of the machine. However, it is not always possible to obtain these tri-axial measurements where noncontact pickups are being employed, since it is not common practice to install vibration-sensing noncontact pickups in the axial direction at each bearing. As a result, it may be necessary to supplement the amplitude-versus-frequency data obtained from radially mounted noncontact pickups with additional modes of display in order to identify certain mechanical problems. One way to supplement the frequency data is to view the time waveform as discussed in the preceding paragraphs. Another technique is to mount two radial noncontact pickups in each bearing, with the axes of the pickups separated by 90°, as shown in Figure 5-44. The signal from one pickup is applied to the horizontal input of an oscilloscope, while the output of the other pickup is applied to the vertical scope input. The resultant display on the CRT will be a plot of the total motion of the shaft within the bearing (see Figure 5-44). Such plots or displays are called Lissajous patterns, and are also referred to as shaft "orbits."

In recent years, it has become common practice to install dual noncontact (X-Y) pickups on critical high-speed turbomachines, as shown in Figure 5-44. Made part of an electronic logic system, they provide redundant, failsafe protection and avoid false shutdowns should one of the pickups fail. These pickups have captured many Lissajous patterns relating to machine malfunctions such as unbalance, misalignment, rubbing, oil whirl, resonance, etc. In the following paragraphs, the general technique of obtaining Lissajous patterns is discussed, followed by a brief coverage of patterns typical of problems often encountered on turbomachines.

Equipment Set-Up for Lissajous Patterns

Assuming that the noncontact pickups have been properly calibrated and installed, the only remaining equipment needed to obtain Lissajous patterns is an oscilloscope. Where permanently installed monitors are being used, the signals necessary to drive the oscilloscope can be obtained directly from the noncontact signal sensors (oscillator demodulators), which are typically installed on or near the machine. Alternatively, the signals can be taken directly from the analyzer receptacles located on the front panel of the monitor instrument, which is usually

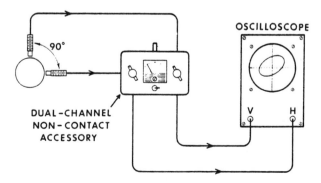

Figure 5-44. Typical setup for displaying Lissajous patterns using non-contact pickups.

located in a control room. If the machine is not being protected by a monitor system, a two-channel, noncontact accessory as shown in Figure 5-44 can be used.

It is preferable if the noncontact pickups are installed with their axes in the true horizontal and vertical directions, as this orientation greatly simplifies interpretation of the pattern observed on the CRT. Vertical motion of the shaft within the bearing would be seen as a vertical trace on the scope, the bottom of the CRT would relate to the bottom of the bearing, etc. However, because of split lines in the bearing housing, it may not be possible to locate the pickups in the true horizontal and vertical directions. In some cases, the pickups may be mounted 45° from true horizontal. This will not affect the shape of the observed pattern, but it will cause the display to be "rotated" 45° on the CRT. To avoid confusion, the operator may wish to mark lines on the scope face representing the true horizontal and vertical axes of the bearing relative to the pickup locations.

Another point to consider when setting up the oscilloscope to observe shaft motion is the relationship between the direction of shaft motion and the resultant direction of the trace on the scope. For example, if a noncontact pickup is mounted vertically, looking down at the shaft, it only seems proper that if the shaft moves upward toward the pickup that this would be seen as a upward movement of the trace on the CRT. However, this may *not* be the case, depending on the polarity of the power supply being used to power the noncontact signal sensor (oscillator demodulator). Nearly all oscilloscopes are designed such that an increase in positive (+) voltage of the input signal [or a decrease in minus (−) voltage] will cause the vertical trace to move *upward* or the horizontal trace to move to the *right*. In practice, when using noncontact pickups to observe shaft motion, the direction that the trace will move on the CRT will depend on two factors:

1. The voltage polarity of the noncontact pickup system.
2. The location of the pickup, i.e. above or below the shaft or to the right or left of the shaft.

To explain the significance of the polarity of the noncontact system, it must be mentioned that some noncontact pickup systems utilize a plus (+) DC voltage to power the oscillator of the signal sensor, while others use a minus (−) voltage supply. To illustrate the effect of system polarity, consider what would happen if a noncontact pickup with positive (+) polarity was installed above the shaft, as illustrated in Figure 5-45A. In this case, an *upward* movement of the shaft would decrease the gap, causing a decrease in the positive gap (signal) voltage, causing the scope trace to move *downward*. This is just the opposite of what is wanted. However, if this same pickup was mounted vertically below the shaft, as illustrated in Figure 5-45B, a *downward* movement of the shaft would again decrease the positive (+) gap voltage, which would cause the scope trace to move *downward* as required. Of course, the same effect will be seen on horizontally mounted pickups as well. If a positive-polarity pickup is mounted to the right of the shaft, as shown in Figure 5-45C, movement of the shaft to the right—toward the pickup—will cause the scope trace to move to the left because of the decrease in positive gap voltage. If this same pickup was mounted to the left of the shaft, Figure 5-45D, then movement of the shaft to the left toward the pickup will cause the scope trace to move left as desired.

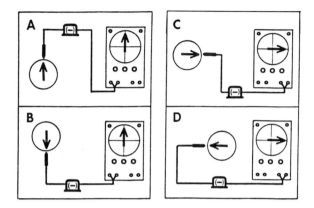

Figure 5-45. Relationship between pickup position, shaft motion and resultant scope trace for non-contact systems utilizing a positive (+) polarity power supply.

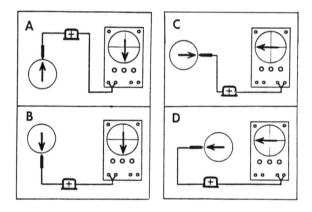

Figure 5-46. Relationship between pickup position, shaft motion and resultant scope trace for non-contact systems using a minus (−) polarity power supply.

Figure 5-46 illustrates the effects of pickup position and the direction of shaft motion on the resultant scope trace when a noncontact system with minus (−) power supply is being used. Essentially, the results are just the opposite of those observed using a system with positive (+) polarity.

Based on this discussion, the following summarizes the points to consider when using noncontact pickups to observe shaft motion on an oscilloscope.

When using noncontact pickups with positive polarity:

1. Vertical pickups mounted above the shaft should have the polarity of the input signal to the ocilloscope *reversed* for a correct indication of vertical shaft motion on the CRT.

2. Vertical pickups mounted below the shaft will read shaft motion correctly on the vertical axis of the CRT.
3. Horizontal pickups mounted to the *right* of the shaft should have the polarity of the input signal to the scope reversed for a correct indication of horizontal shaft motion on the CRT.
4. Horizontal pickups mounted to the left of the shaft will read shaft motion correctly on the horizontal axis of the scope.

When using noncontact pickups with minus polarity:

1. Vertical pickups mounted above the shaft will read shaft motion correctly on the vertical axis of the CRT.
2. Vertical pickups mounted below the shaft should have the polarity of the input signal to the oscilloscope *reversed* for a correct indication of the vertical shaft motion on the CRT.
3. Horizontal pickups mounted to the right of the shaft will read shaft motion correctly on the horizontal axis of the scope.
4. Horizontal pickups mounted to the *left* of the shaft should have the polarity of the input signal to the scope reversed for a correct indication of horizontal shaft motion on the CRT.

As can be seen, careful attention must be given to pickup orientation (relative to how the operator views the display on the CRT) as well as to the polarity of the noncontact pickup system in order to correlate the Lissajous pattern observed on the CRT with the actual position or motion of the shaft relative to the bearing. Failure to do so could lead to false interpretation of the data.

Calibrating the Scope Display

Proper calibration of the horizontal and vertical gain controls of the oscilloscope is important to ensure that the *shape* of the observed pattern accurately reflects the true motion of the shaft, and to make it possible to obtain amplitude data from the scope. While most scopes have separate horizontal and vertical gain controls, using a scope incorporating matched and locked horizontal and vertical gain adjustments can greatly reduce setup and calibration time.

Figure 5-47 shows a field test instrument which can be used to calibrate the oscilloscope horizontal and vertical gain controls.

The Synchronous Reference Pulse

When observing Lissajous patterns on an oscilloscope, it is difficult to obtain frequency information from the display unless some kind of frequency reference, such as a synchronous (1 × rpm) pulse, is superimposed on the display. For example, the nearly circular pattern in Figure 5-48 would normally be considered the result of an unbalance condition where the frequency is 1 × rpm. However, if a 1 × rpm reference pulse was superimposed on the pattern, revealing that shown in Figure 5-49, the interpretation would be quite different. Note in Figure 5-49 that the synchronous pulse appears twice, indicating that this motion of the shaft occurs once for every *two* revolutions of the shaft. Thus, this vibration is apparently occurring at

Figure 5-47. A portable field tester can be used to calibrate the oscilloscope horizontal and vertical gain controls.

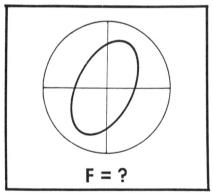

F = ?

Figure 5-48. The Lissajous pattern alone does not provide frequency information.

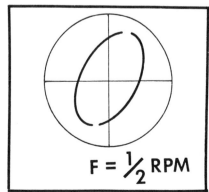

F = ½ RPM

Figure 5-49. Superimposing a reference signal on the Lissajous pattern aids in identifying the frequency of shaft vibration. This shaft motion is occurring at ½ rpm.

one-half rotating speed. Although the example in Figure 5-49 may not be totally realistic, since it would be a rare instance where a vibration at one-half rotating speed was the only frequency present, it does illustrate the value and importance of the synchronous reference pulse for identifying the frequencies of shaft vibration.

The synchronous reference pulse is obtained by installing an electromagnetic pickup, Figure 5-50, or a noncontact pickup observing a protrusion or depression on the rotating shaft, as shown in Figure 5-51. An existing key or keyway can be used to trigger the reference pickup, or a shallow slot can be machined where it will not affect the physical strength of the shaft. The needed pulse can also be obtained by drilling a hole in the end of the shaft, out from the shaft centerline, and then mounting the

Figure 5-50. Electromagnetic reference pickup.

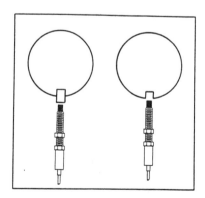

Figure 5-51. A noncontact pickup observing a protrusion or depression will provide the needed speed reference pulse.

reference pickup axially so that the hole passes the tip of the pickup each revolution. Drilling a hole in the end of the shaft will not likely weaken the shaft as a radially drilled hole might in some cases.

The reference pulse can be superimposed on the Lissajous pattern in two ways. First, if the oscilloscope is so equipped, the pulse can be applied to the Z axis (intensity) input. This will produce a blank spot on the pattern, as shown in Figure 5-52. If no Z input is incorporated, the reference pulse can be inserted in parallel with either the horizontal or vertical input. This will result in a spike on the pattern, as shown in Figure 5-49. The latter approach is less desirable, since the reference signal may add unwanted signals to the pattern. If this approach is used, the time waveform of the reference pulse should be observed alone to make sure it is relatively free of spurious signals. Another disadvantage of applying the reference pulse in parallel with a vibration signal is that the reference pulse may be "lost" if the pattern is complex.

Interpreting Lissajous Patterns

While it would be impossible to illustrate and discuss the Lissajous patterns associated with all types of machinery problems, there are many mechanical problems which are readily identified by their characteristic patterns. Several of these are discussed briefly in the paragraphs which follow.

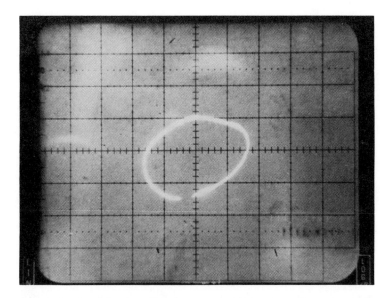

Figure 5-52. Applying the reference pulse to the Z (intensity) axis will produce a blank spot on the pattern.

Unbalance. Unbalance is characterized by a frequency of 1 × rpm and, assuming that no other significant vibrations are present, will reveal a circular or slightly elliptical pattern, as shown in the examples in Figure 5-53. The single reference pulse on the pattern verifies that shaft motion is occurring at a frequency of 1 × rpm.

If the pattern takes on a highly elliptical shape such as the one in Figure 5-54, where the ratio of the major axis to minor axis is relatively high, say 8:1, 10:1, or more, the machine is probably operating on or very near a structural resonance. This could be checked by noting the effect a change in rotating speed has on the pattern shape.

Misalignment. Misalignment will also cause a predominant vibration at a frequency of 1 × rpm in many cases; but unlike unbalance, misalignment will often be accompanied by harmonically related frequencies, including two times, three times, and occasionally higher orders of rotating speed. As a result, significant misalignment may not reveal the circular or slightly elliptical pattern characteristic of unbalance. Instead, misalignment may reveal a pattern shaped like a banana (Figure 5-55). As the higher order frequencies take on greater significance, the pattern may appear as a figure eight or as the other pattern illustrated in Figure 5-56.

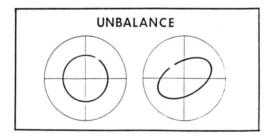

Figure 5-53. Unbalance will typically reveal a circular or slightly elliptical pattern at 1 × rpm.

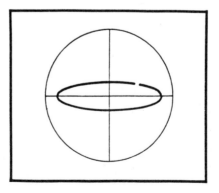

Figure 5-54. A highly elliptical pattern may indicate misalignment, bearing wear, or possible resonance.

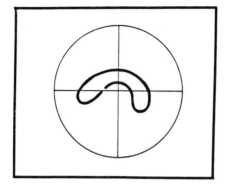

Figure 5-55. Misalignment will sometimes reveal a "banana"-shaped pattern.

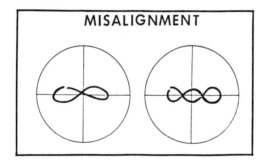

Figure 5-56. Typical patterns resulting from coupling misalignment.

Oil whirl. Oil whirl is characterized by a frequency of vibration slightly less than one-half the rotating speed frequency. In fact, the oil-whirl frequency will normally occur at 44%-48% of rotating speed frequency. The Lissajous pattern of an oil whirl will generally appear as circular or elliptical, with an internal loop, as illustrated in Figure 5-57. The fact that vibration frequency is less than rotating speed frequency is indicated by the *two* reference pulses evident on the pattern. In addition, since oil-whirl frequency is not exactly one-half the rotating speed, the internal loop will appear to rotate as indicated by the dotted pattern in Figure 5-57.

Rubbing. Rubbing between rotating and stationary parts can result in several different patterns, as illustrated in Figure 5-58; and the particular pattern encountered will generally depend on the extent of the rub. In most cases, the reference pulse on the pattern will appear unsteady, although this can also result from other problems, such as mechanical looseness.

A very mild rub, where the rotor only "touches" the stationary part once per revolution, may result in a slight distortion of the circular or elliptical unbalance pattern, as illustrated in Figure 5-58A. Very light rubbing which causes the rotor to

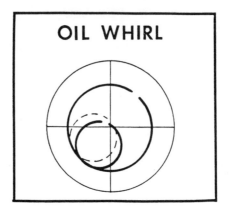

Figure 5-57. Typical oil whirl pattern.

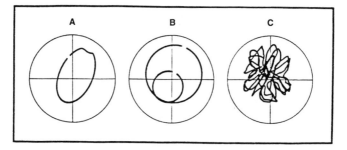

Figure 5-58. Typical Lissajous patterns caused by rubbing.

"hit-and-bounce" can result in submultiple vibration, generally at one-half rpm. This will produce a pattern with an internal loop, Figure 5-58B. This appears similar to an oil-whirl pattern; however, the internal loop caused by the rub will not appear to rotate as it does with oil whirl.

As the rub becomes more severe, the resultant pattern may take on any of a number of configurations, which include harmonic frequencies, random non-synchronous frequencies, and resonant frequencies of various components. Figure 5-58C is a typical pattern resulting from heavy or full rubbing conditions.

Again, it is not possible to present here all the patterns which can be obtained; however, it should be obvious from this discussion that observing the Lissajous patterns can provide important and useful information about the dynamic response of a machine.

Amplitude-Versus-Time Analysis

Most vibration meters, monitors, and analyzers provide connections for, or a receptacle for connecting, a D.C. strip-chart recorder for obtaining hard-copy recordings of vibration or noise amplitude over long periods of time. Typical recorders used with portable instruments are shown in Figures 5-59 and 5-60. The recorder in Figure 5-59 is quite versatile and can be quickly and easily calibrated for use with practically all portable instruments or permanently installed monitors. This recorder also offers 10 selectable chart speeds from ¾ inch (1.9 cm) per hour up to a maximum of 12 inches (30.5 cm) per minute. The recorder in Figure 5-60 is a low-cost unit having a single chart speed of 1 inch (2.54 cm) per hour, although other speeds are available. With this recorder, the recorder stylus writes by striking the pressure-sensitive chart paper once every two seconds.

Figure 5-59. A D.C. strip-chart recorder can be connected to vibration meters, monitors, and analyzers to obtain long-term recordings of noise and vibration.

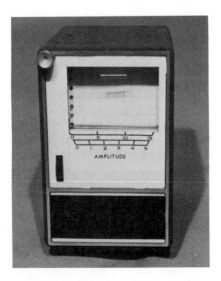

Figure 5-60. Low-cost D.C. recorder.

D.C. recorders are often used with permanently installed vibration monitors. By continually recording vibration amplitude levels, it is possible to determine whether a machinery problem has developed suddenly or gradually over a period of time. Such information may be a significant factor when making the decision to shut down an important machine. For example, if the vibration recording indicates that the vibration has increased very gradually over a long period of time, then a shutdown for correction may be postponed until a more convenient time. However, if the vibration has increased suddenly, plans should be made immediately for detailed analysis and subsequent correction.

Amplitude-versus-time recordings are valuable not only for revealing developing mechanical problems but in diagnosing unusual problems. To illustrate, one paper mill was experiencing repeated premature bearing failures in the press section of a paper machine, for no apparent reason. An amplitude-versus-frequency analysis of the vibration did not reveal any significant mechanical problems which may have been responsible. After exhausting all the normal possibilities, the decision was made to connect a D.C. recorder to the analyzer and record the vibration amplitude for a few days to see if anything could be learned. The next day, an examination of the chart recording revealed several brief periods of extremely high amplitudes of vibration. Since the starting time of the recording and the chart speed were known, it was easy to establish the approximate time of day that each severe vibration occurred. With this information, an investigation was undertaken to determine what might have occurred at these times to cause the high vibration. It was then learned that the periods of high vibration corresponded precisely with the schedule of trains running on a nearby track.

Tape Recording and Playback of Analysis Data

With the availability of portable and battery-operated vibration analysis equipment, most analysis work can be carried out at the machine site. However, circumstances may exist which make "on-line" or in-place analysis impractical. In such cases, it may be necessary to tape record the data for later playback and evaluation. For example, when obtaining plots of vibration amplitude and phase versus machine rpm, it may be necessary to obtain such plots at two, four, or more bearings of the machine and, perhaps, in the horizontal, vertical, and axial pickup directions at each bearing. Further, it may be desirable to obtain comparative filter-in and filter-out amplitude-versus-rpm plots. Of course, to obtain all these data using one analyzer would require that the machine be started and stopped several times. This would be completely out of the question for many machines such as turbogenerators, centrifugal compressors, large fans, etc. A more realistic approach would be to tape record the vibration data from several pickups on a multi-channel recorder and to later play it back through the analyzer in order to obtain all of the data needed for each measurement point.

Other applications where it may be beneficial to tape record vibration data include:

1. Where the machines to be analyzed are in remote locations or have limited accessibility (e.g., those mounted on elevated structures).

2. Where the machines are located in areas of extremely high or low temperature, high noise, or other conditions unsuitable for extended stay of personnel.
3. Where the machine cannot be operated for a sufficient period of time (or under a particular operating condition) to permit on-line analysis.

There are many different models of tape recorders available on the market, ranging from simple, inexpensive, single- or two-channel units up to costly units having 14 or more channels. Recorders will be further classified as either AM or FM. AM recorders are generally considerably less expensive per channel than FM recorders. Moreover, they use standard tape which is readily available, accept a wide range of input sensitivities, and are fairly simple to set up and operate. However, AM recorders do not as a rule have good response to low-frequency vibrations and typically begin to fall off at frequencies below 3000-3600 CPM. Low-frequency data are obtainable with AM recorders; however, the amplitudes will be significantly attenuated. FM tape recorders, by comparison, provide good response to low frequencies—down to 0 CPM (D.C.); but FM units are considerably more expensive per channel and typically operate over a much narrower input sensitivity range. Tape recording and playback of vibration data is fairly simple and straightforward; however, some precautions must be observed to ensure that the final data obtained are accurate.

It is most important that the input signal not be excessive and saturate the tape head. This will cause severe distortion of the vibration data (amplitudes *and* frequency content) on playback through the frequency analyzer. Most recorders provide an input gain control and signal-level meter for each channel. This meter can be monitored while the data are being recorded to make sure the input level is not excessive.

It is possible in many cases to connect the vibration pickup directly to the recorder, as illustrated in Figure 5-61. However, when this is done, the actual vibration levels must first be measured and recorded, or a "calibrated shake" must be recorded on

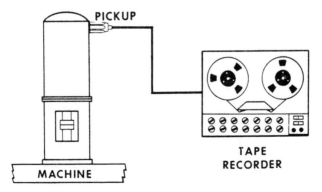

Figure 5-61. If the vibration signal is recorded directly from the vibration pickup, it will be necessary to first record a known or "calibrated" vibration for calibration purposes.

the tape in the beginning, so that the output gain level of the recorder can be calibrated when the tape is played back through the analyzer. This is simply a process of measuring and writing down the overall vibration level being taped and then playing the tape through the analyzer and adjusting the recorder output gain control to achieve the same reading.

When using multi-channel or voice-interrupt recorders, the machine, bearing point, pickup position, overall vibration amplitude, operating conditions, and other facts of interest can be voice recorded for future reference.

Recording signals directly from the vibration pickup is satisfactory as long as the vibration levels and transducer sensitivity combine to produce signals compatible with the input sensitivity requirements of the recorder. Transducer signals are compatible with most AM recorders; however, the input sensitivity of most FM recorders requires that the transducer signal be conditioned (amplified or possibly attenuated) before it can be applied to the recorder. Good-quality vibration and sound-level meters are ideally suited for conditioning signals ahead of the tape recorder. The vibration meter, Figure 5-62, with a multiple-position amplitude range selector, provides the needed amplification or attenuation of the transducer signal and also a direct readout of the overall vibration level for future reference when the tape is played back through the analyzer.

The general setup procedure for recording and playing back vibration data is summarized in the following steps. The recorder manufacturer's operating and calibration instructions should, of course, be carefully studied and followed.

1. Measure and record (manually or voice record) the overall vibration level, machine identification, bearing point, pickup orientation, operating conditions, and other facts of interest. This must be done for each channel being recorded.
2. Connect the vibration signal source (pickup, monitor, signal sensor, vibration meter) to the appropriate recorder input channel. [*Note:* When using an FM recorder with noncontact pickup systems, a large D.C. (gap) voltage is usually

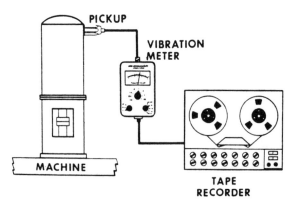

Figure 5-62. Recording from a vibration meter or analyzer provides additional amplitude range and parameter selection.

present with the A.C. (vibration) signal. The D.C. voltage will overload or saturate the tape and thus must be blocked out by inserting a capacitor in series with the input signal to the recorder.]

3. While the vibration signal is applied, note the recorder signal-level meter for the channel being used and adjust the input gain control to achieve an on-scale signal for recording.

4. Record the vibration data for a sufficient length of time for later playback and analysis. The amount of data recorded will depend on the type of equipment available for analysis. For example, if a real-time analyzer will be used, perhaps only a few seconds of recorded data are needed. If the data will be analyzed with a basic analyzer with a manually tunable filter, or if spectrum averaging is anticipated, it may be necessary to record up to three to five minutes of data.

After the vibration data of interest have been recorded, the tape can be rewound and played back through an analyzer for frequency analysis, time waveform analysis, etc., following the general procedure outlined below:

5. Connect the output of the selected recorder channel to the analyzer pickup input receptacle.

6. Select the appropriate mode of readout (displacement or velocity or acceleration). *Note:* If a vibration velocity signal has been recorded, either the displacement or velocity can be analyzed, as selected by the analyzer DISP-VEL selector. If a displacement signal has been recorded, only the displacement can be analyzed; however, the vibration analyzer must be placed in the VELOCITY mode in order to bypass the analyzer integrator.

7. While the recorded signal is applied to the analyzer, adjust the analyzer AMPLITUDE RANGE selector to obtain an up-scale reading on the analyzer amplitude meter. Depending on recorder output sensitivity, it may be necessary to use an amplitude range on playback different from that used when the data were recorded. For example, if the data were recorded with the vibration meter on the 1.0 in/sec range, it may be necessary to have the analyzer on the 0.1 in/sec range in order to obtain a suitable up-scale reading when the signal is played back through the analyzer. If this is the case, it will be necessary to apply a multiplication factor of 10 to the analyzed data in order to obtain their true values.

8. While observing the analyzer amplitude meter, adjust the output gain control for the recorder channel being used in order to obtain a reading on the analyzer amplitude meter equal to the original vibration amplitude recorded in step 1.

9. Analyze the recorded data as required.

10. Repeat steps 5-9 for each remaining channel of recorded data.

Tape-recorded data, once obtained, can be played back to provide amplitude-versus frequency, amplitude-versus-frequency-versus-time, time waveform, or Lissajous pattern (orbit) information. Where it is desirable to record data to obtain amplitude/phase-versus-rpm plots, it will also be necessary to record a 1 × rpm reference voltage pulse from an electromagnetic reference pickup or noncontact pickup on at least one recorder channel. However, when using a multi-channel recorder, it should be determined whether one or two recording heads are to be used for all channels. On many multi-channel units, two recorder heads are used to prevent interference from having too many channels on the same head. If two recorder heads are used, some phase error will occur between the vibration data recorded on one head and a reference pulse which has been recorded on another head. Amplitude/phase-versus-rpm plots can still be obtained for all recorded channels; however, the phase data between channels recorded on different heads cannot be directly compared for relative motion determination due to the phase shift introduced by the recorder. In order to obtain true and directly comparable phase information for all channels, the phase reference pulse should be recorded on one channel of *each* recording head.

As mentioned, tape recording offers many advantages for obtaining analysis data. Vibration data can be obtained simultaneously at many measurement points for a particular operating condition, during startup or coast-down, or for other transient conditions which would be difficult or impossible to analyze on-line.

The data, once recorded, can be played back through an analyzer at leisure, several times if necessary, to obtain the data in many different forms for detailed study and evaluation. Tape recording is also useful for obtaining raw data quickly on machines in remote locations or in adverse environmental conditions.

Despite obvious applications, tape recording vibration data is not free of possible problems which may introduce significant errors leading to false interpretations. Some of the major problems encountered are briefly outlined here:

1. *Signal-to-noise ratio:* The "electrical" noise contributed to the data from the tape recorder will normally be low level compared to most vibration signals of significance. However, when the vibration signals of interest are low, or if a low-sensitivity transducer is being used, recorder electrical noise may contribute false data to the analysis. The extent of recorder noise can usually be determined by recording a "blank" tape, with no vibration signal being applied to the recorder. When the tape is played back through the analyzer, any data obtained is the electrical noise contribution from the recorder and/or tape.

   Tape speed will also affect the amount of electrical noise contributed by the recorder. In general, more noise will be present on tapes recorded and played back at slow speeds such as 1⅞ or 3¾ inches per second. Therefore, higher tape speeds of 7½ or 15 inches/second should be used wherever possible.

2. *Tape speed:* Most commercially available recorders do not provide a means of accurately regulating tape speed. As a result, data are sometimes recorded at one speed but then played back at a slightly different speed—perhaps due to friction, dirt, or moisture in the tape-drive mechanism, or to variations in power supply or battery condition. As a result, the frequencies of vibration obtained when the data are played back may be shifted somewhat, and *precise* frequency determination is impractical.

3. *Automatic gain control:* Some recorders incorporate an automatic gain control (AGC) circuit which tends to balance out or stabilize the input signal level. Such control is useful when voice recording vibration data, but a recorder with

AGC may produce erroneous amplitudes. The AGC tends to attenuate increasing vibration amplitudes and amplify or boost decreasing amplitudes. Therefore, when recording vibration data, be sure to record the data without the AGC mode in operation.

4. *Frequency response:* As mentioned previously, AM recorders tend to attenuate low-frequency signals, generally below 3000-3600 CPM. In any case, it would be well to check the frequency response of the recorder being used to make sure it is capable of reproducing all the data which may be of interest.

In addition to these problems, perhaps the most significant drawback to tape recording data for later playback and interpretation in the lab is the lack of additional insight about the machine and its surroundings which is available when the analysis is done onsite. There are many minor details about a machine, including the type of structure or method of mounting, the proximity of nearby machines which may contribute background vibration, the relationship of piping or ductwork, and other details which may be important to the analysis. Unless a series of photographs are taken to accompany the tapes to the lab, these additional features may not be available to the analyst who evaluates the data—particularly if the taping was performed by someone else. In addition, when the analysis is performed on-site, there are many occasions where studying the normal data suggests that other measurements, perhaps on the structure or piping, should be taken to support one's suspicions about a potential problem. Or, perhaps a particular signature suggests a problem with the transducer or pickup mounting. Such things an be checked quite easily when the analysis is done on-site. If the analysis data are being played back from tape in the laboratory, a second trip to the machine may be necessary to gather the additional data needed. This takes time and may result in unacceptable delays in solving problems with critical plant equipment.

Phase Analysis

Another very useful analysis technique which can help detect and identify machinery problems is phase measurement and analysis. Phase measurements are normally expressed in degrees, where one complete cycle of the vibration equals 360°. Phase can be defined as that part of a cycle (0°-360°) through which one part of a machine has traveled relative to another part or a fixed reference. However, from a practical standpoint, phase measurements simply provide a means of determining the relative motion of various parts of a machine. For example, the two weights in Figure 5-63 are vibrating at the same frequency and displacement; however, weight A is at the upper limit of travel at the same instant weight B is at the lower limit. We can use phase to express this comparison. By plotting one complete cycle of motion of these two weights, starting at the same given instant, it can be seen that the points of peak displacement are separated by a half cycle, or 180°. Therefore, these two weights are vibrating 180° out-of-phase. In Figure 5-64, weight X is at the upper limit at the same instant weight Y is at the neutral position moving towards the lower limit. These two weights are vibrating 90° out-of-phase. In Figure 5-65, weights C and D are vibrating in-phase, or 0° out-of-phase.

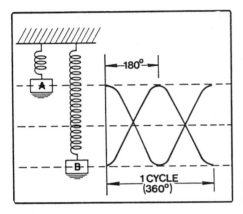

Figure 5-63. Weights vibrating 180° out-of-phase.

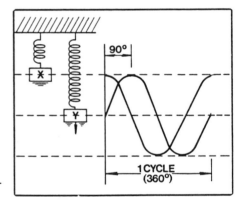

Figure 5-64. Weights vibrating 90° out-of-phase.

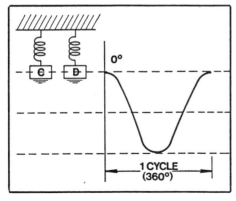

Figure 5-65. Weights vibrating in-phase.

In the paragraphs that follow, various techniques for obtaining phase measurements are presented, along with several examples of practical applications where phase measurement and analysis can aid in problem identification.

Phase Measurement Techniques

There are many techniques which can be used to obtain comparative phase measurements for analysis and balancing, including those using the stroboscopic light, phase meter, and oscilloscope.

Stroboscopic-light phase measurement. High-quality portable vibration analyzers are furnished with a high-intensity stroboscopic (strobe) light which, when triggered by the measured vibration, provides a quick and convenient means of obtaining phase readings for analysis and balancing. To obtain phase readings, a reference mark is placed on the rotor at the end of a shaft or at some other location which can be readily observed. Or, an existing key, keyway, or other distinguishing mark can be used. For accurate phase measurements, an angular reference scaled in degrees (0° to 360°) can be superimposed around the shaft, making it possible to obtain phase readings with minimum error. With the analyzer strobe light being triggered by the machine vibration, the reference mark on the rotor will appear stationary at some angular position. The phase measurement is then obtained by noting and recording the indicated angle from the angular reference. For example, the indicated phase reading in Figure 5-66 is 120°.

Comparative phase readings are obtained by simply moving the *vibration pickup* to other measurement locations but keeping the *same* reference mark and angular reference used for the initial measurement.

Figure 5-66. A stationary angular reference can be used to obtain accurate phase measurements for analysis and balancing.

In some cases it may not be possible to view the end of a shaft, as shown in Figure 5-66. Perhaps only a small portion of the rotor or shaft can be seen at any particular time. For example, when observing the shaft from the side, such as under a coupling guard, it may be that only half of the shaft or coupling circumference is visible. In such cases, a single reference mark placed on the shaft may be out of view for many phase measurements. This problem can often be overcome by putting *two different* phase marks on the shaft diametrically (180°) opposite one another. Of course, the two marks placed on the shaft should be noticeably different to avoid errors in phase comparison. For example, a small (o) might be drawn on one side of a shaft and a small (x) drawn on the other side. While this approach is usually adequate for general phase comparisons needed for analysis, the accuracy of measurements obtained this way may not be sufficient to satisfy some requirements, such as in-place balancing. Where more accurate phase readings are called for, it may be necessary to apply a rotating angular reference such as that illustrated in Figure 5-67, where a piece of tape marked in degrees has been wrapped around the shaft. If the diameter of the shaft is known, the tape can be prepared before the machine is shut down. Simply cut a length of tape exactly equal to the shaft circumference (C = πD). With the tape cut to proper length, divide the tape into equal segments, the number depending on the degree of accuracy required. Dividing the tape into 12 equal parts will provide phase marks every 30°, 24 parts every 15°, 36 parts every 10°, etc. The tape, once divided and marked, is then wrapped completely around the shaft. A stationary reference mark is also needed to obtain accurate readings.

The stationary reference might be the split line between the upper and lower halves of a bearing, or it could be a wire or coat hanger attached to the bearing, base, or other stationary component and bent as necessary to point at the rotating angular reference tape. Accurate phase readings are obtained by noting the angle number on the tape which is in line with the stationary pointer when observed under the stroboscopic light.

The strobe light is very easy and convenient to use for obtaining phase readings where the vibration frequency of interest is equal to 1 × rpm of the machine.

Figure 5-67. Where it is not possible to see the end of the shaft, an angular reference system can be applied to the rotating shaft.

However, if the vibration of interest is occurring at 2 × rpm, the strobe light will flash twice for each revolution of the shaft, and a reference mark will appear at two positions when observed with the light. Obviously, it would be impossible to obtain comparative phase readings with two or more identical reference marks visible under the light. Where it is desired to obtain comparative phase measurements for multiple, sub-multiple or non-harmonically related vibration frequencies, some other technique for phase measurement must be used, such as one using a phase meter or oscilloscope.

Phase measurement with a phase meter. Where it is impractical or unsafe to obtain phase readings with a strobe light, an instrument with a meter for remote phase readout can be used. One such instrument was shown earlier in Figure 5-7. The overlapping 0°-180° and 180°-360° scales used on the phase meter in Figure 5-68 make it possible to obtain very accurate phase data for analysis and balancing. The instrument pictured in Figure 5-69 is the phase readout module incorporated in many

Figure 5-68. Phase meter.

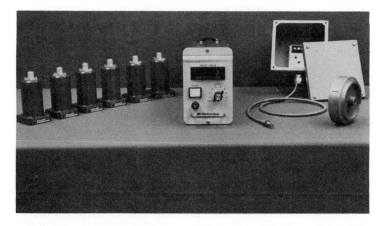

Figure 5-69. This module, used with permanent monitor installations provides a digital readout of the phase of machinery vibration.

vibration monitor systems for monitoring the phase of vibration in addition to amplitude in order to detect significant changes in machinery condition. The monitor phase module provides a direct readout of the measured phase on a digital meter, with accuracy within $\pm 1°$.

To obtain a remote readout of phase requires a voltage reference pulse at the desired vibration frequency. Where it is desired to observe the phase of the vibration occurring at $1 \times$ rpm, a reference pickup such as a photocell, an electromagnetic pickup, or a noncontact pickup is mounted close to the shaft, which has been properly prepared to "trigger" the reference pickup. Since the photocell responds to changes in reflectivity of the target, a common method of preparing the shaft is to wrap black nonreflective tape around the shaft and then paint a white line across the tape or attach a small piece of reflective material such as metal foil to the black tape. Some prefer to paint the target area with flat black paint in lieu of using nonreflective tape. If the shaft is clean and shiny, the proper trigger for the photocell can be provided by simply attaching a small strip of nonreflective tape to the shaft or by painting a small strip with flat black paint. In any case, the objective is to produce an abrupt change in the reflectivity of the target area of the photocell for each revolution of the shaft.

When a noncontact or electromagnetic pickup is used, the target area of the shaft is prepared with an abrupt depression or protrusion, as illustrated in Figure 5-51. An existing key or keyway can be used with excellent results. A satisfactory pulse can also be provided by taping a small piece of key stock to the shaft, using two to three wraps of high-strength tape such as fiberglass reinforced tape. A reference pulse can also be obtained by drilling a small hole axially in the end of the shaft, out from the shaft centerline, and then mounting the reference pickup so that the hole passes in front of the pickup each revolution. Drilling the hole in the end of the shaft will not create stress points or otherwise structurally weaken the shaft, as a radially drilled hole might.

If it is possible to mount a reference pickup on the machine, it is still possible to obtain phase measurements with a remote phase instrument such as the one shown in Figure 5-68. This can be accomplished by using the analyzer or oscilloscope output of another vibration measuring instrument as the reference source. If the vibration is predominant at the desired frequency, a simple vibration meter can be used, as illustrated in Figure 5-70. Simply mount a reference vibration pickup at any convenient location on the machine or structure where a strong signal can be obtained. This reference vibration pickup is then connected to the vibration meter. The analyzer output of the vibration meter is then connected to the reference input of the phase measuring instrument in place of the reference pulse from a photocell or electromagnetic pickup. Adjust the amplitude range selector of the vibration meter until a suitable reference signal is obtained, as indicated by a steady phase reading. Experience indicates that it may be necessary to overdrive the vibration meter by one or two amplitude ranges before an adequate reference signal can be obtained. Apparently, overdriving the amplitude meter "chops" the output reference signal, resulting in a simulated square-wave similar to a normal reference pulse. In any event, overdriving the vibration meter will not damage the instrument and does give the desired results. Of course, the reference vibration pickup *must not be moved* if comparative phase readings are to be made at several measurement locations.

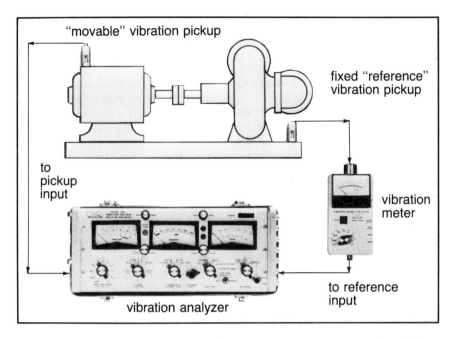

Figure 5-70. If a reference pickup is not available, a vibration meter can be used to provide a reference signal to the vibration analyzer to obtain comparative phase data for analysis. This technique *cannot* be used for balancing, however.

Further, this technique for measuring phase is only usable for phase analysis and *cannot be used for balancing*. Applying a trial weight to the rotor would change the phase of both the phase reference signal and vibration signal. If the vibration is complex or if the vibration frequency of interest is not the predominant vibration, it may be necessary to substitute a vibration analyzer with tunable filter in place of the vibration meter. In this case, simply connect the oscilloscope output of the reference analyzer to the reference input of the phase measurement instrumentation and tune the analyzer filter to the vibration frequency of interest. In this way, comparative phase measurements can be obtained for virtually any frequency of interest—synchronous, multiples, sub-multiples, and even non-harmonic frequencies. This offers a strong advantage over the strobe light for phase measurement, since the strobe technique is limited to vibrations at 1 × rpm.

The 1 × rpm voltage pulse from the reference pickup applied to the analyzer actually serves two purposes. First, the reference pulse automatically tunes the filter of the analyzer to the rotating speed frequency. In this way, should machine rpm change, the filter will automatically adjust to the change to provide consistently accurate amplitude and phase data. Second, the reference pulse is electronically compared to the filtered vibration signal to provide a measure of relative phase. In simple terms, the instrument measures that portion of a cycle by which the reference pulse leads the neutral-to-positive (0 to +) crossover of the vibration signal. This is then converted to a proportional D.C. voltage needed to drive the phase meter. Figure 5-71 illustrates various relationships between the reference pulse and vibration signal and the resulting phase indications.

Unlike the stroboscopic light, the phase meter can be used to obtain comparative phase data for vibration frequencies other than 1 × rpm. All that is needed is a reference signal at the desired frequency. For example, to analyze the phase of vibration for a frequency of 2 × rpm, it is necessary only to put two triggers (protrusions, depressions, etc.) on the rotating shaft 180° apart instead of the usual one trigger. In this way, two reference pulses will be generated for each shaft revolution, the analyzer filter will automatically be tuned to 2 × rpm, and the phase readings noted on the meter will be those for the 2 × rpm vibration. As previously mentioned, where it is necessary to obtain comparative phase data for sub-multiple or non-harmonically related vibration frequencies, a reference vibration pickup and a reference vibration analyzer with tunable filter can be used to provide a reference signal at any desired frequency of machine vibration.

Oscilloscope phase measurement. The oscilloscope is another useful device for obtaining comparative phase measurements and can be used where readings with a strobe light are not possible or when an instrument with remote phase readout is not available. Practically any standard single-trace or dual-trace oscilloscope can be used with standard vibration analysis equipment, as discussed in the paragraphs that follow.

Measuring phase with a single-trace oscilloscope. Obtaining comparative phase measurements with a single-trace oscilloscope requires a voltage reference pulse similar to that required for the analyzer discussed previously. However, the

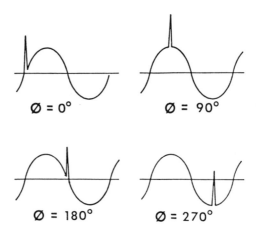

$\emptyset = 0°$

$\emptyset = 90°$

$\emptyset = 180°$

$\emptyset = 270°$

Figure 5-71. Examples of phase measurements using a reference pulse superimposed on the vibration wave-form.

reference pulse can be applied to the oscilloscope in a couple of different ways to measure phase. One way is to switch the oscilloscope to the EXTERNAL SYNC mode of operation and connect the reference signal to the scope EXTERNAL SYNC (horizontal) input. The vibration signal at reference pulse frequency is connected to the oscilloscope VERTICAL input from the analyzer oscilloscope output. The resulting display on the CRT will be one complete cycle of the vibration, and the horizontal position and gain controls of the scope should be carefully adjusted to display the vibration waveform over the full width of the scope graticule. The vertical position of the trace should have been adjusted initially to the center line. The configuration of the vibration waveform on the scope display will depend on the phase relationship between the reference signal and the vibration signal.

By scaling the horizontal axis of the scope from 0° to 360° as shown in Figure 5-72, phase readings can be obtained by simply noting the reading for a common point on the waveform. Normally used for reference is the point of neutral-to-positive (0 to +) crossover. Using this as the reference, the examples in Figure 5-73 show how phase readings of 0°, 90°, 180°, and 270° will appear. The neutral-to-positive crossover point is a common choice and is used here, although some other point such as the neutral-to-minus (0 to −) crossover or peak amplitude could be used just as easily, as long as consistency is maintained.

Another way to obtain phase measurements with a single-trace oscilloscope is to superimpose the reference pulse on the vibration waveform as illustrated in Figures 5-74 and 5-75. The spike which appears on the waveform in Figure 5-74 is the result of connecting the reference signal in parallel with the vibration signal being applied to the scope vertical input. The blanking spot on the waveform in Figure 5-75 results from applying the reference pulse to an oscilloscope Z axis (intensity) input. Either approach will produce the desired results.

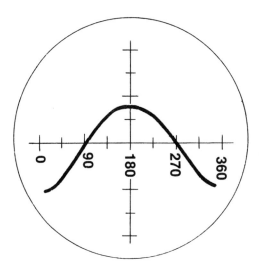

Figure 5-72. The horizontal graticule of the oscilloscope is scaled from 0 deg. to 360 deg. in preparation for phase measurement.

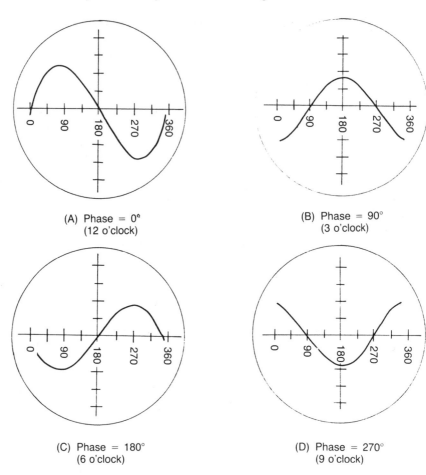

(A) Phase = 0°
(12 o'clock)

(B) Phase = 90°
(3 o'clock)

(C) Phase = 180°
(6 o'clock)

(D) Phase = 270°
(9 o'clock)

Figure 5-73. In the examples above, phase is determined by noting the point on the wave-form which crosses the neutral axis in the positive (+) direction.

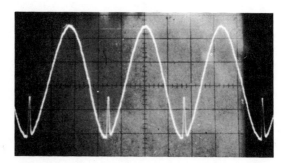

Figure 5-74. By superimposing a reference pulse on the vibration waveform, phase can be determined by noting that portion of a cycle (angle) by which the spike leads (or lags) a common point on the waveform such as the point of neutral to positive cross-over.

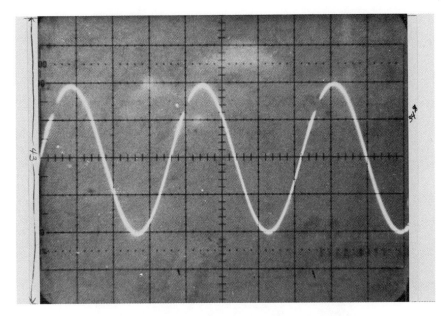

Figure 5-75. Applying the reference pulse to the "Z" (intensity) axis of the oscilloscope will produce a blank spot on the waveform.

With the reference pulse superimposed on the vibration waveform, phase measurements can now be obtained again by noting that portion of the vibration cycle which separates the reference pulse from a common point of reference on the waveform, such as the neutral-to-positive (0 to +) crossover. In addition, the *convention* of phase determination must also be established. Specifically, it must be decided whether the phase angle recorded is the angle by which the reference pulse *leads* or *lags* the neutral-to-positive crossover. Of course, for comparative phase readings, the same convention must be used for all measurements. For the examples given here, the measured phase is the angle by which the reference signal *leads* the neutral-to-positive crossover of the vibration signal. Applying this convention, the phase reading in Figure 5-74 is approximately 260°, and the phase reading for the example in Figure 5-75 is approximately 60°.

A single-trace oscilloscope can be used to obtain comparative phase measurements for practically any vibration frequency of interest as long as a suitable reference pulse is available at the selected frequency. In addition, it is also possible to obtain phase measurements for vibrations occurring at frequencies which are multiples of rotating speed where only a 1 × rpm reference pulse is available. For example, Figure 5-76 shows a 1 × rpm reference pulse superimposed on a 2 × rpm vibration waveform. Note that two complete cycles of the vibration occur for each reference pulse. However, the phase of the 2 × rpm vibration can still be obtained by simply noting the angle by which the reference pulse leads the neutral-to-positive

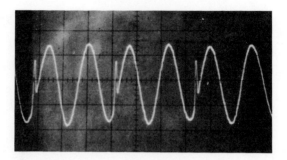

Figure 5-76. The phase of this 2 × rpm vibration is 340 deg.

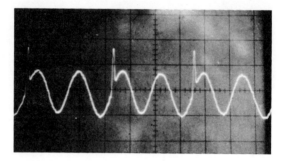

Figure 5-77. The phase of this 2 × rpm vibration is 135 deg.

crossover of the vibration waveform. Thus, the phase of the vibration in Figure 5-76 is approximately 340°; in Figure 5-77 the phase is approximately 135°.

Measuring phase with a dual-trace oscilloscope. In cases where it is not possible to obtain a reference pulse for the vibration frequency of interest for phase measurement with a phase meter or single-trace oscilloscope, it is possible to obtain comparative phase data for two or more measurement locations by using a dual-trace oscilloscope. A dual-trace oscilloscope simply allows two vibration waveforms to be viewed simultaneously on the CRT, as shown in Figure 5-78. The relative phase between the two signals is determined by simply measuring that part of the vibration cycle which separates common points of reference, such as neutral-to-positive crossover points. For example, the two signals displayed in Figure 5-78 are 180° out-of-phase because their points of neutral-to-positive crossover are separated by a half cycle.

Where comparative phase readings will be taken at more than two measurement locations, one of the first two signals should be retained for comparison with all other signals. In addition, the "lead" or "lag" convention for phase determination must also be selected to ensure that all readings are directly comparable.

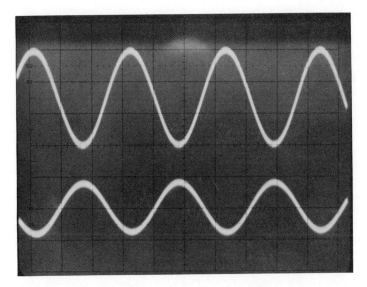

Figure 5-78. A dual-trace oscilloscope can display two vibration waveforms simultaneously for phase comparison. These vibrations are 180 deg. out-of-phase.

Some Practical Considerations for Phase Measurement

There are several factors which can affect the accuracy of phase readings taken for analysis or balancing. Some practical considerations are discussed briefly:

1. The direction of the vibration pickup axis, together with the selected phase reference system, establishes the "fixed reference" needed to take comparative readings. Thus, the direction of the pickup axis should not be changed from one measurement point to the next. Or, if pickup direction is changed, this change must be noted so that the phase readings can be corrected accordingly for comparison. This is true for all phase measurement techniques—strobe light, remote phase meter, or oscilloscope.
2. When using a standard analyzer with manually tunable filter, it is essential that, when used, the filter be precisely tuned for each reading taken. For example, should the machine speed change slightly between readings, causing the analyzer filter to be "detuned," a considerable error in phase measurement may result.
3. When taking comparative phase readings, it is important that all measurements be taken using the same parameter of amplitude readout. For example, switching from displacement to velocity readout will result in a 90° change in the measured phase. Switching from displacement to acceleration (g's) will reveal a 180° phase difference.

Phase Analysis Techniques

The following paragraphs briefly describe several applications where phase measurement and analysis can help detect and identify mechanical problems.

Phase monitoring. Many companies are currently including continuous phase monitoring systems on critical high-speed machines such as turbogenerators and centrifugal compressors for added protection against mechanical failure. The instrument in Figure 5-69 is a phase readout module incorporated in many multi-channel monitor systems. The phase of the vibration being monitored at any point on the machine can be selected for readout on the digital phase meter.

The concept of monitoring phase in addition to vibration amplitude, axial (thrust) position, bearing temperature, and other parameters which reveal machine condition is based on the fact that some machine problems result initially in little or no significant change or increase in vibration amplitude. However, these problems may cause a significant change in the *phase* of the vibration. For example, suppose that yesterday vibration amplitude and phase readings were taken on a machine, and that they indicated an amplitude of 2.0 mils (51 μm) and a phase of reading 90°. Then a new reading was taken at the same location today, indicating an amplitude of 2 mils (51 μm) but a phase reading of 270°. Obviously, the comparative amplitude readings show no increase and would suggest that there has been no change in machinery condition. However, the phase readings show that there has been a substantial (180°) shift in phase which, in this case, represents a 4.0 mil (102 μm) change in vibration amplitude. In this case, the *phase* change detected a significant change in machine condition, possibly due to a thrown rotor blade, cracked rotor shaft, or some other potentially serious problem.

Detecting misalignment versus a bent shaft. When relatively high amplitudes of vibration are detected in the axial direction, the cause may be coupling misalignment, bearing misalignment, a bent shaft, resonance, or possibly unbalance of an overhung rotor. Obviously, when a high amplitude of axial vibration is detected, it would help if further tests could be made to reduce the number of possible causes or to actually identify the problem before the machine is shut down for visual inspection and correction. This is particularly true on critical process and production machinery where only a few hours' downtime can represent tremendous production losses. The following paragraphs outline a simple technique for analyzing the phase of the axial vibration in an attempt to identify the problem source.

The first step is to select a convenient location on the machine where phase readings can be observed. Normally, the end of the shaft or an accessible coupling will serve this purpose; however, it is important that all comparative phase observations be made using the same reference mark.

The objective of our phase analysis for detecting misalignment and bent shafts is actually two-fold:

1. First, to determine how each *individual* bearing is moving axially. In other words, is the bearing twisting, rocking, or simply moving back and forth?
2. Second, to determine how the bearings are moving axially *relative* to one another. Are the bearings vibrating together (in-phase) or in opposition to one another (out-of-phase)?

To meet our first objective, it will be necessary to make several axial phase measurements on each bearing. Normally, four measurement positions for each bearing are recommended, as shown in Figure 5-79. With the analyzer filter tuned to the rpm of the machine, take and record phase measurements with the pickup in the axial direction at each of the four designated points on the bearing. The recorded phase readings will likely reveal one of two possible situations: the phase readings will be noticeably different at the four measurement positions, or the phase readings will all be relatively the same at the four measurement points.

If the phase readings on the bearing are noticeably different at the four measurement points, this indicates that the bearing is "twisting," as illustrated in Figure 5-80. This twisting action is generally caused by a shaft which is bent through

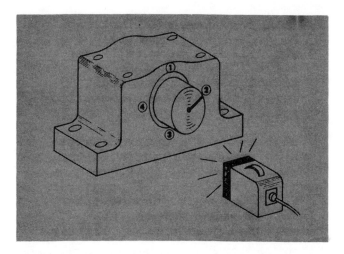

Figure 5-79. Axial phase readings for (4) positions on each bearing are recommended.

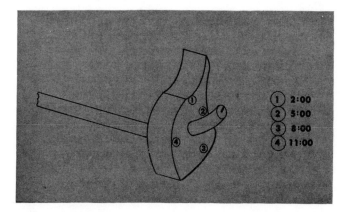

Figure 5-80. Phase readings which are noticeably different at the (4) measurement points indicate that the bearing is twisting.

or very near the bearing. A simple bow in the center of the shaft may not cause this twisting action unless the bow is quite extreme. For machines equipped with antifriction bearings, it is possible to observe a similar twisting motion for a bearing which is cocked on the shaft or in the bearing housing.

If the four phase readings on the bearing are all approximately the same, this indicates that the bearing is vibrating back and forth in a planar fashion, as illustrated in Figure 5-81. This indication alone does not tell us the nature of the problem; therefore, it will be necessary to continue the axial phase measurements at the other bearings of the machine in order to diagnose the cause.

Note that even though our first axial readings revealed a twisting bearing, axial phase measurements should be carried out at the other bearing points to see whether or not any additional problems exist.

After the phase readings have been taken and recorded at the first bearing point, move the pickup to the next bearing of the machine and take a similar set of axial phase readings at four points around the shaft. This procedure is then repeated for each bearing of the machine until axial phase data have been recorded for all points, as illustrated in Figure 5-82.

Again a cautionary note for the analyzing engineer or technician. When the pickup is moved from one bearing to the next, the axis of the vibration pickup could be reversed by 180°. For example, referring to the machine sketch in Figure 5-82, when readings were taken at bearing A, the pickup was probably pointing to the right. However, at bearing B the pickup could have been pointing toward the left. If this is the case, a 180° phase shift has automatically been introduced and the observed phase readings are not directly comparable. Therefore, whenever it is necessary to reverse the direction of the pickup axis to take phase measurements at other bearing points, a 180° correction must be applied to those readings to make them directly comparable with the other readings. If the direction of the pickup axis can be kept the same at all bearing points, no correction factor need be applied.

After axial phase readings have been recorded for all bearings of the machine, the final step is to compare these readings to determine how the bearings are moving

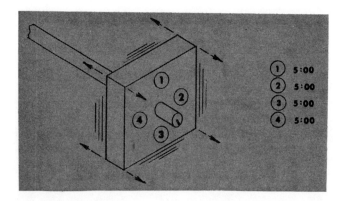

Figure 5-81. Phase readings which are all relatively the same, indicate that the bearing is vibrating axially in a planar fashion.

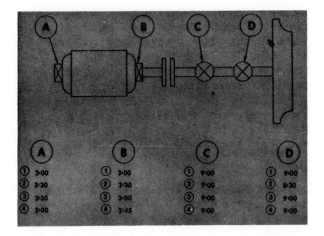

Figure 5-82. Axial phase readings should be taken and recorded for each bearing.

axially, relative to one another. Referring to the example in Figure 5-82, note that the phase readings at bearings A and B are approximately equal. This tells us that these bearings are vibrating in-phase, as a single unit. Likewise, bearings C and D are in-phase. However, bearings B and C are vibrating in opposition to one another. Whenever two adjacent bearings such as B and C are found to be vibrating out-of-phase, this is a strong indication that the originator of the excessive axial vibration is located somewhere in between.

Normally, if a large phase difference is noted between two bearings of direct coupled machinery, such as between B and C, coupling misalignment or a faulty coupling is usually the cause. If the phase difference is noted between two bearings of the same machine, the machine should be checked for a bent shaft or severely misaligned bearings.

If our phase analysis reveals that all bearings of the system are in-phase, the problem may be the result of unbalance—especially on overhung rotors, fans, or blowers. Identical phase readings have also been the result of a foundation which is resonant in the axial direction at a frequency equal to the rpm of the machine.

Detecting reaction forces versus unbalance. When the vibration of a machine occurs at a frequency of 1 × rpm, comparing the phase readings of the horizontal and vertical vibrations can help confirm an unbalance condition and may also help identify other sources of similar vibration such as reaction forces caused by eccentric pulleys and gears, reciprocating forces, or resonance conditions.

A normal unbalance condition produces a force equally applied throughout the 360° rotation of the unbalance heavy spot. This will cause the centerline of the shaft to move in a circular or perhaps somewhat elliptical path, depending on the comparative horizontal and vertical stiffnesses of the machine (see Figure 5-83). In any event, the maximum displacement of the machine in the vertical direction will occur one-quarter revolution, or 90°, before the maximum displacement in the

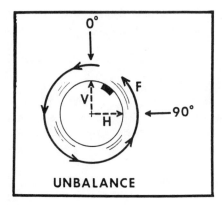

Figure 5-83. Comparative horizontal and vertical phase readings will typically differ by 90 deg. when simple unbalance is the cause.

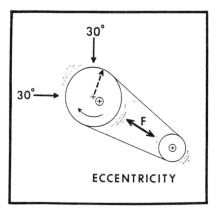

Figure 5-84. Reaction forces generated by eccentricity or reciprocating components may reveal horizontal and vertical phase readings which are the same or differ by 180 deg.

horizontal direction. As a result, a normal unbalance condition will reveal comparative horizontal and vertical phase readings which differ by approximately 90°.

Next, consider the vibration of a machine caused by an eccentric "V" belt pulley, Figure 5-84. In this case, the vibration results from variations in belt tension. The "high side" of the eccentric pulley will cause greater belt tension than the "low side." Thus, as the pulley rotates, the resultant variation in belt tension will cause a vibration at a frequency equal to 1 × rpm of the eccentric pulley—similar to unbalance. However, unlike unbalance, the reaction force caused by the eccentric pulley is not being equally applied throughout the entire 360° rotation of the pulley. Instead, the force is generated in the direction of belt tension, along a line passing through the centerlines of the two shafts. This means that the maximum displacement in the vertical direction might occur at the same time as the maximum displacement in the horizontal direction. As a result, the eccentric pulley in Figure 5-84 might reveal comparative horizontal and vertical phase readings which are the same (or 180° opposite, depending on which side of the bearing the pickup is located for the measurement). In any event, the phase comparison is significantly different from that resulting from a normal unbalance. An eccentric gear would likely produce comparative horizontal and vertical phase readings similar to those caused by an eccentric pulley. The reaction forces caused by eccentric pulleys and gears will result in vibration similar to that caused by unbalance; however, these forces cannot be totally compensated by applying balance correction weights. Attempts to balance out these reaction forces will often result in a reduction of the vibration in one radial direction, with a corresponding *increase* in the other radial direction. That is, balancing to reduce the horizontal amplitude may cause the vertical amplitude to increase. Reciprocating machinery such as compressors or gasoline and diesel engines will sometimes reveal vibration at 1 × rpm because of operational problems such as faulty valves, blow-by, fuel-injector problems, or faulty ignition. Here, also, the vibration will appear similar to unbalance but cannot be totally corrected by balancing. As with reaction forces caused by eccentric pulleys and gears, the inertia forces resulting from operational problems with reciprocating machinery can often be detected by taking comparative horizontal and vertical phase readings as outlined previously for eccentricity.

In addition, it can be noted in Figure 5-85 that operational problems with reciprocating machinery will cause vibration as the result of unbalance torque reactions on the crankshaft. This causes a "rocking" or torsional vibration of the machine about the centerline of the crankshaft instead of the circular or elliptical

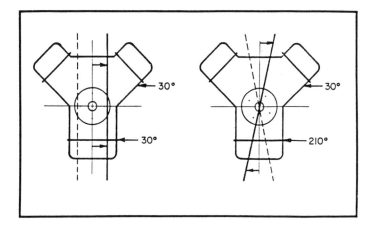

Figure 5-85. Torsional vibration of reciprocating machinery, caused by operational problems, can often be distinguished from simple unbalance by comparing the phase of vibration measured above and below crankshaft centerline.

motion caused by mass unbalance. This torsional type of vibration can often be detected by taking comparative horizontal phase readings above and below crankshaft centerline. Torsional vibration will reveal phase readings which differ by as much as 180°, whereas the circular or elliptical motion of the machine due to mass unbalance will show nearly identical phase readings.

Resonance determination. In the preceding sections discussing amplitude and phase-versus-rpm and polar amplitude-versus-phase plots, several examples were presented where recording the phase of vibration during machine startup or coastdown helped detect and identify machine resonant frequencies. A simple resonance will characteristically plot a 180° shift in phase as machine speed passes through a resonant frequency (see Figure 5-13). If a machine is operating on or very near a structural resonance, say the horizontal bearing resonance, further confirmation of this condition can often be obtained by again comparing the phase of the horizontal and vertical vibrations. As discussed previously, a normal unbalance condition will generally indicate horizontal and vertical phase readings which differ by approximately 90°. However, if the machine is vibrating at a horizontal or vertical resonant frequency, the comparative horizontal and vertical phase readings will likely be the same or 180° opposite, depending on which side of the bearing the vibration pickup is placed for the reading. This apparent 90° "error" in comparative phase is the result of an additional 90° phase "lag" introduced by the resonant condition.

Pinpointing mechanial looseness. Another possible application for phase measurement and comparison is in detecting the source of structural looseness. For example, if all the components of the machine in Figure 5-86 are rigidly coupled together, comparative phase readings for the three measurement locations shown should be practically identical. If the mounting bolt has worked loose, this will likely result in

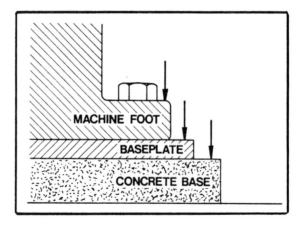

Figure 5-86. Comparative phase readings taken at the three measurement points illustrated may help in locating the source of looseness.

relative motion between the machine foot and baseplate, and comparative phase readings taken on these two components will reveal a noticeable difference in phase. If the grouting has loosened between the baseplate and concrete base, a difference in the phase will likely be measured at these two locations.

Identifying the type of rotor unbalance. In many cases, the solution to a balancing problem can be greatly simplified if the type of rotor unbalance can first be identified. Essentially, there are four types of rotor unbalance: static, couple, quasi-static, and dynamic. Each type of unbalance is identified by the relationship existing between the shaft centerline or shaft axis and the "central principal axis," which is the axis about which the weight of the rotor is equally distributed.

Figure 5-87 illustrates a condition of static unbalance. A single unbalance weight has been placed in the center plane of the rotor, causing the rotor center of gravity to be displaced off the rotating centerline. Note that the central principal axis is displaced parallel to the shaft axis. Obviously, a condition of static unbalance can be totally corrected by making a single weight correction in the center plane of the rotor, or, if this is not possible, by making equal corrections at the end planes of the rotor with the weights in line with one another.

Static unbalance can often be identified by comparing the amplitude and phase of bearing or shaft vibration at the ends of the rotor. A rotor supported between bearings will typically reveal identical amplitude and phase readings measured at the bearings or at each end of the shaft if the unbalance is truly static unbalance. This rule does not apply, however, to overhung rotors.

Figure 5-88 illustrates couple unbalance where equal unbalance weights have been added at the end planes but 180° opposite one another. Although the center of gravity has not been displaced off the centerline, the central principal axis is tilted and intersects the shaft axis at the center of gravity. To correct this unbalance condition it will be necessary to apply equal but opposite correction weights at the end planes of the rotor.

If the unbalance is a couple unbalance, a rotor supported between bearings will most likely reveal equal amplitudes of vibration but phase readings which differ by 180°. Again, this method of detecting couple unbalance does not apply to overhung rotors.

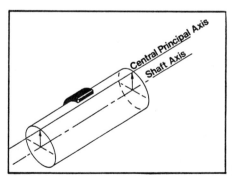

Figure 5-87. Static unbalance.

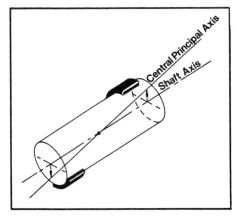

Figure 5-88. Couple unbalance.

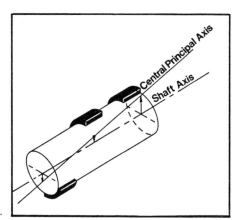

Figure 5-89. Quasi-Static unbalance.

In only a few cases will a rotor have true static or true couple unbalance. Normally, an unbalance will be a combination of static and couple unbalance. Such combinations are further classified as quasi-static and dynamic unbalance.

Quasi-static unbalance is illustrated in Figure 5-89, where the static unbalance component is in line with one of the couple components. With quasi-static unbalance, the central principal axis intersects the shaft axis, but not in the reference plane containing the center of gravity. The amplitude will be noticeably higher at one end of the rotor, and comparative phase readings can be either the same or 180° opposite, depending on whether the central principal axis intersects the shaft axis between the support bearings or outside the bearings.

Theoretically, a quasi-static unbalance can be corrected by unbalancing in a single reference plane—if the proper reference plane can be determined and if it is possible to make the necessary corrections in the selected plane. In most cases this is not practical, so a quasi-static unbalance is generally treated as a typical two-plane problem. However, if the unbalance vibration measured at one end of the rotor is considerably higher than that measured at the other end, a satisfactory balance can often be achieved by simply single-plane balancing the high end.

A dynamic unbalance, Figure 5-90, is perhaps the most common type of unbalance and simply represents a random combination of static and couple unbalance. Comparative phase readings will not be the same or 180° opposite as they would for a static, couple, or quasi-static unbalance. A dynamic unbalance will normally require that balance corrections be made in at least two reference planes.

In the preceding paragraphs, several examples were presented illustrating how phase measurement and comparison can help identify machinery problems. Of course, there are many more applications where phase measurements can aid the vibration analyst. These applications include mode shape determination, which is discussed in the following section.

Mode Shape Analysis. Mode shape analysis is a technique of determining the vibratory or mode shape of a machine, a structure, piping, or other components. To determine the mode shape of a machine or structure, vibration amplitude and phase readings must be taken at numerous measurement points and plotted on a chart to reveal the vibratory shape. This analysis technique is extremely useful for confirming resonance conditions, identifying nodal points, and revealing sources of structural weakness.

To illustrate the technique of determining the mode shape of a structure, suppose that the machine in Figure 5-91 has high amplitudes of vibration in the vertical direction, suggesting a possible resonance or structural weakness. A simple mode shape can be obtained by simply measuring and recording the vibration amplitude in the vertical direction at numerous locations along the structure, as illustrated. The measurement points should be equally spaced to simplify plotting the data. To plot the data, simply prepare a sketch of the structure using a convenient scale and indicate on the sketch the location of each measurement point. At each measurement point on the sketch, draw a vertical line equal in length to the vibration amplitude measured at that point on the structure. Here, also, any convenient scale

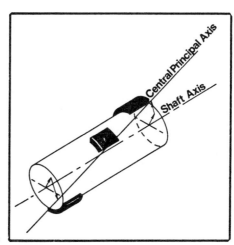

Figure 5-90. Dynamic unbalance.

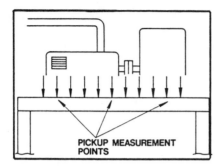

Figure 5-91. To determine the vibratory motion (mode shape) of the structure, measure the vibration amplitude and phase at several points.

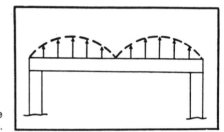

Figure 5-92. Plotting the vibration amplitude measurements to establish the mode shape.

can be used for the amplitude lines and can be selected based on the range of amplitude readings recorded. Once all the amplitude lines have been drawn, the mode shape of the structure can be seen by connecting the ends of the amplitude lines, as shown in Figure 5-92. If the measurements taken indicate the presence of nodal points, which are points of minimum amplitude, it will be necessary to obtain comparative phase readings for the vibration measured on either side of the nodal point in order to determine whether the mode shape appears as shown in Figure 5-92 or as illustrated in Figure 5-93.

The mode shape in Figure 5-93 clearly indicates that the structure is vibrating at the second flexural resonance in the vertical direction because of the characteristic sinusoidal deflection exhibited. Not only has the resonance condition been confirmed, but the nodal points have been identified, as well. This is important, because if the decision was made to change the vertical resonance of the structure by increasing vertical stiffness, if the stiffening members were added to the structure at the nodal points as illustrated in Figure 5-94, little or no improvement might be seen.

The mode shapes in Figure 5-95 illustrate what might be obtained by analyzing a common bearing pedestal and its foundation, perhaps to identify the reason for high amplitudes of bearing vibration in the horizontal direction. In each case, the vibratory motion (mode shape) is indicated by the dotted lines. By examining the bearing on the left of Figure 5-95, it can be seen that the measured vertical amplitudes on the mat or foundation are relatively low, as are the horizontal amplitudes measured on the support pedestal. However, amplitudes measured on the bearing pillowblock show a significant increase with elevation and clearly reveal

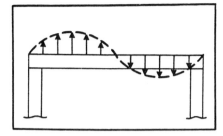

Figure 5-93. Comparative phase readings are needed to determine the true mode-shape—particularly where the amplitude readings suggest nodal points.

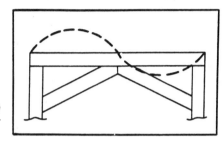

Figure 5-94. Nodal points should be avoided when stiffening the structure to correct resonance.

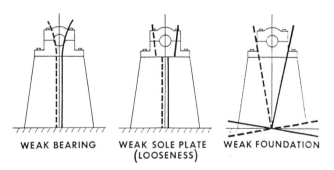

Figure 5-95. Mode shape analysis to help identify sources of weakness of a simple bearing pedestal.

bearing deflection due to bearing weakness and possibly bearing resonance. The solution to this vibration problem is to strengthen the pillowblock bearing; strengthening the foundation or support pedestal would not cause any noticeable improvement.

The mode shape of the bearing in the center of Figure 5-95 also reveals low amplitudes of vibration on the mat and support pedestal, with high amplitudes again measured on the pillowblock bearing. However, in this case, note that the amplitudes on the pillowblock do not increase significantly with increased elevation as they did for the example on the left. Instead, the relatively high amplitudes plotted on the bearing suggest that the pillowback is shifting back and forth on top of the pedestal, suggesting weakness of the sole plate or possible looseness of the bearing mounting bolts.

The mode shape of the bearing on the right of Figure 5-95 reveals a third possible cause of the high horizontal vibration on the bearing. Note in this example that the horizontal amplitudes measured on the support pedestal and bearing show a proportional increase with elevation, and this is caused by the obvious "rocking" of the mat. This would indicate a structural or soil weakness. Attempts to reduce this vibration by stiffening the pillowblock relative to the support pedestal would result in little or no improvement.

Additional examples of mode shape analysis are presented in Figure 5-96, illustrating the application of this technique to vertical pump columns. These examples illustrate the first and second flexural resonances of the column and the effect of a resonant foundation.

A form of mode shape analysis can also be useful when balancing long flexible rotors such as rolls used in paper machinery, Figure 5-97. With the roll spinning in the balancing machine, measurements of roll deflection—amplitude and phase—are obtained at numerous locations along the roll using a vibration pickup and shaft stick. These readings are then plotted in chart form to reveal the mode of roll deflection. Such analysis of roll deflection is useful for selecting the most appropriate planes for applying balance corrections.

Figures 5-95 and 5-96 show examples of simple mode shape analysis presented in two dimensions. This analysis technique can be extended for determining the mode shape characteristics of a structure in three dimensions, as illustrated in Figure 5-98. Simply mark off the structure in a matrix or gridwork and proceed to measure and record the amplitude and phase of the vibration at each marked location. Next, plot

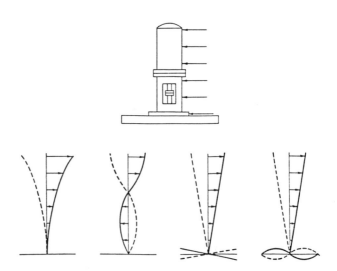

Figure 5-96. Possible mode shapes of a vertical pump and supports.

Figure 5-97. Mode-shape analysis is very useful for finding the mode of deflection (whip) of a long flexible roll, to identify planes for balance correction.

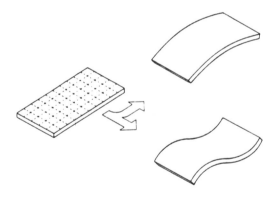

Figure 5-98. Mode shape analysis in three dimensions.

the readings on a scaled, three-dimensioned sketch of the machine and connect the amplitude vectors together as illustrated in Figure 5-98 to reveal the total vibratory shape of the structure.

To summarize, mode shape analysis can be extremely valuable for identifying sources of structural weakness and resonance and is strongly recommended for identifying nodal points *before* structural modifications are made in an attempt to solve a problem of excessive vibration. The time and effort to obtain mode shape information will often result in a clear indication of where structural modifications are needed and may avoid costly and embarrassing trial-and-error approaches.

Special Treatments of Vibration Sources
to Reduce Plant Noise

Basically, there are three ways to make noise: the turbulent mixing of high- and low-velocity air or other fluids; the movement of air over a solid structure; and the vibration of solid structures. Of these three noise generators, the vibration of solid structures is responsible for a large share of industrial noise problems.

Many techniques can be used to control noise generated by a vibrating machine or structure. Complete or partial enclosures, sound barriers, and the control of reverberant sound with absorbing materials are generally effective when properly applied. However, there are many instances where easier and perhaps less costly noise reduction might be achieved by dealing with the problem as a vibration problem and not strictly as a noise problem. For example, Figure 5-99 illustrates comparative noise and vibration analysis (amplitude versus frequency) data obtained from a motor-driven fan. Noise was measured with a microphone located approximately five feet from the fan; vibration was measured with a velocity transducer mounted on the inboard fan bearing. From the data, Figure 5-99, it can be seen that the noise frequencies correlate directly in this instance with the frequencies of mechanical vibration. Of course, identifying specific machinery problems such as unbalance, misalignment, looseness, etc. through vibration measurement and analysis techniques is well documented and generally straightforward. Thus, if the

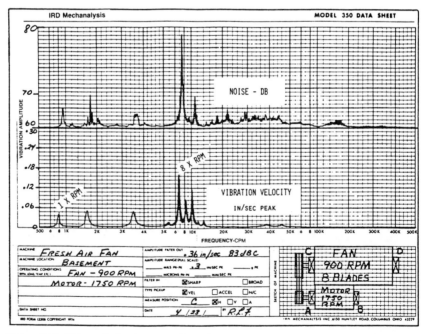

Figure 5-99. This analysis of a motor-driven fan reveals the direct correlation between vibration and noise frequencies.

cause of *vibration* can be identified through analysis and corrected, it is likely that a corresponding reduction in noise will follow. In addition, an added benefit—improved machinery performance—may result.

Obviously, not all noise problems will correlate directly with machinery vibration as clearly as the example given in Figure 5-99; nor can all noise problems be solved through vibration reduction. However, the analysis procedures for confirming or rejecting the feasibility of reducing noise by vibration control are relatively simple and are suggested as a logical, inexpensive first step.

This section presents an approach for dealing with excessive machinery noise through vibration measurement, analysis, and control. Techniques are presented for identifying major noise sources. In addition, suggestions for controlling normal or inherent vibration at the source are outlined, along with methods and devices for controlling "structure-borne" vibration.

Identifying the Noise Source

The chart in Figure 5-100 outlines the various approaches which can be taken to control or reduce excessive noise. Regardless of the approach selected, the first step is to identify the machine or machines which are major contributors to the overall noise level and deal with these first. Little or no reduction in the overall noise level can be achieved until the noise from the major sources is effectively reduced.

Where only one or two machines are operating in a given area, identifying the major noise source may present little problem. In other cases, the major source may seem obvious to the listener, who can readily perceive one machine as being notably

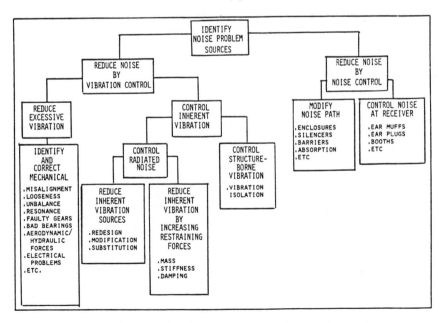

Figure 5-100. Approaches for controlling excessive machinery noise.

"louder." However, it must be remembered that the ear responds differently to various frequencies of sound as well as to variation in amplitude. In other words, the ear can be deceived. Therefore, before expensive noise-reduction schemes are undertaken, the noise source, detected by ear, should be confirmed with actual measurements taken with a sound-level meter, Figure 5-101.

When several machines occupy a noise area, determining the major source(s) of noise becomes more difficult—but not impossible. Of course, one way to measure the relative noise levels of the individual machines is to operate them singly and measure the noise levels directly. However, for many operations it may be impossible to shut down all but the one machine. In such cases, a reasonably accurate measure of the individual noise levels can be obtained using the decibel subtraction chart, Figure 5-102. To use this chart it is necessary to shut down only the machine being studied.

To illustrate the use of the decibel subtraction chart, consider an area containing three fan units having an overall noise level of 97 dBA. To find the noise level of fan 1, shut the fan down and measure the new noise level. Assume this new level is 90 dBA. Next, subtract the new noise level from the overall level (97 − 90 = 7). Referring to the chart, locate this difference along the horizontal scale and, from this point of intersection on the curve, cross horizontally to the scale on the left and read the value indicated. For the example, this value is 1.0. Finally, subtract the vertical scale value from the overall noise level (97 − 1 = 96); the result is the noise level of the machine which was shut down. This procedure is then repeated until the noise level of each individual machine in the area has been established.

Figure 5-101. Suspected noise sources should be confirmed with actual measurements taken with a sound level meter.

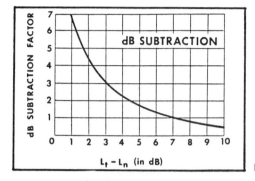

Figure 5-102. dB subtraction chart.

Figure 5-103. Vibration readings taken at each bearing of the machine are needed to determine machine condition.

Determining Machinery Condition

After the major noise sources have been determined in order of their significance, the next step is to measure and analyze machinery vibration and noise, beginning with those machines having the highest noise levels.

Overall or unfiltered vibration measurements should first be taken at each bearing of the machine to determine general machinery condition. Readings are taken on or as near the bearings as possible, because it is through the bearings that the vibration forces are transmitted. These readings of *overall* vibration amplitude can be taken using a vibration/sound-level meter, as illustrated in Figure 5-103.

Since the pickups normally used to measure vibration are sensitive only in one direction, and since different machinery troubles will often cause more vibration in one direction than another, the vibration readings should be taken with the pickup placed in the horizontal, vertical, and axial directions. Horizontal and vertical readings are take perpendicularly to the shaft, whereas axial readings are taken parallel to the shaft.

The overall vibration amplitude readings taken at each bearing should be recorded on a data form such as the one shown in Figure 5-104, along with a sketch of the machine and other pertinent data, including noise level measurements taken at selected locations around the machine. Note that the overall vibration readings have been taken in units of vibration velocity (inches/second peak or mm/sec peak). Experience has indicated that overall vibration velocity measurements provide the best indication of vibration severity over the frequency range of 600 to 60,000 CPM (10 Hz to 1000 Hz), which covers most vibration frequencies found on rotating machinery. Vibration amplitude measured in displacement (mils or microns peak-to-peak) and acceleration (g's peak) may also be used to indicate vibration severity; however, when displacement or acceleration is used, it is usually necessary to know the vibration frequency as well, in order to determine severity. Displacement measurements are generally used for low-frequency vibration, below 600 CPM, where large displacements cause high stresses. On the other hand, vibration acceleration is often used for high frequencies, above 60,000 CPM, where vibratory forces may be high even though displacement and velocity may be small. The initial vibration readings taken on the machine, compared with the machinery manufacturer's recommended values, will provide a reasonable evaluation of machinery condition. In the absence of manufacturer's recommendations, the vibration velocity values in Figures 5-105 and 5-184 can be used for overall readings taken on the machine structure or on bearing caps. When displacement or acceleration is measured, the charts in Figures 5-106, 5-107 and 5-183 can be used to cross-reference with frequency to determine vibration severity.

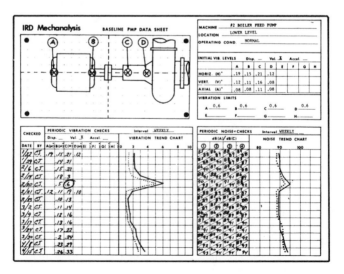

Figure 5-104. Typical form for recording overall vibration and noise-level readings.

Vibration Velocity
(Inches/Second Peak)

0 - .005 in./sec	Extremely Smooth
.005 - .01 in./sec	Very Smooth
.01 - .02 in./sec	Smooth
.02 - .04 in./sec	Very Good
.04 - .08 in./sec	Good
.08 - .16 in./sec	Fair
.16 - .32 in./sec	Slightly Rough
.32 - .64 in./sec	Rough
Above .64 in./sec	Very Rough

Figure 5-105. Velocity table.

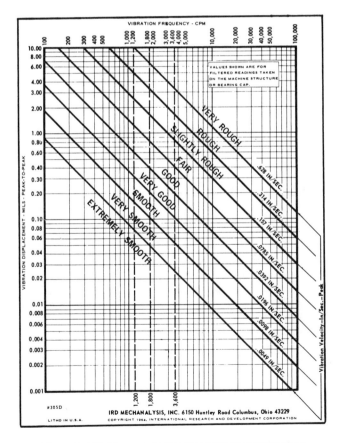

Figure 5-106. General machinery vibration severity chart.

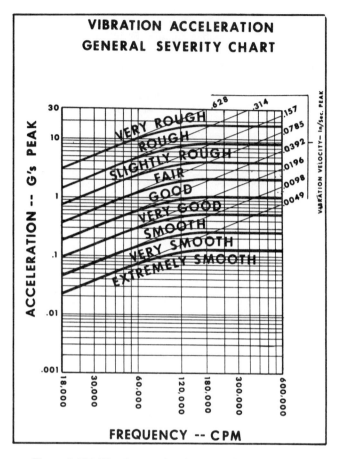

Figure 5-107. Vibration acceleration general severity chart.

The overall vibration readings should provide a reasonable estimate of machinery condition. Of course, if there are no significant mechanical or operational defects, the overall vibration readings will be low. However, if the readings are high, or at a level where further reduction seems feasible, then a complete analysis of the vibration and noise can be carried out to identify the specific causes of vibration.

With mechanical problems detected and diagnosed by vibration measurement and analysis, needed corrections can be carried out to reduce the vibration, hopefully with a corresponding reduction in noise as well. Corrections may include realigning bearings and couplings, correcting mechanical looseness, or replacing defective bearings or worn gears. If the problem is unbalance, the correction can normally be made in place using the vibration analyzer and standard balancing techniques. Figure 5-108 shows the vibration and noise analysis data taken from a direct-drive 1750-rpm fan. From the data, there appears to be little correlation between the initial vibration and noise frequencies: the initial vibration is predominant (.59 in./sec) at 1 × rpm because of fan unbalance, whereas the noise frequencies of significance are 2 × rpm,

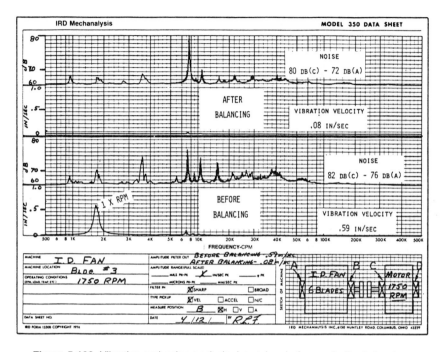

Figure 5-108. Vibration and noise-analysis data taken before and after balancing.

4 × rpm, 6 × rpm, and 10 × rpm. However, it is not uncommon to find a fundamental low frequency of vibration responsible for excessive noise at higher, harmonically-related frequencies. Therefore, the first attempt to reduce noise in this case was to balance the fan. This was accomplished, resulting in a reduction of overall vibration from .59 in./sec (15mm/sec) to .08 in./sec (2mm/sec). The follow-up noise analysis reveals that balancing did, in fact, reduce the noise amplitudes at frequencies of 1 × rpm, 2 × rpm, 6 × rpm, and 10 × rpm. However, the noise at 4 × rpm was unaffected. The result of balancing was a noise reduction from an initial 82 dBC and 76 dBA to 80 dBC and 72 dBA, or a 2-dBC and 4-dBA reduction.

Figure 5-109 illustrates another example of machinery noise and vibration analysis data. These data were taken on a speed-reducer gear driven by a 1750-rpm, four-pole induction motor. The vibration analysis clearly shows excessive vibration (.7 in./sec) at a frequency equal to 1 × rpm of the motor; however, the axial vibration at this frequency was equally high, indicating misalignment between the motor and gear box, not unbalance. The initial noise analysis data revealed very little noise at the 1 × rpm frequency, but did reveal high amplitude noise at a frequency of approximately 160,000 CPM, corresponding to gear-meshing frequency (i.e. 1750 rpm × 90 teeth = 157,500 CPM).

The gear box and motor were realigned, which reduced the vibration from .7 in./sec (18 mm/sec) to .1 in./sec (2.5 mm/sec). After realignment, a follow-up noise analysis disclosed a substantial reduction of the noise gear-meshing frequency, with an initial 92 dBC and 92 dBA reduced to 75 dBC and 73 dBA after alignment.

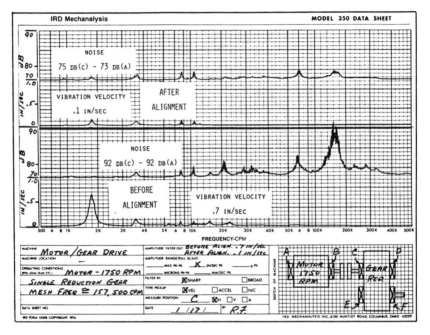

Figure 5-109. Vibration and noise-analysis data taken before and after coupling alignment.

Another problem which illustrates the relationship between machinery vibration and noise involved a 3600-rpm centrifugal pump, where an unusually high amplitude of vibration and noise was detected at a frequency of 21,600 CPM. This frequency was equal to the product of the number of vanes on the pump impeller (6) times the rotating speed (3600), indicating that hydraulic forces were responsible. Normally, hydraulic forces are inherent in the operation of a pump and seldom cause excessive vibration unless some part of the system or supporting structure happens to be in resonance at that particular frequency. However, in this case, the usual tests for resonance were carried out, but no natural frequencies were found corresponding to the objectionable 21,600-CPM vibration and noise frequency.

Next, the pump was disassembled for inspection and it was discovered that the radial clearances between the impeller and two diffusers cast in the pump housing diametrically opposite one another differed by approximately .4 in. (10mm) thus creating an imbalance of hydraulic forces. This condition was corrected by machining both diffusers to equalize radial clearance and diffuser profile. The result was a substantial reduction of both vibration and noise.

Controlling Normal Vibration

Of course, conditions do exist where a machine, identified as a major noise source, reveals very low levels of vibration, indicating that no significant mechanical or operational problems are present. In such cases, there may be little or no correlation between the vibration and noise analysis data, such as demonstrated in Figure 5-99, or no apparent cause-and-effect relationship, as demonstrated in Figure 5-108 and 5-109. This is not unusual because, it must be remembered, vibration readings are taken primarily at the bearings of a machine and in specific directions, whereas noise measurements taken with a microphone are the result of noise coming from all parts of the machine and from virtually all directions. For example, a fan may show very little vibration measured at the bearings. However, the fan housing and ductwork may be vibrating due to inherent aerodynamic pulsations, and because of the large surface area of the housing, even a small amplitude of vibration can result in high noise levels. In addition, machinery vibration may be transmitted to the surrounding structure, causing large wall, floor, or ceiling surfaces to vibrate. Again, the amplitudes of vibration may be quite small; however the large vibrating surfaces tend to create a loud-speaker effect with very little applied energy.

As can be seen, normal or low levels of machinery vibration can still produce excessive noise. And, the resultant noise can be radiated directly from the machine and/or the result of structure-borne vibration. Of course, techniques and materials for controlling radiated noise will differ from those used to control structure-borne vibration. Therefore, before corrective measures can be selected, it will be necessary to determine which problem exists.

Determining whether noise is predominately radiated or structure-borne can be difficult—especially on large machines where both problems may exist. In such cases, further analysis of noise and structural vibration should be carried out before undertaking extensive modifications.

Radiated noise can usually be detected by measuring the noise level at various distances from the machine and noting a gradual steady decline in noise amplitude as the distance is increased. In a "free field" (i.e. a sound field having no reverberant influences), the noise level will decrease six dB with each doubling of distance from the source. Of course, few plant areas will be free from reverberant effects, so this much change will not likely be seen in actual practice. Obviously, if the noise amplitude increases, or if no decrease in noise amplitude is measured at increased distances from the machine, there is a good possibility that the problem is structure-borne. This possibility should be further studied by actually analyzing the vibration of suspected structural members such as walls, floors, decks, ceilings, etc. For example, the analysis data in Figure 5-110 were obtained from a second-floor conference room of a two-story office building. The vibration analysis data were obtained by attaching the pickup directly to one wall of the conference room while the noise analysis data were picked up by a microphone located in the middle of the room. Although the vibration amplitudes were all very low compared to general machinery standards, the frequencies of vibration corresponded directly with the predominant noise frequencies. This vibration was ultimately traced to air-conditioning equipment located in the basement of the building. The vibration and noise at approximately 900 CPM was due to a 900-rpm fan supplying fresh air to the conference room. This vibration was being transmitted through the ductwork leading from the fan to the conference room. The vibration at 3600 rpm corresponds to 2 × rpm of an 1800-rpm, 6-cylinder reciprocating chiller compressor. Vibration predominant at 2 × rpm is normal for reciprocating machines because of inherent

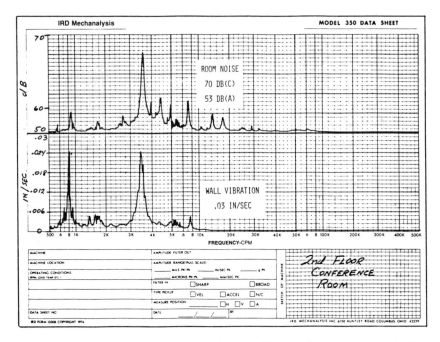

Figure 5-110. Vibration and noise analysis of a second-floor conference room reveals direct correlation between noise and structure.

inertia forces of reciprocating components. The vibration from the compressor at 3600 CPM was being transmitted to the conference room through rigid piping leading to a cooling tower located on the roof of the building.

In some instances, the amplitude of a particular frequency of machinery-related vibration will be higher on a structural member than measured at the machine. This would indicate the possibility of a structural resonance condition, where even a small amount of applied energy can result in high levels of vibration and noise. Of course, plots of overall noise amplitude versus rpm can be made to determine resonance problems contributing to excessive noise. Resonance conditions detected in this way can be corrected by changing the operating speed of the machine so that the exciting frequency no longer coincides with the natural frequency of the structure. Or, the natural frequency of the structure can be changed by increasing or decreasing structural mass or stiffness. If these solutions are impractical, isolating the structure from the vibration source is another possible solution.

Having identified the noise problem as radiated or structure-borne vibration, we can now investigate additional methods of control.

Controlling Radiated Noise

The control or reduction of normal low-level vibration to reduce noise radiated from the machine usually requires some addition or modification to the machine.

Solutions can range from simple, inexpensive modifications to costly redesign projects. In any event, there are only two choices:

1. Increase the inherent forces which oppose or restrict vibration. These forces are stiffness, inertia, and damping.
2. Substitute, modify, or redesign the machine or selected components to minimize vibration forces.

Increasing stiffness by adding braces and structural supports can reduce noise radiated by vibrating panels when the fundamental vibration and noise frequencies are relatively low. When the vibration and noise frequencies are high, increasing stiffness by adding a few braces or supports becomes less effective. Of course, increasing the mass of the vibrating member, perhaps by increasing wall thickness, will also reduce the vibration and radiated noise. However, when dealing with large vibrating surfaces such as large machine housings or ductwork, increasing the mass or stiffness is generally less practical and less economical than conventional damping or acoustic lagging techniques.

Damping materials are readily available in many forms, including brush-on, spray-on, and trowel-on liquids and pastes. These are usually asphalt, latex, or epoxy materials. In addition, damping can be achieved by cementing flexible sheets of rubber, glass fiber, leaded vinyl, and other materials directly onto the vibrating surface. These materials, also, are commercially available in many forms. The heavy limp-mass of materials such as leaded vinyl is particularly effective for reducing vibration and radiated noise. When these materials are properly applied to the vibrating surface, damping is substantially increased. In addition, the added weight of these materials, usually one to two lbs/ft^2 (4.5 to 9 kg/m^2) effectively increases mass to further reduce vibration and noise. These materials are also an effective acoustic barrier and reduce the transmission of internally generated noise.

The selection of the most appropriate damping or sound-deadening materials is usually based on temperature requirement; resistance to water, oil, or other chemicals; fire resistance; cost; etc.

In addition to damping materials applied to vibrating surfaces, it has also been demonstrated that certain lubricants, such as those containing molybdenum disulfide, have improved damping and load-bearing properties to minimize impact forces generated between mating gears, cams, and followers, or on similar applications involving heavily loaded industrial machinery.

If it is not possible or practical to reduce vibration and radiated noise by increasing stiffness, mass, or damping, minimize the vibration forces by selective substitution or modification of various machine components:

1. Substitute helical gears for straight-cut (spur) gears.
2. Substitute nylon, plastic, or sintered gears for steel gears.
3. Use "V"-belt drives in place of gear drives.
4. Reduce free-play clearances in cam drives, in oscillating linkage arrangements, and in similar applications in order to minimize impact forces.
5. Use larger, lower-rpm machines in place of smaller high-speed machines.

6. Use hydraulic presses in place of purely mechanical stamping presses.
7. Substitute rotating shears for square shears.
8. Use step dies or progressive operation dies in place of single-operation dies.

In some cases, it may be possible to make design modifications to existing equipment to reduce vibration forces which have been identified through analysis. To illustrate, a petrochemical plant was experiencing excessive noise from a particular model centrifugal pump. An analysis of the noise revealed that the disturbing frequency was equal to the product of pump rpm times the number of impeller vanes. However, the vibration at the pump was not considered excessive at this frequency, indicating that no major mechanical or operational problems were responsible. Upon examining the design of the pump, it was concluded that the hydraulic forces causing the noise were due to the fact that the leading edges of the impeller vanes were square, as were the leading edges of the diffusers, thus creating a slapping action or abrupt pulsation each time a vane passed a diffuser. To reduce these abrupt pulsations, a "V" notch was machined in the leading edge of each diffuser to provide a more gradual, progressive impulse. The result was a substantial noise reduction. (See page 432 for additional information.)

Controlling Structure-Borne Vibration

There are many situations where the problem is not the noise radiated by the machine but machinery vibration being transmitted into the structure, resulting in vibration of walls, floors, ceilings, etc. This structure-borne vibration can be transmitted over large distances and regenerated as noise at locations far from the machine.

Structure-borne vibration and noise is controlled by vibration isolation. Vibration isolation involves using resilient materials such as rubber, cork, felt, or fiberglass pads; coil springs; flexible joints in piping or conduit; or other devices which prevent the vibration from being transmitted into the structure. Basically, a properly selected resilient isolator material placed between the machine and structure prevents the transmission of vibration energy to the surrounding structure.

The selection of vibration isolation equipment requires extreme care. If isolators are chosen arbitrarily, it is possible to actually amplify structural vibration instead of reduce it. This fact is revealed by the transmissibility chart, Figure 5-111, which presents a fundamental relationship essential to effective vibration isolation. The vertical scale represents the percent of vibration transmitted, and the horizontal scale is the ratio F_f/F_n, where F_f is the forcing frequency of machinery vibration and F_n is the natural frequency of the installed isolator.

According to the chart, no isolation is achieved for F_f/F_n ratios less than 1.4. Therefore, the goal in vibration isolation is to select a spring or other isolation material that will provide a natural frequency much lower than the forcing frequency of machinery vibration. Of course, the first step in accomplishing this goal is determining the forcing frequency or frequencies. Normally, the forcing frequency will be the vibration occurring at rotating speed (rpm); however, a vibration analysis should be performed to determine whether or not any other significant frequencies are present. For example, coil springs are widely used to isolate low-frequency

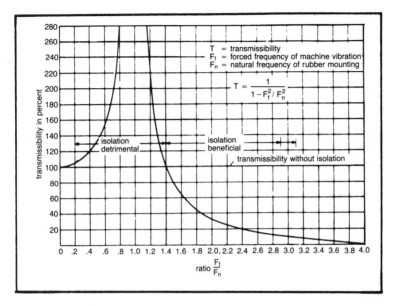

Figure 5-111. Transmissibility chart.

vibration, but they are generally ineffective for controlling high-frequency vibration. Therefore, if the vibration analysis reveals both low and high frequencies, it may be necessary to combine a high-frequency isolator such as a felt or rubber pad in series with a coil spring in order to achieve satisfactory isolation.

Once the forcing frequencies of vibration are known, the next step is to select an isolator to be installed between the machine and supporting structure. The isolator manufacturer or supplier needs to know the vibration amplitudes and frequencies involved as well as other details about the machine, including its size, weight, load or weight distribution, location within the building, plus any adverse environmental factors such as heat, moisture, oil, chemicals, etc. Based on this information, the isolator manufacturer or supplier will be able to recommend the isolators best suited for the job.

Another important factor to consider when isolating machinery vibration is the possibility of vibration being transmitted into the structure through rigid "flanking paths," such as rigid piping or electrical conduit. One can go to a lot of trouble to specify and install isolators under the machine; however, if flanking paths such as piping or electrical conduit are ignored, all efforts may be wasted because these paths can completely "short circuit" the isolators. For this reason, it is vital that all rigid connections to the machine be treated as well. In the case of piping and conduit, flexible connectors can be installed as shown in Figure 5-112. In some cases, piping may be left rigidly connected to the machine and isolated from the structure by using resilient pipe hangers, such as the one illustrated in Figure 5-113. Note in Figure 5-113 that a high-frequency isolator disc is used in conjunction with a coil spring to effectively isolate both low and high vibration frequencies.

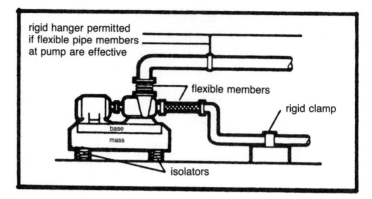

Figure 5-112. Piping, ductwork, or conduit rigidly connected to the structure must be isolated from machinery vibration with flexible connectors.

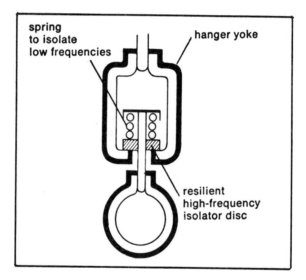

Figure 5-113. Piping which is rigidly connected to a machine can be isolated from the structure by using resilient hangers.

Summary

These sections have discussed various techniques for reducing machinery noise by reducing or controlling vibration. The methods presented included the detection, analysis, and correction of mechanical faults such as unbalance, misalignment, and looseness; the control of normal or inherent vibration resulting in radiated noise; and the control of structure-borne vibration.

Of course, not all noise problems can be solved by vibration control. In some instances, the problem must be treated as both a vibration problem as well as an acoustic (noise) problem. However, working to control vibration can substantially reduce noise levels to make further reduction through acoustic treatment less demanding and consequently less costly. And, machines with low levels of vibration do provide longer trouble-free service.

Data Interpretation

In the preceding sections various techniques for observing and displaying the vibration characteristics of a machine were discussed and illustrated. Now, the vibration characteristics associated with common machinery problems will be discussed in detail. Vibration amplitude-versus-frequency signatures, time waveform displays, and other methods of data presentation are included.

Making the Comparison

After the vibration data have been obtained, the next step is to compare the readings of significance with the characteristics of vibration typical of various types of trouble. The key to this comparison is frequency. A frequency comparison is made on the basis of the rotating speed or speeds of the machine parts. Figure 5-114 lists the frequencies of parts in terms of rpm and possible cause of the vibration. The trouble referred to will be associated with the part whose rpm is some multiple of the vibration frequency, and this comparison should identify the part causing the trouble.

The vibration identification chart in Figure 5-115 lists the most common causes of vibration likely to be encountered, together with the amplitude, frequency, and strobe-picture characteristics for each cause. The "remarks" column provides helpful information about any peculiar characteristics that will help pinpoint the trouble.

Since many machine troubles have similar characteristics, and since several troubles may be present in a machine simultaneously, it is often necessary to choose between several likely possibilities. One useful technique is to examine the relative probability of the occurrence of machine faults and start with the most probable fault first. The vibration and noise identification chart in Figure 5-116 originally proposed by turbomachinery consultant John S. Sohre in 1968 at the ASME Petroleum Mechanical Engineering Conference in Dallas, Texas, will assist in this task. This chart represents a comprehensive listing of the most common machinery problems encountered and provides a relative probability-rating number system which indicates which trouble is most likely in a given set of circumstances. Where no probability number appears, this indicates that the specific fault will not exhibit the amplitude or frequency characteristics shown. The number 10 indicates that the fault will always show the characteristics described. The number five indicates that the

VIBRATION FREQUENCIES AND THE LIKELY CAUSES

Frequency In Terms Of RPM	Most Likely Causes	Other Possible Causes & Remarks
1 x RPM	Unbalance	1) Eccentric journals, gears or pulleys 2) Misalignment or bent shaft — if high axial vibration 3) Bad belts if RPM of belt 4) Resonance 5) Reciprocating forces 6) Electrical problems
2 x RPM	Mechanical Looseness	1) Misalignment if high axial vibration 2) Reciprocating forces 3) Resonance 4) Bad belts if 2 x RPM of belt
3 x RPM	Misalignment	Usually a combination of misalignment and excessive axial clearances (looseness).
Less than 1 x RPM	Oil Whirl (Less than ½ RPM)	1) Bad drive belts 2) Background vibration 3) Sub-harmonic resonance 4) "Beat" Vibration
Synchronous (A.C. Line Frequency)	Electrical Problems	Common electrical problems include broken rotor bars, eccentric rotor, unbalanced phases in poly-phase systems, unequal air gap.
2 x Synch. Frequency	Torque Pulses	Rare as a problem unless resonance is excited
Many Times RPM (Harmonically Related Freq.)	Bad Gears Aerodynamic Forces Hydraulic Forces Mechanical Looseness Reciprocating Forces	Gear teeth times RPM of bad gear Number of fan blades times RPM Number of impeller vanes times RPM May occur at 2, 3, 4 and sometimes higher harmonics if severe looseness
High Frequency (Not Harmonically Related)	Bad Anti-Friction Bearings	1) Bearing vibration may be unsteady — amplitude and frequency 2) Cavitation, recirculation and flow turbulence cause random, high frequency vibration 3) Improper lubrication of journal bearings (Friction excited vibration) 4) Rubbing

Figure 5-114. This chart lists the vibration frequencies normally encountered and the most likely causes for each frequency.

characteristics indicated for a specific fault will occur approximately 50% of the time. For example, looking at the vibration characteristics of unbalance, it can be seen that the number 10 appears in the predominant frequency column of 1 × rpm. This indicates that the vibration frequency of unbalance will always be 1 × rpm. In the predominant amplitude column, it can be seen that the predominant amplitude of unbalance will be in the horizontal direction 50% of the time, in the vertical direction 40% of the time, and in the axial direction 10% of the time. Further, the probable location of unbalance will be the shaft or rotor 90% of the time and caused by the bearings only 10% of the time, perhaps as the result of eccentricity of rolling-element bearings such as ball bearings or roller bearings.

As can be noted from the vibration and noise identification charts, there are many mechanical and electrical problems which can generate very similar vibration characteristics, and a particular mechanical problem may develop different vibration characteristics under varying circumstances. The following paragraphs discuss the common mechanical problems and will supplement the information available from the identification charts.

VIBRATION IDENTIFICATION

CAUSE	AMPLITUDE	FREQUENCY	PHASE	REMARKS
Unbalance	Proportional to unbalance. Largest in radial direction.	1 x RPM	Single reference mark.	Most common cause of vibration.
Misalignment couplings or bearings and bent shaft.	Large in axial direction 50% or more of radial vibration	1 x RPM usual 2 & 3 x RPM sometimes	Single double or triple	Best found by appearance of large axial vibration. Use dial indicators or other method for positive diagnosis. If sleeve bearing machine and no coupling misalignment balance the rotor.
Bad bearings anti-friction type	Unsteady - use velocity measurement if possible	Very high several times RPM	Erratic	Bearing responsible most likely the one nearest point of largest high-frequency vibration.
Eccentric journals	Usually not large	1 x RPM	Single mark	If on gears largest vibration in line with gear centers. If on motor or generator vibration disappears when power is turned off. If on pump or blower attempt to balance.
Bad gears or gear noise	Low - use velocity measure if possible	Very high gear teeth times RPM	Erratic	
Mechanical looseness		2 x RPM	Two reference marks. Slightly erratic.	Usually accompanied by unbalance and/or misalignment.
Bad drive belts	Erratic or pulsing	1, 2, 3 & 4 x RPM of belts	One or two depending on frequency. Usually unsteady.	Strob light best tool to freeze faulty belt.
Electrical	Disappears when power is turned off.	1 x RPM or 1 or 2 x synchronous frequency.	Single or rotating double mark.	If vibration amplitude drops off instantly when power is turned off cause is electrical.
Aerodynamic hydraulic forces		1 x RPM or number of blades on fan or impeller x RPM		Rare as a cause of trouble except in cases of resonance.
Reciprocating forces		1, 2 & higher orders x RPM		Inherent in reciprocating machines can only be reduced by design changes or isolation.

Figure 5-115. Vibration identification chart.

IRD Mechanalysis, Inc.
Vibration and Noise Identification Chart

Causes of Vibration
(RELATIVE PROBABILITY RATINGS: 1 THRU 10)

PREDOMINANT FREQUENCIES — **PREDOMINANT AMPLITUDE** — DIRECT / PROBABLE LOCATION

Category	Cause	<40%	40-50%	50-100%	1×RPM	2×RPM	Higher Multiples	½RPM	¼RPM	Lower Multiples	Odd Frequencies	Very High Freq.	Horizontal	Vertical	Axial	Rotor (Shaft)	Bearings	Casing	Foundation	Piping	Couplings
UNBALANCE	Initial Unbalance			10									5	4	1	9	1				
	Shaft Bow–Lost Parts			10									5	4	1	9	1				
MISALIGMENT LOOSENESS AND DISTORTION	Misalignment				4	5	1						3	2	5	8	1	1			
	Mechanical Looseness					8	1				1		5	4	1		3	2	2	2	1
	Clearance Induced Vibration	1	8	1									5	4	1		7	1	1		1
	Foundation Distortion		2		5	2					1		5	4	4		3	1	1	1	
	Case Distortion	←1→		8		½	½						5	4	1	9	1				
	Seal Rub	1	1		2	1	1		1	1	1	1	4	3	3	8	1	1			
	Rotor Rub (Axial)	←2→		3	1	1			1	1	1		4	3	3		7	1	2		
	Piping Forces				4	5	1						3	2	5	8	1	1			
BAD BEARINGS AND JOURNALS	Journal & Bearing Eccentric				8	2							5	4	1	9	1				
	Radial Bgr. Damage	1→			4	2						2	4	3	3		7	2	1		
	Thrust Brg. Damage	9→										1	3	2	5		6	2	2		
	Bearing Excited Vibration	←10→											5	4	1		5	2	2	2	
	Unequal Brg. Stiff. Horiz/Vert					9 @CR							5	4	1		4	3	3		
GEARING AND COUPLINGS	Gear Inaccuracies						2				2	6	5	3	2	8	1	1			
	Coupling Inaccuracies				1	8	1						4	3	3		7	2			1
CRITICALS	Critical Speed			10									5	4	1		6	4			
	Rotor & Brg. Sys. Critical			10									5	4	1		7	3			
	Coupling Critical			10									4	2	4		1	1			8
	Overhang Critical			10									5	4	1		7	1			2
RESONANCE	Resonant Vibration			10									4	4	2	2	1	2	3	2	
	Sub–Harmonic Resonance							←10→					3	3	4	2	2	2	2	2	
	Harmonic Resonance						←10→						4	4	2	2	1	1	3		
	Casing Resonance			8	1		1						5	4	1		4	4	1	1	
	Support Resonance			8	1		1						5	4	1		2	5	2	1	
	Foundation Resonance			8	1		1						4	3	3		1	4	4	1	
	Torsional Resonance				4	2	2				2		Torsion.				1	4	4		1
MISCELLANEOUS BASIC CAUSES	Bad Drive Belts			←10 Belts→									4	3	3		5	3		2	
	Reciprocating Forces				3	5	2						3	6	1		5	3	1	1	
	Aero./Hydr. Forces				2		6					2	5	4	1		4	3	2	1	
	Friction Induced Whirl	8	1	1									5	4	1		8	2			
	Oil Whirl		10										5	4	1		8	2			
	Resonant Whirl		10										5	4	1	2	2	2	2	2	
	Dry Whirl											10	4	3	3	4	2	2	1		1
ELECTRICAL	Rotor Not Round			10									5	4	1	9	1				
	Rotor/Stator Misalignment			10									4	3	3	8	2				
	Elliptical Stator Bore			10									5	4	1	8	2				
	Defective Bar			10									5	4	1	9	1				
	Bent Rotor Shaft			10									3	2	5	9	1				
	Rotor Not Elect. Centered			10									3	2	5	6	4				

NOISE RADIATION

Mechanical and Electrical Defects—are noise sources which appear initially as vibration and are later transferred into air-borne noise. Mechanical Noise may be associated with Fan/Motor Unbalance; Bearing Noise; Alignment; Duct and Panel Flutter—Oil Canning Effect; Flutter of dampers, blades, vanes, tubes and support as well as structural vibration. **Electrical Noise**—may be due to Electrical Energy transformation:
1) Magnetic Forces—A function of flux densities, number and shape of poles or slots and air gap geometry.
2) Random Electrical Noise—Brushes, electrical arcing, sparks, etc.

Figure 5-116. Vibration and noise identification chart. (See explanation on page 398 of how to use this chart.)

Phase (No. of Reference Marks)	Low Freq. "Hum"	Loud "Roar"	"Rumble"	Periodic "Beat"	High Pitch "Whine"	Very High Loud "Scream"	Very High "Squeal"	Ultrasonic	REMARKS
(1)	8	2							Most common cause of vibration whose amplitude is proportional to the amount of unbalance. May be aggravated by or may produce complications such as seal rubs, bearing failures or resonances. (Overhung rotors may show relatively high Axial Vibration).
	8	2							
(1) (2) (3)	4	4	2						Misalignment appears as a large axial vibration. Use dial indicators or other methods for positive diagnosis. May porduce friction or deflection forces which cna be severe. **Looseness** creates many problems. Small amount may allow violent vibration. Looseness in bearings may be mistaken for oil whirl. Usually accompanied by unbalance and/or misalignment. **Distortion** causes vibration indirectly by generating misalignment, causing internal rubs or uneven bearing contact. **Piping Forces & Foundation Distortion** often cause resonance problems. Rubs are characterized by the presence of many frequencies all over the spectrum often ultra-sonic. Produce "Hot spots" resulting in bent shaft, bearing cavitation and resonances.
(2) Erratic	8	1	1						
Erratic	6	2		2					
Erratic	1	5	3	1					
Erratic	1	7	1	1					
Erratic	2	5		1		1	1		
Erratic	3	5		2					
(1) & (2)	3	4	3						
(1)	1	9							In the case of Anti-friction Bearing failures, very high frequencies will be noted with the bearing responsible being the one at the point of the largest high frequency vibration. Journal eccentricity relating to gears appears largest in line with gear centers. On motors or generators vibration disappears when power is turned off. On pumps and blowers, improvement may be accomplished by balancing. Velocity measurements are recommended when analyzing for Anti-friction bearing failures.
Erratic	2	4	1		1	1	1		
Erratic	8	1	1						
Erratic	6	1		3					
Changing	1	6	2	1					
Erratic	2	1	1	2	2	1		1	Misalignment is prime cause of gearing failures. Pitting, scuffing & factures from non-uniform loading results. Couplings are susceptible to be both misalignment and torsional forces. Friction whirl/low damping also contribute.
Erratic		5			5				
180° Chg	5	3	2						For practical purposes, the terms "Natural Frequency", "Resonance" and "Critical Speed" are synonymous. Minute unbalances cause large shaft deflections due to centrifugal force at critical speed. Differs from resonant Vibration in that the shaft does not vibrate "Back and forth" but rotates in an ever increasing bow, assuming equal radial damping. Shaft will bend rather than fail from fatigue as in the case of resonance. A critical may be improved by balancing. Resonance may be improved by internal damping.
Changing	5	3	2						
Changing	2	4	2	2					
Changing	5	4	1						
Erratic	4		3	3					**Resonance**—Only amplifies vibrations from other sources, cnnnot generate vibration. Can create highly dangerous situations by amplifying normal vibration in rotating machines or from pulsations in piping. May cause rotors or bearing abnormalities such as **Resonant Whirl**. **Torsional Vibration** is not usually externally since motion is superimposed on the rotation similar to the action of a washing machine agiator. Failures may occur without warning unless gearing is involved resulting in noise; also bearing and case vibration. Special transducers usually required. **Torsional Resonant** frequencies coinciding with electrical frequencies can become very serious.
Rocking	8			2					
			4	2	4				
	2		2	6					
	2		2	6					
		1	8	1					
	1	2	2	3	1	1			
(1)—(2) Erratic	1	1	3	5					**Bad Belts**—Strob light will freeze faulty belt. Cure is matched belt sets, equal tnesion & correct alignment. **Recip. Forces**—Inherent in reciprocating machines—can only be reduced by design changes or isolation. **Aero-Hydro Forces**—Occur usually at N°. of impeller blades X RPM. Random Pulses may produce related resonance. **Friction Whirl**—Sometimes called "Hysteresis Whirl". Rare but violent. Cause: Rotor passes thru critical; angle between unbal & shaft "High Spot" swings 180° with friction damping also 180° out of phase. Frequency of vibration always at actual rotor critical speed. **Oil Whirl**—Caused by shaft being pushed around in bearing clearance by oil pressure wave. Frequency ½ shaft speed less 2%–8% due to friction effects.
Erratic	8	2							
(1) or Multiple	3	2	1	2	2				
Erratic	6		2	2					
Erratic	6		1	3					
Erratic	6		2	2					
Erratic							2	8	
(1)* or	8	2							*Phase at synchronous frequency. Electrical causes of vibration will show up at 60 & 120 Hz (1 & 2 X line frequency) and disappear quickly when power is turned off. A "Slip-beat" vibration may occur at slip speed times number of poles. "Beat Frequency" relates to more than one machine operating at nearly the same speed. Mechanical defects may be detected with conventional indicating methods. **Defective Bar**—Break bar connection, energize one phase with low voltage and turn rotor by hand. Current surge will indicate broken bar. Check air gaps.
Rotating	5	2	3						
Double	6	2	2						
Mark	2	6	2						
(1)	8	2							
(1)	3	3	4						

Aerodynamic—May be related to vortex shedding, pressure pulsations, windage, etc., and create both broad and narrow band noise. **Broad Band**—a) Fan blades, vanes, obstructions, supports in the air stream. b)Mechanical Rotation—Integal fans, belts, slots, etc. c) Abrupt changes in direction of flow or cross-section of ducts (Rumble). Differing flow velocities in adjacent streams—flow separations such as boundry layer effects, compression effects, etc. **Narrow Band**— a) Resonances—Organ-piping effects, vibrating strings, panels, structural members, etc. b) Sharp edge vortex effect– air columns excited by blow (whistle). c)Mechanical Rotation -Siren effects, slots, holes, vanes, grooves on rotating parts.

Impactive—Created by the forceful contact of one body or element with another such as the noise produced by a dropped hammer a thunder clap, sonic boom, etc. Tooth impact in gearing may be audible as well as the slap of faulty drive belts. Impact noises may occur so rapidly that special high speed recording techniques must be used to distinguish the periodic impact from the unpredictable transient. Areas with many impact generators will have a steady state "Drone" resulting from the accumulation of many impact "Peaks"

Figure 5-116. Continued.

HOW TO USE THE CHART (Figure 5-116)

1. Calculate or determine expected vibration and noise frequencies of main rotating machine components.

2. Measure vibration and noise at various measuring points.

3. Using the tunable filter of the analyzer, carefully tune through the appropriate frequency range of the analyzer recording the amplitude and frequency at which significant components are detected.

4. Generate a Machinery Signature by making a "Narrow Band" analysis on each frequency range of interest. Obtain signatures of both vibration and noise as required.

5. Study the Machinery Signature and carefully examine each significant vibration component (Amplitude vs Frequency).

6. Relate all significant vibration or noise components to rotational speeds of machines main components.

7. Follow the appropriate vertical columns on the chart looking for machine troubles that show high probablity ratings.

8. Perform further detailed analysis as required to complete the identification of all sources of vibration and noise in the machine in question

Reference:
J. S. Sohre, "Operating Problems with High Speed
Turbomachinery Causes and Correction", ASME
Paper, presented at the ASME Petroleum Mechanical
Engineering Conference, September 23, 1968,
Dallas, Texas.

Figure 5-116. Continued.

Vibration Due to Unbalance

The horizontal, vertical, and axial vibration signatures presented in Figure 5-117 illustrate typical amplitude-versus-frequency analysis data resulting from an unbalance condition. It can be noted that the predominant vibration occurs at 2200 CPM, corresponding to the 2200-rpm fan speed. Since the amplitude of vibration in the axial direction is relatively low compared to the radial amplitudes, a bent shaft or misalignment is not indicated. It can be noted that small amplitudes of vibration are present at 2, 3, and 4 × rpm; however, when a significant unbalance is present, the appearance of small amplitudes at the harmonic frequencies is common and does not necessarily indicate any unusual problems such as mechanical looseness.

Normally, the largest amplitudes of vibration due to unbalance will be measured in the radial (horizontal or vertical) direction. However, unbalance of an overhung rotor like that shown in Figure 5-118 will often result in a high amplitude in the axial direction as well, perhaps as high as, or in some cases higher than, the radial amplitudes. When high amplitudes of axial vibration are encountered, the axial phase analysis technique described earlier under Phase Analysis can be carried out to confirm whether the problem is unbalance or misalignment.

A rotor mounted between bearings and having a substantial couple unbalance may also exhibit high amplitudes of axial vibration at the rotating speed frequency. If comparative amplitude and phase readings reveal a couple unbalance condition (see

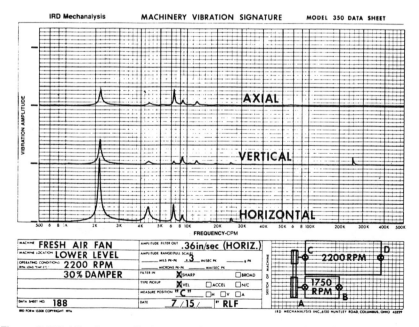

Figure 5-117. Vibration amplitude versus frequency data recorded with an X-Y recorder clearly identifies fan unbalance.

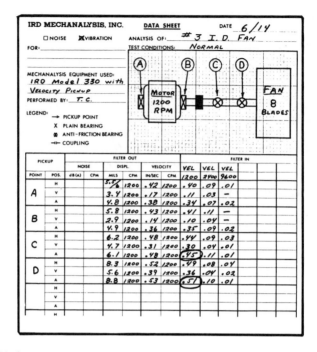

IRD MECHANALYSIS, INC.   **DATA SHEET**   DATE 6/14

☐ NOISE ☒ VIBRATION   ANALYSIS OF: #3 I.D. FAN

FOR: _____   TEST CONDITIONS: NORMAL

MECHANALYSIS EQUIPMENT USED: IRD Model 330 with Velocity Pickup

PERFORMED BY: T.C.

LEGEND: → PICKUP POINT   X PLAIN BEARING   ⊗ ANTI-FRICTION BEARING   ⊣⊢ COUPLING

(Diagram: A B C D — MOTOR 1200 RPM — FAN 8 BLADES)

| PICKUP | | NOISE | | DISPL. | | VELOCITY | | VEL | VEL | VEL | | | |
POINT	POS.	dB(A)	CPM	MILS	CPM	IN/SEC	CPM	1200	2400	9600			
A	H			3.5/6	1200	.42	1200	.40	.09	.01			
	V			3.4	1200	.17	1200	.11	.03	—			
	A			4.8	1200	.38	1200	.34	.07	.02			
B	H			5.8	1200	.43	1200	.41	.11	—			
	V			2.9	1200	.14	1200	.10	.04	—			
	A			4.9	1200	.36	1200	.35	.09	.02			
C	H			6.2	1200	.48	1200	.44	.09	.03			
	V			4.7	1200	.31	1200	.30	.04	.01			
	A			6.1	1200	.48	1200	.45	.11	.01			
D	H			8.3	1200	.52	1200	.49	.08	.04			
	V			5.6	1200	.39	1200	.36	.04	.02			
	A			8.8	1200	.53	1200	.51	.10	.01			
	H												
	V												
	A												
	H												

Figure 5-118. Overhung (outboard) rotors will often reveal high amplitudes of axial vibration with unbalance.

Figure 5-88), balancing the rotor will likely reduce both the radial and axial amplitudes of vibration. Again, however, an axial phase analysis should be carried out to confirm that no misalignment exists.

Although unbalance is relatively simple to diagnose, there are several problems which can result in vibration characteristics very similar to those of normal unbalance. For example, the eccentric pulley in Figure 5-84 will produce vibration at a frequency of 1 × rpm, with amplitudes predominant in the radial directions. The vibration signatures may appear identical to those of normal unbalance; however, the vibration resulting from the reaction forces of the eccentric pulley on the belts cannot be totally compensated by balancing. While the vibration forces caused by the eccentricity condition may be reduced by balancing in, say, the horizontal direction, the vibration amplitude typically increases in the vertical direction. Correspondingly, balancing for the vertical vibration would result in an increase in the horizontal amplitude.

One clue which may suggest that the vibration is actually the result of eccentricity and not unbalance can be obtained by comparing the horizontal and vertical phase indications. A normal unbalance condition will generally reveal horizontal and vertical phase readings which differ by approximately 90°, whereas the reaction force generated by an eccentric pulley may reveal phase readings which are identical or which differ by 180°. Although this is not always a positive means of detecting eccentricity, such comparative phase readings indicate that a visual inspection of the pulley is in order.

When balancing, problems with eccentricity can also be avoided by carrying out balancing solutions simultaneously for the horizontal and vertical readings. In other words, simply obtain original amplitude and phase readings for both the horizontal and vertical pickup positions. Apply a trial weight and measure the resultant horizontal and vertical amplitude readings. Then solve the balancing problem for both the horizontal and vertical data and compare the results. If the horizontal and vertical solutions indicate that the same balance correction is required, then it is highly probable that the problem is, in fact, unbalance. However, if the calculated horizontal and vertical solutions differ considerably, then the problem is not unbalance and the machine should be checked for other mechanical problems such as eccentric pulleys or gears, looseness, etc. It was mentioned in the section on time waveform analysis that a single chipped, broken, or deformed gear tooth could result in vibration at a frequency of 1 × gear rpm, and that the vibration signatures could appear quite similar to those resulting from a normal unbalance. It is quite likely that vibration resulting from a defective gear tooth would result in comparative phase measurements of the horizontal and vertical vibration similar to those illustrated in Figure 5-84. It is unlikely that the horizontal and vertical phase readings from the defective tooth would be 90° out of phase, as they would be with a normal unbalance. Further, it is unlikely that a balance correction would reduce both the horizontal and vertical amplitudes. When difficulty in balancing is encountered, observing the vibration time waveform on an ocilloscope can help identify the problem as an eccentric or damaged gear (see Figure 5-33).

Vibration Due to Insufficient Shrink Fit

Figure 5-119 is extremely interesting for two reasons:

- It serves as an example of how the analyst could reach the wrong conclusion if he would look at one single vibration signature only.
- It demonstrates the effective use of accelerometer-based vibration monitoring and analysis techniques.

Prior to making the record of the compressor speed change behavior pattern, the conclusion was reached that the amplitude peak at 2 × RPM, or 2 × 240 cycles-per-second, indicated severe misalignment between the output shaft of the gear and the input to the high-pressure (H.P.) compressor. Fortunately, on this steam-turbine driven compressor train the speed was reduced to 235 cycles-per-second (14,100 CPM) and a second vibration signature superimposed on the first one. This second signature showed the vibration amplitude peak to shift to 1 × RPM, while the 2 × RPM or 2 × 235 cycles-per-second amplitude was significantly reduced. Similar speed change patterns were recorded at other locations in the machinery train and the traces obtained at 14,100 CPM operation superimposed on those obtained at the

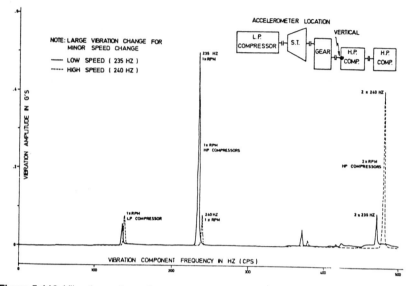

Figure 5-119. Vibration pattern of compressor with one impeller having insufficient shrink-fit (Ref. 1).

14,400 CPM speed. With the characteristic large vibration change for a relatively minor speed change (300/14,400 or approximately 2 percent) appearing on the indicated side of the high-pressure compressor only, it was clearly established that one or more impellers had been assembled on the shaft with insufficient shrink fit.

Vibration Due to Compound Problems

Vibration signatures of before versus after repair of a motor-gear-compressor train are superimposed in Figure 5-120. Again, the analyst chose to use accelerometers for data acquisition. He marked the predominant frequencies (1 × RPM, RPM × impeller vanes, RPM × thrust bearing pads) and showed the gear in damaged versus undamaged condition. He then observed poor coupling hub fit versus proper interference fit, and finally, the signature effects of the compressor inboard bearing with excessive clearance versus normal clearance.

The vibration presented in the horizontal, vertical, and axial signatures in Figure 5-121 might be interpreted as a normal unbalance problem because the predominant vibration occurs at 1 × rotor rpm. However, the vibration occurring at a frequency of 2 × rpm is also significant. As mentioned previously, when a significant unbalance

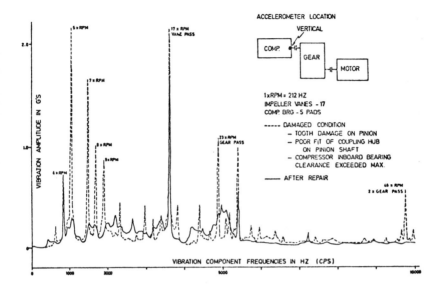

Figure 5-120. Vibration pattern of damaged vs. repaired compressor train (Ref. 1).

is present, it is not unusual to find that some vibration at the harmonic frequencies (2 × rpm, 3 × rpm, etc.) will also appear. As a general rule, as long as the amplitudes of the harmonic frequencies in the radial directions are less than 50% of the fundamental (1 × rpm) frequency, the problem can generally be regarded as simple unbalance. However, when the amplitude of a harmonic frequency such as the 2 × rpm vibration in Figure 5-121 exceeds 50% of the fundamental frequency, then mechanical looseness is indicated. The analysis data in Figure 5-121 suggest that the machine should be checked for loose mounting bolts, a loose rotor, excessive bearing clearances, structural cracks, etc. before an attempt is made to balance. If a significant mechanical looseness does exist, extreme difficulty in balancing may be encountered and the solution will very likely be only temporary.

Vibration Due to Misalignment

Misalignment is an extremely common problem. In spite of self-aligning bearings and flexible couplings, it is difficult to align two shafts and their bearings such that no forces exist which will cause vibration. Misalignment, even with flexible couplings, results in two forces, axial and radial, which cause axial and radial vibration. This is true even when the misalignment is within the coupling flexibility limits. The size of

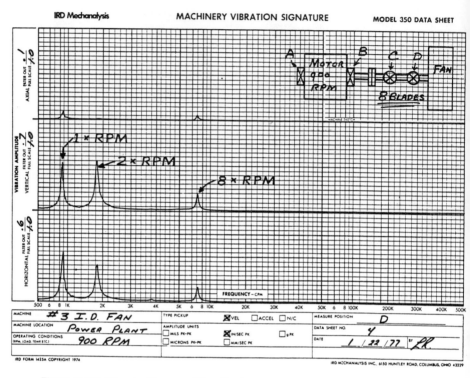

Figure 5-121. The significance of the vibration at 2 × rpm suggests possible looseness.

the forces and therefore the amount of vibration generated will increase with increased misalignment. A bent shaft acts very much like angular misalignment, so its vibration characteristics are included here.

The significant characteristic of misalignment and bent shafts is that vibration will be noted in both the radial and axial directions. As a result, comparative axial vibration is the best indicator of misalignment or a bent shaft. In general, whenever the amplitude of axial vibration is greater than one-half (50%) of the highest radial vibration (horizontal or vertical), then misalignment or a bent shaft should be suspected. Basically, there are two types of misalignment:

1. Angular misalignment, where the center lines of the two shafts meet at an angle.
2. Offset misalignment, where the shaft centerlines are parallel but displaced from one another.

As indicated in Figure 5-122, angular misalignment primarily subjects the driver and driven machine shafts to an axial vibration at the same frequency as shaft rpm. This figure is, of course, a simplified example of coupling with a single pin

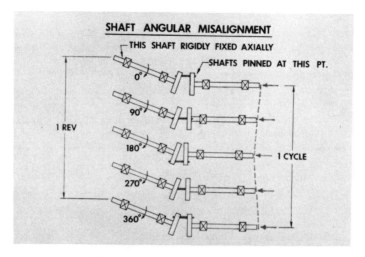

Figure 5-122. Simple model of angular misalignment.

SHAFT OFFSET MISALIGNMENT

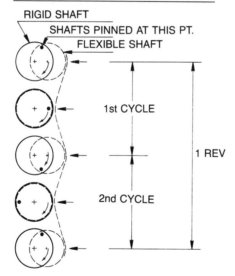

Figure 5-123. Simple model of offset misalignment.

connection. The concept, however, is applicable to actual operating couplings, and such axial vibration commonly indicates a misaligned machine.

Offset misalignment, illustrated in Figure 5-123, produces primarily a radial vibration at twice the rotational frequency of the shaft. Again, this is a simplified example using a single pin connection across the coupling, but it illustrates how a 2 × rpm vibration can be generated by offset misalignment. It can also be readily seen how a multi-pin or multi-tooth coupling could generate proportionately higher frequencies. It should also be mentioned that a twice-per-revolution vibration commonly indicates a misaligned machine. The vibration analysis data in Figure 5-124 are the result of misalignment between the motor and fan. Although the predominant vibration occurs in the axial direction at a frequency of 1 × rpm, note that a significant 2 × rpm vibration is also evident in the axial direction.

The vibration signature in Figure 5-125 was obtained from a vertically-mounted noncontact pickup measuring relative shaft vibration. This vibration was also the result of misalignment between the gear box and centrifugal compressor. Here, also, the predominant vibration occurs at a frequency of 1 × rpm, but with a very substantial vibration at 2 × rpm as well.

The analysis data in Figure 5-126 are another example of the vibration resulting from coupling misalignment. These data, obtained on a 150-megawatt turbogenerator unit, show the predominant vibration in the axial direction at frequencies of 1 and 3 × shaft rpm.

Referring to the misalignment diagrams in Figures 5-122 and 5-123, it can be seen that the radial vibrations resulting from misalignment will occur predominantly in the direction of misalignment. In other words, if the coupling halves are offset vertically, the predominant radial vibration will likely occur in the vertical direction. As a result, the radial vibration of misaligned couplings will often be somewhat directional in nature, similar to the vibration generated by an eccentric pulley or eccentric gear. This, in part, explains why the Lissajous patterns characteristic of

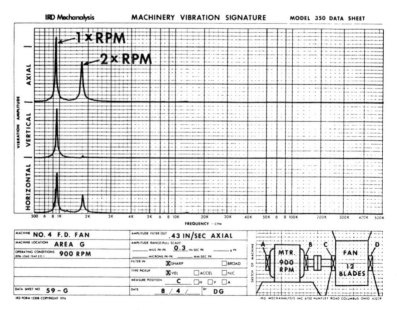

Figure 5-124. This vibration analysis data is typical of misalignment or possibly a bent shaft.

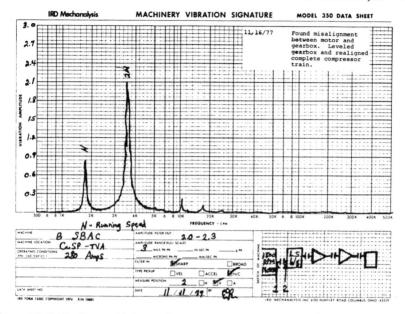

Figure 5-125. Misalignment between the gear case and centrifugal compressor resulted in a predominant vibration at 2 × rpm.

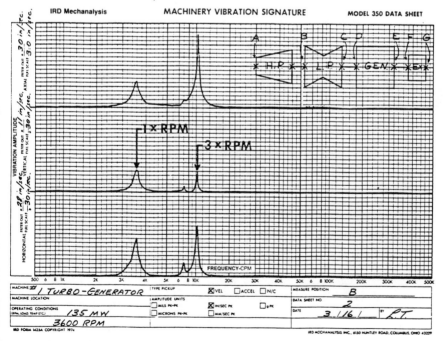

Figure 5-126. Misalignment between the turbine and generator of this turbo-generator unit produced a predominant vibration at 3 × rpm.

misalignment will have a banana, eliptical, or figure-eight configuration, as illustrated earlier in Figures 5-55 and 5-56.

In addition to the vibration frequencies of 1, 2, and 3 × rpm normally encountered with misalignment, higher frequencies of vibration are sometimes found, particularly on machines equipped with ball bearings and gears. For example, analysis of a direct-coupled motor/fan unit disclosed the predominant vibration in the axial direction at a frequency of 7 × rpm, measured on the outboard bearing of the motor. This vibration was traced to coupling misalignment, and the vibration generated at 7 × rpm was the result of the motor bearing having seven rolling elements.

Misalignment involving gear drives can also result in amplitudes of vibration at gear-mesh frequency. For example, referring to the analysis data in Figure 5-127, it can be noted that a high-frequency vibration is present on the gear box, indicating possible gear problems. However, note that the axial vibration ocurring at motor

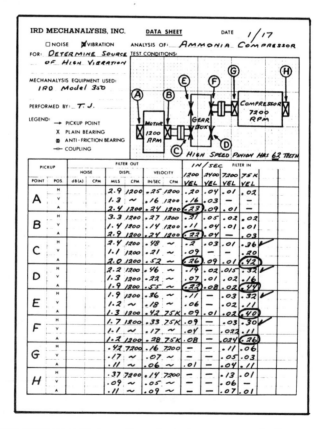

Figure 5-127. Misalignment between the motor and gear case may be the cause of this gear-related vibration.

rpm frequency is also relatively high on the gear box and motor, suggesting that misalignment may be the source of trouble. In this case, misalignment should be corrected first, as this may also eliminate the high-frequency gear vibration. In other words, the gear vibration may simply be the response of the gears to the excessive axial forces of misalignment.

Of course, there are many reasons for misalignment to develop, including the settling of foundations, piping strains, and variations in machinery growth due to changes in temperature. When misalignment is detected, all of these possibilities should be checked to help ensure a satisfactory solution to the problem. The reader is encouraged to consult applicable texts discussing the use of noncontact pickups, hot-alignment monitors (see Volume 1 of this series), and other devices used for measuring machinery growth at operating temperatures.

Torque Lock

There are many types of couplings available that are designed with flexibility to permit minimal amounts of offset and angular misalignment. However, flexible couplings which have been improperly selected or maintained can contribute to vibration. For example, gear-type flexible couplings may exhibit torque lock, which causes the coupling to behave as a rigid coupling. If the coupling is properly sized and lubricated, the coupling will perform normally. However, if the coupling is grossly underrated or if the internal components are not properly lubricated, the coupling can virtually lock up due to the driving torque between the driver and driven units. When this happens, misalignment which may normally be insignificant can result in excessive vibration.

A condition of torque lock can sometimes be detected by noticing that the amplitude and phase readings are not always repeatable from one run of the machine to the next. The coupling locks in a different angular position for each start and stop of the machine. Therefore, to help determine whether torque lock is the source of misalignment, simply obtain an amplitude and phase measurement for the axial vibration. Then, shut down the machine and allow it to come to a complete stop. When the machine is started and brought back to the original operating speed, temperature, and load conditions, a new amplitude and phase reading should be taken for comparison with initial measurements. If a significant change in the axial vibration amplitude and/or phase is noted, the machine should be shut down for visual inspection of the coupling.

Bearing Misalignment

Conditions of misalignment can exist which do not involve a coupling. The misalignment of a bearing with its shaft is one example. In the case of a misaligned sleeve-type bearing, Figure 5-128, no vibration will result unless there is also unbalance. With unbalance, a radial vibration will be present as well as an axial vibration. Both result from the reaction of the misaligned bearing to the forces due to unbalance. The real cause of this vibration is unbalance, and both the axial and radial readings will be reduced when the part is balanced.

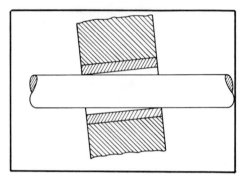

Figure 5-128. Misalignment of a sleeve-type bearing with its shaft will only cause axial vibration if accompanied by unbalance.

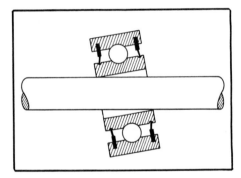

Figure 5-129. When an anti-friction bearing is misaligned with its shaft, axial vibration will occur whether unbalance is present or not.

When an antifriction bearing is misaligned with a shaft (Figure 5-129), axial vibration will exist even when the part is balanced. The vibration may occur at a frequency of 1, 2, or perhaps 3 × rpm, or may occur at a frequency equal to the product of the number of rolling elements in the bearing times shaft rpm. Proper bearing installation will eliminate this vibration.

When the vibration readings indicate a misalignment condition and yet no coupling misalignment is indicated, it is possible that the machine is being distorted and that the bearings misaligned as the result of improper machine mounting. If the feet of the machine are not on the same plane, or if the machine has been bolted to an uneven base plate, tightening the mounting bolts will cause distortion and bearing misalignment. If such a condition is suspected, the outboard mounting bolts of the machine can be loosened slightly, one at a time, to check the effect on the measured vibration. If loosening one of the bolts results in a significant change in the measured vibration, it is likely that the machine has a "soft foot," and it may be necessary to remachine the base plate or install shims to eliminate distortion when the mounting bolts are tightened.

Another condition of misalignment resulting in high axial vibration is the misalignment of sheaves and sprockets used in "V"-belt and chain drives. The angular and offset misalignment conditions illustrated in Figure 5-130 not only result in destructive vibration but in accelerated wear on pulleys, sprockets, chains, and drive belts.

Vibration Due to Eccentricity

Eccentricity is another common source of machinery vibration and exists whenever the shaft rotating centerline is not coincident with the rotor geometric centerline. Figure 5-131 illustrates several common examples of eccentricity.

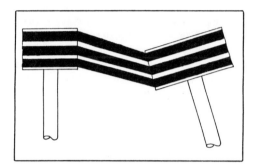

Figure 5-130. Angular or offset misalignment of "V" belt pulleys causes axial vibration and accelerates belt wear.

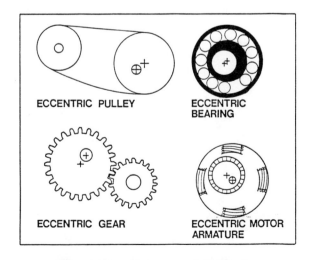

ECCENTRIC PULLEY

ECCENTRIC BEARING

ECCENTRIC GEAR

ECCENTRIC MOTOR ARMATURE

Figure 5-131. Typical sources of eccentricity.

Actually, eccentricity is a common source of unbalance, resulting in more weight on one side of the rotating centerline than on the other side. For example, with the antifriction bearing illustrated in Figure 5-131, the bore of the inner race of the bearing is not concentric with the geometric centerline of the inner race. This will introduce an apparent unbalance in the part mounted in the bearing. However, by balancing the rotor, the forces causing the vibration will be compensated for. It is for this reason that balancing a rotor in its own bearings is recommended. In addition, care must be taken to ensure that the position of the bearing inner race on the shaft does not change, because the eccentricity of the bearing race is compensated by balancing correction weights on the rotor. If the relationship changes, then the condition may be worse than if no balance correction had been made.

Although eccentricity is a source of unbalance which can be corrected in many cases by routine balancing techniques, eccentricity can also result in reaction forces which cannot be corrected by balancing. For example, the eccentric gear illustrated in Figure 5-131 produces reaction forces because of the cam-like action against the mating gear. The largest vibration will occur in the direction of a line through the centers of the two gears at a frequency equal to 1 × rpm of the eccentric gear. The resultant vibration will appear similar to unbalance, and yet the reaction forces cannot be totally compensated for by balancing. In addition to the vibration generated at 1 × rpm, the eccentric gear may also result in a high-frequency vibration at the meshing frequency of the gear teeth. However, because gear tooth clearances are continually changing as the eccentric gear rotates, the vibration generated at gear-mesh frequency will have its amplitude modulated at 1 and, in some cases, 2 × rpm. This modulated amplitude of gear-mesh frequency vibration would appear as illustrated in Figure 5-132 when viewed on an oscilloscope in time waveform.

Eccentricity of the "V"-belt pulley in Figure 5-131 will result in reaction forces due to variation in belt tension as the pulley rotates. As discussed earlier under "Phase Analysis," the largest vibration resulting from the eccentric pulley will occur in the

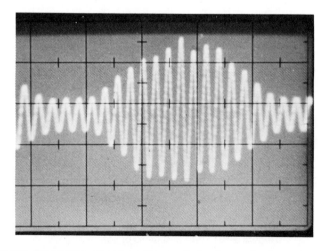

Figure 5-132. Eccentricity may cause the amplitude of gear vibration to be modulated.

direction of belt tension at a frequency equal to 1 × rpm of the eccentric pulley. Since the forces are directional, the resulting vibration cannot be totally corrected by balancing.

Eccentricity of other rotors, such as motor armatures, fan and blower rotors, pump impellers, turbine rotors, compressor rotors and others, can also cause vibration at 1 × rpm similar to unbalance. In the case of an eccentric motor armature, a 1 × rpm force is generated between the armature and stator because of unequal magnetic attraction between the eccentric armature and motor poles. Although it may be possible to compensate for this condition by balancing for a particular motor load condition, increasing magnetic field strength by increasing motor load may result in increased vibration. Eccentric fan, blower, pump, and compressor rotors may also respond to balancing for a given load condition. In these cases, unequal aerodynamic or hydraulic forces act against the rotor, and changing the load condition of the machine will likely produce an increase in the vibration at 1 × rpm. In general, if it is noted that the 1 × rpm vibration changes with variations in load condition, the rotor should be inspected for possible eccentricity.

Vibration Due to Mechanical Looseness

As noted in the vibration and noise identification chart, Figure 5-115, mechanical looseness typically results in a vibration at a frequency of twice the rotating speed (2 × rpm), but may also result in higher-order frequencies such as 3, 4, 5, or perhaps even 6 × rpm. This vibration may be the result of loose mounting bolts, excessive bearing clearance, a crack or break in the structure or bearing pedestal, a rotor which is loose on the shaft, or some other loose machine component. Electric motors, generators, and alternators having loose rotor bars, loose rotor windings or loose stator windings have been known to cause vibration at harmonics of torque-pulse frequency. Normal torque-pulse vibration occurs at a frequency of twice synchronous frequency, and where synchronous frequency is 60 Hz, or 3600 CPM, the torque-pulse vibration will occur at 7200 CPM. A loose rotor bar or stator winding excited by this torque-pulse vibration may reveal a predominant vibration at perhaps twice torque-pulse frequency, or 14,400 CPM.

The amplitude of vibration resulting from mechanical looseness will often be unsteady, and observing the rotor with the stroboscopic light will generally reveal an erratic image. Comparing the amplitude and phase of vibration at various points on the machine or structure as illustrated in Figure 5-86 can often help detect the source of mechanical looseness. Any significant difference in amplitude or phase measured at adjacent components of the machine or structure clearly indicates relative motion and a source of mechanical looseness.

Of course, the vibration characteristic of mechanical looseness will not occur unless there is some other exciting force such as unbalance or misalignment to cause it. However, when there is excessive looseness, even relatively small amounts of unbalance or misalignment can result in large amplitudes of looseness vibration. Looseness simply allows more vibration to occur than it would otherwise. For example, the vibration analysis data in Figure 5-133 show the vibration resulting from excessive clearance in a sleeve-type bearing. It can be seen that significant amplitudes of vibration occur at frequencies of both 1 and 2 × rotor rpm. For comparison purposes, Figure 5-133 also includes vibration analysis data obtained

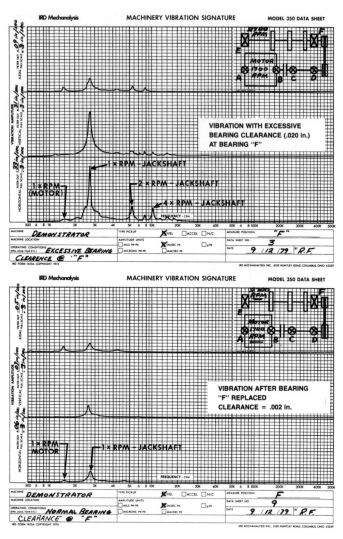

Figure 5-133. Correcting the problem of excessive bearing clearance not only eliminated the vibration at 2 and 4 × rpm, but significantly reduced the vibration at 1 × rpm as well.

after the defective bearing was replaced with one having proper radial clearances. It can be seen that eliminating excessive bearing clearance reduced not only the vibration at 2 × rpm but also the vibration at 1 × rpm.

The nature of mechanical looseness and the reason for the characteristic vibration at 2 × rpm or at higher harmonic frequencies can be explained by referring to the sequence in Figure 5-134. Illustrated is an unbalanced rotor mounted in a bearing with loose mounting bolts. In Figure 5-134A, the heavy spot of unbalance has rotated to the six o'clock position, where the unbalance force is directed downward. This tends to force the bearing down against the pedestal. In Figure 5-134B, the heavy spot has rotated to a position of 12 o'clock, and the resultant unbalance force is now in the upward direction. This upward force tends to lift the bearing off the pedestal, as shown. In Figure 5-134C, the heavy spot has rotated around to a position of three o'clock, and in this position the upward lifting force of unbalance is zero. Therefore, the bearing will simply drop against the pedestal. This action produces two applied forces for each revolution of the shaft. One force is applied by the rotating unbalance, and a second force is applied when the bearing drops against the pedestal. Therefore, the vibration frequency resulting from this looseness will be 2 × rpm.

The sequence in Figure 5-135 illustrates how higher-order frequencies such as 3, 4, or 5 × rpm can be generated by mechanical looseness.

Of course, there will always be some clearances inherent in every machine, and it is normal that some vibration will occur at 2 × rpm and at higher-order frequencies whenever a significant unbalance or misalignment is present. However, mechanical looseness should be suspected whenever vibration severity at the higher-order frequencies is more than one-half the vibration severity occurring at rotating-speed frequency. In addition, where extreme difficulty is encountered when attempting to eliminate the vibration by balancing or realignment, a visual inspection should be made to detect possible looseness conditions.

Vibration Due to Defective Antifriction Bearings

Rolling-element bearings having flaws on the raceways, rolling elements, or cage will usually cause a high-frequency vibration. Actually, a defective bearing will not generally cause a single, discrete frequency of vibration but, instead, may generate several frequencies simultaneously. For example, the vibration analysis in Figure 5-136 was taken on a machine with a faulty ball bearing. Note from the XY plot that

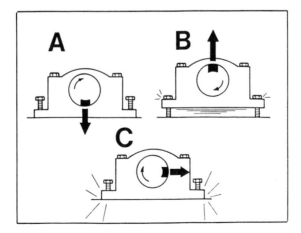

Figure 5-134. Basic mechanism of mechanical looseness resulting in harmonically related vibration frequencies.

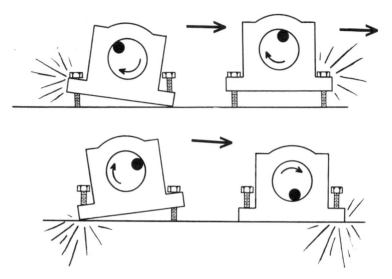

Figure 5-135. Looseness which allows the machine to rock and bounce as illustrated by this sequence can produce vibration at 2 × rpm, 3 × rpm, 4 × rpm or even higher harmonics.

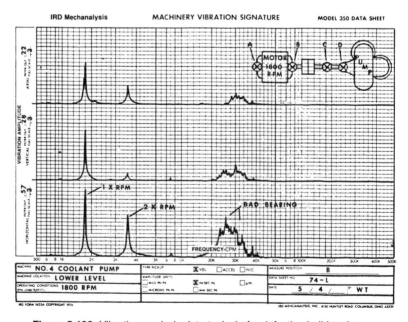

Figure 5-136. Vibrating analysis data typical of a defective ball bearing.

there are actually several high frequencies generated by the faulty bearing. The vibration signature in Figure 5-136 further suggests that the high-frequency bearing vibration is somewhat random or unsteady. Thus, observing the rotating shaft with the stroboscopic light will probably not show a stationary image as it would for vibration caused by unbalance, misalignment, or gears which occurs at even multiples of shaft rpm. In most cases, the bearing vibration is not steady and the frequency meter and amplitude meter can be observed to twitch or fluctuate slightly.

The high frequencies and occasional unsteady or twitching frequencies normally encountered with faulty rolling-element bearings can be explained by examining the nature of the exciting forces generated by a defective bearing. The vibration frequencies generated by a defective bearing will either be "rotational" frequencies or component resonant frequencies.

The rotational frequencies generated by a defective bearing do not occur at multiples of shaft rpm. However, they will be steady-state frequencies which can often be readily identified through analysis. These frequencies include:

1. Ball or roller rpm.
2. The ball-passing frequency for the inner bearing race.
3. The ball-passing frequency for the outer bearing race.

In many cases, the steady-state bearing rotational frequencies can be calculated from the dimensions of the bearing components, including the diameters of the balls and of the inner and outer races. Of course, the actual frequencies may not necessarily agree precisely with calculated frequencies because of ball sliding and the fact that the actual ball path may be different from the calculated raceway diameters. If the dimensions of the bearing are known, the steady-state rotational frequencies of vibration generated by the bearing can be calculated using the equations presented in Figure 5-137.

If the exact dimensions of the bearing components are not known, the ball-passing frequencies for the inner and outer race can be estimated. As a rule of thumb, the ball-passing frequency for the inner race of a bearing will be roughly 60% of the number of rolling elements multiplied by the shaft rpm. For example, suppose that a bearing has 12 balls and is being used in a machine rotating at 3600 rpm. The rule of thumb states that 60% of the rolling elements would pass over a given spot on the inner race for each revolution of the shaft, and 60% of 12 balls is 7.2 balls, which means that the ball-passing frequency for the inner race is 7.2 balls per revolution. Multiplying this by the rotating speed of 3600 rpm provides a ball-passing frequency for the inner race of 25,920 CPM.

To approximate the ball-passing frequency for the outer race, it has been generally noted that approximately 40% of the rolling elements will pass a fixed point on the outer race for each revolution of the shaft. Therefore, for a bearing having 12 balls, 4.8 balls will pass a point on the outer race for each shaft revolution. If the shaft is rotating at 3600 rpm, the ball-passing frequency for the outer race will be approximately 17,280 CPM. Of course, this technique for calculating the ball-passing frequency of the inner and outer races provides only a rough approximation of these frequencies; the actual ball-passing frequencies may differ significantly, depending upon bearing configuration. Knowing the rotational bearing frequencies of ball rpm

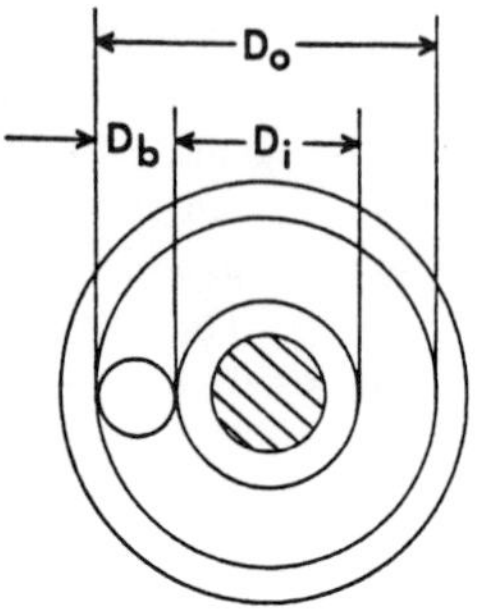

Defect of Cage or Ball: $F_{(cage)} = \dfrac{D_i}{D_i + D_o} \times RPM$

Defect of Ball: $F_{(ball)} = \dfrac{D_o}{D_b} \times \dfrac{D_i}{D_i + D_o} \times RPM$

Defect of Inner Race: $F_{(inner)} = \dfrac{D_o}{D_i + D_o} \times M \times RPM$

Defect of Outer Race: $F_{(outer)} = \dfrac{D_i}{D_i + D_o} \times M \times RPM$

Where D_i = Diameter of inner race

D_o = Diameter of outer race

D_b = Diameter of ball

M = Number of balls in bearing

RPM = Shaft RPM

$F_{(\ )}$ = Frequency in CPM of the defect

Figure 5-137. Calculating bearing vibration frequencies.

and the ball-passing frequencies for the inner and outer race can often be useful in analyzing the vibration from rolling-element bearings and may provide additional clues as to why the bearing failed. For example, if the bearing has developed a defect because of excessive unbalance, the inner race will typically be the first bearing ·component to show signs of deterioration. This is because the unbalance forces will have a constant angular relationship with a fixed position on the inner race. By comparison, if the bearing is being damaged because of background forces or misalignment with another machine component, the bearing outer race will more likely show initial signs of deterioration. If the rolling elements themselves are the first components in the bearing to fail, the problem may be due to improper lubrication, overheating, or electric currents passing through the bearing because of improper grounding. In some cases, the rotational frequencies of interest can be detected through routine frequency analysis. However, since bearing vibration can often be complex, consisting of many frequencies, spectrum averaging with a real-time spectrum analyzer may be required to enhance the rotational frequencies while suppressing the somewhat random frequencies of component resonances. Observing the bearing vibration in time waveform on an oscilloscope can also help detect specific bearing frequencies. The time waveform presented in Figure 5-138 was obtained from a bearing having a known defect on the inner raceway. The spikes generated by the rolling elements of the bearing passing over the inner race defect can be clearly seen, and since the time period of the display is known, the frequency of this vibration can be easily determined for comparison with the calculated

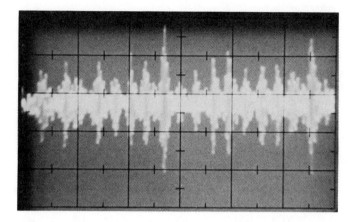

Figure 5-138. Time wave-form analysis of this bearing having a defect on the inner raceway clearly shows the vibration "spikes" generated by the balls passing the defect.

rotational frequencies of the bearing. The amplitude of the vibration generated appears somewhat modulated due to the fact that the defect on the inner race is rotating and thus continually changing position relative to the vibration pickup attached to the bearing.

Of course, it is easy to visualize how multiple defects in the inner and outer raceways or more than one defective rolling element can result in a multitude of high frequencies.

In addition to the rotational frequencies generated by the bearing, the momentary impacts between the rolling elements and bearing raceways will excite the natural frequencies of the various bearing components. The vibration generated by these impacts is much like that generated by a bell or tuning fork. Every object, including the components of a bearing, has its own unique natural frequency, and a flaw in the bearing can produce the intermittent, impact type of force which will cause the various parts of the bearing to vibrate at their respective natural frequencies. Normally, the natural frequencies of these parts will be quite high compared to the rpm of the machine. As a result, the vibration frequencies measured from a defective bearing are also characteristically high. Further, it is unlikely that the natural frequencies of the various bearing components will be exact multiples of shaft rpm. Thus, the frequency of bearing vibration will probably not be a direct multiple of shaft rpm. Since there are many parts, including the inner race, outer race, rolling elements, bearing cage, bearing housing, and rotor shaft, which may vibrate from the impact of a bearing fault, it is likely that several vibration frequencies will be excited simultaneously and to varying degrees. As a result, the high-frequency bearing vibration will most likely be somewhat random or erratic.

Of course, there are other machinery problems which will cause erratic high-frequency vibration similar to that generated by a defective bearing. Such problems include cavitation in pumps, rubbing, defective gears, combustion, and flow turbulence. However, if the vibration is the result of a defective bearing, the problem can usually be diagnosed because the vibration generated by the bearing is not readily transmitted to other points on the machine. Therefore, the bad bearing is usually the one nearest the point where the greatest vibration of this type occurs. Referring to the analysis data in Figure 5-139, note that the amplitude of the high-frequency vibration is high at one point on the machine but barely measureable at other points. This clearly points to the bearing as being the source of this vibration.

Severe misalignment, excessive thrust, and other problems will sometimes result in a high-frequency bearing vibration which is not the fault of the bearing. For example, in one instance a high-frequency vibration was detected on the bottom bearing of a 900-rpm motor driving a vertical pump. In this instance, it was noted that the vibration frequency was steady at 12,600 CPM, or 14 x motor rpm, and that the amplitude was high in the axial direction. Because of the high amplitude of vibration, the motor was removed and the bearing replaced. However, after the motor was reinstalled, a vibration check revealed that the high axial vibration at 12,600 CPM was still present. Further investigation revealed that the motor mounting flange was being distorted when the mounting bolts securing the motor to the pump flange were tightened. The vibration frequency generated in this case was equal to the number of rolling elements in the bearing (14) times motor rpm. Proper shimming to avoid distortion corrected the bearing misalignment problem and eliminated the high-frequency vibration.

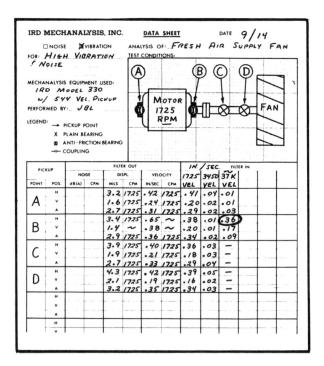

Figure 5-139. The high frequency vibration generated by the defective bearing is not readily transmitted to other bearing points.

It should be remembered that bearings are among the most precisely made devices available, and they generally do not fail unless other problems are encountered. A careful check of your analysis data for other difficulties such as unbalance, misalignment, excessive background vibration, and other problems should be made and the problems corrected accordingly, particularly where premature bearing failures have been a problem. In addition, an inspection of the defective bearing can often uncover the cause of failure. An inspection of the bearing may reveal problems such as:

1. Lack of lubrication.
2. Overheating, perhaps due to excessive lubrication.
3. Corrosion of the raceways or rolling elements due to contamination.
4. Electrical pitting of the rolling elements or raceways due to electric currents passing through the bearing. This is the result of improper grounding of the machine.
5. Excessive thrust.
6. Excessive preload due to improper fit.

For additional information, see Chapter 3, under "Rolling-Element Bearing Failures and Their Causes."

Vibration Due to Drive Belts

V-type drive belts are popular for power transmission because they have good capacity to absorb shock and vibration. In addition, for many applications, V-belts offer relatively quiet operation compared to chain or gear drives. However, V-belts can be the source of objectionable vibration, especially on machine tools where very low levels of vibration must be maintained.

Vibration problems associated with V-belts are generally of two types: belt reaction to *other* disturbing forces in the equipment, and vibration due to actual *belt* problems.

V-belts are often blamed as the source of vibration because the flexible strands between the pulleys are sometimes seen to whip and flutter. Since belt vibration is more visible than the vibration of other parts of the machine, and because the belts are usually the easiest part of the machine to change, belt replacement is often one of the first attempts to correct the vibration problem. However, it may be that the belt is reacting to other disturbing forces in the machine. For example, excessive unbalance, eccentric pulleys, misalignment, or mechanical looseness may result in belt vibration which is readily visible. Thus, the belt may simply be an indicator of other disturbances in the equipment. Therefore, before replacing drive belts, an analysis should be made to determine the true nature of the problem.

The frequency of the vibration is the *key factor* in determining the nature of belt vibration. If the belt is simply reacting to other disturbing forces in the machine, such as unbalance or eccentricity of pulleys, the frequency of belt vibration will likely be the same as the disturbing frequency. The belt is simply amplifying or exaggerating these other disturbing forces. If this is the case, that part of the machine which is actually generating the disturbing forces will appear to stand still under the strobe light of your analyzer. With multi-belt drives, it is important that all belts have equal tension. If one or more belts are slack while the other belts are under proper tension, the slack belts may cause excessive vibration even from very minor disturbing forces. This condition will also cause belt slippage and accelerate belt and pulley wear. Vibration from actual *belt* defects will usually occur at frequencies which are direct multiples of belt rpm. The normal frequencies found are 1, 2, 3, and 4 × belt rpm. The particular frequency encountered will depend on the nature of the belt problem as well as on the number of pulleys and idlers over which the belt must pass.

The rpm of a belt can easily be determined if the length of the belt is known, as well as the pitch diameter and rpm of one of the pulleys. Use the following formula to find belt rpm:

$$\text{belt rpm} = \frac{\text{pulley dia.} \times 3.14}{\text{belt length}} \times \text{pulley rpm}$$

Belt defects that cause vibration at frequencies equal to direct multiples of belt rpm include cracks; hard spots, soft spots or lumps on the belt faces; and pieces or chunks which have broken off. A belt which is crooked, having taken a set shape during packing and storing, may cause high vibration on lightweight equipment until it has had an opportunity to become limber. In addition, a V-belt may vary in width. This causes the belt to ride up and down in the pulley grooves, creating vibration because of variations in belt tension.

Regardless of the problem, belt defects can be readily distinguished from other disturbing forces in the machine. The vibration frequency will occur at direct multiples of belt rpm; therefore, the belt will appear to stand still under the strobe light. The best way to detect belt vibration is to apply the pickup on the bearing housing perpendicularly to and in the direction of belt tension. Belt defects normally result in a higher amplitude in the direction parallel to belt tension.

In some cases, the vibration amplitude resulting from faulty belts will be unsteady. This is particularly true on multiple-belt installations, where the belts may slip to varying degrees so that the faults on the belts will at one time add to one another and another time subtract from one another. The net result will often be an amplitude which increases and then decreases in a periodic or cyclic manner.

The extent of belt slippage on multi-belt installations can be readily observed under the strobe light. Simply shut the machine down and mark a line with white chalk across the belts. Then, with the machine operating at normal speed and your analyzer filter tuned to 1 × belt rpm, observe the white chalk marks under the strobe light. If the belts are slipping relative to one another, the white marks on the individual belts will also be seen to move relative to one another.

Improper belt tension, pulley misalignment, mismatched belts, or excessive load and horsepower requirements which cause belt slippage may also produce a high-frequency noise and vibration due to the friction generated as the belt rubs over the pulleys. The result is usually a distinct "squeal" or "chirp."

In summary, smooth operation can be obtained from V-belt drives by following a few simple precautions:

1. Make sure belts are in good condition.
2. Be sure the number and size of belts used will meet the horsepower and load requirements of the equipment.

3. With multiple belt installations, use belts which are a "matched" set in order to obtain equal tension on all belts.
4. Make sure pulleys and sheaves are round and accurately aligned with one another.
5. Check for wear of pulley grooves. Excessive wear may allow the belt to ride in the bottom of the groove, causing slippage and poor efficiency.
6. Be sure belts are properly installed and adjusted to proper tension as recommended by the belt manufacturer.
7. Finally, keep other disturbing forces in the machine to a minimum.

Special Techniques for Monitoring Bearing Condition

As just mentioned above, bearings are among the most precisely made devices available. Hence, the vibration levels characteristic of a rolling-element bearing in good condition will typically be very low compared to other sources of vibration in the machine, such as unbalance, misalignment, hydraulic forces, etc. Moreover, the vibration amplitudes encountered in the early stages of bearing failure may also be relatively low compared to amplitudes of other vibrations in the machine. As a result, when taking overall velocity or acceleration measurements for monitoring machinery condition, an increase in the overall vibration due to a defective bearing may not become apparent until bearing deterioration has become critical. For this reason, measuring or monitoring the overall vibration velocity or acceleration may not provide sufficient warning of developing bearing problems. The vibration meter in Figure 5-140 provides for measurements of normal vibration displacement, velocity, and acceleration and also includes special provisions for measuring

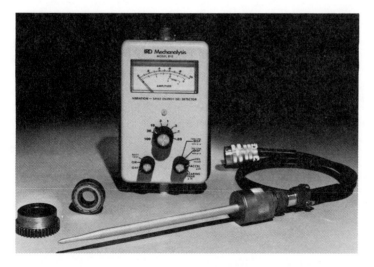

Figure 5-140. This portable Vibration Meter includes provisions for measuring "Acceleration-Spike Energy" for improved sensitivity to bearing and gear defects. (IRD Model 810).

spike-energy vibration, which is characteristic of defective bearings and gears. A similar vibration meter (IRD Model 820) incorporating spectrum display and hard-copy features was introduced in 1981 and has been quite successful in monitoring rolling element bearing condition.

Spike energy is the measurement of the ultrasonic microsecond-range pulses caused by impacts between bearing elements which have microscopic flaws. Special spike energy circuits have been designed to detect not only this pulse amplitude but also the rate of occurrence of the pulses and the amplitude of the high-frequency broad-band vibratory energy associated with bearing defects. These three parameters of pulse amplitude, pulse rate, and high-frequency random vibratory energy are electronically combined into a single quantity called "g-SE," which is a measurement of bearing condition. The term g-SE (acceleration units of spike energy) has been selected to indicate that this type of measurement is more comprehensive than g-Peak or g-RMS.

As illustrated in Figure 5-138, a defect on the bearing raceways or rolling elements causes intermittent impacts between the bearing components. It was also mentioned that these impacts, or spikes, excite the natural frequencies of the various bearing components. The instrument in Figure 5-140 utilizes this characteristic of bearing condition. The transducer used to detect the vibration is an accelerometer having a mounted natural frequency of approximately 27,000 Hz (minimum), or 1,260,000 CPM. Of course, the natural frequency of this accelerometer is well above those vibration frequencies generated by mechanical problems in the machine, such as unbalance, misalignment, electrical problems, aerodynamic and hydraulic forces, etc. As a result, the only sources of vibration which would excite the natural frequency of the accelerometer are the impact or spike forces generated by a defective bearing or gear tooth. So that the instrument does not respond to other vibration frequencies when measuring this spike energy, a high-pass filter is included in the instrument to reject all the vibrations generated at frequencies below the accelerometer natural frequency. Consequently, increases in vibration due to other problems such as unbalance or misalignment will not result in an increase in the spike energy vibration. In this way, the spike-energy mode is designed to respond to the impact-type forces that are characteristic of defective bearings. This approach allows bearing condition to be accurately checked and monitored, regardless of the complexity of overall machinery vibration.

Vibration Due to Plain (Journal) Bearings

Problems with plain bearings which result in high levels of vibration are generally the result of excessive bearing clearance, improper bearing load, or lubrication problems. A plain bearing with excessive clearance may allow a relatively minor unbalance, misalignment, or some other vibratory force to result in mechanical looseness or pounding. The resultant vibration frequency may be at 2 times, 3 times, or some higher multiple of shaft rpm. In this case, the bearing is not the actual cause of the vibration; it simply allows more vibration to occur than would exist if bearing clearances were correct. A wiped bearing can often be detected by comparing the horizontal and vertical amplitudes of vibration. Machines securely mounted to a rigid foundation or structure will normally reveal a slightly higher amplitude of vibration in the horizontal direction. In several instances where the amplitude of vibration in the vertical direction appeared unusually high compared to the horizontal amplitude, a wiped bearing was found to be the cause. A wiped bearing with excessive clearances, which allows the shaft to change position within the bearing, may also result in misalignment. For example, the vibration signature in Figure 5-141 was obtained on a high-speed centrifugal compressor, motor driven through a gear box. In this case, the output bearing of the high-speed shaft of the gear box had been wiped because of water in the lube oil, and yet the resultant vibration of the shaft relative to the bearing, measured with a noncontact pickup, shows the predominant frequencies of 1 and 2 × rpm characteristic of misalignment.

Plain or journal bearings with excessive clearance have also been known to cause rubbing. In one case reported, the bearing of a centrifugal compressor had worn significantly. This allowed the shaft to rub on the seal, producing a predominant vibration at one-half rpm. As noted in the vibration and noise identification chart, Figure 5-115, rubbing may result in any one of a number of possible vibration frequencies, ranging from less than rotating speed to very high frequencies.

Oil Whirl

Oil whirl is another problem associated with plain or journal-type bearings. This vibration has been known to occur primarily on machines equipped with

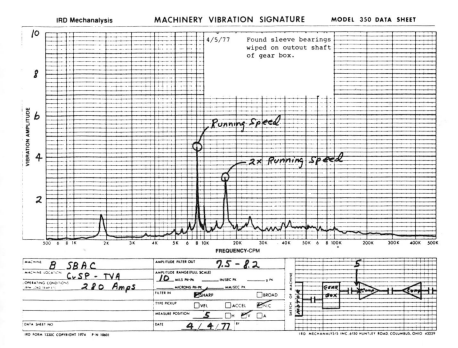

Figure 5-141. Water in the lube oil caused the out-board bearing of the high-speed gear shaft to wipe, thus creating excessive bearing clearance. The result was high vibration at 1 × rpm and 2 × rpm.

pressure-lubricated sleeve bearings and operating at a relatively high speed, normally above the critical speed of the rotor. Oil-whirl vibration is often quite severe but is easily recognized because the frequency is slightly less (3% - 8% less) than one-half the rpm of the shaft. For example, the vibration analysis data in Figure 5-142 show an oil whirl detected on a 40-megawatt steam-turbine generator rotating at 3600 rpm. In this case, the predominant vibration of oil whirl occurs at a frequency of approximately 1400 CPM.

The mechanism of oil whirl can be explained by referring to the diagram in Figure 5-143. Under normal operation, the shaft of the machine will ride up the side of the bearing slightly, as shown. The shaft, operating at an eccentric position from the bearing center, draws oil into a "wedge" to produce a pressurized load-carrying film. This oil film is made up of molecules, and those oil molecules adjacent to the shaft tend to stick to the shaft and rotate at shaft rpm. However, those oil molecules adjacent to the bearing tend to adhere to the bearing, which is not rotating. As a result, the oil between the shaft and bearing will be in shear and will tend to rotate at a speed that is the average between shaft speed and bearing rpm, which of course is zero. As a result, the average rotating speed of the oil film between the shaft and bearing is approximately one-half shaft rpm, and if friction losses are considered, the average speed of the oil film will be slightly less than one-half shaft rpm.

The fact that all oil-lubricated sleeve bearings will have an oil film rotating at slightly less than one-half shaft rpm makes all such machines susceptible to oil whirl.

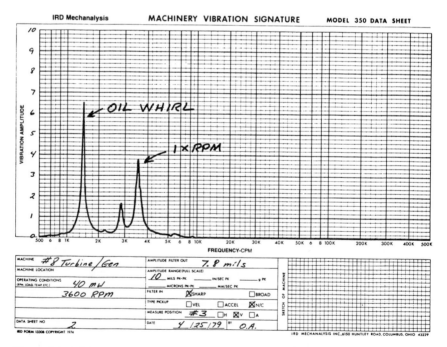

Figure 5-142. This analysis data shows a severe oil whirl condition on a 3600 rpm turbogenerator.

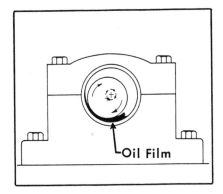

Oil Film

Figure 5-143. The "average" rotating speed of the oil film will be slightly less than one-half shaft rpm due to friction losses.

However, the force generated by this rotating oil film is normally small when compared to other sources of static and/or dynamic loading of the bearing. Nevertheless, the force of the rotating oil film can become the predominant force, in which case the oil film simply pushes the shaft around within the internal clearance of the bearing. This is oil whirl. The shaft is simply forced to whirl within the bearing in a manner similar to a boat which is being pushed by a wave of water.

Oil whirl can often be attributed to improper bearing design. For example, if the static loading of the shaft journal against the bearing is too small, then the force generated by the rotating oil film may be the predominant force, in which case the machine will be highly susceptible to oil whirl. Excessive bearing wear is another problem which can contribute to oil whirl. The tendency of a machine to develop oil whirl depends in part on the extent of eccentricity of the shaft within the bearings. If the shaft was perfectly centered in the bearing, the machine would not be susceptible to oil whirl. However, as a bearing wears, the shaft will ride more and more eccentrically, with increased bearing wear. The machine will thus become more susceptible to oil whirl. Therefore, if the machine has operated satisfactorily in the past and has only recently developed oil whirl, the bearings should be checked for excessive wear.

An increase in lube-oil pressure or viscosity may also increase susceptibility to oil whirl. Therefore, when oil whirl is encountered, the pressure, viscosity, and temperature of the lube oil should be checked to make sure they meet manufacturer's specifications. In some cases a temporary solution for oil whirl is to increase the temperature of the lubricant, which lowers its viscosity. Increasing the loading on the bearing by introducing a slight unbalance or misalignment has also been effective as a temporary solution to oil whirl.

On occasion, a machine which would normally be completely stable will exhibit oil-whirl vibration. This can occur when an external source transmits vibration to the machine through the foundation or piping. If this background vibration occurs at just the right frequency, i.e. the probable oil-whirl frequency, oil whirl will likely occur.

This condition is referred to as "externally excited whirl." For example, a steam-turbine generator rotating at 3600 rpm would have a probable oil-whirl frequency around 1700-1750 CPM. Of course, there are many four-pole induction motors whose operating speed is within this range. Consequently, externally excited oil whirl is not uncommon. Similarly, a normally stable machine may be excited into oil whirl by a foundation or piping which is vibrating in resonance at a frequency equal to the probable oil-whirl frequency. The resonant vibration of the piping or foundation may be the result of pulsations or flow turbulence.

Whenever the vibration characteristic of oil whirl is found, a complete vibration survey of the installation, including background sources, foundation, and related piping, should be made to determine the true cause.

Several special bearing configurations are available to minimize the possibility of oil whirl. Some of these designs are illustrated in Figure 5-144. The axial-groove bearing is normally used on smaller machines such as light gas turbines and turbochargers. The three-lobed bearing provides improved bearing stability against oil whirl. The three individual bearing surfaces generate pressurized oil films that center the shaft within the bearing. By minimizing the eccentric position of the shaft within the bearing, the rotor will be less subject to oil whirl. Axial grooves are sometimes included at the intersection of the lobe segments to increase whirl resistance.

The tilting-pad bearing is a common choice on larger high-speed industrial machinery such as centrifugal compressors or high-speed centrifugal pumps. In a manner similar to the lobed bearing, each segment or tilting pad provides a pressurized oil wedge which tends to center the shaft in the bearing.

Although the bearing configurations illustrated in Figure 5-144 are specifically designed to minimize the possibility of oil whirl, oil-whirl vibration has occurred even on machines equipped with these special bearings. However, the bearing may not be totally at fault. In one case reported, a floating mechanical seal in a centrifugal compressor had bound in its housing and was carrying the load of the shaft instead of allowing the tilting pad bearing to perform this duty. The result was an oil-whirl condition because the seal was actually serving as a bearing.

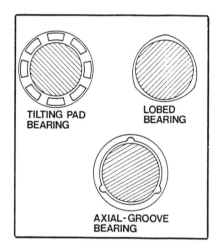

TILTING PAD
BEARING

LOBED
BEARING

AXIAL-GROOVE
BEARING

Figure 5-144. These and other special bearings have been designed to improve stability and resistance to oil whirl by minimizing the eccentricity of the shaft within the bearing.

Dry Whirl

Improper lubrication of a plain journal bearing can also cause vibration. If the bearing lacks lubrication, or if the wrong lubricant is used, the result may be excessive friction between the rotating shaft and stationary bearing. This friction excites vibration of the bearing and other related parts of the machine in a manner similar to the vibration generated by simply rubbing a moistened finger over a pane of glass. This vibration is called dry whirl. The vibration resulting from dry whirl is generally high frequency and thus will often produce the distinctive squeal normally associated with a dry bearing. The vibration frequencies generated are not likely to occur at direct multiples of shaft rpm. Therefore, they will give no definite image under the stroboscopic light. In this respect, the vibration from dry whirl is similar to the vibration caused by a faulty rolling-element bearing.

Whenever the vibration characteristic of dry whirl is encountered, the lubricant, lubrication system, and bearing clearances should be inspected. This condition has been found on bearings having excessive as well as insufficient clearance.

Vibration Due to Resonance

The mechanism of resonance has been covered earlier in some detail. Designing a machine installation, including the structure, piping, ductwork, etc., such that there are no natural frequencies coincident with any significant exciting force generated by the machine is extremely difficult. As a result, resonance is a very common problem throughout industry.

There are many ways to confirm whether or not a part is vibrating at resonance. Amplitude and phase-versus-rpm plots (Bodé plots) obtained during the startup or coast-down of a machine positively identify the resonance frequencies by the characteristic peak amplitude and 180° shift in the phase of vibration (see Figure 5-13 and 5-16). This technique was discussed in some detail in the section on data acquisition.

If an instrument with a tracking filter is not available for obtaining plots of amplitude and phase versus machine rpm, the resonance frequencies can be determined by viewing the long-term time waveform during machine startup or coast-down, as illustrated in Figure 5-37.

Evidence of possible resonance can also be obtained from the frequency analysis data as illustrated in Figure 5-119, where the horizontal vibration amplitude at the rotating speed frequency appeared unusually high when compared to the vertical and axial amplitudes. High amplitude ratios such as this strongly suggest the possibility of resonance.

Another simple yet effective way to confirm whether or not a part is vibrating in resonance is the "bump" test. With the machine shut down and a vibration pickup held or attached to the machine, simply bump the machine or structure with a force sufficient to cause it to vibrate. Since an object will undergo free vibration at its natural frequency when bumped or struck, the natural frequency generated in this way will be indicated on the analyzer frequency meter. The analyzer filter must be in the "out" position for this test. If the vibration diminishes very quickly, it may be necessary to bump the machine several times in succession in order to sustain free vibration long enough for it to register on the analyzer frequency meter. Modern real-time spectrum analyzers, as shown in Figures 5-5 and 5-8, provides an instantaneous display of the vibration amplitude-versus-frequency data. These are ideal instruments for determining natural frequencies.

In many cases when a bump test is performed using a standard analyzer, bumping the machine may excite more than one natural frequency. When this occurs, the frequency meter may not be able to lock on to the resonant frequencies excited. If a real-time analyzer is not available, the natural frequencies excited can still be determined in other ways. One way is to view the results of the bump test in time waveform on an oscilloscope. For example, the time waveform in Figure 5-145 is the result of a bump test where two resonant frequencies have been excited. Evidence that more than one natural frequency has been excited by this bump test is indicated by the unusual decay characteristics of the free vibration. In this case, the two natural frequencies excited are nearly harmonically related. These frequencies can be determined by knowing the time period covered by the waveform and by simply counting the number of complete cycles which occur over a certain period, such as one second. This technique of determining frequency from the time waveform was discussed earlier in the section on time waveform analysis.

Another technique for determining the natural frequencies of a structure, where several natural frequencies may be excited by bumping the machine, is to perform a routine frequency analysis using the standard vibration analyzer. With the machine shut down, simply set up the analyzer as though you were going to perform an amplitude-versus-frequency analysis. Then, while the machine is being bumped repeatedly, simply tune the analyzer filter over the various frequency ranges to identify all the natural frequencies being excited. Recording the amplitude-versus-frequency data of the bump test on an XY recorder greatly simplifies this analysis procedure and makes it possible to identify specific resonant frequencies more accurately.

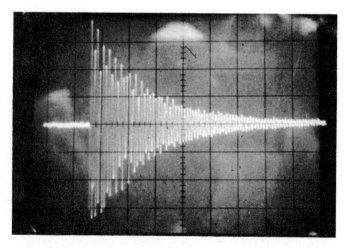

Figure 5-145. Time wave-form analysis of impulse to structure.

Of course, if the natural frequencies of the machine or structure identified as the result of the bump test or amplitude/phase-versus-rpm (Bodé) plots are the same as the exciting frequencies noted during machine operation, then a condition of resonance does exist. If a resonance problem is encountered, there are several ways it can be corrected. One way is to change the frequency of the exciting force so that it no longer coincides with the natural frequency of the machine or structure. This is normally accomplished by either increasing or decreasing machine rpm. If the exciting frequency cannot be changed, the problem can be correlated by actually changing the natural frequency. This can be accomplished by either increasing or decreasing the mass (weight) or stiffness of the machine or structure. Increasing stiffness will increase the natural frequency, whereas increasing the mass will decrease the natural frequency. Before structural modifications are made to change the mass or stiffness of the machine or structure, a thorough analysis of the system is required to determine precisely what portion of the machine or structure is in resonance. This can normally be accomplished by performing a mode shape analysis as discussed in the section on data acquisition (see Figures 5-91 through 5-98). The mode shape analysis not only helps identify the particular spring-mass system in resonance but also identifies nodal points which should be avoided when structural modifications will be made to stiffen the machine support.

If it is not possible to separate the exciting and natural frequency by changing machine rpm or the mass or stiffness characteristics of the machine or structure, another possible solution is to create an antinode by the addition of a dynamic absorber. This technique was discussed in the data acquisition section under "Amplitude and Phase Versus rpm," and is illustrated in Figure 5-21.

Of course, resonance could be avoided by minimizing the exciting force, and balancing to lower than normal levels will sometimes be effective in controlling a resonance problem. However, in many cases, attempting to eliminate the exciting force by balancing and alignment may prove extremely difficult. The best solution to a resonance problem is to separate the natural frequencies from the exciting frequencies.

Vibration Problems with Specific Machinery

In this section, the interpretation of vibration data continues with a discussion of problems encountered on specific types of rotating machinery and their associated vibration characteristics.

Centrifugal Pumps

In addition to the problems of unbalance, misalignment, defective bearings, and resonance which have already been discussed, centrifugal pumps have some unique problems which deserve discussion. These include:

1. Hydraulic forces.
2. Cavitation.
3. Recirculation.

Hydraulic forces. Most centrifugal pumps will exhibit some vibration due to inherent hydraulic forces, and this vibration will generally occur at a frequency equal to the number of impeller vanes multiplied by shaft rpm. This vibration is simply the result of pressure pulsations within the pump, created whenever an impeller vane passes a stationary diffuser. If the impeller is centrally located within the pump housing and is properly aligned with the pump diffusers, the hydraulic pulsations will be balanced and the amplitudes of vibration from inherent hydraulic forces will be minimal. However, problems can occur which result in excessive hydraulic forces.

On page 386 we described the case of a 3600-rpm centrifugal pump which had an unusually high amplitude of vibration at a frequency of 21,600 CPM. This frequency was equal to the product of the number of vanes on the pump impeller (six) times the rotating speed of 3600 rpm, indicating that hydraulic forces were responsible. The usual tests for resonance were carried out, but no natural frequencies were found corresponding to the objectionable 21,600-CPM vibration frequency. When the pump was disassembled for inspection, it was discovered that the radial clearances between the impeller and diffusers cast in the pump housing diametrically opposite one another differed by approximately .4 inch (10 mm). Apparently, this was creating an unbalance of hydraulic forces. The condition was corrected by machining both diffusers to equalize radial clearances and diffuser profile, and vibration significantly reduced as a result.

Experience shows, however, that high pump noise and/or vibration will sometimes exist in pumps whose rotating elements are fully concentric with stationary diffuser or cutwater components. As a rule of thumb, the vane tip to volute cutwater clearance should be no less than 6% of the impeller diameter.

The analysis data in Figure 5-146 were obtained from a centrifugal boiler feed pump operating at approximately 4700 rpm. This analysis was made because of a substantial increase in pump vibration, and it can be noted that the predominant vibration was found to be 28,000 CPM, which is 6 × rpm. This corresponds directly to the number of vanes on the first-stage impeller, indicating that hydraulic forces are responsible for this vibration. Since the vibration represented a sudden increase, it was felt that the pump should be shut down and opened for a visual inspection. This inspection revealed that a first-stage stationary component had seized to the shaft. Replacing the component with a new one of adequate clearance virtually eliminated the vibration at the frequency of 6 × rpm.

The vibration analysis data in Figure 5-147 were obtained on another centrifugal boiler feed pump similar to the one in Figure 5-146. Here again, the analysis was performed as the result of a substantial increase in pump vibration. In this case, the predominant vibration was found at 35,500 CPM, which is 7 × rpm of the pump and corresponds directly to the number of impeller vanes on the second, third, and fourth pump stages. Here again, it was felt that the pump should be shut down for visual inspection, and in this case the inspection disclosed that three out of the four welds which secured the second-stage diffuser to the channel ring had broken free. Apparently, this allowed the diffuser to shift slightly relative to the impeller, resulting in an increase in hydraulic forces. Had the pump been allowed to continue in operation and the fourth weld broken loose, extensive damage to the pump would undoubtedly have resulted.

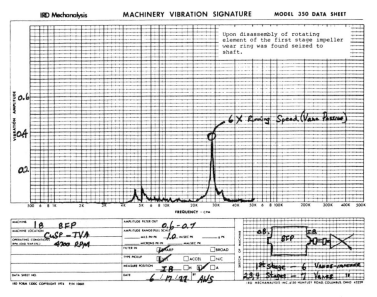

Figure 5-146. A wear ring seized to the shaft caused this vibration at 6 × rpm, corresponding to 6 vanes on the first stage impeller.

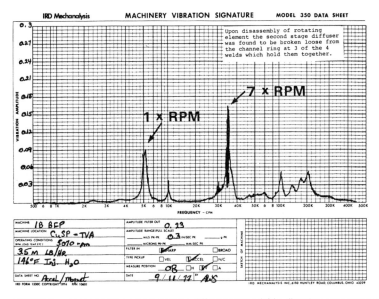

Figure 5-147. Looseness of the second stage diffuser caused this vibration at 7 × rpm, corresponding to the 7 vanes on impeller stages 2, 3 and 4.

Cavitation. Cavitation is a somewhat common problem with centrifugal pumps and results whenever the pump is operated below its designed capacity or with inadequate suction pressure. Since the pump is "starved," the fluid coming into the pump will literally be pulled apart in an attempt to fill the void which exists. This creates pockets or cavities of nearly perfect vacuum which are highly unstable, and which collapse or implode very quickly. Due to their impactive nature, these implosions excite the local natural frequencies of the pump housing, impeller, and other related pump parts, and since these implosions may occur at random intervals at various locations within the pump or piping, the resulting vibration will be random in amplitude and frequency. Thus, a vibration caused by cavitation in a pump may cover a broad frequency range, where individual amplitudes and frequencies are constantly changing. The analysis data in Figure 5-148 are typical of the vibration due to cavitation. The distinct spikes at 3600 and 7200 CPM represent steady-state vibration at 1 and 2 × pump rpm, possibly due to some unbalance or misalignment. However, the vibration occurring between 30,000 CPM and 100,000 CPM is random, with no steady-state amplitude and frequency characteristics.

Cavitation can be quite destructive to the internal components of a pump, and a visual inspection of a pump which has been subjected to cavitation will often reveal metal loss similar to erosion. In one case reported, a pump which had been subjected

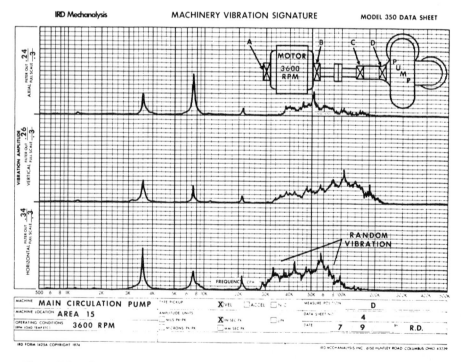

Figure 5-148. Cavitation, recirculation and flow turbulence normally cause random noise and vibration.

to minor cavitation had progressively shown a decrease in operating efficiency. When the pump was disassembled for visual inspection, it was found that the hub was all that remained of the pump impeller. The impeller vanes had been completely eroded away by cavitation.

A distinctive noise often accompanies cavitation. Mild cavitation may have the sound of sand being pumped, whereas a more severe cavitation may sound as though gravel was being passed through the pump. In cases of extremely severe cavitation, it may sound as though rocks are being passed through the pump.

In some cases, where a fluid undergoes a substantial pressure drop at a valve, in the pump, or at abrupt changes in piping diameter, gases dissolved in the fluid may be released, or the liquid may boil. This condition is also called cavitation and has the same random vibration amplitude and frequency characteristics. A pump or valve subjected to this form of cavitation may show the same evidence of wear or erosion, since the unstable gas bubbles tend to implode upon contact with the impeller or pump housing.

Recirculation. Recirculation manifests itself similar to cavitation and normally occurs on pumps which have a large impeller eye diameter-to-outlet diameter ratio. These pumps are generally high suction specific speed pumps which cannot be operated too far from the Best Efficiency Point (BEP). Recirculation, as the term implies, is reverse flow and can cause localized heating and vaporization of pumpage. This reverse flow in the pump and the mixing of fluids moving in opposite directions result in random vibration similar to cavitation. Although the vibration characteristics of recirculation are similar to those of cavitation, recirculation will not likely cause the wear or erosion of pump components in a manner similar to true cavitation. However, the vibration resulting from recirculation can result in damage to mechanical seals, bearings, impeller vanes, and related pump components.

Fans and Blowers

Fans and blowers will often exhibit excessive vibration due to aerodynamic forces and flow turbulence in a manner similar to that already described for centrifugal pumps.

Aerodynamic forces. Fans and blowers will inherently have some vibration due to aerodynamic forces. This vibration results from the fan blades striking the air and will occur at a frequency equal to the number of fan blades times fan rpm. Normally, the amplitudes of vibration resulting from aerodynamic forces will be low and no cause for concern. When excessive vibration at the aerodynamic frequency is encountered, a common cause is resonance of some part of the machine or structure, and the checks for resonance, described previously, should be carried out to determine the resonant part.

If it is confirmed that a condition of excessive aerodynamic vibration is not due to resonance, the fan should be carefully checked for obstructions which may disturb the smooth flow of air through the fan. For example, high amplitudes of aerodynamic vibration are sometimes encountered on cooling-tower fans. Many of these fans consist of a drive motor, mounted outside the fan stack, coupled to the fan gear box by means of a long torque tube or drive shaft. This torque tube can obstruct the smooth flow of air through the fan, and an aerodynamic pulsation is generated each time a fan blade passes over the torque tube. The result is often excessive vibration at the aerodynamic frequency, and the distance between the blade path and torque tube may need to be increased to minimize these aerodynamic pulsations.

On centrifugal fans, excessive vibration at the aerodynamic frequency can sometimes result if the fan rotor is positioned eccentrically in the fan housing. Therefore, this should be checked if the problem cannot be traced to resonance.

Vibration due to aerodynamic forces can also occur at a frequency equal to 1 × fan rpm and will appear similar to normal unbalance. This aerodynamic unbalance will result if the fan blades do not have the same track or pitch. If the fan operates under a constant aerodynamic load, the force of aerodynamic unbalance can be compensated by following normal balancing procedures. However, changing the fan load will often produce a corresponding change in the vibration at 1 × rpm. For example, a centrifugal fan was balanced with the access doors in the fan housing removed to facilitate the addition and movement of trial weights. After the fan had been satisfactorily balanced, the access doors were replaced. Unfortunately, this produced a significant change in aerodynamic conditions, and the result was a significant increase in unbalance vibration. In this case, it was necessary to balance the fan operating under its normal aerodynamic conditions. If a fan must operate smoothly over a broad range of aerodynamic loads, it may be necessary to check and correct for significant variations in blade track or pitch before smooth operation can be guaranteed.

Flow Turbulence. Vibration due to flow turbulence is common on fans and blowers and is the result of variations in pressure or in velocity of the air passing through the fan or associated ductwork. Anything that disrupts the smooth flow of air through the system, such as sharp right-angle turns or changes in the cross-sectional area of the air passage, will generally result in flow turbulence. For example, the original ductwork design in Figure 5-149 will likely result in flow turbulence. The air flow approaching the right-angle turn has inertia and wants to continue moving in a straight line. However, the right-angle turn acts as an obstruction to the smooth flow

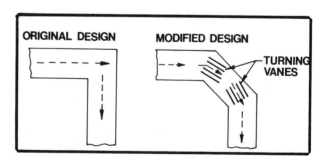

Figure 5-149. Excessive vibration and noise due to flow turbulence in air ductwork may require re-design to provide a more gradual change in flow direction.

of air, resulting in turbulence at this point. The modified design of the ductwork in Figure 5-149 will exhibit less turbulence by providing a more gradual change in flow direction. Where it is necessary to change the diameter or cross-sectional area of the ductwork, this change should be made gradually, perhaps over several feet, and not abruptly, as illustrated in Figure 5-150. The vibration due to flow turbulence in fans and blowers will generally be random, low-frequency vibration, as illustrated in Figure 5-151. Typically, the frequencies excited may range from 50 CPM up to 1800 or 2000 CPM and represent the natural frequencies of the machine, structure, and ductwork.

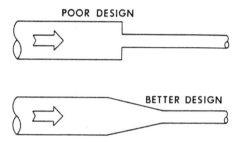

Figure 5-150. Avoid abrupt changes in duct and piping diameter to prevent turbulence and cavitation.

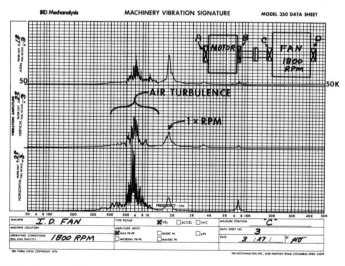

Figure 5-151. The erratic low frequency (400–1000 cpm) vibrations are resonances being excited by flow turbulence in the fan ductwork.

Gear Drives

Gear problems are usually easy to identify because the vibration normally occurs at a frequency equal to gear-meshing frequency, i.e. the number of gear teeth times the rpm of the faulty gear. Figure 5-152 is typical of analysis data resulting from gear vibration. It can be seen that the vibration frequency in this case is equal to the product of the number of gear teeth times the rpm of the high-speed pinion gear. In complex gear drives where several meshing frequencies are possible, examining drawings or blueprints of the gear box to determine the rpm and number of teeth on the various gears may be required in order to identify which gear or gears are most likely at fault.

Common gear problems which result in vibration at gear-meshing frequencies include excessive gear wear, improper adjustment of backlash, gear-tooth inaccuracies, faulty lubrication, and gear eccentricity. In addition to actual gear problems, the vibration characteristic of gears may occur as the result of other disturbing forces in the machine, such as misalignment or a bent shaft. For example, referring to the analysis data in Figure 5-153, note that a high-frequency vibration is present on the gear box, indicating possible gear problems. However, also note that the axial vibration occurring at motor rpm frequency is also relatively high on the

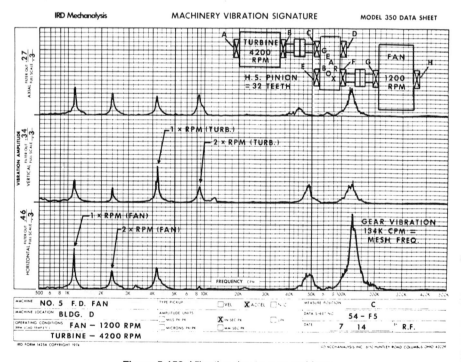

Figure 5-152. Vibration due to gear problems.

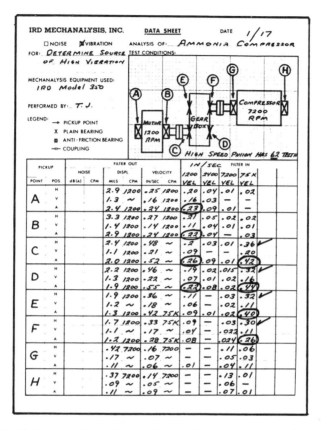

IRD MECHANALYSIS, INC.   **DATA SHEET**   DATE 1/17

☐ NOISE ☒ VIBRATION   ANALYSIS OF: *AMMONIA COMPRESSOR*
FOR: *DETERMINE SOURCE* TEST CONDITIONS:
OF HIGH VIBRATION

MECHANALYSIS EQUIPMENT USED:
IRD Model 350

PERFORMED BY: *T. J.*

LEGEND:
→ PICKUP POINT
X PLAIN BEARING
⊕ ANTI-FRICTION BEARING
⊣⊢ COUPLING

MOTOR 1200 RPM — GEAR BOX — COMPRESSOR 7200 RPM

HIGH SPEED PINION HAS 62 TEETH

PICKUP		NOISE		DISPL		VELOCITY		1200 VEL	2400 VEL	7200 VEL	75K VEL	
POINT	POS	dB(A)	CPM	MILS	CPM	IN/SEC	CPM					
A	H			2.9	1200	.25	1200	.20	.04	.01	.02	
	V			1.3	~	.16	1200	.16	.03	—	—	
	A			2.4	1200	.24	1200	(.23)	.09	.01	—	
B	H			3.3	1200	.27	1200	.21	.05	.02	.02	
	V			1.4	1200	.14	1200	.11	.04	.01	.01	
	A			2.9	1200	.24	1200	(.22)	.04	—	.03	
C	H			2.4	1200	.48	~	.2	.03	.01	.36	
	V			1.1	1200	.21	~	.09	—	—	.20	
	A			2.0	1200	.52	~	(.26)	.09	.01	(.42)	
D	H			2.2	1200	.46	~	.19	.02	.015	.32	
	V			1.3	1200	.22	~	.07	.01	.02	.16	
	A			1.9	1200	.55	~	(.22)	.08	.02	(.44)	
E	H			1.9	1200	.36	~	.11	—	.03	.32	
	V			1.2	~	.18	~	.06	—	.02	.11	
	A			1.3	1200	.42	75K	.09	.01	.02	(.40)	
F	H			1.7	1200	.33	75K	.09	—	.03	.30	
	V			1.1	~	.17	~	.04	—	.022	.11	
	A			1.2	1200	.28	75K	.08	—	.024	(.26)	
G	H			.42	7200	.16	7200	—	—	.11	.06	
	V			.17	~	.07	~	—	—	.05	.03	
	A			.11	~	.06	~	.01	—	.04	.11	
H	H			.37	7200	.14	7200	—	—	.13	.01	
	V			.09	~	.05	~	—	—	.06	—	
	A			.11	~	.09	~	—	—	.07	.01	

Figure 5-153. Misalignment of couplings and bearings is a common cause of excessive gear vibration.

gear box and motor, suggesting that misalignment may be the source of trouble. In this case, misalignment should be corrected first, as this may also eliminate the high-frequency gear vibration.

In addition to a misalignment of the gear case with the driver or driven equipment, internal misalignment of the gear-box bearings can also cause gear vibration. Such bearing misalignment may be the result of case distortion from mounting the gear case on a warped baseplate. This can easily be checked by loosening the mounting bolts slightly, one at a time, while observing the effect on the measured vibration, or by putting a dial indicator on the mounting foot and watching for a change as the bolt is loosened. Gear cases have also been known to distort with age because of stress relieving, thermal effects, or other internal forces. When this occurs, it will be necessary to line-bore the bearings to reestablish proper alignment.

High amplitudes of axial vibration on straight-cut (spur) gears generally indicates misalignment. However, relatively high axial vibrations are common on helical gears where the normal gear load is not perpendicular to the shaft axis but includes an axial as well as a radial component.

Careful examination of the vibration analysis data obtained on gear drives can sometimes provide additional clues to the nature of the specific problem. For example, an eccentric gear or a gear drive subjected to excessive misalignment will often reveal side-band frequencies around the gear-mesh frequency. The side-band frequencies typically encountered are those at gear-mesh frequency plus or minus rotational frequency. To illustrate, the analysis data in Figure 5-154 were obtained on a gear drive with severe misalignment. The pinion gear has 12 teeth and rotates at 1750 rpm, giving a gear-mesh frequency of 21,000 CPM. This gear-mesh frequency vibration is clearly the predominant vibration indicated on the analysis data in Figure 5-154. However, since the gear vibration results from excessive misalignment, the gear-frequency vibration is actually modulated by the vibration at rotating-speed frequency, producing side-band frequencies at gear-mesh frequency plus or minus gear rpm. Therefore, the side-band frequencies indicated on the analysis signature in Figure 5-154 are evident at 19,250 CPM and 22,750 CPM.

If the predominant vibration due to misalignment is 2 × shaft rpm, additional side-band frequencies at gear-mesh frequency plus or minus 2 × rpm may also be present. In any event, when an analysis of gear vibration reveals these side-band

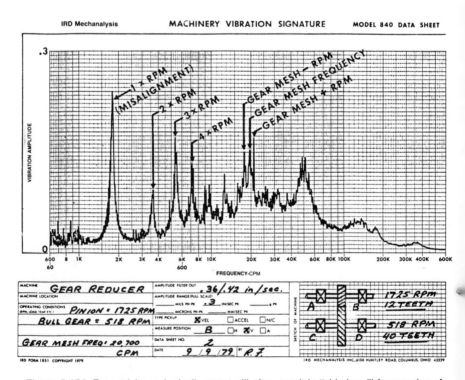

Figure 5-154. Eccentricity and misalignment will often result in "side-band" frequencies of gear-mesh plus and minus gear rpm.

frequencies, a careful check should be made for misalignment, a bent shaft, or an eccentric gear.

If the problem is excessive gear clearance due to gear wear or improper adjustment of gear backlash, the side-band frequencies just discussed will not likely appear on the analysis data. However, excessive vibration will still be indicated at gear-mesh frequency. If gear clearances are within tolerance and the gears are properly lubricated, the transfer of the load from one gear tooth to the next will be a smooth rolling action. However, when gear clearances are excessive, gear mesh will occur in the form of tooth impacts. The result, of course, will be increasing vibration at the gear-mesh frequency. As gear tooth clearances increase even more, the initial tooth impact may cause the gear tooth to bounce within the clearances available, resulting in vibration frequencies at harmonics of gear-mesh frequency. This is 2 ×, 3 ×, or perhaps even higher multiples of gear-mesh frequency. To illustrate, the vibration signature in Figure 5-155 was obtained on a gear drive where the pinion gear having 12 teeth was rotating at 1750 rpm. The gear-mesh frequency of 21,000 CPM is the predominant vibration in this case; however, the second and third harmonics of gear-mesh frequency, 42,000 CPM and 63,000 CPM respectively, are also quite evident. These harmonic frequencies are the result of excessive tooth clearance or backlash. Note the absence of significant side-band frequencies.

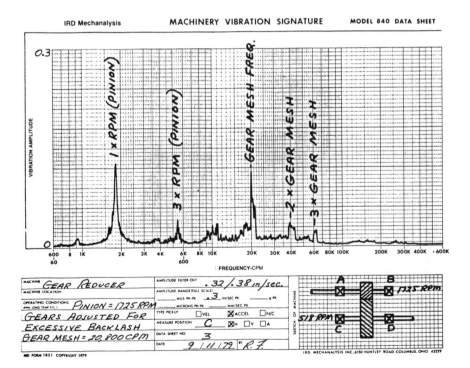

Figure 5-155. The presence of vibration at multiples of gear-mesh frequency is normally the result of excessive gear wear or excessive back-lash.

Gear vibration amplitude and frequency data may be erratic in some cases. This erratic vibration normally occurs with gears operating under very light loads, where the load may randomly shift back and forth from one gear to the next. The impacts which occur when the load is shifted will excite the natural frequencies of the gears, bearings, and other associated machine components, along with gear-meshing frequencies and their harmonics. On occasion, problems associated with gears and gear drives will produce vibration at a frequency not equal to gear-meshing frequency. For example, if a gear has only one broken, cracked, or deformed tooth, the vibration at 1 × gear rpm may result. Viewing the vibration in time waveform on an oscilloscope will help distinguish this problem from unbalance because of the spike-like signal caused by the faulty gear tooth. This was illustrated in Figure 5-33 under "Data Acquisition" in the section on time wave form analysis. Of course, if more than one tooth is defective, vibration frequencies at multiples of gear rpm may result. When the vibration data indicates that only one or a few gear teeth are defective, the gear box should be inspected for loose objects which can get into the gear teeth and cause further damage. The failure of one or only a few gear teeth may also result from an excessive load on the gear teeth during initial startup.

A somewhat unusual vibration frequency sometimes encountered on gear drives is a submultiple of gear-mesh frequency. This vibration is generally the result of an eccentric gear, misalignment, or possibly a bent shaft. Any one of these conditions would result in variations in tooth clearances for each revolution of the gear. As a result, the amplitude of gear-mesh frequency vibration may appear modulated, as shown in Figure 5-156. Since the gear-tooth impacts may only be excessive during a portion of each revolution of the gear, the resulting vibration will have a frequency less than the meshing frequency of the gear. Although this frequency will be less than gear-mesh frequency, it will still be a multiple of gear rpm, and thus the shaft will appear to stand still under the strobe light.

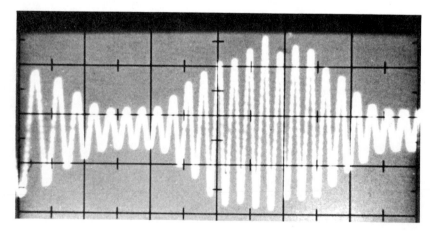

Figure 5-156. Eccentricity or misalignment may cause the amplitude of gear vibration to appear modulated when displayed in time wave-form.

Electrical Problems

Vibration of electrical machinery such as motors, generators, and alternators can be either mechanical or electrical in origin. Mechanical problems, including unbalance, misalignment, defective bearings, and looseness, have already been discussed in some detail. Vibrations caused by electrical problems are normally the result of unequal magnetic forces acting on the rotor or stator. These unequal magnetic forces may be due to open or shorted windings, a broken rotor bar, unbalanced phases, unequal air gap, and similar problems. Generally, the frequency of vibration resulting from these electrical problems will be 1 × rpm and thus will appear similar to unbalance. A common way to check for electrical vibration is to observe the change in vibration amplitude the instant electrical power is disconnected from the unit. This check must be made with the analyzer in the "Filter Out" position. If the vibration disappears the instant power is shut off, the vibration is likely due to electrical problems. If this is the case, conventional electrical testing procedures can be carried out to pinpoint the true cause of vibration. On the other hand, if the vibration amplitude decreases only gradually after power is disconnected, the problem is more likely mechanical in nature. Perhaps an even better indication of the contribution of electrical problems to the vibration of the machine is provided by observing the vibration in time wave form when power is disconnected. This is illustrated in Figure 5-39 and discussed in the data acquisition section under "Time Waveform Analysis."

Electrical problems with induction motors will often cause the amplitude meter to swing or pulsate cyclically. In addition, a reference mark on the motor shaft observed with the analyzer strobe light may be seen to swing back and forth. This pulsating vibration common with induction motors will either be a single frequency whose amplitude is being modulated, or it will be a "beat" between two frequencies of vibration which are very close together. If the nature of the pulsating vibration can be determined, this can significantly help identify the specific problem, as discussed in the following paragraphs.

Armature-related problems. Typical problems associated with the rotor or armature of the induction motor which cause electrical vibration include:

1. Broken rotor bars.
2. Open or shorted rotor windings.
3. A bowed rotor.
4. An eccentric rotor.

The pulsating vibration amplitude characteristic of all of the listed armature problems is a single vibration frequency whose amplitude is modulated with time. As a result, an armature-related problem with an induction motor will generate vibration which in time waveform will appear as shown in Figure 5-157. To explain how this vibration is generated, consider a motor armature having one broken rotor bar (see Figure 5-158). In this case, a two-pole motor is illustrated for simplification. The stator windings of the motor in Figure 5-158 are being energized by an alternating (AC) current.

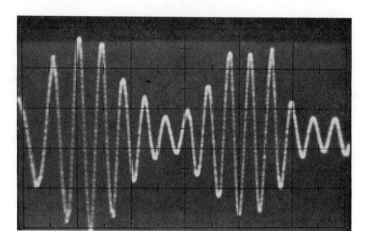

Figure 5-157. Armature problems with induction motors, such as broken rotor bars or a bowed or eccentric rotor may cause a single frequency of vibration whose amplitude is modulated, causing a noticeable pulsation in amplitude.

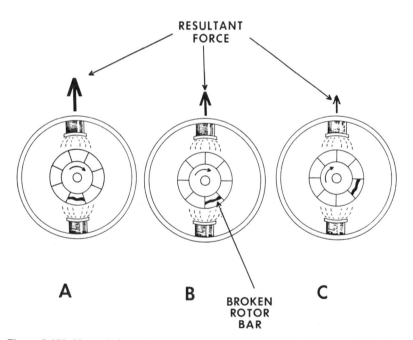

Figure 5-158. Magnetic forces between the armature and stator will vary periodically as the defective rotor bar "slips" relative to the rotating magnetic field of the stator.

In Figure 5-158A, the broken rotor bar is aligned with the lower motor pole at the exact instant the motor pole is energized by the peak current. The upper motor pole is also being energized at peak current at this same instant, but with a polarity opposite that of the lower pole. Assuming that the motor poles are identical, the magnetic fields acting on the armature will be equal, but opposite in polarity. However, since the lower rotor bar of the armature is broken, the magnetic forces between the armature and motor poles will not be equal, and a reaction force between the armature and stator will be generated.

When the stator windings are again energized with the peak A.C. current (Figure 5-158B), the broken rotor bar will not be perfectly aligned with the lower pole as it was in Figure 5-158A. This is due to the magnetic field of the stator pole rotating at a frequency exactly equal to, or an exact submultiple of, the A.C. current powering the motor. The motor armature itself is rotating at a frequency slightly *less* than this electrical frequency. This lag in the rpm of the armature is referred to as "slip," and while the magnetic field may be rotating at 3600 rpm, the armature may be rotating at only 3550 rpm. Therefore, there is a 50-rpm slip. As a result, the next time the lower motor pole is energized at peak current, as shown in Figure 5-158B, the broken rotor bar will not be perfectly aligned as it was in Figure 5-158A; instead, the broken rotor bar will lag by one-fiftieth of a revolution. As a result, the unbalanced magnetic attraction between the motor poles and armature will not have as great a reaction force in Figure 5-158B as it did in Figure 5-158A. The third time the lower motor pole is energized at peak current, the broken rotor bar will lag behind even further, as shown in Figure 5-158C. Since the broken rotor bar in Figure 5-158C is even further from the energized motor pole, the unbalanced magnetic forces acting on the rotor will have even a lower amplitude than was the case in either Figure 5-158A or B. This rotating armature with broken rotor bar will produce a single frequency of vibration whose amplitude is modulated at a rate equal to the difference between the rotating speed of the magnetic field and the rotating speed of the armature. The vibration frequency generated in this case is the rotating speed of the magnetic field and not the rotating speed of the armature.

Stator-related problems. Electrical problems in the stator of an induction motor can also result in vibration with a pulsating amplitude. However, in this case the pulsation is the result of a beat between two separate frequencies of vibration which are very close together. Common stator-related problems which can be expected include:

1. Open or shorted stator windings.
2. Unequal air gap.
3. Unbalanced phases.

In the case of a stator-related problem, producing a vibration whose amplitude pulsates cyclically requires that two frequencies of vibration be present. One of these vibration frequencies may be the result of some unbalance or misalignment occurring at the rotating speed (rpm) of the armature. The other vibration required is an electrical vibration which occurs at the rotating speed of the magnetic field powering the motor. If any of the listed stator problems are encountered, a mechanical vibration will occur at the rotating speed of the magnetic field. Since the mechanical and electrical vibrations are relatively close in frequency, their amplitudes will alternately add and subtract at a rate equal to the difference between their frequencies. The result will be a noticeable steady pulsation, or beat, of the vibration amplitude.

Observing the pulsating vibration in time waveform on an oscilloscope can be useful in identifying the beat frequency characteristic of stator-related induction motor problems. The time waveform will appear as illustrated in Figure 5-159. Note that the phase relationship between the two individual vibration frequencies is constantly changing, producing a resultant vibration whose amplitude increases and decreases in a periodic fashion.

If an oscilloscope is not available for viewing the time waveform of the pulsating vibration, another way to determine whether the vibration is a single frequency whose amplitude is modulated or whether it consists of two frequencies in beat, is to simply expand the horizontal frequency axis when plotting the vibration signature on an XY recorder. This is illustrated in Figure 5-160. Note that when the normal horizontal frequency scale is used, the vibration at 1800 cycles per minute appears as a single frequency. However, when the 600-6000 CPM frequency range was expanded over the entire width of the graph paper, the expanded signature revealed that this vibration at approximately 1800 CPM actually consisted of two frequencies very close together.

Torque pulses. Electric motors have inherent vibration due to torque pulses. Torque pulses are generated as the rotating magnetic field of the motor energizes the stator poles. Since each motor pole is essentially energized twice for each cycle of A.C. current, the vibration resulting from torque pulses will be two times the A.C. line frequency powering the motor. Thus, if an A.C. line frequency is 60 Hz (60 cycles per second), or 3600 CPM, torque-pulse frequency will be 7200 CPM. This

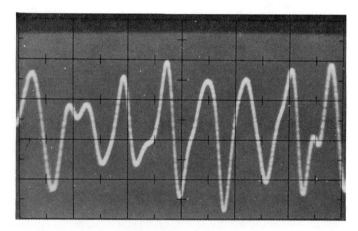

Figure 5-159. This pulsating amplitude is the result of two different frequencies—one is unbalance at 1 × rpm and the other is electrical at the rotating speed of the magnetic field.

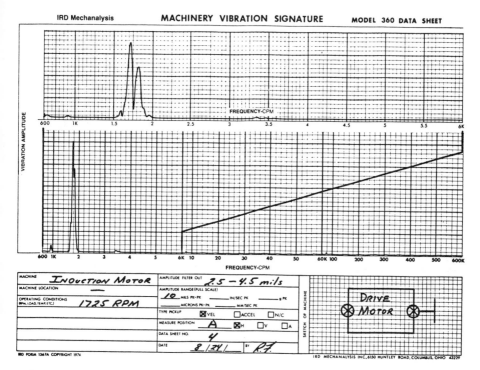

Figure 5-160. Whether a pulsating amplitude is a single frequency with modulated amplitude or two close frequencies in "beat" can often be determined by expanding the frequency scale.

vibration is rarely troublesome except where extremely low vibration levels are required, or if the torque pulses excite a resonance condition in the machine or structure. Torque pulses have also been known to excite loose rotor bars and loose stator windings at frequencies of 2, 3, and even 4 × torque-pulse frequency.

Magnetic interference. Measuring the vibration on large A.C. motors or alternators sometimes presents a problem because of the alternating magnetic fields inherent with this type of machinery. Such magnetic fields can induce a signal in velocity-type pickups at a frequency equal to that of the A.C. field. The amplitude reading which results from the induced signal is actually a "false" reading, which has nothing to do with the condition of the machine. Of course, the strength of the induced signal will depend on the strength of the magnetic field where the pickup is placed.

The presence and approximate influence of a magnetic field can be easily checked using a velocity pickup and analyzer instrument. Connect the vibration pickup to the analyzer just as you would for measuring machine vibration. Next, suspend the vibration pickup by its cable in the area where vibration readings are normally made. Hold the pickup as steady as possible, but do not touch the machine with the pickup.

To measure the amplitude of the magnetic field, carefully tune the analyzer filter to A.C. line frequency and note the amplitude reading. This is the signal caused by the magnetic field.

Ideally, the pickup should be installed in a magnetic shield. Figure 5-161 depicts this shield—actually a mu-metal can available as an option—on an independent Velocity Alarm Module for monitoring small rotating machinery. This particular module is a self-contained velocity alarm indicator and is easy to install and replace. The device mounts directly on the machine. Installation requires absolutely no cable.

No larger than a 6 oz (~160 ml) can, the device is intended for general-purpose equipment monitoring and consists of a velocity transducer, a light-emitting diode (LED) indicator, a battery, adjustable trip points, and test points for reading the velocity level. The Velocity Alarm Module gives an alarm indication at the machine. The LED flashes when the adjustable present level is exceeded.

Turbomachinery Problems

The most common problems encountered on turbomachines, including centrifugal compressors, high-speed centrifugal pumps, turbogenerators, steam and gas turbines, etc., include:

1. Unbalance
2. Misalignment
3. Oil whirl
4. Rubbing

Figure 5-161. Magnetically shielded, self-contained Vibration Alarm Module on a small screw air compressor (Courtesy Bently-Nevada).

5. Looseness
6. Aerodynamic/hydraulic forces

The vibration characteristics associated with all of these problems have been discussed in some detail. Additional problems which may be encountered on turbomachinery include:

1. Internal friction (hysteresis) whirl
2. Aerodynamic cross-coupling
3. Surging
4. Choking (stone-walling)

Friction-induced (hysteresis) whirl. Although rare, hysteresis, or internal friction-induced whirl, is sometimes encountered on turbomachines operating above the first critical speed of the rotor. A rotor which operates above critical speed will tend to deflect or bow in a direction opposite the unbalance heavy spot. As a result, internal friction damping of the rotor (hysteresis damping), which normally works to restrict deflection, will be out of phase, and this internal damping will further deflect the rotor. This condition is normally kept in check by damping provided by the bearings. However, if stationary damping is low, perhaps because of improper bearing lubrication, or if internal rotor damping increases, perhaps because of a poorly lubricated coupling, a friction-excited or hysteresis whirl may occur. This condition has also been traced to excessive interference fit of turbine wheels on the shaft.

The vibration excited by hysteresis whirl is similar in many respects to oil whirl described earlier; however, hysteresis whirl will always occur at a frequency equal to the first rotor/bearing critical speed. As indicated by the vibration and noise identification chart in Figure 5-115, the frequency of friction-induced whirl will occur between 0% and 40% of rotor rpm 80% of the time, between 40% and 50% of rotor RMP 10% of the time, and between 50% and 100% of rotor rpm 10% of the time. To illustrate, if a rotor operates at 3600 rpm and the first bearing/rotor critical speed is 2200 CPM, hysteresis whirl will occur at a frequency of 2200 CPM. Since the frequency of oil whirl will always be slightly less than one-half rotor rpm (42%-47%), hysteresis whirl can usually be distinguished from oil whirl. However, extremely severe vibration problems have been encountered when the frequency of hysteresis whirl coincided with the oil-whirl frequency.

Sometimes hysteresis whirl can be confirmed by changing machine rpm and noting the effect on vibration frequency. If the problem is oil whirl, changing machine rpm will produce a corresponding change in the oil-whirl vibration frequency as well. However, if the problem is hysteresis whirl, changing the rpm of the machine will not change the hysteresis-whirl vibration frequency, since this frequency is determined only by the first rotor/bearing system critical speed.

Whenever hysteresis whirl is confirmed, bearing conditions, including the lubrication system, should be checked thoroughly. The rotor should also be checked for any evidence of rotor or seal rubbing and overly tight interference fits. Another likely cause of hysteresis whirl is excessive friction in gear-type couplings. Improper coupling lubrication, coupling wear, and undersized couplings have also been responsible for exciting hysteresis whirl.

Aerodynamic cross-coupling. Aerodynamic cross-coupling is occasionally encountered on turbines and centrifugal compressors operating above the first rotor critical speed, and it generally results from eccentric rotation of the rotor caused by rotor bow or deflection. In the case of a centrifugal compressor, the layer of air or other gas being compressed will have a rotating speed less than that of the rotor, similar to that of the rotating oil film in a plain bearing. And, if the rotor is slightly bowed, the layer of rotating gas between the rotor and machine housing will produce a torque reaction on the rotor, causing the rotor to whirl at the rotating speed of the gas layer. This is illustrated in Figure 5-162. The frequency at which this whirl occurs can vary from one machine to the next. The vibration may have the frequency characteristics of oil whirl or hysteresis whirl when the lowest natural frequency of the rotor bearing system is excited. In most cases, the vibration frequency will be less than the rotating speed frequency. Cases have also been reported where sub-harmonic resonant frequencies at ½, ⅓, or perhaps ¼ × rotor rpm have been excited.

Since rotor whirl generated by aerodynamic cross-coupling is excited by the compressed air or gas, it seems logical that this condition would be affected by machinery load. In general, the machine will more likely experience this condition under heavily loaded conditions, and changing the load of the machine to determine its effect on the vibration can be useful in diagnosing this problem.

Surging. This is a common problem encountered on high-speed centrifugal and axial-flow compressors and occurs when the compressor is operated outside design limits. Typically, a compressor is designed to deliver air or other gas over a specified volume (mass-flow) range and at a specified pressure ratio. These requirements are met by selecting rotor speed, the number of compressor stages, blade configurations, and other factors. The compressor manufacturer can supply performance characteristic curves showing the range of stable operation in terms of pressure ratio, mass flow, and rpm; attempting to operate the unit outside the design range can result in excessive vibration and machine damage.

The problem of surge occurs when, for a particular operating speed, the discharge pressure (pressure ratio) is too high *or* mass flow (volume) is too low relative to design conditions. This results in a reversal of gas flow in the compressor. In the initial stages of surge, the flow reversal may occur only in the boundary layers of the

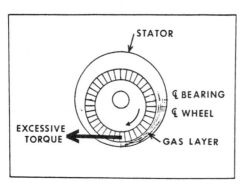

Figure 5-162. An eccentric or bowed rotor can result in aerodynamic torque causing the rotor to whirl.

rotor blades; however, at full surge, the gas flow reverses its direction and flows from the discharge to the inlet.

The vibration characteristics resulting from compressor surging can vary, depending on the extent of the problem. In cases of mild surge, a noticeable increase in the vibration at blade-passing frequency can usually be detected. This frequency is the product of the number of rotor blades times the rpm of the rotor. In other cases, multiples of blade-passing frequency may also be detected. When a full surging condition is encountered, the result may be a high amplitude of random, erratic vibration usually covering a broad frequency range. This is caused by the turbulent flow within the compressor exciting the various natural frequencies of the rotor wheels, rotor blades, diffuser blades, casing, shaft, and other components. Of course, if this condition continues, extensive damage to the compressor can result.

Choking (stone-walling). The problem of choking, or stone-walling, in a compressor is essentially the opposite of surging, but again is the result of attempting to operate the unit outside its design range. Choking occurs when discharge pressures are too low. When discharge pressures are low, velocities are high; and when flow velocity in the diffuser section approaches Mach 1, a turbulent or circulating flow between the blades will occur, which has the effect of blocking the flow of gas. When this occurs, a noticeable drop in efficiency and pressure ratio can be seen, along with an increase in vibration due to the turbulent flow within the compressor. The vibration characteristics of choking will be essentially identical to those encountered with surging, and a check of other operating parameters such as pressure, mass flow, etc. may be necessary to identify which problem it is.

Shaft Crack Detection*

Cracks result from the coupled action of tensile loads, variable stresses, and plastic strain, eventually causing micro- and macro-structural discontinuities. The crack-causing mechanisms of primary importance include high-cycle fatigue, low-cycle fatigue, creep, and stress corrosion cracking. Turbomachinery rotors usually operate in complex stress conditions. The shafts are subject to forced vibrations of a wide frequency spectrum (due to unbalance, constant force fields, etc.) and self-excited vibrations (mostly due to fluid dynamic actions in bearings and seals). Particularly unfavorable dynamic conditions occur at transient processes during all startups and shutdowns of the machines. Hostile environmental conditions (temperature gradients, water, or steam stimulating corrosion) significantly precipitate the crack nucleations and accelerate their growth. Direct ultrasonic and electrical crack-detection methods are very useful and powerful, but it may be extremely difficult to apply them successfully in field operating conditions. In

---

*Adapted by special permission from "Shaft Crack Detection", presented at Seventh Machinery Dynamics Seminar, NCR, Edmonton, Canada, October 1982; by Dr. Agnieszka Muszynska (Bently-Nevada Corporation, 1982) and "Detecting Cracked Shafts At Earlier Levels," ORBIT, July 1982, by Don Bently.

particular, for some complex rotor structures, shaft-crack detection by nondestructive testing does not unambiguously indicate whether a crack exists, because multiple wave reflections due to the geometric complexity of the structure can cause high levels of noise.

Vibration monitoring gains more and more applications in shaft-crack detection. A shaft crack causes some specific dynamic phenomena during machine operation. Observing the appearance of these phenomena and, in particular, the evolution of their growth can help detect shaft cracks. The observation should be made with noncontacting shaft-displacement measuring probes. Shaft riders are likely to miss the relevant data because they cannot interrogate the shaft without disturbing its performance. Since cracked-shaft signal excursions manifest themselves at very low speed (half of first balance resonance), accelerometers and velocity transducers are often out of proper amplitude and phase, thus disqualifying them for most shaft-crack observations.

The data from vibration tests and rotor dynamic analysis provide a basis for fracture-mechanics calculations of fatigue-crack growth. This information can help determine when to shut down the machine. The dynamic phenomena associated with a shaft crack originate from decreased shaft stiffness in the direction of the crack. The stiffness decreases in only one direction, causing the following nonsymmetry in the rotating shaft:

1. The stiffness coefficients in two perpendicular directions will be different because nonequal geometric moments of inertia in the crack-affected cross section (classical "nonsymmetry" of the rotating shaft—Figure 5-163).
2. The stiffness axis will differ from the axis of rotation, first due to the immediate crack, and second due to the resulting shaft bow (Figure 5-164).

This last aspect causes an elastic unbalance excitation synchronous with the shaft rotational velocity, and is of greater magnitude than the dynamic change observed because of different stiffness coefficients in two perpendicular directions.

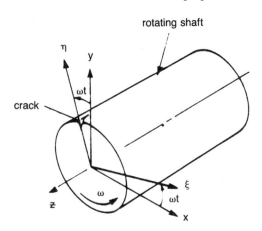

Figure 5-163. Stiffness coefficients in cracked shafts.

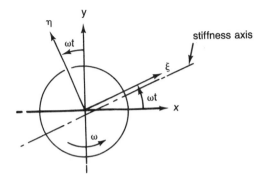

Figure 5-164. Stiffness axes in cracked shafts.

In some cases, shaft-crack events have announced themselves by an increase in shaft radial vibration in the order of one peak-to-peak mil (0.025 mm) per hour. Upon examination, the shafts showed lateral cracks of 60 percent depth-to-diameter ratio.

However, most companies would certainly like to know that a crack exists *before* it penetrates 60 percent into the shaft. There are two methods of detecting shaft cracks sooner. The first method is on-line documentation of the variations in amplitude and in phase vectors of radial vibration. The second method is the recording and subsequent analysis of radial vibration on startups, coast-downs, and heat spin tests of turbines and generators. While on-line vibration-based crack-detection methods are not intended to replace conventional nondestructive testing techniques employed with the rotor at rest, they are an important means of defect diagnosis during actual machine operations.

On-Line Crack Detection

On-line crack detection involves plotting each shaft displacement motion point monitored: vertical and horizontal shaft displacement (shaft displacement with respect to the housing and shaft displacement with respect to the free space, i.e. dual probe, if available). The analysis requires making a polar plot of the rotative displacement motion vector as a function of time, megawatts, horsepower, field current, steam condition, or other factors which influence the shaft rotative speed response vector.

Plotted over a period of time and load, a "shotgun pattern" of the normal rotative-speed vibration vector distribution will be established. Typical polar plot histories are shown in Figures 5-165 and 5-166.

In the historical polar plot of each bearing, a lateral crack will appear as a shaft bow. The shaft bow, in turn, will exhibit a vector pattern outside this range.

This on-line crack-detection method should yield a reliable indication of a crack much earlier than the 60 percent range at which machines are currently shut down. Cracks in the order of 40 percent may be spotted using this historical sampling technique. Surely the indication will precede gross vibration and, therefore, buy time before shut down on gross vibration level may become necessary.

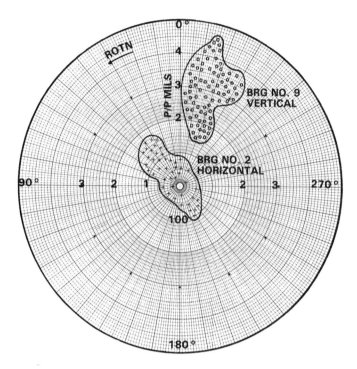

Figure 5-165. Typical "shotgun" pattern of rotative motion vectors.

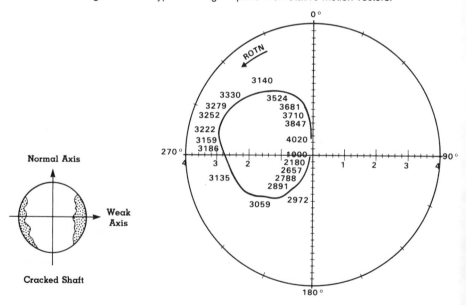

Figure 5-166. Typical 2×-rotative speed vibration vector response of cracked shaft.

In making this measurement, the machine train must be equipped with at least vertical and horizontal shaft relative-displacement probes at each bearing, and ideally with dual probes to measure shaft absolute displacement as well. A digital vector filter is best suited to resolve the rotative speed ($1 \times$) component of each displacement motion. This can be accomplished by using the data acquisition and analysis instrumentation described later in this chapter and shown in Figures 5-167 through 5-171. Portable instruments may also be utilized to measure rotational speed vectors. The historical polar plot for each radial displacement vibration probe can be recorded by hand or by a host processor.

A typical graph of the expected behavior of the twice rotative-speed vibration vector is shown in Figure 5-166.

Startup and Shutdown Vibration
Recording and 2X Data Reduction

In the early history of rotor mechanics, the behavior of asymmetric shafts was observed. Over many years, more than 50 papers have appeared on this subject. It is well known that a lateral crack produces asymmetric spring restraint, and that if the rotor is loaded by a soft preload, such as a horizontal machine with soft gravity preload, the rotor responds to this twice-per-turn reaction of gravity by producing a special Mathieu equation effect generally called the "gravity critical."

Three steps are required to perform this second method of crack detection. First, during each startup and shutdown of the machine train, use an FM tape recorder or an equivalent digital system to document the once-per-revolution, or Keyphasor signal.

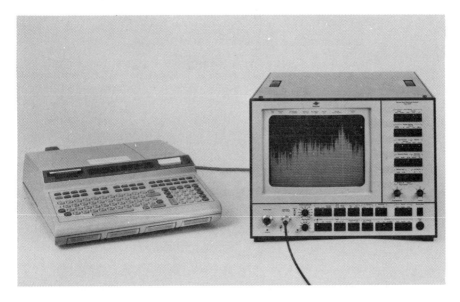

Figure 5-167. Desk-top computer and FFT Analyzer used for machinery analysis on major international pipelines. (Source: Bruel and Kjaer).

Also, the radial shaft-displacement signals, both vertical and horizontal, must be recorded to determine the shaft relative and absolute displacements.

Then document the rpm region of the first self-balance resonance speed (first critical) of the span that each transducer is observing. (Documentation of the second self-balance resonance speed may also be useful.)

Finally, using a data terminal or system as shown in Figures 5-167 through 5-171, set the digital vector filter for 2 × and study each radial vibration channel in the rpm

Figure 5-168. Real-time display terminal (Source: Zonic Corp.).

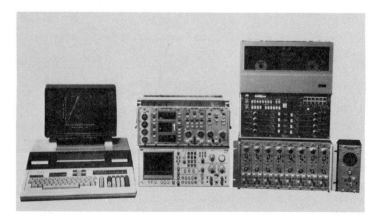

Figure 5-169. Dedicated machinery diagnostic system (Source: Bently-Nevada Corp.).

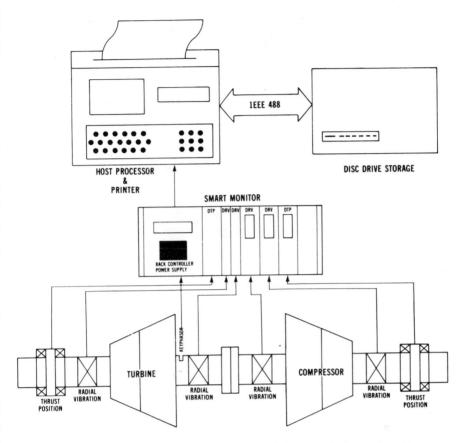

Figure 5-170. Integrated machinery diagnostics and data acquisition system (Source: Bently-Nevada Corp.).

range of 40%-60% of the documented self-balance resonance. (For example, if the first self-balance resonance is 1500 rpm, run 2× rotative-speed response in the region of 600-900 rpm.)

If the shaft is asymmetric due to a crack, vertical and horizontal polar plots will reveal a 2× rotative-speed gravity critical response. Be sure that a foundation or piping resonance of the rotor system is not in this same rpm range by observing the casing response along the machine train. The best data will probably be obtained during coast-down because of the absence of cold thermal bows and "hot" running alignment.

This 2× gravity critical should be plotted in polar form. With the normal damping of a rotor system, lateral cracks of as little as 10% depth-to-diameter ratio are readily detected. This is at a much earlier crack stage than other tests can accomplish under actual installed operating conditions.

Three warnings regarding the second method are appropriate. First, a crack at a purely vertical shear point (no bending moment) may not be observed by either

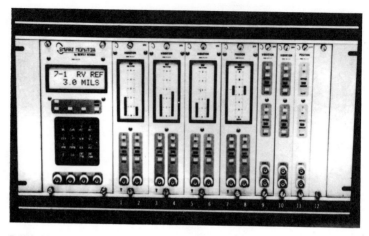

Figure 5-171. Monitoring portion of the integrated system shown in Figure 5-170 (Source: Bently-Nevada Corp.).

method. Second, this method is very sensitive to damping. Either a crack or poor damping may cause complete rotor destruction at gravity critical and/or at first self-balance resonance speed. Use shaft bow limit clearance bearings at the center of the rotor to ensure safety during spin pit tests. Third, the $1 \times$ motion (due to unbalance) will influence the size and orientation of the $2 \times$ polar plot. Good trim balance is desirable for best results.

In summary, it is recommended that the online polar plotting method of the $1 \times$ radial-displacement motion vector history be instituted for plants desiring the best possible equipment protection. This makes it possible to see deviations in the unbalance of the rotor system which may be caused by a propagating lateral shaft crack.

All startups and shutdowns should be documented and analyzed for the gravity critical effect showing up on a good $2 \times$ polar plot.

As compared to other categories of machine malfunctions, cracked shafts are very rare. A broken shaft, however, can be just as dangerous as a fire in an oxygen machine. Any machine with a suspected cracked shaft should be shut down and operated only under dire emergency production conditions.

Continuous Monitoring and Diagnostics

Because an expanded treatment of continuous monitoring and diagnostic methods is outside the scope of this text, only an introductory overview will be given.

Continuous monitoring and diagnostic methods use the very same principles of vibration analysis which have been explained in the preceding pages. These continuous methods are attractive because they promise to give advance warning of impending failures and may allow root-cause determination with greater accuracy than achievable with spot or after-the-fact analysis methods.

Continuous monitoring systems range in complexity from straightforward and proven to involved and prototype-like. Most of the proven continuous monitoring systems use both an FFT (Fast Fourier Transform) analyzer and a calculator. Combining a modern FFT analyzer and desk-top calculator systems as shown in Figures 5-167 and 5-168 allows automatic spectrum comparisons and thus the detection of significant changes in machine condition. At the same time, the analyzer-calculator system is insensitive to small changes which can be disregarded. Depending on type and model, the calculators include a cassette recorder or floppy disc for mass storage of reference spectra or other measured points of interest.

One of the major advantages of the analyzer/calculator combinations is the efficient way in which large quantities of data can be handled and stored. A typical cassette could accommodate reference data from several hundred measurement points. Additional data can be stored by simply changing the cassette. Also, standard, proven software is available to perform many of the calculations and conversions which are useful for observing machine behavior and tracking signal trends.

Other dedicated diagnostic systems consist of a desk-top computer with specific software formats, an integral thermal printer/plotter and tape cartridge drives, a digital vector filter, and a modern spectrum analyzer. Figure 5-169 shows these components.

A third integrated machinery-monitoring, diagnostics, and data-acquisition system is depicted in Figures 5-170 and 171. Traditional machinery monitoring modules which provide such functions as readout, alarms, and analog outputs are here supplemented by trend plots and estimated time-to-danger calculations. A record is kept of maximum, minimum, and typical values for each channel to allow presentation of current data in relevant historical perspective. Integrated systems such as this one give amplitude and phase of the rotative-speed vibration in continuous fashion. In conjunction with a host processor and available dedicated software, the analyzing engineer can readily obtain polar and Bodé plots and spectrum, orbit, and time-base plots. Moreover, the host processor enables long-term storage of machinery condition data. This historical data is of value whenever decisions have to be made or status reports generated.

Vibration of Reciprocating Machines

Routine vibration amplitude-versus-frequency analysis techniques applied to reciprocating pumps, compressors, and gasoline and diesel engines are generally quite effective for diagnosing mechanical problems such as rotating unbalance, misalignment, looseness, etc. However, reciprocating machines will frequently have inherent vibrations which are the result of inertia of the reciprocating components plus varying pressures on the pistons which cause torque variations. The vibrations resulting from these inherent reciprocating forces often have frequency characteristics similar to those associated with more common mechanical problems. The vibration frequencies normally encountered are those at 1 and 2 × rpm; however, higher-order frequencies are also common with some designs, depending on the number of pistons and their relationships with one another. For example, a six-cylinder, four-cycle engine will have three power pulses for each revolution of the crankshaft, and this will cause vibration at a frequency of 3 × rpm. On the other hand, an eight-cylinder engine with four power pulses per revolution will show a vibration at a frequency of 4 × rpm.

Other multiple vibration frequencies of crankshaft rpm are also common on reciprocating units. For example, the vibration signatures in Figure 5-172 were obtained on a truck-mounted, six-cylinder, four-cycle diesel engine operating at 2000 rpm. Looking at the vibration velocity signature, it can be seen that many harmonic (multiple) vibration frequencies are present. These frequencies at multiples of crankshaft rpm are common on reciprocating machines and, if amplitudes are low, may not indicate an abnormal condition. Of course, since so many harmonically related frequencies are present, the possibility of one or more of these frequencies exciting a resonance condition in the machine or structure is high.

The analysis data in Figure 5-172 will also show that vibration is present at a frequency of one-half rpm. In addition, vibrations are also present at half-order harmonics of 1½, 2½, 3½, 4½, 5½, etc. These half-order related vibration frequencies are common on reciprocating compressors and engines that work on a four-stroke (720°) cycle. These units will typically have a camshaft rotating at one-half crankshaft rpm, which can further contribute vibration at a frequency of

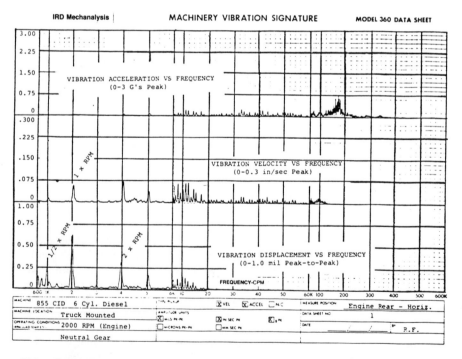

Figure 5-172. This vibration analysis of truck engine vibration discloses numerous multiple (and sub-multiple) frequencies of engine rpm—common on reciprocating equipment.

one-half rpm. In general, however, four-cycle engines and compressors will inherently have some vibration at the half-order frequencies; and, again, the presence of low amplitudes of these vibration frequencies does not necessarily indicate problems. However, an excessive amplitude can result from operational problems with one or more cylinders. For example, the vibration signature in Figure 5-173 was obtained on a V-8 engine with one cylinder misfiring because of a defective spark plug. Note that the predominant vibration occurs at a frequency of one-half rpm. This is due to the absence of a power pulse once for every two revolutions (720°) of the crankshaft. The analysis in Figure 5-173 also shows a comparative vibration signature obtained after the misfiring condition was corrected, and it can be seen that the vibration at one-half rpm has been eliminated. Of course, if two or more cylinders experience similar operational problems, then higher multiples of the one-half rpm vibration frequency may occur. Therefore, when high amplitudes are found at frequencies of ½, 1½, 2½, etc. × rpm, each cylinder should be checked for possible ignition, compression, and carburetion (fuel-injection) problems.

It should be mentioned that operational problems will only generate vibration at the half-order frequencies on those units which operate on a four-stroke (720°) cycle. On two-stroke (360°) cycle engines and compressors, similar operational problems will generate vibration only at even multiples of crankshaft rpm.

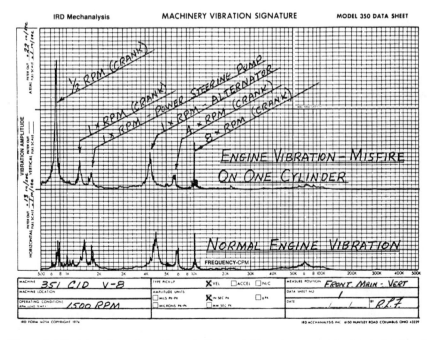

Figure 5-173. A 4-cycle engine with one cylinder misfiring shows a predominant vibration at ½ rpm.

Cases have been reported where the half-order related frequencies inherent on four-cycle units have excited resonant conditions. In one case reported, the predominant vibration measured on a diesel-generator set was at a frequency of 4½ × rpm. Again, the vibration at 4½ × rpm will be inherent, but was excessive in amplitude in this case only because coincidence with a natural frequency of the machine resulted in resonance.

Although some vibration at frequencies of ½ and 1 × rpm plus their harmonic frequencies is common on reciprocating machines, excessive amplitudes at these frequencies may also indicate mechanical or operational problems. The common mechanical problems such as unbalance and misalignment and their vibration characteristics have already been discussed. Typical operational problems which can cause excessive vibration at higher-order frequencies include excessive wear of rod and main bearings, piston slap, valve clash, blow-by (worn rings), compression leaks, faulty ignition, leaking valves, and faulty carburetion or fuel injection. On liquid-cooled engines, leakage of coolant into the combustion chamber can also cause misfiring and vibration. Also, worn cam bearings and worn timing gears and chains can cause high frequencies of vibration.

Worn Connecting-Rod Bearings

Loose or excessively worn rod bearings will often be characterized by a noticeable increase in noise and vibration, particularly during deceleration of the engine. The predominant vibration frequency will typically occur at 2 × rpm of the crankshaft, since bearing impacts will be generated each time the connecting rod (piston) changes direction, and this occurs when the piston reaches top-dead-center (T.D.C.) and again at bottom-dead-center (B.D.C.) for each crankshaft revolution. If more than one connecting rod is defective, then vibrations at frequencies higher than 2 × rpm will likely occur.

Worn Crankshaft Main Bearings

In many cases, the noise and vibration caused by defective main bearings will be most apparent during acceleration or deceleration of the engine, since variations in bearing loads will be greatest under these conditions. However, if bearing clearances become excessive, the noise and vibration may be detected under all operating conditions. The predominant frequency associated with excessive clearances in main bearings will often be equal to or multiples of power-pulse frequency. Thus, on a six-cylinder engine the vibration may occur at a frequency of 3 × rpm, resulting from three power pulses per crankshaft revolution. In addition, higher multiples of power-pulse frequency, such as 6, 9, or 12 × rpm, may also show a significant increase. By comparison, on an eight-cylinder engine the vibration may occur at multiples of 4 × rpm because of the four power pulses per crankshaft revolution.

Piston Slap

Piston slap caused by excessive piston-to-cylinder clearance will have vibration characteristics similar to those caused by excessive clearances in connecting-rod bearings, except that the piston slap vibration will most likely be predominant under heavy load or acceleration conditions, when pressures acting on the piston will be greatest. This problem is also characterized by a noticeable metallic knocking noise coming from the engine. Typically, piston slap will cause vibration at a frequency of 2 × rpm of the crankshaft, although increases in vibration at ½ and 1 × rpm are also possible.

Unbalanced Inertia Forces

Unbalanced reciprocating (inertia) forces are sometimes encountered on pumps, compressors, and engines with a characteristic vibration frequency of 2 × rpm accompanied by higher-order frequencies of lesser amplitude. On modern four-, six-, and eight-cylinder engines, the primary and secondary forces and moments are reasonably well balanced. However, if during a major overhaul a piston or connecting rod has been replaced with one whose weight differs considerably from that of the original equipment, the resultant unbalanced inertia forces can cause excessive vibration.

Because operational problems and reciprocating forces will often cause vibration with frequency characteristics similar to those generated by unbalance, misalignment, and looseness, it may be necessary to obtain additional analysis information in order to distinguish one problem from another. Additional details about the vibration can be obtained by such techniques as phase analysis, time waveform analysis, and mode shape analysis.

Mode shape analysis techniques can be most useful for identifying sources of mechanical looseness. Phase analysis techniques are also helpful for detecting looseness conditions and for confirming misalignment problems. Techniques for distinguishing between unbalance and the torsional vibration caused by operational problems and reciprocating forces using comparative phase indications of the vibration measured above and below crankshaft centerline were also discussed earlier under "Phase Analysis Techniques," (see Figure 5-85).

One clue that will often help distinguish operational problems from unbalance or misalignment is that operational problems tend to create unequal reciprocating forces and thus will usually show a much greater increase in vibration in a direction parallel to the reciprocating motion, but they may show only a small increase in vibration in a direction perpendicular to this motion. Mechanical problems such as unbalance or misalignment will normally show proportional increases in vibration in all radial directions.

If it is concluded that rotating (mass) unbalance is the most likely problem with a reciprocating machine, a balancing technique which can help avoid problems involves working out balance solutions simultaneously for readings obtained in both the horizontal and vertical measurement positions. In other words, obtain original (0) unbalance amplitude and phase measurements for both the horizontal and vertical directions. Then, apply a trial weight to the rotor and obtain resultant (0 + T) readings for both the horizontal and vertical directions. After the data have been collected, calculate the required balance correction using the horizontal data and also using the vertical data. If the problem is truly one of unbalance, the horizontal solution should indicate the same required correction as that obtained for the vertical solution. If this is the case, the correction can be made with confidence that *both* the horizontal and vertical vibrations will be corrected. On the other hand, if the horizontal and vertical solutions disagree significantly, the vibration is *not* due to simple unbalance. Instead, the vibration is probably the result of reciprocating forces which cannot be totally corrected by balancing. If this is the case, the horizontal and vertical solutions should be examined to see if a reasonable compromise can be reached to minimize the radial amplitudes.

Additional information about reciprocating machinery can also be obtained by observing the vibration in time waveform. For example, Figure 5-174 shows a long-term time waveform of the vibration measured on the cylinder head of a six-cylinder reciprocating chiller compressor. This time waveform display clearly shows the repetitive pulses of vibration caused by the opening and closing of the intake and discharge valves. The periodic vibration pulses from this valve action are the result of open-intake, close-intake, open-discharge, and close-discharge. By carefully comparing the vibration time waveforms for each cylinder, it is likely that leaking or sticking valves could be identified by noting unusual or dissimilar pulse characteristics. Further, by superimposing a 1 × rpm reference pulse from a electromagnetic or noncontact pickup on the time waveform, it should be possible to actually identify the specific valve or cylinder at fault. Viewing the time waveform can also be useful for studying internal problems such as piston slap or loose rod and main bearings, since these problems will produce short-duration impact-type vibration and not the sinusoidal vibration typical of unbalance or misalignment (see Figure 5-33).

Engine Analyzers for Reciprocating Machinery

Performance analyzers contribute to the economical operation of engines and compressors by showing which equipment is operating efficiently, which is not, and the reason why it is not. Performance analysis obviously cannot eliminate all engine problems. What it can do is alert reciprocating engine and compressor operators to problems and potential problems before serious equipment damage occurs.

The condition of reciprocating machinery is revealed by efficiency calculations involving horsepower, pressure versus time relationships, pressure versus swept volume, ignition and vibration waveforms, or acquisition of data containing clues as to ignition malfunction, valve leakage, worn bearings, and other mechanical problems.

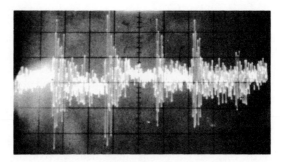

Figure 5-174. Time wave-form analysis of reciprocating compressor shows vibration impulses due to opening and closing of inlet and discharge valves.

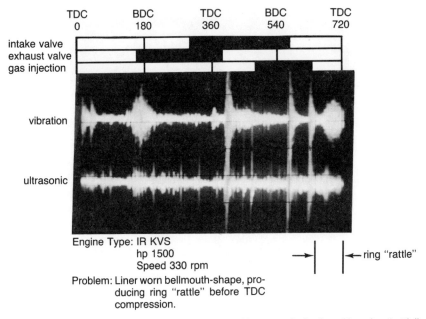

TDC	BDC	TDC	BDC	TDC
0	180	360	540	720

intake valve
exhaust valve
gas injection

vibration

ultrasonic

Engine Type: IR KVS
hp 1500
Speed 330 rpm

→ ←ring "rattle"

Problem: Liner worn bellmouth-shape, producing ring "rattle" before TDC
compression.

Figure 5-175. Analyzer signatures of a gas engine with worn cylinder liner. Note ring "rattle" before TDC compression. (Source: Beta Machinery Analysis, Ltd.)

In the mid-1960s, electronic performance analyzers became commercially available that can analyze higher-speed equipment better than can be analyzed with older instrument types. Not only are pressure/time and pressure/volume displays available on its oscilloscope readout, but vibration, ignition, and ultrasonic traces can be displayed also. Various connecting rod length-to-stroke ratios can be more easily accommodated, and changes in pressure scaling are feasible. Because of this relative versatility, many of these electronic performance analyzers are used in the petrochemical and gas pipeline industry.

Typical problems diagnosed by one such analyzer* are shown in Figures 5-175 through 5-182.

Figure 5-175 shows both the vibration (low frequency) and ultrasonic (high frequency) signatures of a gas engine with piston ring "rattle" due to a worn cylinder liner.

A more severe problem is illustrated in Figures 5-176 A through C. Progressing from normal electronic signal traces in Figure 5-176A to incipient defect indications in Figure 5-176B, we finally observe severely notched cylinder port bridges in Figure 5-176C.

Low-frequency vibration and high-frequency ultrasonic signals are compared for a small gas engine operating normally in Figure 5-177A and experiencing cylinder scuffing in Figure 5-177B.

---

* Beta Machinery Analysis, Ltd., Calgary, Alberta, Canada.

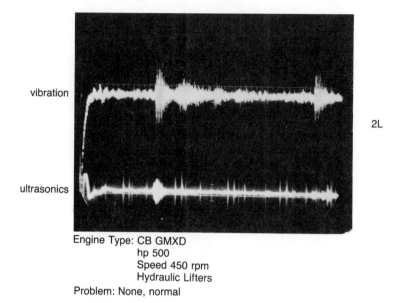

Engine Type: CB GMXD
hp 500
Speed 450 rpm
Hydraulic Lifters
Problem: None, normal

Figure 5-176A. Gas engine analyzer signature with normal port bridges.

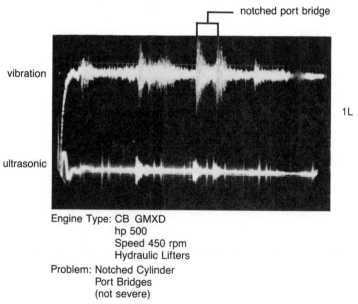

Engine Type: CB GMXD
hp 500
Speed 450 rpm
Hydraulic Lifters
Problem: Notched Cylinder
Port Bridges
(not severe)

Figure 5-176B. Gas engine with slightly notched cylinder port bridges.

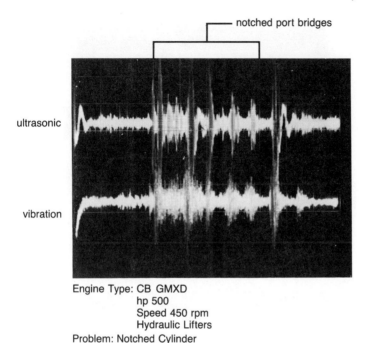

notched port bridges

ultrasonic

vibration

Engine Type: CB GMXD
hp 500
Speed 450 rpm
Hydraulic Lifters
Problem: Notched Cylinder
Port Bridges
(severe)

Figure 5-176C. Gas engine with severely notched cylinder port bridges.

Figure 5-178 depicts the oscilloscope traces for cylinder 1R with worn rocker arm bushings, and for cylinder 2R whose piston and cylinder liner are badly scuffed.

Power cylinder pressure versus time displays are shown in Figures 5-179 and 5-180. Figure 5-179A indicates power cylinder unbalance. The underfueled No. 1 cylinder shows wide fluctuation of peak pressures, typical of a lean condition. Adjusting the fuel balancing valve raised peak firing pressures to average levels and reduced peak pressure deviations, as shown in Figure 5-179B.

Misfiring and "soft" firing by excessive scavenging air pressure led to significant deviations in peak cylinder pressures, Figure 5-180A. Reducing the scavenging air pressure resulted in steadier peak pressures, as illustrated in Figure 5-180B.

Pressure-time and pressure-volume curves of a 14-inch diameter reciprocating compressor cylinder are shown in Figures 5-181 and 5-182, respectively. The discharge valve on the head end of this cylinder had failed. With engine load remaining normal, there was no significant change in discharge temperature. However, a significant reduction in interstage pressure was noted.

In summary, vibration and pressure-volume analysis techniques can be applied effectively to reciprocating machines to detect and identify mechanical *and* operational problems. However, since many problems encountered on reciprocating

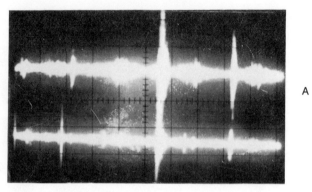

Engine Type: IR SVG
hp 400
Speed 400 rpm
Problem: None, normal

Figure 5-177A. Gas engine with normally worn cylinder walls. (Source: Beta Machinery Analysis, Ltd.)

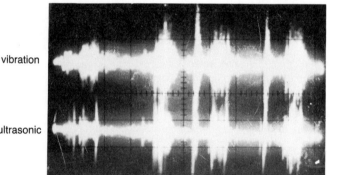

Engine Type: IR SVG
hp 400
Speed 400 rpm
Problem: Cylinder scuffing

Figure 5-177B. Gas engine with scuffed cylinder walls. (Source: Beta Machinery Analysis, Ltd.)

machines have similar vibration characteristics, further investigation may be necessary in order to pinpoint the exact cause. Analysis and identification of defects can be greatly enhanced if prior vibration history or baseline signatures can be obtained on units which are known to be operating in satisfactory condition. This will help identify the normal or inherent vibration frequencies and their corresponding amplitudes. Then, when vibration increases do occur, a new signature can be obtained and compared with the original baseline signature to quickly reveal which frequency components have changed. This approach can greatly simplify the analysis of reciprocating machinery vibration, which in many cases is inherently complex.

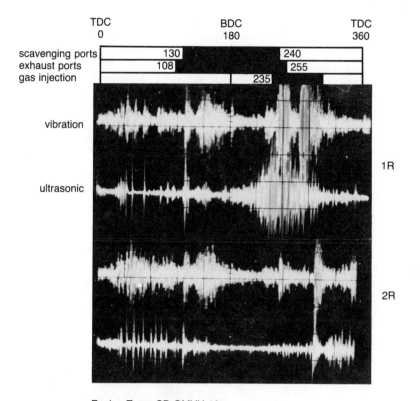

Figure 5-178. Analyzer traces for gas engine with cylinder 1R exhibiting worn rocker arm bushings, and cylinder 2R exhibiting badly scuffed piston and cylinder liner. (Source: Beta Machinery Analysis, Ltd.)

Establishing Safe Operating
Limits for Machinery

The establishment of safe operating limits for major machinery requires a knowledge of machine sensitivity and failure risk. Single-train, critical machines are usually in the spotlight because their shutdown—scheduled or unscheduled—can cause costly interruptions in production and unit output. Moreover, serious failures of major machinery may incur burdensome repair expenditures even for a relatively wealthy owner company.

(text continued on page 473)

Power Cylinder Pressure Time Display

Right Bank

Cylinder No.   1   4   3   2   5

13° BTDC

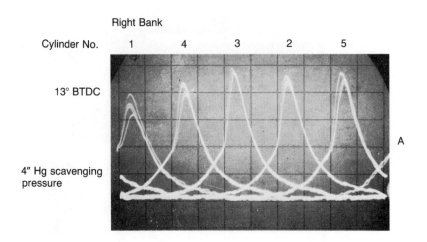

A

4″ Hg scavenging
pressure

Figure 5-179A. Gas engine with cylinder no. 1 underfueled. (Source: Beta Machinery Analysis, Ltd.)

After balancing

13° BTDC

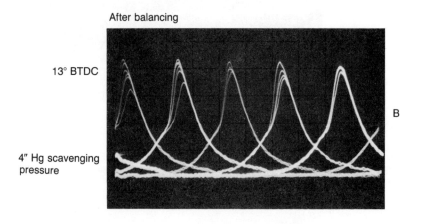

B

4″ Hg scavenging
pressure

line separation = 150 psi

Engine Type: CB GMVH 10
          hp 700
          Speed 330 rpm

Problem: Power Cylinder unbalance. Underfueled No. 1 cylinder shows wide fluctuation of
        peak pressures typical of a lean condition.
        Adjusting fuel balancing valve raised peak firing pressure to average levels and
        produced reduced deviation in peak pressures.

Figure 5-179B. Gas engine, properly balanced. (Source: Beta Machinery Analysis, Ltd.)

Power Cylinder Pressure Time Display

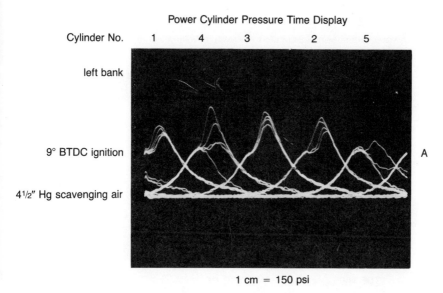

1 cm = 150 psi

Figure 5-180A. Misfiring in gas engine caused by excessive scavenging air pressure. (Source: Beta Machinery Analysis, Ltd.)

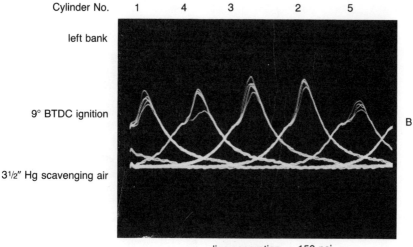

line separation = 150 psi

Engine Type: CB GMVH 10
      hp 700
      Speed 330 rpm

Problem: Lean mixture causing excessive misfiring and soft firing. Note steadier peak pressures when scavenging air pressure is reduced.

Figure 5-180B. Gas engine exhibiting much steadier pressure after scavenging air pressure was reduced. (Source: Beta Machinery Analysis, Ltd.)

Compressor Data

Note: The pressure shown adjacent to the Pressure-Volume curves are based on the dead-weight markers.

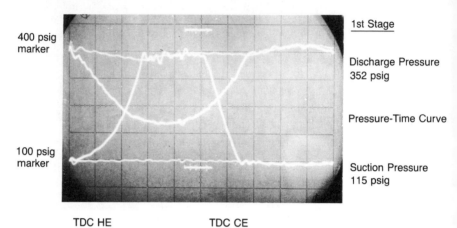

400 psig
marker

100 psig
marker

TDC HE   TDC CE

1st Stage

Discharge Pressure
352 psig

Pressure-Time Curve

Suction Pressure
115 psig

Figure 5-181. Pressure-time curve for reciprocating compressor cylinder with failed discharge valve. (Source: Beta Machinery Analysis, Ltd.)

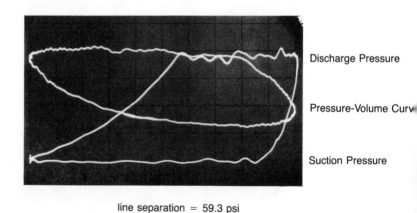

Discharge Pressure

Pressure-Volume Curve

Suction Pressure

line separation = 59.3 psi
HE IHP = 182   CE IHP = 205
HE Vol. Eff. = 0%   CE Vol. Eff. = 85%

Compressor: IR KVG 14″ cylinder diameter
    Speed 330 rpm

Problem: Failed discharge valve on head end. No significant change in discharge temperature but a significant reduction in interstage pressure. Engine load remained normal.

Figure 5-182. Pressure-volume curve for reciprocating compressor cylinder showing significant reduction in interstage pressure. (Source: Beta Machinery Analysis, Ltd.)

Some safe operating limits are firmly set and require neither debate nor discussion. Regardless of the critical nature of a potential shutdown or its impact on plant output, when the suction drum level rises to the point where liquid is about to enter into a centrifugal compressor, or when the lube oil supply pressure at the bearings dips to zero, the machine *must* be shut down. But what about high vibration? How high is still acceptable?

In 1977, S. M. Zierau of Essochem Europe ventured to put years of troubleshooting experience into a set of "limit guidelines." He prefaced these guidelines by stating that the decision to shut down vulnerable turbomachinery trains calls for risk evaluation by a team of process-operations, maintenance, and machinery-oriented technical staff personnel. The machinery engineer must bring to this risk-assessment meeting pertinent vibration baseline and available trend data. He should be prepared to explain the machine's sensitivity, possible problem causes, and available repair and remedial options. He should also give his expert view on the probable consequences of allowing anticipated defects to progress further.

Knowing the machine's sensitivity is of paramount importance. All machines are certainly not created equal. Experiencing six mils of shaft vibration at 5000 rpm may be high, but still tolerable for one machine, whereas another machine experiencing the same vibration amplitude at 5000 rpm may literally start to come apart at the joints. A qualified machinery engineer must have a feel for the probability of his machine belonging to one category or the other.

Similarly, we would expect the process operations person to know the value of unit downtime to the plant. For the low-loss case it would not be appropriate to take high risks by operating a defective machine beyond conservative limits, while for the high-loss case additional risk taking may be deemed appropriate as long as supplementary protective measures are put into effect.

In Figures 5-105 through 5-107, we had given general vibration guidelines for general turbomachinery. These guidelines are approximate at best and can be exceeded when the specific machine behavior in its physical environment of structural supports and its response to load, speed, and temperature changes are known. It is important to understand that experiences of successful operation at a higher than maximum guideline level are not transferable to other machines, since such a level usually reflects a specific installation peculiarity. However, recording such incidents is valuable because it provides needed experience background. Again, we need to say that there is no easy method for establishing such "custom-tailored" maximum allowable levels, independent of the sophistication of the analysis equipment used. Such limits require knowledge of one's machine, confirmative field testing, experience, and risk judgment. In any event, Mr. Zierau's experience led to the development of Figures 5-183 and 5-184, which are offered as a supplement to the many charts available from machinery vendors.

Finally, a few observations on skills requirements. Process plant engineers and managers are sometimes under the impression that sophisticated analysis equipment and/or computerized machinery monitoring alleviates the need for experienced analysts. This impression may well stem from the need to justify the expensive and often elaborate equipment, and one way chosen was to overemphasize actual and potential benefits. However, while signal or spectrum analyzers and trend/trip data from computer systems provide valuable information which eases the job of the

(text continued on page 476)

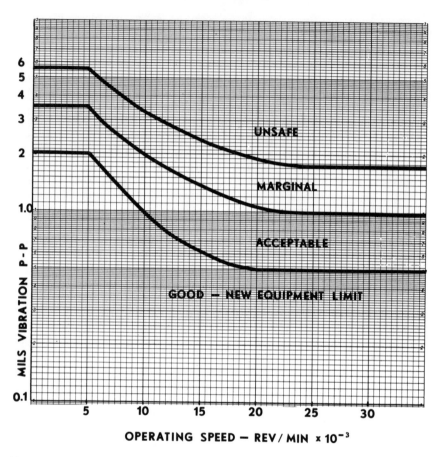

Notes:
1. Operation in the "unsafe" region may lead to near-term failure of the machinery.
2. When operating in the "marginal" region, it is advisable to implement continuous monitoring and to make plans for early problem correction.
3. Periodic monitoring is recommended when operating in the "acceptable" range. Observe trends for amplitude increases at relevant frequencies.
4. The above limits are based on Mr. Zierau's experience. They refer to the typical proximity probe installation close to and supported by the bearing housing and assume that the main vibration component is $1 \times rpm$ *frequency*. The seemingly high allowable vibration levels above 20,000 rpm reflect the experience of high-speed air compressors (up to 50,000 rpm) and jet-engine-type gas turbines, with their light rotors and light bearing loads.
5. Readings must be taken on machined surfaces, with runout less than 0.5 mil up to 12,000 rpm, and less than 0.25 mil above 12,000 rpm.
6. Judgment must be used, especially when experiencing frequencies in multiples of operating rpm on machines with standard bearing loads. Such machines cannot operate at the indicated limits for frequencies higher than $1 \times$ rpm. In such cases, enter onto the graph the predominant frequency of vibration instead of the operating speed.

Figure 5-183. Turbomachinery vibration limits proposed by experienced machinery engineer.

MACHINE TYPE	1xRPM	2xRPM	3xRPM	4xRPM	1xVane Pass	2xVane Pass	1xGear Mesh Freq.	2xGear Mesh Freq.	1xBlade Pass	2xBlade Pass
					SPEED HARMONICS					
Blowers, Up to 6,000 RPM Maximum	.50	.40	.25	.25	.10	.050				
Centrifugal Compressors										
Horizontal	.25	.20	.15	.15	.10	.050				
Barrel	.15	.10	.10	.10	.05	.025				
Steam Turbines										
Special Purpose	.30	.25	.15	.15					.10	.05
General Purpose	.50	.40	.25	.25					.10	.05
Gas Turbines and Axial Compressors	.50	.40	.25	.25					.10	.05
Centrifugal Pumps										
Between Brgs.	.25	.20	.15	.15	.10	.05				
Overhung Type	.50	.40	.25	.25	.10	.05				
Electric Motors	.25	.20	.15	.15						
Gear Units					Intermed. Freq. 1,000-3,000 Hz					
Parallel, Spec.Purp.	.25	.20	.15	.15	.15		.10	.05		
Parallel, Gen. Purp.	.50	.40	.25	.25	.15		.10	.05		
Epicyclic	.15	.10	.10	.10			.10	.05		
	1xRPM	2xRPM	1xLOBE Pass	2xLOBE Pass	3xLOBE Pass	4xLOBE Pass	5xLOBE Pass			
Screw Compressors	.25	.20	.20	.20	.20	.20	.20			

NOTE: (1) The significance of vane, blade and lobe pass frequencies is not yet fully understood. More field data must be evaluated to arrive at universally meaningful maximum levels.

(2) Vane, blade, lobe pass and gear mesh frequency amplitudes vary with load and/or speed changes. The actual sensitivity should be part of the data base.

Figure 5-184. Maximum allowable vibration limits as a function of speed harmonics (all measurements are in inches per second).

technical support engineer, the engineer's skill is still needed to make use of this information. Manufacturing excellence will not be achieved through black boxes, but will continue to be the result of skill applied by all levels of plant personnel. Computer prediction of faults and action recommendations are only possible on clearly defined repeat problems or by remaining at conservative guideline levels. Repeat problems are generally rare, and conservative guideline levels are unacceptable for high-outage-loss units. In other words, we need the sophisticated tools to *assist* the skilled personnel in optimizing plant operations, but skilled personnel are needed more than ever to take advantage of the additional information now available.

Chapter 6

Generalized Machinery Problem-Solving Sequence

In our introductory chapters we developed the need for a uniform approach to problem solving, progressing from the specifically machinery-related "troubleshooting" to a more generally applicable "Professional Problem Solver's Approach (PPS)." This direction was prompted by the fact that:

1. Machinery failure analysis and troubleshooting do not only entail cause identification.
2. Follow-up action is almost always necessary.
3. Our goal is future failure and trouble prevention.

Consequently, we may occasionally be looking at a chain of one problem-solving event after another as part of a single problem situation.

We also alluded to so-called executive processes in thinking and problem solving as part of the Professional Problem Solver's Approach. They are the systematic techniques common to all problem-solving activities. Five suitable techniques will be discussed*:

1. Situation analysis
2. Cause analysis
3. Action generation
4. Decision making
5. Planning for change

These techniques are parts of what we earlier identified as problem-solving sequences. Table 6-1 is an overview of what will be presented in this chapter.

---

*Copyright © T. J. Hansen Company, 6711 Meadowcreek Drive, Dallas, TX. Adapted by permission.

Table 6-1
Problem-Solving Sequence

LOGICAL QUESTION	What's the problem?	Why did it happen?	What are some answers?	Which one is best?	How can I avoid trouble?
TECHNIQUE OR PROCESS	Situation Analysis	Cause Analysis	Action Generation	Decision Making	Planning for change and avoiding surprises
END RESULT	Identify areas requiring my action	Discover the critical cause/ effect relation- ship	Create some new solutions	Select the action which best meets our objectives	Set up actions to: • avoid trouble • reduce damage • capitalize on key opportu- nities

Situation Analysis

Situation analysis considers the following three steps:

1. Separating the elements of a problem.
2. Assigning priorities.
3. Action planning.

Separating the elements. The most difficult part of the separation step is accepting that there may be more than one situation to deal with and more than one action required. After all, who wants to solve three problems if you only have to solve one?

Big problems can rarely be solved easily or quickly, because they are truly multiple problems. So when partial answers are proposed, they get rejected because they don't solve the whole problem. For example, the statements "This large motor suffers from too many starts, the operators need more training, and nobody knows from one moment to the next whether or not the machine is needed" are probably too general to be handled in one step. They are all candidates for situation analysis.

Steps in separating. Take for example "the operators need more training":

1. Define the key words and ask, for example: What kind of training? What is happening (or not happening) that tells you they need more training?

2. List the effects of the problem. Are these some troubles caused by specific lack of training or instruction?
3. Make each separation a complete sentence with modifiers, if possible. An example might be: "Several operators have complained that they cannot find proper instructions on how to start the machine."
4. Write everything down where you can see it. An easel or an overhead projector is good for this.

Figure 6-1 shows several good examples of the separating process.

EXAMPLE I

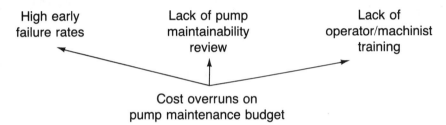

In this case there are three problems which can now be attacked separately. Also the "high early failure rates" seem to be most important.

EXAMPLE II

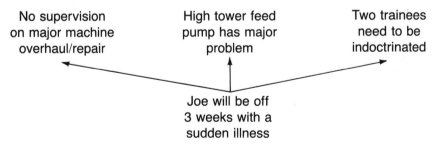

This example shows separating by effects. Unless the question, "What effects will Joe's absence have?" is asked, these potential problems may become unwelcome surprises.

Figure 6-1. The separating process.

Assigning priorities. If we have trouble agreeing on priorities, our objectives may be unclear. When we are in accord about our destination, how we get there is an easier decision. Objectives and goals ought to be clearly stated and understood.

Here are some priority-setting methods:

1. Do the quickest or easiest first.
2. Pick the items that have been waiting the longest.
3. Do some you know you can complete.
4. Do the most urgent ones.
5. Let the boss do the priority setting.
6. Make a list and begin at the top.
7. Take each item in order from the pile or in-basket.

Most of us have at least a subconscious method for deciding what is worth doing and when to do it. A good technique should save time, be conscious, and be repeatable. Three questions will help set priorities in most situations:

1. What is the objective here?
2. What is it worth to fix the problem?
3. When do I have to act to be effective?

For potential or future problems, there is a fourth question:

4. How likely is it to happen?

Action planning. The heart of action planning is active personal intent. *You* are going to *do something.* (Either act, recommend, or drop it because you are wasting time.) If you are not going to act on the results, analyzing is frustrating and nonproductive.

Recommending, for example, is taking action. Not everyone gets to be the final decision maker. Ask: "Am I the recommender, the decision maker, the implementor, or a combination? What is my role in the situation?" Ask these questions: "If I am the recommender, should the person implementing my recommendation/decision be consulted or involved?" If the answer is yes, add some structure and visibility to your process. Communications will be better, and you will have a better shot at success.

A list of actions is shown in Table 6-2. The kind of action taken depends on the situation. But you should be conscious of your action, the purpose of your action, and what triggers the action.

What kinds of action are you most often involved in?

1. Corrective
2. Innovative
3. Opportunistic
4. Interim
5. Adaptive
6. Replacement
7. Selective
8. Preventive
9. Contingent

Table 6-2
Types of Action

Triggering Situation	Action Types	Purpose
Something goes wrong. (Inlet Guide Vanes—IGVs—in forced-draft fan jammed because operating temperature exceeds design temperature of IGV shaft bearings.)	Corrective	Removes the cause and returns to standard performance. A fix, (i.e., install bearings suitable for temperature).
Something goes wrong. (IGV in forced-draft fan jammed)	Innovative or Opportunistic	Extends the cause of improved performance in one area and spreads it around. Raises the standard (i.e., review if IGVs needed at all).
Something goes wrong. (IGV in forced-draft fan jammed)	Interim	Attacks symptoms or effects to buy time. A *patch* (i.e., fix in wide-open position & use by-pass damper for modulation. Fix it for now!)
Something goes wrong. (IGV in forced-draft fan jammed)	Adaptive	Attacks the effects to allow normal or near normal results. Allows you to live with the cause. A more permanent patch (i.e. keep repairing, learn to live with it)!
A *need*. (Sudden coupling Failure)	Replacement or Selective	Fills the void while meeting a set of objectives (i.e., find custom shop with quick turnaround).
Recognition that something in a planned action may go wrong.	Preventive	Removes future causes of trouble. Reduces risk. It is set up and activated ahead of time.
Recognition that something in a planned action may go wrong.	Contingent	Reduces or eliminates bad effects of future problems. Allows continued progress toward goal, even though trouble occurs. Set up ahead of time, but not activated until the problem happens or is imminent.

If you chose the first four, you are concerned with problems and opportunities. Situation analysis and cause analysis should be helpful tools. Adaptive, Replacement, and Selective actions involve deciding. If you chose those, you'll find decision making useful. Preventive and Contingent actions are associated with planning for change. Table 6-2 shows appropriate examples.

An action plan shows what *you* intend to do. Here are the elements:

1. The priority concern—what you are taking action against. This could be a piece of the problem, its cause, or one of its effects.
2. The cause of that concern, if known.
3. The "fix," or action, you plan. List several—be imaginative.
4. Who is responsible—names, which should include the Clout, the Data Source, the Worker, the Reporter (one of these people should be *you*).
6. Your specific actions.
7. The timing. This should include start, report, and completion times. Dates and times only—not "immediately" or "right away."

In other words, what action, against which area, with whom, and by when.

An action plan often includes both short-term and long-term actions. Troubleshooters frequently need to take interim or short-term action against symptoms or effects, then they can take long-term actions to seek and cure the cause. For instance, sometimes machinery distress symptoms must be temporarily treated—like fractional-frequency oil whirl in a centrifugal compressor—before attacking the root of the problem.

Visibility techniques. Frequently, the troubleshooter has to explain a machinery malfunction, distress situation, or component failure to an audience not familiar with the technical details in question. Visibility techniques will help in this case.

Components of a situation can be displayed or made visible several ways. The "cloud technique" is commonly used. At least three others are worth discussing. They may even be more useful:

1. *Force Field**
   - The force field is useful in showing the positive and negative elements affecting a situation.
   - Either the driving forces can be strengthened or the restraining forces reduced to improve things.
   - Again the main purpose is to support action to make decisions more rational.

---

*Although Force Field is discussed here as a visibility technique, there are whole planning processes built around it. It can be a system all its own.

Action usually comes from identifying which elements you wish to attack so the current situation line can be moved to the right. Your question should be "What items here can I act against?" and "How?"

2. *Pictorial*

Often it is better, cleaner, and faster to show a diagram or a picture of how something is (or is not) working (see Figure 6-2). This is especially true if:

- You are explaining a situation to people who do not understand the technical terms or system.
- You are showing an expert how the situation exists. The expert can then compare the present state with how it should be and attack the problem at its causes.

3. *Stair-Step*

Problem

(cause)
Problem

(cause)
Problem

Root Cause

It is helpful most of the time to keep digging for the root cause of the trouble and solve it there. The stair-step method is excellent for showing a long cause-effect relationship. It is also good for showing where you are attacking the problem. The

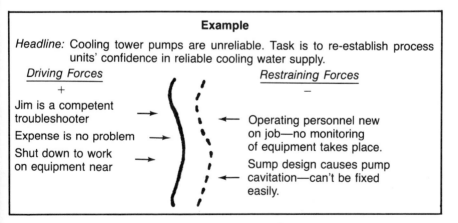

Figure 6-2. Example of the force-field method.

thinking that goes on behind this is that distinguishing between cause and effect is no more than chopping a chain of happenings at some convenient point and calling what goes before the chop "cause," and what comes after the chop "effect." If there are numerous "causes," you ask "which, if any of these can I reduce or eliminate?"

Figures 6-3 through 6-5 are convenient worksheets for using any of the previous techniques.

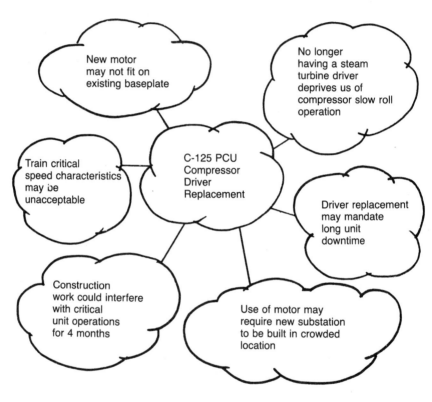

ACTION PLAN

Priority concern	Cause	Fix or other action	Responsible persons and roles	Your actions	When
Interruption of unit operations	Work activity on C-125 slab	Surround with fence	H. G. Smith to coordinate and execute	Sign work order	3/22/83
Driver replacement may mandate long downtime	Unforeseen difficulties with mounting dimensions	Obtain certified drawings for all equipment	P. C. Monroe	Reduce his remaining workload	3/29/83

Figure 6-3. Action plan worksheet—preliminary.

Cause Analysis

You need cause analysis if:

1. Something is wrong and you are not sure why.
2. It is high priority.
3. *You* intend to take action. (If not, do not bother.)
4. Knowing the cause will help you come up with a better solution.

ACTION PLAN

Priority concern	Cause	Fix or other action	Responsible persons and roles	Your actions	When
Interruption of unit operations	Work activity on C-125 compressor floor slab	Erect 8 ft high wooden board fence	H. G. Smith	Sign work order. Notify unit supervisors	3/22/83
Driver replacement may mandate excessive downtime	Unforeseen technical and dimensional interference problems	Arrange with HQ-staff for alternate source of unit product (contingency plan)	Primary: P. C. Monroe Secondary: E. D. Visser	Communicate with HQ product coordination personnel	3/23/83
Train critical speed problems	Torsional critical speed coincident with train speed	Locate Holset-type couplings	Lou Rizzo to tabulate all available options	Authorize advance funding for computer study	4/4/83
New substation located in congested part of plant	Equipment density too high	Install special fire monitors. Obtain variance from safety committee	T. L. Hernandez	Confirm in writing to TLH. Set up meeting with safety coordinator	4/4/83 4/4/83
Lack of slow roll capability may cause rotor to bow	High temperature process gas surrounds compressor rotor	Change operating procedures. Develop clutch system.	G. F. Moeller P. C. Monroe	Request opinion Request feasibility study	3/29/83 3/23/83
Long lead time for motor delivery	Manufacturers too busy. Strikes.	Develop 5 possible sources. Negotiate penalties with Union.	F. D. Corpute J. D. Crumb	Authorize contact Initiate via letter	3/22/83 5/17/83
Project gets cancelled	Economic downturn	Negotiate cancellation charges	J. D. Crumb	Request work to commence	5/17/83

Figure 6-4. Action plan worksheet—final.

FORCE FIELD

Headline:

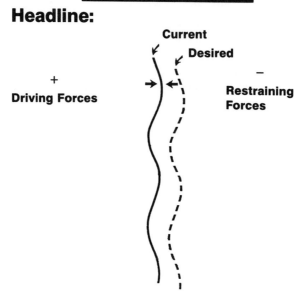

Figure 6-5. Force-field method worksheet.

Process steps. These steps require you to:

1. State the problem (headline or problem statement).
2. Describe the problem (What is wrong and what is not wrong with it?).

Specifications	Problems −	Non-Problems +
Identify Locate Timing Dimensions		

3. State differences which seem significant. Also mention relevant changes.
4. Develop possible causes or hypotheses from Step 3.
5. Test causes against the known facts about the problem to determine the most probable cause. Try to shoot it down. (Test destructively.)
6. Devise a next step to assure you are correct.
7. Return to Action Plan.

Seeking the cause of a problem is like being a detective: you want to know "who did it." The cause of most trouble is evident either by observation or by asking a simple question: "Why did this happen?"

In the few cases where it is necessary and you can take better action knowing the cause, a proven method to gather and analyze information is needed. That is what this section is about.

Remember, you are discussing a past event. Your only limitations are your ability to discover the facts and to process them accurately. It is important to proceed systematically.

A pitfall to avoid. If you intuitively feel you know the cause of the trouble and then set out to prove you are correct, two things—both bad—will happen. People may screen out information and your judgment will be biased. You will have, therefore, compounded your only two limitations: those of discovering and of processing the data. Instead, you must actively try to *disprove* any suggested cause-effect relationship. If it is really true, it will stand the test.

Step 1—the problem statement. Stating the problem is not easy. You need to headline the trouble without making it more difficult by introducing extra emotion because when things go wrong, people are emotional enough.

The problem statement should show *what is having trouble, and what is wrong with it.* For example, "drive motors are burning out." In the case of people, state the unwanted behavior and who is exhibiting that behavior. Do not evaluate or place blame. Usually, you know what is expected. If it is not clear or generally known, put it in the problem statement. Then you have a third piece—*how it should be.* For example, "drive motors are wearing out after six months" *may* be a problem, but there should be some more information. When we discover the design specs call for a two-year life, we *know* we have trouble. Restated, the problem is:

> Drive motors (the object) are wearing out after six months (the defect) when they are supposed to last two years (the expected performance).

If you had some typical problems each week, you might list them as problem statements in tabular format:

What Is the Expectation?	What Is Really Happening?
1.	
2.	
3.	

Test the value of your tabulation by asking:
1. Does the way you stated each problem tell both what (or who) you are having trouble with and what the trouble is?
2. If something is wrong, do you know specifically how it should be?
3. Will it be more clear if you reword the statement?

Step 2—problem description. This is the most critical step—information gathering. And you need to gather information not only about the problem that *is*, but you must also describe the problem that *is not*.

A good set of questions and a format are helpful. You will not get answers to all the questions, and they will not all be relevant. But you can often save considerable time and effort with a good problem specification.

Let us take the easy part first—describing the problem you have:

Identity. Ask, "What's the thing that's affected? What's wrong with it? Important: make certain you separate these two. For example, if the problem *is* cold flow (rolling and peening) of #1 gear set, the *is not* or non-problem portion could be listed as Broken Teeth or Interference Wear. Even better would be: *is* Cold Flow, *is not* Broken Teeth, and *is* #1 gear set, *is not* #2 and #3 gear sets.

Location. Ask, "Where is the trouble found? Any special part? Where geographically? Where is the process?"

Time. Try to keep this to time and dates. Ask, "When did you first notice this? Did it happen before? What has been the pattern since then? Is there any cycle or period?"

Size. Ask, "When it occurs, is it progressive? Are all objects affected? Is the whole thing affected? Is there additional information observed at the site of the problem?"

Now we want to turn each of these questions around and note similar problems which you might have but do not:

Identity. Ask, "What similar objects, products, symptoms, etc., are there where there is no problem like this? Are there other things we might have expected to go wrong, but have not? Is there one that works?"

Location. Ask, "What other similar items, operating units, machines, etc., have we? Any trouble there?"

Time. On the surface, this looks simple. Do *not* write down "no other time" or similar statements. If the problem occurs Monday a.m. and Wednesday p.m., say so, but additionally, state that it does not happen Monday p.m., Tuesday, Wednesday a.m., Thursday, or Friday. Also, ask, "When would we have expected such trouble?"

Size. If the *whole* item is affected, then it is not *partially* damaged. Also, if *all* items in a defective package are bad, it is not just the top or bottom items.

A pitfall to avoid. Do not build information into your description or specification for the purpose of supporting your pet cause. Remain objective and *ask questions*. Otherwise, you will have a piece of information looking for a home.

Step 3—analyze differences which seem significant; also relevant changes. The first question to ask is, "What is different about where we have a problem from where we do not have a problem?" (Now you see one of the reasons for describing the problem you *do not have*.) Next you ask, "Has anything changed with this element? When?"

Step 4—possible causes (hypotheses). For *each* significant difference under Item 3 ask, "What about this difference or change could have caused the trouble?"

Step 5—test causes against both (−) and (+) (minus & plus) problems and non-problems. In testing, you want to shoot down possible causes that do not fit.

A pitfall to avoid. It is tempting here to "prop up" a possible cause that you fancy. *Do not do it.* If it really did cause the trouble, it will fit the facts.

Questions to ask. Does this cause fully explain both the (−) and (+) of the problem description? What additional data do I need? What conditions do I have to check out?

Step 6—devise a next step to assure you are correct. Asking a few questions will usually do this, although a field test or thorough lab test may be called for.

Step 7—return to action plan. We now return to the action plan under cause. The Fix or Action should be better for having discovered the cause.

About Day-One Deviations

Definition: A deviation from planned, designed, or expected standard. It has been present since startup. It never has worked out.

Actions:

  a. Use the minus (problem)/plus (non-problem)/difference approach:

Reality	Expected (+)	Difference (−)
Actual construction	Design specs	
Actual use	Designed use	
Actual operating conditions	Operation assumptions in design	
Actual location	Location initially assumed	
Actual implementation (timing, staffing, budget, etc.)	Implementation (startup assumptions)	

  b. Find one somewhere that is working or that has worked properly. What is different? Seek out similar situations, equipment, objectives, elsewhere. Any trouble? What is different?
  c. Ask these questions:
  Has anyone else attempted to solve this problem? What approach did they take? Did it work? What did they do differently?
  Has this system, equipment, technique, material, etc., been used elsewhere with success? What is different?

When looking for the cause of the troubles relating to people, ask:

1. What *behavior* am I seeing that troubles me?	1. What things are going well? (What things are not problems?)
2. Who is exhibiting that behavior?	2. Who in a similar situation or position is *not* showing the troublesome behavior?

Your goal: try to find some factors *under your control* that contribute to the problem. Look for environmental rather than personal factors, since you can manage environment or arrangements. People manage themselves.

Another valuable use of the expected/reality pairing of information is with problems or conflicts with people. If you carefully state what you expect and what the other individual expects, often the source of the trouble is evident.

The same is true for describing reality, or what is really happening. Two people observing the same event or set of facts may have totally different feelings about it. They may even see two different realities.

Communication Techniques

In other parts of this book, we are making the point that our success as machinery failure analysts and troubleshooters is highly dependent on how well we communicate with others. Examples of successful problem solving show time and again how we depend on the art of communication because frequently we are just not "there" when things go wrong: other people have to supply the background information, the subtleties and details of a trouble-causing event. In the following, we look at some of the more important aspects of communication.

Roles. Successful communication about a problem involves three roles and any number of individuals. The roles or positions are:

1. *Data source*—The person, people, or records that contain data about the problem.
2. *Analyst*—The individual or group of individuals who will gather the information and draw conclusions or recommendations.
3. *Decision maker*—The person or group who will take action.

These roles may be overlapped. The person who knows the details may analyze and take action. Then again, the data source may answer questions and let others analyze and recommend. For any productive communication to occur, there must be willingness and ability. It is critical to recognize these as a pair. One alone will not do.

Willing. If individuals who know the facts about a trouble are threatened or put off by your manner or your technique, they may not give you all the information or they may give you false information. Either way, communication breaks down because they were unwilling.

Able. There are some things you can do to increase the willingness of others to supply the information you need. Your best chances of success will come if you first assume people are doing the best they can. In other words, they are not out to specifically undermine your efforts. Therefore, you can assist by:

1. Finding the person who has the information.
2. Tell them how you will use the information.
3. Ask specific, direct questions to get the selected information you need.

The first is self-explanatory. Someone who does not know the facts or cannot get them is clearly unable to help you. The second will help defuse some emotional hesitancy (defensiveness). If you can show you are objective and avoid jumping to quick conclusions, you will get more cooperation. There is a big difference in seeking the true cause(s) of trouble and seeking merely to place blame.

Ideally the data source, analyst, and decision maker share the same method and concern for the problem. Not always so. The data source almost always provides screened information (that is what good process questions are designed to do). If you are not asking people precise, specific questions, they will give the information *they* consider important. Also some data may be withheld to avoid blame.

An analyst will often be hampered by a lack of time (demand for a quick solution) and bombardment by speculations. Therefore, an analyst will give you only the information *he* thinks is important or necessary. In other words, an analyst will select or screen information based on: (1) what he thinks you want, (2) what he knows, (3) what he thinks, and (4) what he feels. Let us look at these one at a time:

1. *What he thinks you want.* The questions you ask have a lot to do with this. Also, the use you make of the information tells the other person something. You have a lot of control. *You* decide what questions to ask, and you can show the other individual what you are going to do with the data. Crisp, fact-oriented questions in a logical order are necessary. You also need structure to place the answers in.
2. *What he knows.* People usually know more about a situation than they are aware of. Here again, crisp directed questions are necessary. Further, a critical consideration is reaching the *correct people*. Individuals who know or can find out the critical facts.
3. *What he thinks.* People will often have speculations or hypotheses about a problem. Premature judgment turns people off. Write it *all* down.
4. *What he feels.* If the others involved know your method, they will be more eager to work with you. Questions to discover cause can sound quite aggressive (they are). And they can turn off people's feelings if they do not know where you are going.

Pattern Matching

When there is a major disaster, for instance a machinery wreck, it is critical to discover as soon as possible what happened. Often, as with many problem situations, there is no clue that allows us to understand immediately what went on before the wreck occurred—all that remains is the physical evidence that something has gone wrong. We cannot get accurate accounts of what was happening at the time. All we see are the results.

Pattern matching is a useful tool in such instances. It calls for openness or divergence of thought when speculating or hypothesizing about the cause. And it calls for skillful logical processing of the data in testing.

Here is how it works:

1. Examine the evidence (firsthand if possible).
2. Do some "Hypothetical What-Iffing" (HWI) Do not judge or evaluate at that point!
3. Beginning with the top of the list, or anywhere you like, build your test model. You do this on the cause analysis work sheet (Figure 6-8) by asking: "If _____caused the trouble/accident/etc., what item(s) would we see evidence on? What items would we see *none* on?" And so on down the list of questions from the specification sheet. To do this you will need some experience or expertise. If you do not have it, seek someone who does. In the absence of expertise (perhaps the problem has not happened before), you can picture what you would expect to find.
4. Return to the actual evidence. Again, on-site is best.
5. Look for what you would expect if your hypothesis were correct.
6. When you have misfit, go to another "Hypothetical What-If."

This should help solve any mystery you encounter. Whether it involves machinery, people, process, natural occurrences, or a combination, some evidence either physical or behavioral will remain. We conclude our discussion of cause analysis by presenting Figures 6-6, 6-7 and 6-8 as working aids.

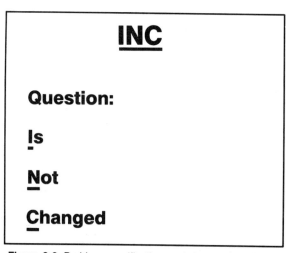

Figure 6-6. Problem specification worksheet—short form.

PROBLEM SPECIFICATION
WORKSHEET

Instructions: Fill in the answers to the specifying questions as
precisely as you can.

DIMENSION	PROBLEMS −	NON-PROBLEMS +
ID	**1a. What item specifically has or shows trouble?**	**1b. Are there other similar items? What are they?** **Are they affected?**
	2a. What is wrong with it?	**2b. What similar or related trouble could we be seeing?** **Are we?**
TIME	**3a. When did it happen? (as precisely as possible)**	**3b. What were other possible or likely times?**
PLACE	**4a. Where exactly did it happen on the item?**	**4b. What parts of the item are unaffected?**
	5a. Where was/is the item with the problem located?	**5b. Are there other locations? Are they having trouble?**
	6a. Did the problem happen in a certain geographical location?	**6b. Where could it have happened but didn't?**
SIZE	**7a. When the trouble happens how much is affected?**	**7b. Is some portion consistently not involved?**
	8a. Is there a pattern or a cycle to the trouble? What is it?	**8b. Is this usual?** **Is there another pattern, cycle, or lack of one you would expect?**

Figure 6-7. Problem specification worksheet—long form.

I. Problem Statement or Headline

| II. DESCRIBE | | III. ANALYZE | IV. POSSIBLE CAUSES | V. QUESTIONS |
PROBLEM −	NON-PROBLEM +	What's different or changed with the—? (Problem Areas)	What about this could have caused the trouble?	Questions for missing infor-mation or con-firmation of possible cause
Existing condi-tions that are off standard or unwanted	Related existing conditions, things, people that aren't off standard		W H E N ?	
ID				
TIME				
PLACE				
SIZE				

Based on this analysis, what caused this trouble?

Figure 6-8. Cause analysis worksheet.

Action Generation

Too often problems are looked at or defined only one way. We are in a hurry to do something, so we grab the first statement of the problem and run. This limits the number and the range of possible solutions. A complaint frequently heard from top management is, "My folks aren't considering enough different answers to problems."

There are some helpful techniques. Understanding them is not difficult. But using them effectively requires some familiarity and practice. Thinking of new ideas is fun. And it can be channeled in a productive way. To do it you need:

1. Some people willing to work with you as resources.
2. Visibility—either a flipchart, blackboard, or someplace to write things down in view of the others.
3. Adequate time to work on the problem.
4. An agreed-to method including role assignments.

The essential roles are:
1. *Client*—the person with the responsibility for action.
2. *Recorder*—The individual who writes the solutions, questions, etc. This should not be the client (too much power of omission in the pen).
3. *Chairperson*—This person keeps the discussion on track, watches the time, and especially guards against premature evaluation of any sort.

If it is your problem, you must take responsibility for the outcome or direction of the meeting. As the client, you must define the problem or opportunity. A method proven successful has been popularized by the Synectics Corporation of Cambridge, Massachusetts. They suggest eight steps.

Step 1—briefly define the task. Tell the group what is wrong. Give them a brief history, and tell them what you expect in terms of group output. If you know cause/effect relationships, show them. Try to present the problem in some non-verbal way as well. Pictures, diagrams, clouds, stair-steps, etc., are ways to do this.

Do not over-define the problem. Remain non-judgmental. Avoid stifling the group with a list of "we've already tried."

It is helpful here to restate the problem in several ways. For example, a problem stated as: "The inlet guide vane bearings of the forced draft fan are seized" could be restated:

"How to get operations to live with it."
or
"How to enlist operations cooperation in the effort to try to run with inlet guide vanes positioned permanently at one setting."

The client picks one problem to work on. The rest are saved for possible action later.

Step 2—goal wishing. This is a free-thinking approach to solutions. It is totally without regard to feasibility.

Opening the doors—First, let us discuss getting the ideas started. Some of the more important things to do and avoid:

a. *Do* go first for quantity.
b. *Avoid* evaluating (looking for quality)—that comes later.
c. *Do* include some other people.
d. *Avoid* assumptions about limits, restrictions, salability, etc.
e. *Do* write everything down.
f. *Do* have fun.
g. *Do* loosen up your brain a little before starting.

A couple of other aids are beginning statements with "I wish . . ." and building on other's comments or ideas.

Step 3—narrowing. The client picks one or more directions or wish statements that hold some intrigue.

Step 4—excursion and force fit. There are several ways to do this:
a. *Key Words.* You select unrelated key words like "circus," "landscaping," "árt," "sports," etc., and again let your mind open up with examples that relate to the selected direction (from Step 3). Then you relate these to problem solutions.
b. *Role Playing.* You might ask them to assume a different identity. For instance, if the problem has to do with motivating first-line supervisors, you might ask your associates to assume they are first-line supervisors, half of whom are "turned on" the other half of whom are disgruntled. Now they are in a frame of mind to invent ways to stay motivated or to get that way. These are then worked into possible solutions.
c. *Imaging.* This is only slightly different from role playing. The difference is the use of objects along with or instead of people. For example, if expensive tools were being stolen and solutions were needed, you might have part of the people imagine they were prisoners (the tools) and the prison was the machine shop. The other people would be the guards. Half would generate ideas on how to escape. The guards, of course, would devise all possible ways to keep them confined. From this you look for new security measures.

Step 5—potential solution. From the wishes and the excursions, the client again narrows to a possible solution. It need not be feasible or viable yet.

Step 6—Pro/con or itemized response. This is the initial evaluation step. *First* you list the good points, the pros, about the idea. Stretch for as many as you can. *Second,* the client states the cons or concerns. These must be proved and worded as "how to" statements.

For example, "Costs too much," might be reworded, "How to do the job with less-expensive materials" or "How to recover some of the $." These "how to" statements are more action-oriented rather than purely negative shoot downs.

Steps 7—actions to reduce concerns. This again is creative and open. However, you are really trying to come up with ways to make the idea fly now. You do this by removing obstacles or concerns.

New ideas often start out as totally ridiculous. As concerns are addressed, they get a little more realistic until finally they reach a critical point: the threshold of "tryability."

Step 8—next steps. If the concerns have been adequately addressed, the client should be ready to say what is next.

Summary

As can be seen, this is a constant divergent/convergent process. We begin by expanding the number of problem statements, then narrow to one. We proceed with multiple wishes or approaches, then narrow to one, and so on.

There are some other keys about the group and the technique:

1. No more than seven people.
2. Include people who have knowledge of the problem.
3. Someone must be willing to take action.
4. Visibility is critical.
5. Method must be agreed to.
6. Chairperson assigned.
7. Strive for freshness. (Creativity is not necessarily a new idea, but one that is new to you.)

The core step in solving problems is developing appropriate action. Indeed, supporting the action step is the primary reason for any other analysis, and perhaps the only sound reason.

A novel, new, or creative solution may also be worth a lot in terms of:

1. Support by fellow employees and management.
2. Interest.
3. Motivation of workers.
4. Competitive advantage.
5. Reducing other problems.
6. Allowing previously unusable techniques to become feasible.

Being effective with creative thought involves letting go—not an easy thing to do. It involves using the right side of the brain—something we are not used to. Only by doing this can we become balanced thinkers and balanced problem solvers.

Decision Making

Deciding is choosing a course of action. Frequently, it is in this area where the machinery failure analyst and troubleshooter has to "show his stuff." Since all action involves risk, we believe that a good process should help to reduce this risk.

Usually the machinery troubleshooter will be involved at the point where everybody asks the question "Where do we go from here?" We are talking about followup decisions.

The Followup Decision Routine depicted in Figure 6-9 can be used to determine appropriate action steps after a machinery component failure has occurred. It begins with the decision as to whether or not the observed failure mode turned up within its standard life expectancy. Standard life expectancy for a given failure mode is the time period in which we, quite often subjectively, expect this failure mode to appear. A few simple examples may illustrate the process.

Let us start with the failure of a rolling-element bearing in a refinery pump. After verifying that this failure mode occurred within the standard life expectancy of the bearing, we move down to the question. Since rolling-element bearings have a finite life, we reason that such a failure can be predicted, but not prevented, by appropriate PM techniques. This decision leads to the next question: Could modification of maintenance ensure earlier recognition of defect? If we answer yes, we are directed to periodic monitoring, inspection, repair, or overhaul of the pump. If the answer is no, we proceed to additional questions which, when answered, allow us to identify appropriate followup steps.

A second example is given to again illustrate how decision trees can be used. This time we are dealing with a gear coupling serving a 3000-hp steam turbine-driven refrigeration compressor rotating at a nominal speed of 5000 rpm. The train was taken out of service when shaft vibration reached excessively high levels after approximately two years of operation. Experience tells us that this failure mode occurred within the standard life expectancy of the coupling. Also, the reviewer decides that we are dealing with a failure mode—electric spark discharge erosion—which is neither predictable nor preventable by PM methods. Judging the failure to be totally preventable, the reviewer is directed to two additional questions. Could modification of maintenance or modification of operation prevent failure? Answering no, he arrives at the last question: Could modification in design prevent failure? This time the answer is affirmative and several possible followup action steps are finally identified for the reviewer. One of these, improved protection, would actually have to be implemented in the form of leakoff brushes for electostatic currents.

The process just developed will suffice for most of our day-to-day decision making. In the following, we would like to deal with the more critical and unusual decisions, such as "select best way to get back on line after a machinery wreck," or "decide whether or not to continue running after a major vibration problem on a critical machine has appeared." The systematic part of decision making included both the correct steps (doing the right things) and following them in the correct order (doing things right). Decision making is a step in the total problem-solving continuum, and the fourth item in it coming up for discussion. As we remember:

1. What is the problem?
2. Why did it happen?
3. What are some solutions?
4. What is the best answer, and how can I present it?

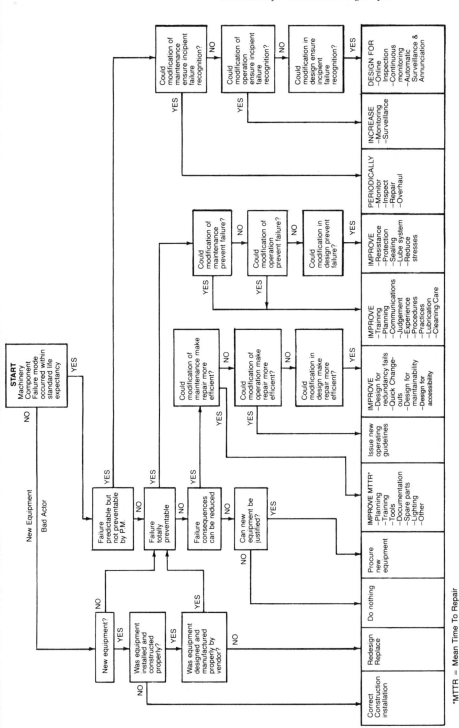

Figure 6-9. Followup decision tree.

*MTTR = Mean Time To Repair

Often people in decision or action situations do not realize they are trying to solve a problem. Others involved in the decision or who need to approve the action may be expecting something to be fixed. As a result, the chosen action is rejected or fails to meet expectations. To help avoid this, you should ask, "what problem are we trying to solve?" Here is an example:

	First	Second
Choose best action to fix inlet guide vanes in F.D. fan	Ask: Why do they have to work? State objectives: 1. Combustion air control for furnace 2. Energy savings 3. Startup needs	Seek and analyze solutions that meet needs and constraints

A good decision-making process must be usable individually or in a group. It is almost always desirable to have some visibility—to write down some of your logic.

There are seven steps to the process. Depending on the situation, some steps will receive more of your attention than others:

Step 1—The decision statement (your mission).
Step 2—Objectives or criteria—what you want to accomplish or avoid.
Step 3—Ranking of objectives. Which are mandatory? Rank order the others, and give them a relative weight.
Step 4—Possible actions—different people, methods, machines, etc., which might meet your objectives.
Step 5—Find the best fit (usually done with a matrix).
Step 6—Assess risk (can be incorporated into Step 2).
Step 7—Take best choice back to the Action Plan.

Three Pitfalls to Avoid

1. Sometimes you will hear, "I don't have time to think things through or talk them over. I've got to make split-second, on-the-spot decisions." Even in tight situations, it *never* hurts to examine quickly objectives, alternatives, and risk.
2. Do not confuse decision making with following a plan where the decisions have been made. The follower of a program is not really deciding, the person who wrote it made the decisions.
3. We are talking about *choosing,* making a choice from several possible actions. Probably you are going to change something—a method, a machine, a person in a new job. As in any change, you would like to reduce uncertainty. Also, you would like your ideas to be willingly accepted or "bought." Notice, it was not said you want to *sell* them. Acceptance is vital when others will be implementing or using your ideas. Sometimes those who are "sold" have not really accepted and may fight the implementing.

Step 1—the decision statement. A decision statement should allow a good selection of alternatives. And, it should reflect decisions already made. For example, "Select new centrifugal pump" is too narrow if what you wanted is to increase pressure from one pressure level to a higher one and deliver a certain quantity of fluid. Another example, "Select new gear coupling" eliminates all other types of couplings. The point is you may be limiting your choices to one kind of coupling when other ideas could better meet your objectives. Also you should ask, "What problem are we trying to solve?" During World War II, the military strategists were stymied trying to decide "How can we avoid the disastrous losses from German U-boats?" The convoy system was helping, but the situation in England was critical. Finally, some bright person suggested the most urgent problem was, "How to get top-priority goods to Great Britain." Flying them in became an obvious alternative. Unusual? Far fetched? Not at all. Failing to seek out the problem to be solved, the reason the action must be taken, can easily result in totally misdirected solutions.

Step 2—objectives. This is a critical step. It is also difficult. Many decision criteria are not conscious. "I don't know what I want, but I'll know it when I see it," is not an irrational statement. It is the legitimate remark of someone who has unconscious objectives difficult to describe, write down, or communicate. Even with oneself! Recognizing some gut-feel objectives probably enter everyone's process, there is considerable value in making your objectives visible and conscious (not necessarily *public*—but conscious). If you went through the action generation step, the "pro" and "con" approach provides some input. You already have an idea of what you like and dislike in the solutions offered.

These can be written in selection criteria form. For example, "a reliable component" (a pro) becomes: "Make best contribution to system reliability." This expression can be used to measure several possibilities.

"How to avoid operator errors" (a con) becomes: "Cause least (or no) machinery related incidents," which can again help in selection.

A simple and always useful question is: "What do we want to consider?" For example, reliability, operating and maintenance staff, time, energy, space, efficiency, etc. Develop your objectives from such a list. The worksheet "Do This Before Deciding" (Figure 6-10) is a series of questions to help you state your objectives. Let us consider an example and follow it through all seven steps. Here is the situation: you are involved in a task group assigned to recommend the best course of action to reduce early failures of pumps in a petrochemical complex. After some discussion, the decision statement or mission is determined:

Decision Statement—Select best method to reduce early failures of repaired pumps.

To set our objectives, we consider the needs of the organization, specific problems our recommendation is to address, and any constraints or directives that have been imposed on us. Using the worksheet, we came up with this list of objectives (see Figure 6-10):

    a. Achieve minimum average runtime of one year after repair.

    b. Minimize Mean-Time-To-Repair (MTTR).

DO THIS BEFORE DECIDING

This will be your decision statement

What is the headline of the decision you have to make?

Does it reflect the reason for this decision or the problem you're trying to solve?

These will be objectives. They will be used as selection criteria.

List at least three things you especially want your choice to accomplish—benefits you expect.

List at least one thing you wish to avoid.

Who will be affected by or will implement this decision?

What are at least three things *they* would like for this to do?

What resources are to be spent?

What are the limits on these, if known?

Figure 6-10. "What to do" before deciding.

    c. Minimize impact on operations.
    d. Minimize cost of repair (overtime).
    e. Minimize additional supervision.
    f. Maximize personnel safety.

We are now ready for Step 3.

Step 3—Ranking of objectives. First determine which objectives are absolutely mandatory (which means if the solution does not meet them, you throw out the solution). These objectives are used as a screen and, therefore, need recognizable limits. In the example, the achievement of a minimum average run time of one year after repair is probably in the mandatory category. The other objectives have relative importance. Numerical weights are helpful and almost any set of numbers will do. A one, two, three scale (low, medium, high) is easy to use. A 10-point scale is good for complex decisions with a larger number of objectives.

For our example, we have used a three-point scale. We went through the list assigning each objective high, medium, or low importance. Then we converted the highs to three, mediums to two, and lows to one.

Now we have obtained:

    A. *Decision Statement*—Select best method to reduce early failures of repaired pumps.

    B. *Objectives*

Achieve Minimum Average Run Time of One Year	Mandatory
Maximize personnel safety	3
Minimize impact on operations	3
Minimize additional supervision	2
Minimize MTTR	2
Minimize repair cost (overtime)	1

A pitfall to avoid. It is important to agree on objectives before selecting your final action. There is a good reason to identify objectives first. Possible actions are possible choices, possible solutions. Like political candidates, they often have backers. Naturally, these backers are potential enemies of all ideas except theirs. If you have good agreement on the objectives *prior* to considering these actions, the chances of win/lose confrontations are reduced.

Step 4—possible actions. Possible actions may come from:

    a. A meeting called for that purpose: What factors affect pump repair quality?
    b. Vendors asked to propose solutions to a mechanical problem that is recurring and affecting repair quality.
    c. Solicited or unsolicited solutions to a problem—five or six suggestions from mechanics and/or operating technicians.

A pitfall to avoid. "There's really only one way to do this," should merit our immediate attention. If objectives are examined, there is almost always at least one other approach.

Possible actions should be stated as completely as is necessary to clearly understand the various choices. If you fix one up or improve it, you need to write in beside the action how you improved it. For example, if additional mechanical-technical help is being considered, you might add: "Find way to release mechanical-technical service personnel from other duties."

Getting back to our example, let us imagine we settled on many possible actions or recommendations.

    a. Incorporate formal failure analysis step for every repair.
    b. Introduce training sessions for mechanics.

c. Select "teams" of top mechanics to work on pumps only.
d. Advise operations of any repair compromises made due to lack of sufficient time to repair.
e. Introduce repair quality checklists.
f. Define what is expected of mechanics and communicate this expectation.
g. Critique bad workmanship on a positive note.
h. Plan to have mechanics present at first startup after repair and develop consistent startup procedures.
i. Review spare parts policy to avoid stock-outs and "make do" compromises.
j. Subcontract some or all pump repairs—with one-year guarantees on parts and workmanship.

The final matrix would look something like the one in Figure 6-11. The content of this matrix is further explained on page 508.

Step 5—find the best fit. Now we decide—at least tentatively:

a. Each possible action is measured against the mandatory objectives on a "go or no-go" basis.
b. Then measure the possible actions against each other and rank them on how well they meet each objective. One way is to give each a number showing how it

HEADLINE OF DECISION:

OBJECTIVES		POSSIBLE ACTIONS			
	WEIGHT				

$\Sigma W \cdot \Sigma \overset{1}{\underset{PA}{N}} = \Sigma T$

Figure 6-11. Decision-making worksheet.

ranked. If you have three, then the best gets three; if there are six possible actions, then give the best one six, etc. Ties are permitted.

A choice rarely fits all conditions perfectly. There will be some trade-off. However, the mandatory conditions *must* be met, or the action is dropped from consideration.

Step 6—*assess risk*. Risk can be considered after the choice is made. Asking the question, "What'll happen if we go with choice A?" will sometimes highlight potential trouble. This step is emotionally difficult. We have just built up a choice, a direction. We have *decided*. Now we are asking ourselves to reconsider. Not easy. There is an emotional release even when the answer is adverse—we are no longer under the strain of indecision. Still, risk must be considered. Anything else is avoiding reality. There are two ways to do this:

a. Look at potential risks of the actions and build them into your objectives with a weight. For example, "Subcontracting makes us dependent upon too many extraneous circumstances beyond our control." It becomes "Subcontract must maximize insurance against extraneous changes—bond, etc."

b. The other way is to ask, "What about this choice is risky?" Then determine if it can be overcome, or if it is serious enough to warrant dropping the choice in favor of the next best alternative.

Step 7—*take best choice of the action plan*. The selected alternative comes back in under Fix or Action.

Fine tuning and details

a. *The irrational part of deciding.* There is also an irrational portion to most decisions. All of us have intuitive feelings about what to do. Some we listen to and some we do not. Whether we act on these "gut feelings" may depend on how well they have worked in the past. If a new solution is especially appealing, but does not seem to meet your objectives, perhaps your objectives need looking at. There generally are reasons why the creative new idea looks good, and these reasons can be put into words. Your "public" objectives may not be the real ones. Ask, "What do I really like about this new idea?" Then list the reasons as objectives. Show *why* this answer is the best. This will help sell your idea, and everyone will feel good knowing why they decided the way they did. Or, in the case of a lone decider, you will know why you picked this choice. There is no guarantee of success, but you will be more confident.

b. *Making recommendations.* All recommendations are risky. The only way to be sure you have made the correct choice is to have a controlled laboratory situation. Even then, you cannot be sure you have tested all options. The best solution may never be found. When you are recommending or selling you can bet your listener is comparing your solution with one or more others. But how are they being measured? This is the major rule in recommending: *discover the presentee's objectives.* Remember, little is sold in this world. Successful transactions involve people *buying* things, and they buy for their reasons, not yours. Therefore, you must find out what those reasons are. Some options are to ask "what do you want?" show a possible list of selection criteria and have them respond or rank order, ask what they especially like in your idea, ask what would make it better, ask what they like about other ideas. The point is, you probe. You help them think it through. Many people do not actually know what they want on a conscious level, and if they do not know, they cannot tell you. Everyone wins if you help them think this through. Once you have the criteria or list of objectives, you need to rank them, possibly assigning relative weights. You need to agree on how these will be measured. Then it is only a matter of comparing the options or actions to see which best fits the criteria.

c. *Recommendation variation.* There is a variation which works a little better when you need to sell your idea, when you are not sure you are deciding among the best choices, or when you would like to highlight areas which are especially good or bad about the possible actions.

Now listen carefully. The difference is subtle. Here is how it works:

a. Arrange the Possible Actions across the top like before.
b. Add the criteria or objectives down the left side.
c. Give each criterion a weight (at least high/medium/low).
d. Now check the Possible Actions against each of the criteria. How well do they do? Good? So-so? Poorly? Give them some points. The simplest is three for good, two for so-so, one for poor.
   If you assigned numerical weights to your criteria, you can multiply to get the final weighted score.
e. *Improving the actions.* Wrong idea won? Either there are some important criteria you have not stated or you need to strengthen your favorite. Look at the areas where your pet idea fared poorly. How can you improve it? When you are in a group situation, the hardest part is to stick with the objectives and their weights. For example, you are deciding which person to promote. There are several mandatory criteria, one of which is: "must have technical degree." Jackie's name comes up. Jackie does not have the necessary degree, but is a favorite of one of the key executives. If you decide to keep Jackie in the running, you *must* remove the mandatory nature of that criterion for *all* candidates, or else Jackie is dropped from consideration. A positive way to use this is to figure out a way around the requirement such as making it a condition that Jackie enroll in a course of study to get the degree and in the meantime receive the necessary technical support. To conclude this discussion of the decision process, we are including Figures 6-12 and 6-13 as working aids.

Planning for Change

A good plan should contain problem-preventing steps. If risks are high, you also need protection as part of your plan. The process to do this *must* be quick and simple. It can be done by asking a couple of mental questions:

1. What could go wrong?
2. What can we do about that?

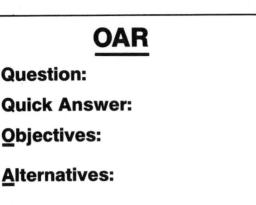

Figure 6-12. Decision-making worksheet—short form.

In cases where the exposure of money, people, time, or other resources is high and critical, the plan should be written.

Where Problems Come From

When a certain concentration or combination of factors occurs, we have trouble. Avoiding unwelcome surprises means discovering these factors or conditions. That is not enough, though. The person who can observe the conditions needs to know what to look for—both warnings and symptoms. In addition, someone must be willing and able to take the right action.

If you are to avoid or minimize trouble effectively, you need a good process. And, the simpler it is, the more likely it will be effective. Unused safety devices prevent few accidents.

The process is simple. There are six parts:

1. Statement of plan (what, where, when, and how much?)
2. Chronological steps.
3. Separate potential problems (think of everything that will change and ask "What could go wrong?")
4. Pick two or three with highest priority. Assign priority with these questions:
   a. What is it worth to solve?
   b. When must I act to be effective?
   c. How likely is it to happen?
5. Make an Action Plan for the highest priority items.
6. Update your plan with the Action Items.

Gathering, organizing the information, and setting up the system is not easy. The concept is simple, but it is emotionally difficult. Common sense tells you that removing likely causes of problems reduces the probability of their occurring. Since most problems have interactive causes, you can cut down trouble by attacking the various conditions which act together.

HEADLINE OF DECISION: SELECT BEST METHOD TO REDUCE EARLY FAILURES OF REPAIRED PUMPS

OBJECTIVES	WEIGHT Must	POSSIBLE ACTIONS Formal FA	Repair QC	No Stock-Outs	Subcontract
Achieve Minimum Average Run Time of One Year		Contributing	Contributing	Contributing	Unknown
Maximize Personnel Safety	3	Perhaps: End Result 3 / 9	Improvement 4 / 12	No Impact 1 / 3	Perhaps 2 / 6
Minimize Impact on Operations	3	Follow-up Required 2 / 6	No Impact 3 / 9	Positive 4 / 12	Subcontractor Needs to be called 1 / 3
Minimize Additional Supervision	2	No Impact 3 / 6	Need Supv. 1 / 2	No Impact 4 / 8	Needs follow-up & co-ordination 2 / 4
Minimize MTTR	2	Extra Time Required 1 / 2	Need extra time 2 / 4	Eliminate waiting time 4 / 8	Could take longer (week-ends) 3 / 6
Minimize Repair Cost	2	No additional cost 4 / 8	Extra Time = $$ 3 / 6	Increase in cost 2 / 4	Higher cost 1 / 2
ΣW		T 31	T 33	T 35*	T 21

$$\Sigma W \cdot \Sigma \overset{1}{\underset{PA}{N}} = \Sigma T \qquad 12 \cdot 10 = 120 \qquad 120$$

$$\left.\begin{array}{l}31\\33\\35\\21\end{array}\right\} \;120$$

*Highest total score = Preferred Action

Figure 6-13. Decision-making worksheet—example of long form. Each of the possible actions is given a value ranging from 4 (high) to 1 (low). This value is multiplied by the weight of each given objective and the resulting products added up to give the total (T) for each possible

A working system of problem prevention has immense value. Operating technicians, mechanical-technical service personnel, and the machinery trouble-shooter and failure analyst usually gain increasing confidence when they apply this system.

Details of the System

1. *The Plan*
   - What is going to change?
   - Where will it be done?
   - When will it begin and end?
   - How much is involved?
   - What is your part—are you in charge, an observer, or a participant?

2. *Steps*

   List them in order of occurrence. What has to be done first, second, etc. Put in times and dates. This is not much detail. If you are going to plan at all, you should do at least this much.

3. *Problems or Opportunities*

   List things that might go wrong *without regard to whose fault or control*. Also any positive surprises that may occur. Here are the questions to ask:
   - What could go wrong?
   - What trouble have we had before?
   - Are we dealing with new people, equipment, technology, vendors, supplies, etc.?
   - Anything where we have no experience?
   - Where is timing or scheduling tight?
   - Is this event dependent on another?
   - Are we counting on people not under our control? Other departments? Other companies?
   - Would trouble for us benefit anyone else? Are they in a position to hinder us? To withhold or be less than enthusiastic with support?

List them in the same way you separated items under Situation Analysis.

4. *Assign Priorities*
   - Value—what is it worth to avoid or capitalize on this situation?
   - Timing—when must you act (or when must action be taken) to be effective. If you want to order a standby item in case your schedule is delayed, and the lead time is 30 days, you obviously must act at least 30 days ahead. This may seem obvious, but why are there so many crash programs and emergency expedite orders in our industry? The key here is "when timing is really tight, plan backward."

- How likely is this to happen? Has it happened before? What incentive do others have to make sure it does not happen? The more likely it is to occur, the more you need to consider actions to prevent (if a trouble) or to take advantage (if an opportunity)

5. *Action Planning*

Use the same Action Planning format as before. The priority concerns are clearly just that—items which are critical in importance, timing, and likelihood of occurrence.

Under causes, the concept of contributions is helpful.

Contributors

List the conditions or factors which make this problem happen. Some of them merely increase the chances of trouble. Be *sure* to list those that increase probability. List the factors, events, or conditions that could cause it, indicate its approaching, or raise the probability of its happening. Questions to ask:

- What usually happens just before this problem?
- What conditions make this problem likely? Possible?
- What key material, people, timing, etc., *must* be correct for all to go well? Is it all in order?
- What tells us trouble is imminent? Is happening? Has happened?
- Are those indicators functioning? Are they monitored?
- What caused trouble last time? Any other time?

Fix or Other Actions

a. *Before Actions.* These are actions to prevent trouble and are specifically designed for each condition factor. The idea is to reduce the probability of its happening, or to prevent it entirely if possible. Frankly, there are few perfect preventives. Inoculations, total physical isolation from risk, total removal of a vital link (for example, no fire in the absence of oxygen) are some examples. Total removal of risk is often prohibitively expensive. So, we reduce the probability to an acceptable risk level and monitor the key indicators of trouble. Examples of Before Actions are:

Problems	Contributors	Before Actions
Large motor is damaged when transported to electrical shop	Suffers traffic accident	Provide front & back escort
	Falls off float	Check & doublecheck tie-down
Large motor fails	Dirt in windings	Clean and inspect regularly.

b. *When actions.* These are arranged ahead of time to activate when the indicators show the problem has happened or is imminent. They are your contingency plan. They are like insurance. They do not keep the trouble from happening, but they reduce the damage. Effects of the problem are their target. Examples of When Actions are:

Problem	Contributors	When Actions
Large motor suffers winding failure	Dirt & overload	Enact contingency plan, i.e., use reserve copper bar stock for windings.

Note that When Actions include the specific symptoms or warning which tells you to begin the action. Also, both Before and When Actions are appropriate in connection with serious problems.

6. *Update your plan with the Action Items*

You have accomplished nothing until the plan itself is altered. If you pick out some actions designed to prevent problems as you decide to change the timing or sequence in your plan, *go back and physically make the change.*

7. *Assign responsibilities and see that the plan gets executed*

Figures 6-14 and 6-15 show appropriate examples of a "planning for change" sequence. Figure 6-16 may be used as a convenient working aid for future planning work.

PLANNING FOR CHANGE

What's going to change?

Replace 4500 HP motor with new 5500 motor - Booster compressor driver.

Where will it be done?

On Ethylene Cracker machinery deck

When will the change begin and end?

Start   3/20/82
Finish   4/04/82

How much is involved?

Remove existing motor - Install new motor.

What's your part in this?

Troubleshooter and overall coordinator.

What's the plan?

Review Foundation Drawings	1/15/81
Specify New Motor	2/15/81
Make Vendor Selection	3/15/81
Order New Motor	3/20/81
Review Vendor Drawings & Approve	5/01/81
Review New Coupling Arrangement	8/15/81
Review New Motor Storage Arrangemt.	9/10/81
Remove Old Motor	3/20/82
Install New Motor & Coupling	3/25/82
Start-up	4/05/82
Evaluate Performance	4/06/82
Full Load Performance	5/15/82

What's the most critical step?

No one step is most critical. (Therefore, look at whole thing.)

Figure 6-14. Planning for change example.

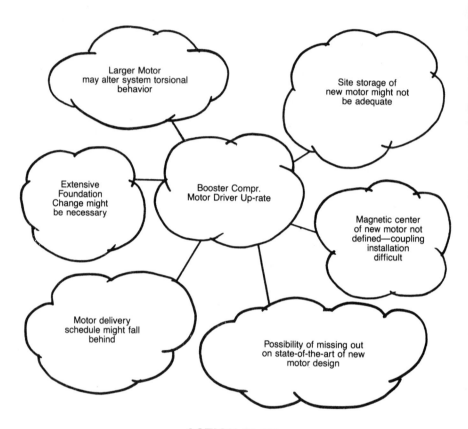

ACTION PLAN

Priority concern	Cause	Fix or other action	Responsible persons and roles	Your actions	When
Larger motor chg. system torsional	1. Different mass-elastic character	Define torsional criticals early	DFM/ vendor	Assist	3/15/81
	2. Wrong couplg. selection	Specify	Vendor	Review	4/01/81

Figure 6-15. Action plan example.

PLANNING FOR CHANGE

What's going to change?

Where will it be done?

When will the change begin and end?

How much is involved?

What's your part is this?

What's the plan?

What's the most critical step?

Figure 6-16. Planning for change worksheet.

A tip. The people who suffer from the problem are ideal to enlist for Before Actions. They are motivated. For example, prior to a long sailing voyage, the captain asks the crew, "Anyone here who can't swim?" When Charlie raises his hand, the captain says, "Ok, Charlie, you're in charge of leaks!"

A final word. Reducing the probability of trouble tomorrow is one of the best reasons to spend part of today seeking the cause of yesterday's problems.

References

1. Edwards, B., *Drawing on the Right Side of the Brain,* Houghton-Mifflin Co., Boston, 1979.
2. Stein, M.I., *Stimulating Creativity, Volume 1: Individual Procedures, Volume 2: Group Procedures,* Academic Press, New York, 1974-5.
3. Gallwey, W.T., *The Inner Game of Tennis,* Random House, New York, 1974.
4. *The Journal of Creative Behavior,* The Creative Education Foundation, Buffalo, New York.
5. Parnes, S.J., *Guide to Creative Action,* Charles Scribner's Sons, New York, 1977.
6. Parnes, S.J., *This Magic of Your Mind,* The Creative Education Foundation, Inc., Buffalo, NY 1981.

Chapter 7

Statistical Approaches in Machinery Problem Solving

In the preceding chapters it was shown how machinery failure and distress behavior can be influenced by appropriate design techniques. Ideally, we would like to verify low failure risk during the specification, design, installation, and start-up phases. However, experience shows that good, basic design alone may not prevent failures. Failure statistics in earlier chapters demonstrated clearly that many machinery component failures result from equipment operation at off-design or unintended conditions. Our failure analysis and troubleshooting activities will thus often be post-mortem and commence after installation and start-up. The maintenance phase has then begun, and failure analysis and trouble-shooting will become an integral part of that phase.

Machinery maintenance activities, together with failure analysis and trouble-shooting, can be regarded as failure fighting processes. Failure fighting calls for appropriate strategies in order to be successful. However, before we can apply any strategies, we must gather as much information as possible about failure modes and failure frequencies. Suitable statistical approaches will accomplish this task and therefore, have to be part of the failure fighting arsenal.

This chapter discusses statistical approaches and other important issues in machinery maintenance as they relate to failure analysis and troubleshooting.

Machinery Failure Modes and Maintenance Strategies

Most petrochemical plants have large maintenance departments whose sole reason for being is to take care of various and numerous forms of plant deterioration. Typical forms of plant deterioration are machinery component failure events, malfunctions, and system troubles.

As in most areas of human endeavor, we would like to perform process machinery maintenance in a rational and planned manner. Planning in turn could imply that the object of our efforts also behaves in a predictable way, i.e., it responds favorably to planned maintenance actions. Is that always true in machinery?

To answer this question satisfactorily, we must examine how machinery component failure modes appear and behave as a function of time. This discussion concerns the concept of machinery component life.

Life Concepts

Failure is one of several ways in which engineered devices attain the end of their useful life. Table 7-1 shows the possibilities.

Since most machinery failure analysis takes place at the component level, one must define what constitutes life attainment in connection with component failure modes. The term "defect limit"[1] is used in this context and needs to be explained first.

Defect limit describes a parameter which we are all familiar with. For instance, original automotive equipment tires often have built-in tread wear indicators. These are molded into the bottom of the tread grooves and appear as approximately ½-inch wide bands when the tire tread depth is down to $1/16$ inch. The wear indicators are "defect limits."

Other examples of defect limits are clearance tolerances of turbomachinery assemblies which are observed and documented during machinery overhauls. But not all defect limits are obvious. Mechanical shaft couplings are an example of a difficult-to-define-limit where we wish we had something like a tread wear indicator to determine serviceability. Another example, but connected with a random failure mode, is the decision point at which a leak from a mechanical seal or a vibration severity value on a pump is considered excessive and deserving of maintenance attention. Here, the defect limit defies ready quantification and, more often than not, depends on subjective judgment.

One noted machinery troubleshooter[2] refers to this phenomenon as "user tolerance" to the distress symptoms exhibited by machinery equipment. In his discussion of flow recirculation in centrifugal pumps, he describes a pump designer's view as to how users determine acceptable minimum flow for centrifugal pumps. He

Table 7-1
Reaching End of Useful Machinery Life

Way in Which the End of Useful Life Is Reached		Example
Failure	Slow	Mechanical seal leakage
	Sudden	Motor winding failure
Obsolescence		Recip. steam pump/engine
Completion of mission		Oil mist
		Packaging
		Wear pads
Depletion		Electric battery

states while some users will accept that an impeller subject to cavitation may have to be replaced every year, others will not tolerate impeller replacements in the same service more frequently than every three or four years. Similarly, noise and pulsation levels perfectly acceptable to one user are totally unacceptable to another.

In any case, defect limit defines the degree of failure mode progress that would initiate such maintenance activities as replacement, repair, more thorough or more frequent checks, or perhaps a calculated risk decision to continue without resorting to these conventional maintenance activities.

"Life" of a machinery component can now be defined as the time span between putting a machinery component into service and the point in time when the component attains its defect limit. "Standard Life" is the average lifetime acceptable to any particular plant failure analyst or troubleshooter. Have you arrived at defect limits and therefore "standard life" of *all* the failure modes encountered on machinery components in your plant? Are you willing to live with them? Or are you looking for improvement?

These questions remind us that we need to set defect limits objectively, albeit just for our own plant. Two implications seem obvious: (1) if we do not, we have no baseline from which to improve, and (2) if we allow subjective judgment to prevail, we will most likely experience unpleasant surprises, because those judgment calls will range from the conservative to the optimistic. The conservative judgment will reduce component life arbitrarily and the optimistic judgment call will expose us to unjustifiably high risks. A team approach to documented defect limit setting seems therefore appropriate.

From experience we know that some failure modes occur rather often, for example, "wear of coke crusher rotor fingers" in tar service. Other failure modes are hardly ever encountered; for instance, electrical pitting (FM: erosion) on surfaces of spherical rollers caused by current passage through the bearing. Obviously, different failure modes have different frequencies of occurrence. In the past, we have generally used mean-time-between-failure (MTBF) to express these differences. What we are describing is the probability of machinery failure and breakdown events as a function of operating time. While this is of some interest to all of us, it is of particular concern to the maintenance failure analyst and troubleshooter. No doubt at one time or another we all have to grapple with the realization that some machinery failure modes appear slowly and predictably while others occur randomly and unpredictably. In any event we have probably encountered a mix of both types of failures.

The Swedish mathematician W. Weibull developed mathematical models to describe characteristics of component failures. In "Weibull functions" we find a constant β which characterizes the life cycle of a given group of components. Figure 7-1 shows these Weibull functions. In its simplest form, the Weibull relationship looks like this:

$$\text{failure rate} = at^{(\beta-1)} \qquad (7\text{-}1)$$

where:

$\quad a$ = scale factor.
$\quad \beta$ = Weibull slope.
$\quad t$ = time in service.

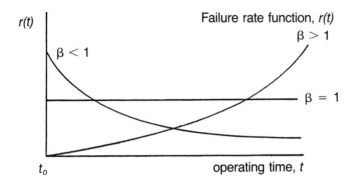

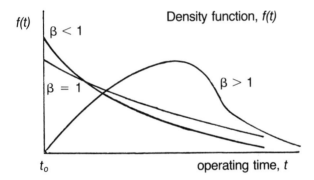

Failure Rate, r(t)

Consider a test where a large number of identical components are put into operation and the time to failure of each component is noted. An estimate of the failure rate of a component at any point in time may be thought of as the ratio of a number of items which fail in an interval of time, say, a week to the number of the original population which were operational at the start of the interval. Thus the failure rate of a component at time t is the probability that the component will fail in the next interval of time given that it is good at the start of the interval, i.e., it is a conditional probability.[3] A noteworthy relationship is

$$r(t) = \frac{1}{\text{MTBF}}$$

Other terms for failure rate are hazard rate, force of mortality, and failure intensity.

Density Function or Frequency Function, f(t)

The density function, similar to a histogram, shows the percentage of failing components during a time interval relative to the total population operating at time t_o.

Figure 7-1. Weibull functions.

Where β is equal to 1.0, the analysis describes random failures such as experienced in ball bearings and many other mass-produced machinery components before the effects of wear-out set in. Those failures occur with constant probability during component operating time.

A very useful example in this context is the failure experience of mechanical shaft seals. A classic study[4] shows how to use the Weibull function to describe the nature of seal failures in a petrochemical plant. The Weibull indices (β) published in the study suggested that seal failures were largely a mixture of "infant mortality," i.e., seal defects, fitting errors, and random failures caused by the inability to maintain design assumptions (dry running and excessive vibration).

Other examples are the failure experience on diesel engine control units shown in Figure 7-2 and diesel engine pistons, Figure 7-3.[5]

Another important contribution to the proper understanding of random failure experience of machinery components was made by Bergling.[6] He applied a β of 1.1 for rolling bearings based on numerous laboratory tests and demonstrated that, if a bearing is removed and inspected after operation and no signs of fatigue as manifested by

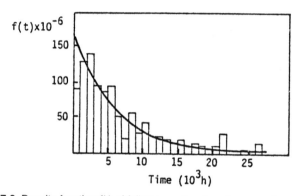

Figure 7-2. Density function *f(t)* of failures on diesel engine control units (Ref. 5).

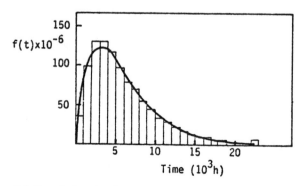

Figure 7-3. Density function *f(t)* of failures on diesel engine pistons (Ref. 5).

spalling, are detected, then this bearing is practically just as serviceable from a fatigue point of view as a brand new bearing. Figure 7-4 shows probable remaining life for used, undamaged rolling bearings. We see from this that if $L_{10\ 128}b$ 28 years, the bearing after 40 years of operation can be expected to function for a further period corresponding to approximately 83% of the basic rating life. There is no reason to discard a bearing because of age.

The condition $\beta > 1.0$ describes machinery component wear-out modes which occur with increasing probability over operating time. Carbon brushes on electric motors are one example. Most wear- and aging-related failure modes fall into this category.

Since our discussions have mainly been at the machinery component level, we should at this point ask how life performance of machinery assemblies, units, and systems could be evaluated. What life should be expected for our machinery systems? Figure A-1 in Appendix A shows typical life spans for machinery equipment and related systems. Figure 7-5 shows life dispersions for different industrial products compared with those of human beings in industrial countries. This information provides a greater understanding of life concepts.

To conclude our discussion of machinery life concepts, a popular view of failure rate dependence from operating time of machinery and other systems should be considered. If the failure rate is plotted as a function of time, a curve, known as the "bathtub curve" is generated as shown in Figure 7-6.

In this curve three conditions can be distinguished: (1) early failures, (2) random failures, and (3) wear-out failures. One would be hard pressed to construct an actual curve like this for a given machinery system. However, it has been used to give an

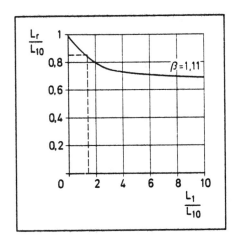

L_{10} = basic rating life
L_1 = attained life at a certain period of time
L_r = remaining life at a certain period of time
β = dispersion exponent

Figure 7-4. Probable remaining life for used, undamaged anti-friction bearings (Ref. 6).

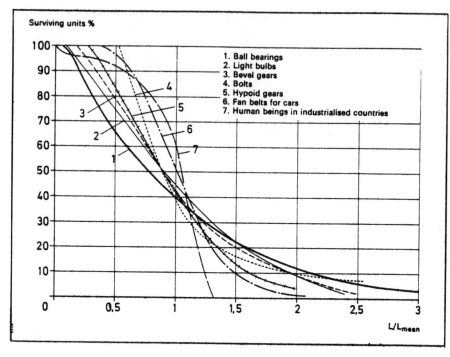

Figure 7-5. Life distribution for various industrial products and for human beings (L = Life) (Ref. 6).

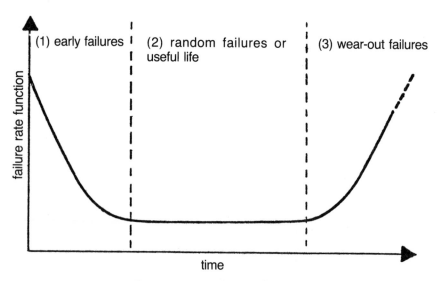

Figure 7-6. Bathtub curve concept.

overall picture of the life cycle of many systems, particularly complex systems. The closest approximation is the life cycle of a petrochemical process unit with compression and pumping machinery as an integral part of it.

Condition 1 describes the early time period of the system by showing a decreasing failure rate over time. It is usually assumed that this period of "infant mortality" is caused by the existence of material and manufacturing flaws together with assembly errors. Condition 2, the area of constant failure rate, is the region of normal running. Here, failure occurs as a consequence of statistically independent factors. This period is also termed "useful life," during which time only random failures will occur. Condition 3 is characterized by an increase of failure rate with time. As described previously, here failures may be due to aging effects. Also, for a good percentage of lubrication-related failures, the failure rate increases with time period.[5]

The broad applicability of the bathtub concept at the system level has unfortunately been postulated as an ideal-failure or life-cycle model all the way down to individual component levels. We agree with J. R. King[7] that the too easy representation of failure rate behavior has often been a deterrent to effective product improvement. Quick debugging procedures to remove infant mortality have been overemphasized, and the random *times to failure* in the useful life period have erroneously been understood to mean that the *causes of failure* are random. Real progress in process machinery reliability can only be made by diligently isolating individual failure modes and their causes and then finding ways to eliminate or minimize them permanently.

Machinery Maintenance Strategies

Different failure modes and their behavior in time are met with different maintenance strategies. Machinery maintenance strategies are:

1. Preventive maintenance
2. Predictive maintenance
3. Breakdown maintenance
4. Bad actor management

Preventive Maintenance

All over the world the term "preventive maintenance" means "periodic" maintenance.[1] Preventive maintenance in this context should therefore be understood as a periodic or scheduled activity which has as its objective the direct prevention of failure modes or defects. In its simplest form this activity entails periodic lubrication, coating, impregnating, or cleaning of machinery components to increase life spans.

Some failure modes respond very well to periodic preventive maintenance, whereas others do not. Moreover, quite often deterioration cannot be completely eliminated even in those failure modes that do respond to periodic preventive maintenance. For instance, by periodically renewing the lube oil in a pump bearing (if one has not decided to use superior oil mist lubrication) one will effectively reduce wear (FM) of the bearings by postponing or even preventing the failure modes "age" and "contamination" of the lube oil itself. Another example is the periodic cleaning of large pipe-ventilated motor drivers of process gas compressor trains where excessive winding dirt contamination (FM) and, more important, insulation failures (FM) can be effectively prevented by this maintenance activity. We presented an example related to this maintenance action in Chapter 4.

Predictive Maintenance

The other maintenance activity is periodic inspection, followed by replacement or overhaul if incipient defects are detected. These actions do not directly reduce the deterioration rate but indirectly control the consequences of accidents, breakdowns, malfunctions, and general trouble. We refer to this maintenance activity as *predictive maintenance*.

While preventive maintenance should look after the predominantly time-dependent failure modes, predictive maintenance addresses the randomly and suddenly occurring failure modes as far as possible by searching for them and by effecting timely repairs. Through the use of this maintenance technique, we expect not to prevent the failure mode, but to influence the consequences of the unexpectedly occurring defect. An appropriate example would be the inspection and change of a major compressor face-type oil seal where random heat checking (FM) has been observed over the years.

Ideally, predictive maintenance strategy should dictate a continuous search for defects, i.e., continuous monitoring of machinery condition and performance coupled with continuous feedback. Usually, however, depending upon a chemical plant's management philosophy and cost-benefit considerations based on failure severity and risk evaluations, the monitoring task will be shared to a varying degree among vigilant operators or maintenance people and sophisticated onstream monitoring equipment. In reality, most predictive maintenance activities are somewhat periodic and involve the observer's senses of seeing, hearing, feeling, and smell.

Breakdown Maintenance

Many people have the tendency to be overly optimistic about the possibilities of machinery preventive maintenance. We believe that there are several limitations to machinery preventive maintenance (PM) concepts.

First, the limitations set by random failure events. Random machinery failure events, according to our definition, could occur with equal probability in time. They do not appear periodically and we conclude from this that PM, or periodic maintenance, will be ineffective in that case. Continuous monitoring is the only resort.

Second, the life dispersion of machinery components. Even time-dependent failures are not all that predictable. They do not appear after absolutely equal operating intervals, but after very dissimiliar time periods. Consider the life dispersion of mechanical gear couplings on process compressors. Obviously, they are components subject to wear. If we conclude that their MTBF (mean-time between failure), or mean-time-between-reaching-of-detect-limit is 7.5 years, it is possible to have an early failure after 3 years and another after 15 years. The assumptions are that there are no undue extraneous influences, such as excessive misalignment, lubricant loss, etc. The longest life span in this case will be five times as long as the shortest life span. Figure 7-7 describes this relationship.

According to H. Grothus,[8] most time-dependent failures show even larger life dispersions. Accordingly, we must consider ratios of maximum to minimum of 4:1 and even 40:1. Appendix A, Figure A-1 provides additional insights into this phenomenon. As a rule of thumb, it can be said that relative life dispersion increases with the absolute value of MTBF. This means that wear items with relatively short life expectancy such as rider rings on reciprocating compressors, will have a comparatively smaller dispersion than components such as gear tooth flanks, which can be expected to remain serviceable for long periods of time.

The third limitation of predictive maintenance in process machinery is that in order to inspect we have to shut down and open up. Who is not familiar with the risk of this procedure? The Canadian Navy uses the phrase "Leave well enough alone." There have been numerous studies that show how machinery troubles started to mount right after predictive maintenance inspections had made a disassembly of the equipment necessary. Figure 7-8 shows how trouble incident frequencies can increase contrary to expectations after inspection and overhaul.

We conclude from this that a certain amount of breakdown maintenance of our machinery will always be necessary, even if we succeeded to strike that fine balance

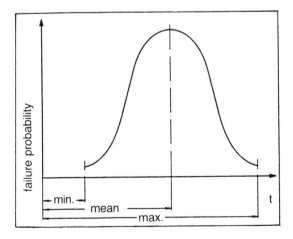

Figure 7-7. Failure probability (density) and minimum–maximum life of machinery components (Ref. 8).

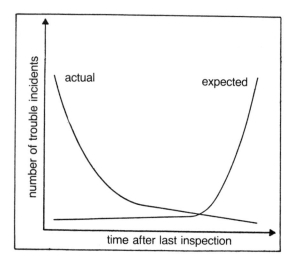

Figure 7-8. Number of trouble incidents per specified operating hours as a function of time after last inspection and overhaul (Ref. 8).

between too much predictive/preventive maintenance and too little. There will be failure modes that will not respond to periodic servicing, such as lube-oil changes, nor can they be detected by some form of inspection or monitoring. For instance, we have hardly any possibility of predicting a motor winding failure, the insulation breakdown of a feeder cable for a large compressor motor driver, or pump shaft fracture caused by fatigue.

Breakdown maintenance is here to stay! It is the maintenance activity necessary to restore machinery equipment back to service after failure modes developed that were:

1. Bad actor identification and tracking.
2. Failure analysis and documentation.
3. Follow-up.
4. Organizational aspects.

Bad Actor Management

Bad actor management addresses that highly variable portion of our process machinery population that wants to behave as though it did not belong.

Bad actors among our process pump population, for instance, sometimes prove Pareto's law, i.e., 20% of all pumps cause 80% of the trouble. All or some of the causes previously identified in connection with machinery component failure modes can give rise to bad actor behavior. Some of the predominant causes are:

1. Not suited for service due to:
   a. Wrong design assumptions.
   b. Incorrect material.
   c. Change in operating conditions.
2. Maintenance:
   a. Improper repair.
   b. Design and/or selection errors.

Bad actor management is preoccupied with the following necessary problem-solving steps[9]:

1. Bad actor identification and tracking.
2. Failure analysis and documentation.
3. Follow-up.
4. Organizational aspects.

Since a large portion of this book concerns itself with steps 2, 3, and 4, we would like to limit the discussion at this point to step 1, namely, bad actor identification and tracking.

Bad Actor Identification and Tracking

As in reliability work, a prerequisite for bad actor identification is the knowledge of machinery MTBF by type, unit, and service. The next step is to decide what time to failure (TTF) is acceptable to us vis-à-vis a machinery break-down, keeping in mind that our resources will be limited to allow us to investigate *all* items that deviate from an acceptable or expected TTF. So let us make sure we do not set our sights too high at the beginning. Even though this sounds complicated, do we not all know our bad actors? It is necessary to institute this kind of decision step covering *all* failures. Later in our program we will not be able to tell our bad actors from acceptable ones quite that easily. Figure 7-9 shows the methodology of this process.

The third step is to proceed to the component level because "the devil is hiding in the detail."* At this point, Table 7-2 will be a good aid in decision making. It shows machinery component failure modes commonly encountered in machinery failure analysis together with suggested standard life values, Weibull indices (β), and responsiveness to preventable or predictable maintenance strategies. Finally, Figure 7-10 is a tracking sheet that will allow the analyst to again document bad actor failure modes on the component level.

In conclusion, bad actor identification and tracking is the first step in the painstaking effort to eliminate recurring and costly machinery failures. Later we will deal with all other necessary steps that will allow us to reach our objective.

In the two following sections we are going to take a look at some useful statistical tools that will help us to define failure experience trends in a given machinery population. *(text continued on page 532)*

---

*Old German proverb.

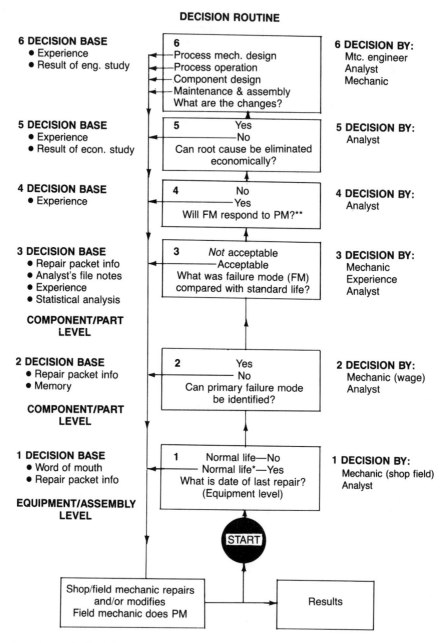

DECISION ROUTINE

6 DECISION BASE
- Experience
- Result of eng. study

6
—Process mech. design
—Process operation
—Component design
—Maintenance & assembly
What are the changes?

6 DECISION BY:
Mtc. engineer
Analyst
Mechanic

5 DECISION BASE
- Experience
- Result of econ. study

5 Yes
—No
Can root cause be eliminated
economically?

5 DECISION BY:
Analyst

4 DECISION BASE
- Experience

4 No
—Yes
Will FM respond to PM?**

4 DECISION BY:
Analyst

3 DECISION BASE
- Repair packet info
- Analyst's file notes
- Experience
- Statistical analysis

**COMPONENT/PART
LEVEL**

3 *Not* acceptable
—Acceptable
What was failure mode (FM)
compared with standard life?

3 DECISION BY:
Mechanic
Experience
Analyst

2 DECISION BASE
- Repair packet info
- Memory

**COMPONENT/PART
LEVEL**

2 Yes
No
Can primary failure mode
be identified?

2 DECISION BY:
Mechanic (wage)
Analyst

1 DECISION BASE
- Word of mouth
- Repair packet info

**EQUIPMENT/ASSEMBLY
LEVEL**

1 Normal life—No
Normal life*—Yes
What is date of last repair?
(Equipment level)

1 DECISION BY:
Mechanic (shop field)
Analyst

START

Shop/field mechanic repairs
and/or modifies
Field mechanic does PM

Results

** PM = Periodically inspect, clean, lubricate & repair
* Normal life = Expected life for individual machine

Figure 7-9. Machinery failure analysis methodology.

Table 7-2
Primary Machinery Component Failures

Failure Mode	Weibull Index*	Standard Life*	Maintenance Strategy	
			Preventive	Predictive
Deformation				
Brinelling	1.0	Inf		●
Cold flow	1.0	Inf		●
Contracting	2.0	Inf	●	
Creeping	2.0	Inf		
Bending	1.0	Inf		●
Bowing	1.0	Inf		
Buckling	1.0	Inf		
Bulging	1.0	Inf		
Deformation	1.0	Inf		
Expanding	1.0	Inf		
Extruding	1.0	Inf		
Growth	1.0	Inf		
Necking	1.0	Inf		
Setting	2.0	Inf		
Shrinking	2.0	Inf		
Swelling	3.0	Inf		
Warping	1.0	Inf		
Yielding	1.0	Inf		
Examples:				
Deformation of springs	1.0	Inf		●
Extruding of elastomeric seals	1.0	4.0Y		●
Force-induced deformation	1.0	Inf		
Temperature-induced deformation	2.0	Inf		●
Yielding	1.0	Inf		
Fracture/Separation				
Blistering	1.0	Inf		
Brittle fracture	1.0	Inf		
Checking	1.0	Inf		
Chipping	1.0	Inf		
Cracking	1.0	Inf		
Caustic cracking	1.0	Inf		
Ductile rupture	1.0	Inf		
Fatigue fracture	1.0	Inf		
Flaking	1.0	Inf		
Fretting fatigue cracking	1.0	Inf		
Heat checking	1.0	Inf		
Pitting	1.0	Inf		
Spalling	1.0	Inf		
Splitting	1.0	Inf		

*From References 3-5, 8 and author's estimates.

(table continued on next page)

Table 7-2 (cont.)

Failure Mode	Weibull Index*	Standard Life*	Maintenance Strategy	
			Preventive	Predictive
Examples:				
Overload fracture	1.0	Inf		
Impact fracture	1.0	Inf		
Fatigue fracture	1.1	Inf		●
Most fractures	1.0	Inf		
Change of Material Quality				
Aging	3.0	5.0Y	●	
Burning	1.0	Inf		
Degradation	2.0	3.0Y	●	
Deterioration	1.0	Inf		
Discoloration	1.0	Inf		
Disintegration	1.0	Inf		
Embrittlement	1.0	Inf		
Hardening	1.0	Inf		
Odor	1.0	Inf		
Overheating	1.0	Inf		
Softening	1.0	Inf		
Examples:				
Degradation of mineral oil-based lubricant	3.0	1.5Y	●	●
Degradation of coolants	3.0	1.0Y	●	●
Elastomer aging	1.0	4.0-16Y		●
'O'-Ring deterioration	1.0	2.0-5Y		●
Aging of metals under thermal stress	3.0	4.0Y		●
Corrosion				
Exfoliation	3.0	2.0–4.0Y	●	
Fretting Corrosion	2.0	3.0Y	●	
General Corrosion	2.0	1.0–3.0Y	●	
Intergranular Corrosion	2.0	1.0–3.0Y	●	
Pitting Corrosion	2.0	1.0–3.0Y	●	
Rusting	2.0	0.5–3.0Y	●	
Staining	2.0	0.5–3.0Y	●	
Examples:				
Accessible Components	2.0	2.0-4.0Y	●	●
Inaccessible Comp.	2.0	2.0-4.0Y		
Wear				
Abrasion	3.0	0.5–3.0Y		●
Cavitation	3.0	0.5–3.0Y		●
Corrosive Wear	3.0	0.5–3.0Y		●
Cutting	3.0	0.5–3.0Y		●
Embedding	3.0	0.5–3.0Y		●

*From References 3-5, 8 and author's estimates.

Table 7-2 (cont.)

Failure Mode	Weibull Index*	Standard Life*	Maintenance Strategy	
			Preventive	Predictive
Erosion	3.0	3.0Y		●
Fretting	3.0	2.0Y		●
Galling	3.0	2.0Y		●
Grooving	3.0	2.0Y		●
Gouging	3.0	2.0Y		●
Pitting	3.0	1.0Y		●
Ploughing	3.0	1.0Y		●
Rubbing	3.0	3.0Y		●
Scoring	3.0	3.0Y		●
Scraping	3.0	0.5–3.0Y		●
Scratching	3.0	3.0Y		●
Scuffing	3.0	1.0Y		●
Smearing	3.0	1.0Y		●
Spalling	3.0	0.5–16Y		●
Welding	3.0	0.5–3.0Y		●
Examples:				
Non-lubed relative movement	3	1.0Y		●
Contaminated by	3	3.0M		●
lubed sleeve bearings	3	4.0-16Y	●	●
Spalling of antifriction bearings	1.1	16.0Y		●
Displacement/seizing/adhesion				
Adhesion	1.0	Inf		
Clinging	1.0	Inf		
Binding	1.0	Inf		
Blocking	1.0	Inf		
Cocking	1.0	Inf		
Displacement	1.0	Inf		
Freezing	1.0	Inf		
Jamming	1.0	Inf		
Locking	1.0	Inf		
Loosening	1.0	Inf		
Misalignment	1.0	Inf		
Seizing	1.0	Inf		
Setting	1.0	Inf		
Sticking	1.0	Inf		
Shifting	1.0	Inf		
Turning	1.0	Inf		
Examples:				
Loosening (locking fasteners)	1	Inf		
Loosening (bolts)	1	Inf		●

*From References 3-5, 8 and author's estimates.

(table continued on next page)

Table 7-2 (cont.)

Failure Mode	Weibull Index*	Standard Life*	Maintenance Strategy	
			Preventive	Predictive
Loosening (nuts)	1	Inf		●
Misalignment (process pump set)	2	1.5-3.0Y		●
Seizing (linkages)	1	Inf	●	●
Seizing (components subject to contamination or corrosion)	1	Inf		●
Shifting (unstable design)	1	Inf		
Leakage				
Joints with relative movement	1.5	3.0M-4.0Y		
Joints without relative movement	1	16.0Y		●
Mechanical seal faces	0.7-1.1	0.5-1.5Y		
Contamination				
Clogging	1.0	Inf		
Coking	2.0	0.5–3.0Y		●
Dirt accumulation	2.0	0.5M–3.0Y		●
Fouling	1.0	Inf		●
Plugging	1.0	Inf		●
Examples				
Fouling gas compressor	3	1.5-5.0Y	●	●
Plugging of passages with moving medium	1	Inf		●
Plugging of passages with non-moving medium	1	Inf		
Conductor Interruption				
Flexible cable	1	Inf		
Solid cable	1	Inf		
Burning through insulation				
Motor windings	1	16Y	●	
Transformer windings	1	16Y	●	

Legend: Inf = Infinite
       M = Month(s)
       Y = Year(s)
       ● = Yes. Absence of (●) indicates that break-down maintenance is appropriate strategy.

*From References 3-5, 8 and author's estimates.

BAD ACTOR TRACKING	FAILURE ANALYSIS			NO: 023
DESCRIPTION: P-101 , Hightower Feedpump				
INFORMATION	FAILURE	FAILURE	FAILURE	FAILURE
DATE	80/11/05	81/01/15	81/05/10	
COMPONENT / PART	SEAL	SEAL	SEAL	
VENDOR DRWG.NO.				
ELEMENT/PART	Secondary sealing O-ring Part # 13	Same Part # 13	(3)Springs Part # 5	
FAILURE MODE	Swelling	Swelling	Breakage	
TTF *	?	1.5M	5.0M	
CORRECTIVE ACTION	Replaced	Changed to Viton	Corrected internal alignment	
BAD ACTOR PART (Check)	?	✓		
FAILURE CAUSE	Changed pump service	Changed pump service	Internal Mis-alignemnt due to flange loading	
FAILURE RESISTANCE	O-ring material	O-ring material	Pump casting strength	
ANALYSED BY: J.Pickel		DATE: 81/05/30		

REVISON NO.	REMARKS	DATE
1		
2		
3		
4		

* Time-To-Failure

Figure 7-10. Bad actor tracking.

Hazard Plotting for Incomplete Failure Data*

Plotting and analysis of data must take into account the form of the data. Failure data can be complete or incomplete. If failure data contain the failure times of all units in the sample, the data are complete. If failure data consist of failure times of failed units and running times of unfailed units, the data are incomplete, are called "censored," and the running times are called "censoring times." If the unfailed units all have the same censoring time (which is greater than the failure times), the data are singly censored. If unfailed units have different censoring times, the data are multiply censored.[10]

Complete data result when all units have failed. Singly-censored data result in life testing when testing is terminated before all units fail. Multiply-censored data result from:

1. Removal of units from use before failure.
2. Loss or failure of units due to extraneous causes.
3. Collection of data while units are still operating.

The following are examples of situations with these reasons for multiply-censored data:

1. In some life-testing work, units are put on test at various times but are all taken off together; this is done to terminate testing by some chosen time to meet a schedule.
2. In medical followup studies on the length of survival of patients after treatment, contact with some patients may be lost or some may die from causes unrelated to the disease treated. Similarly, in engineering studies, units may be lost or fail for some reason unrelated to the failure mode of interest; for example, a unit on test may be removed from a test or destroyed accidentally.
3. In many engineering studies, field data are used which consist of failure times for failed units and differing current running times for unfailed units. Such data arise because units are put into service at different times and receive different amounts of use.

Probability plotting is commonly used for graphical analysis of failure data. Knowledge of probability plotting is not essential to understanding the hazard plotting method presented here. Probability plotting can easily be used for complete data and for singly-censored data but is difficult to use for multiply-censored data. One simple expedient that is used to plot multiply-censored data is to ignore all failures that occurred after the earliest censoring time and plot only those that occurred before the first censoring time. This procedure may be acceptable, if there are relatively few failures after the first censoring time and thus little information is lost by excluding them from the plot. However, if there are relatively many failures after the earliest censoring time, considerable information is lost. Moreover, if the earliest censoring time is in the lower tail of the distribution and only the failure times

---

*Adapted by permission of Dr. Wayne Nelson, G. E. Research and Development, and of the American Society for Quality Control.

before that time are used, the distribution function must be estimated by considerable extrapolation beyond the earliest censoring time. Then the estimate of the distribution function may be considerably in error; that is, a theoretical distribution that agrees with the data below the earliest censoring time may differ considerably from the true distribution function above the earliest censoring time. This is a risk of extrapolating beyond the data that are plotted. For these reasons, it is preferable to plot all failure times occurring both before and after the earliest censoring time to make full use of the information in the data.

The method presented here of plotting data on hazard paper achieves full use of all failures and gives the same information but with less effort. Hazard papers are shown here for the exponential, Weibull, normal, log normal, and extreme value distributions.*

In the foregoing, failure is described in terms of time to failure. In various situations other measures of the amount of exposure before failure may be relevant, for example, cycles to failure, miles to failure, etc. Also, failure is used here in a general sense to denote any specific deterioration in performance of a unit. If units are put into service at different times, the time in use for each unit is figured from its starting time.

The following sections contain a detailed presentation of the hazard plotting method. The next section contains step-by-step instructions on how to make a hazard plot of multiply-censored data. The instructions are illustrated with an example based on actual failure data. The section after that shows how to interpret and use a hazard plot to obtain engineering information on the distribution of time to failure. This section also contains plots and analyses of simulated data from different known theoretical distributions; this permits us to assess the performance of the hazard plotting method. The last section contains additional practical advice on how to use hazard paper. Appendix B contains a presentation of theory underlying hazard plotting and descriptions of the theoretical distributions for which there are hazard papers.

How to Make a Hazard Plot

Step-by-step instructions are given here for making a hazard plot of multiply-censored failure data. Data used to illustrate the hazard-plotting method are shown in Table 7-3. These data for 70 diesel generators consist of hours of use on 12 generators at the time of fan failure, and hours on 58 generators without a failure. The ordered times to failure for the failed fans and the running times for the unfailed fans are shown in Figure 7-11. The engineering problem was to determine if the failure rate of the fans was decreasing with time. Information on the failure rate was to be used to aid in arriving at an engineering and management decision on whether or not to replace the unfailed fans with a better fan to prevent future failures. There was the possibility that the failures had removed the weak fans, and the failure rate on the remaining fans would be tolerably low. Steps in the hazard-plotting method follow with the fan failure data used as an example:

(text continued on page 536)

---

*These hazard plotting papers can be obtained from TEAM, Box 25, Tamworth, NH 03886.

Table 7-3
Generator Fan Failure Data and Hazard Calculations

n(K)	Hours	Hazard	Cumulative Hazard	n(K)	Hours	Hazard	Cumulative Hazard
1 (70)	4,500*	1.43	1.43	36 (35)	43,000		
2 (69)	4,600			37 (34)	46,000*	2.94	18.78
3 (68)	11,500*	1.47	2.90	38 (33)	48,500		
4 (67)	11,500*	1.49	4.39	39 (32)	48,500		
5 (66)	15,600			40 (31)	48,500		
6 (65)	16,000*	1.54	5.03	41 (30)	48,500		
7 (64)	16,600			42 (29)	50,000		
8 (63)	18,500			43 (28)	50,000		
9 (62)	18,500			44 (27)	50,000		
10 (61)	18,500			45 (26)	61,000		
11 (60)	18,500			46 (25)	61,000*	4.00	22.78
12 (59)	18,500			47 (24)	61,000		
13 (58)	20,300			48 (23)	61,000		
14 (57)	20,300			49 (22)	63,000		
15 (56)	20,300			50 (21)	64,500		
16 (55)	20,700*	1.82	7.75	51 (20)	64,500		
17 (54)	20,700*	1.85	9.60	52 (19)	67,000		
18 (53)	20,800*	1.89	11.49	53 (18)	74,500		
19 (52)	22,000			54 (17)	78,000		
20 (51)	30,000			55 (16)	78,000		
21 (50)	30,000			56 (15)	81,000		
22 (49)	30,000			57 (14)	81,000		
23 (48)	30,000			58 (13)	82,000		
24 (47)	31,000*	2.13	13.62	59 (12)	85,000		
25 (46)	32,000			60 (11)	85,000		
26 (45)	34,500*	2.22	15.84	61 (10)	85,000		
27 (44)	37,500			62 (9)	87,500		
28 (43)	37,500			63 (8)	87,500*	12.50	35.28
29 (42)	41,500			64 (7)	87,500		
30 (41)	41,500			65 (6)	94,000		
31 (40)	41,500			66 (5)	99,000		
32 (39)	41,500			67 (4)	101,000		
33 (38)	43,000			68 (3)	101,000		
34 (37)	43,000			69 (2)	101,000		
35 (36)	43,000			70 (1)	115,000		

*Denotes failure.

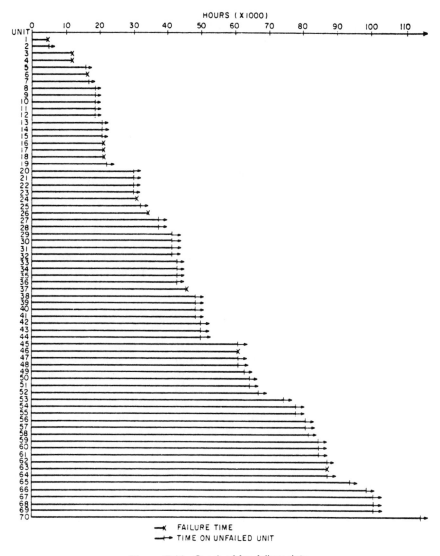

Figure 7-11. Graph of fan failure data.

1. Suppose that the failure data on n units consist of the failure times for the failed units and the running (censoring) times for the unfailed units. Order the n times in the sample from smallest to largest without regard to whether they are censoring or failure times. In the list of ordered times the failures are each marked with an asterisk to distinguish them from the censoring times. This marking is done in Table 7-3 for the generator fan failure data. If some censoring and failure times have the same value, they should be put into the list in a well-mixed order.

2. Obtain the corresponding hazard value for each failure and record it as shown in Table 7-3. The hazard value for a failure time is 100 divided by the number K of units with a failure or censoring time greater than (or equal to) that failure time. The hazard value is the observed conditional probability of failure at a failure time, that is, the percentage $100 (1/K)$ of the K units that ran that length of time and failed then. This observed conditional failure probability for a failure time is readily apparent in Figure 7-11, where the number of units in use that length of time and surviving is easily seen. If the times in the ordered list are numbered backward with the first time labeled n, the second labeled $n - 1, \ldots$ and the nth labeled 1, then the K value is given by the label. For the example, these K values are shown in parentheses in Table 7-3. Three figure accuracy for K is sufficient. Note that the hazard value of a failure is determined by the number of failure and censoring times following it; in this way, Hazard plotting takes into account the censoring times of unfailed units in determining the plotting position of a failure time.

3. For each failure time, calculate the corresponding cumulative hazard value which is the sum of its hazard value and the hazard values of all preceding failure times. This calculation is done recursively by simple addition. For example, for the generator fan failure at 16,000 hours, the cumulative hazard value 5.93 is the hazard value 1.54 plus the cumulative hazard value 4.39 of the preceding failure. The cumulative hazard values are shown in Table 7-3 for the generator fan failures. Cumulative hazard values can be larger than 100%.

4. Choose a theoretical distribution of time to failure and use the corresponding hazard paper to plot the data on. Various hazard papers are shown in the figures. The theoretical distribution should be chosen on the basis of engineering knowledge of the units and their failure modes. On the vertical axis of the hazard paper, mark a time scale that includes the range of the sample failure times. For the generator fan failure data, Weibull hazard paper was chosen, and the vertical scale was marked off from 1000 to 100,000 hours, since the range of failure times for the data is from 4500 to 87,500 hours. This is shown in Figure 7-12.

5. Plot each failure time vertically against its corresponding cumulative hazard value on the horizontal axis of the hazard paper. This was done for the 12 generator fan failures and is shown in Figure 7-12.

6. Check whether the failure data on the hazard plot follow a reasonably straight line. This can be done by holding the paper up to the eye and sighting along the data points. If the plot is reasonably straight, draw a best-fit straight line through the data points. This line is an estimate of the cumulative hazard function of the distribution that the data are from. This was done in Figure 7-12

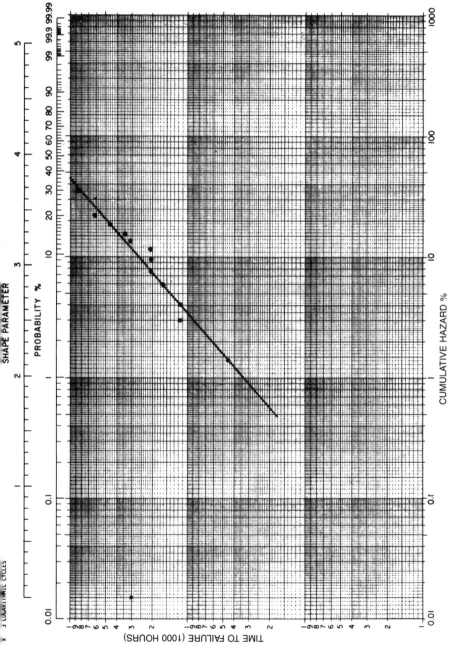

Figure 7-12. Weibull hazard plot of generator fan failure data.

for the generator fan failure data. The straight-line estimate of the cumulative hazard function is used to obtain information on the distribution of time to failure. It provides estimates of the cumulative distribution function and parameters of the distribution of time to failure. If the data do not follow a reasonably straight line, then plot the data on hazard paper for some other theoretical distribution to see if a better fit can be obtained. Curvature of a data plot such as shown in Figure 7-13 indicates poor fit.

It may happen that data plotted on two different hazard papers give a reasonably straight plot on each. For practical purposes, it usually does not matter much which plot is used to interpolate within the range of the data. If extrapolation beyond the range of the data is necessary, the choice of which plot to use must be determined by engineering judgment.

The generator fan failure data were also plotted on exponential, normal, and log normal hazard paper (see Figures 7-13, 7-14, and 7-15). The exponential plot is a reasonably straight line; this indicates that the failure rate is relatively constant over the range of the data. The probability scale on this exponential hazard plot is crossed out, since it is not appropriate to the way the data are plotted; this is explained later. The normal plot is curved concave upward; this indicates that a normal distribution is not appropriate to the data. The log normal plot is a reasonably straight line. The log normal σ parameter of a line fitted to the data is approximately 0.7. For σ values near 0.7, a log normal distribution has a relatively constant failure rate over a large percentage of the distribution, particularly in the lower tail. For prediction purposes, the exponential, Weibull, and log normal plots are all comparable over the range of the data. For prediction to times beyond the data, it is necessary to decide which distribution is best on the basis of engineering knowledge.

The hazard plotting method, like all other methods for analyzing censored failure data, is based on a key assumption that must be satisfied in practice if it is to yield reliable results. It is assumed that if the unfailed units were run to failure, their failure times would be statistically independent of their censoring times. That is, there is no relationship or correlation between the censoring time of a unit and the failure time. For example, suppose that a certain type of unit has two competing modes of failure which can be identified after failure, but there is interest in only one of them. Then failure of a unit by the other mode might be regarded as a censoring of the time to failure for the mode of interest. The two times to failure that a unit potentially has may not be independent, but correlated, for physical reasons. If this is so, the hazard-plotting method fails to give valid results on the distribution of time to failure for the mode of interest.

The theoretical basis for the hazard plotting method and paper is given in Appendix B. Some readers may wish to refer to Appendix B at this point before proceeding to the next section which covers how to use and interpret hazard plots.

How to Use Hazard Plots

The line that is fitted to data on a hazard plot is used to obtain engineering information on the distribution of time to failure. Methods of obtaining such

(text continued on page 542)

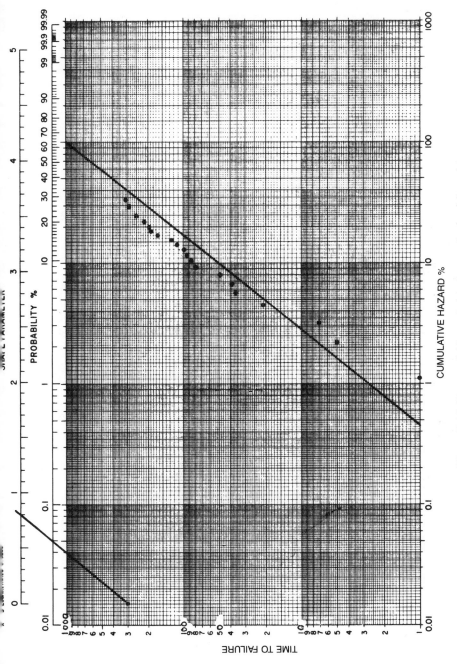

Figure 7-13. Exponential hazard plot of generator fan failure data.

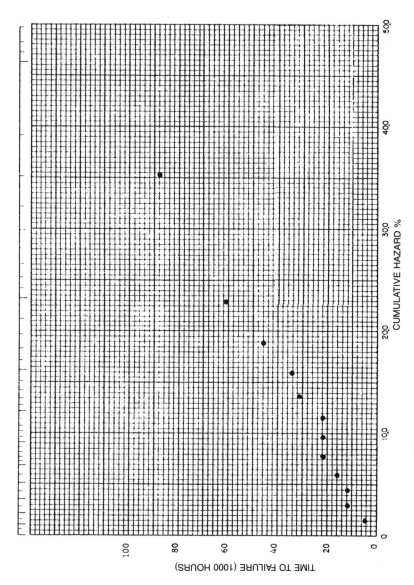

Figure 7-14. Normal hazard plot of generator fan failure data.

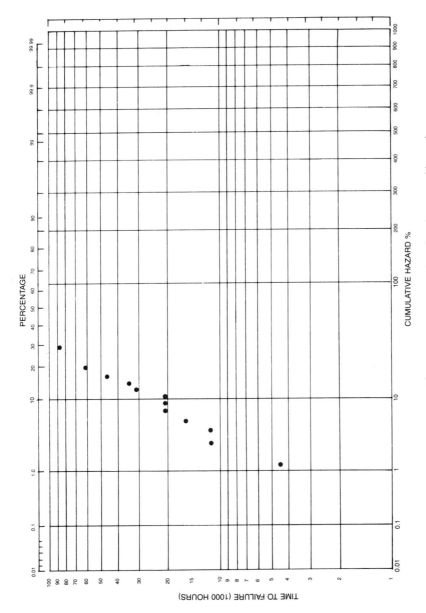

Figure 15. Generator fan failure data plotted on log normal hazard paper.

information are given here and are illustrated by the fan failure data and simulated data. The methods provide estimates of distribution parameters, percentiles, and probabilities of failure. The methods that give estimates of distribution parameters differ slightly from the hazard paper of one theoretical distribution to another and are given separately for each distribution. The methods that give estimates of distribution percentiles and failure probabilities are the same for all papers and are given first.

Estimates of Distribution Percentiles and Probabilities of Failure

It is shown in Appendix B that for any distribution the cumulative hazard function and the cumulative distribution function are connected by a simple relationship. The probability scale for the cumulative distribution function appears on the horizontal axis at the top of hazard paper and is determined from that relationship. Thus, the line fitted to data on hazard paper provides an estimate of the cumulative distribution function. The fitted line and the probability scale are used to obtain estimates of distribution percentiles and probabilities of failure. This procedure is explained next.

Suppose, for example, that an estimate based on a Weibull fit to the fan data is desired of the fifth percentile of the distribution of time to fan failure. Enter the Weibull plot, Figure 7-12, on the probability scale at the chosen percentage point, 5%. Go vertically down to the fitted line and then horizontally to the time scale where the estimate of the percentile is read and is 14,000 hours.

An estimate of the probability of failure before some chosen specific time is obtained by the following. Suppose that an estimate is desired of the probability of fan failure before 100,000 hours, based on a Weibull fit to the fan data. Enter the Weibull plot on the vertical time scale at the chosen time, 100,000 hours. Go horizontally to the fitted line and then up to the probability scale where the estimate of the probability of failure is read and is 38%. In other words, an estimated 38% of the fans will fail before they run for 100,000 hours.

Given next is an estimate of the conditional probability that an unfailed unit of a given age will fail by a certain time. Such estimates for units operating in the field are useful for planning. An example of this is in the next paragraph. Suppose that an estimate based on the Weibull plot is desired for the probability that the unfailed fan with 32,000 hours will fail by 40,000 hours. Enter the Weibull plot on the time scale at 32,000 hours. Go horizontally to the fitted line and then down to the cumulative hazard scale where the value is read as a percentage and is 13%. Similiarly, obtain the corresponding cumulative hazard value for 40,000 hours, which is 16.8%. Take the difference between the two values, that is, $16.8 - 13 = 3.8\%$. Enter the plot on the cumulative hazard scale at the value of the difference, and go directly up to the probability scale to read the estimate of the conditional probability of failure, which is 3.7%. For small percentages, cumulative probabilities and corresponding hazard values are approximately equal; this can be seen on the hazard papers. Thus, one could use the difference, if it is small, of the two cumulative hazard values as an approximation to the conditional probability of failure; this is 3.8% here.

Conditional probabilities of failure can be used to predict the number of unfailed units that will fail within a specified period of additional running time on each of the units. For each unit, the estimate of the conditional probability of failure within a specified period of time (8000 hours here) must be calculated. If there is a large number of units and the conditional probabilities are small, then the number of failures in that period will be approximately Poisson distributed* with mean equal to the sum of the conditional probabilities, which must be expressed as decimals rather than percentages. The Poisson distribution allows us to make probability statements about the number of failures that will occur within a given period of time.

It should be emphasized that the estimation methods presented previously apply to any hazard paper and, in addition, to a nonparametric fit to data obtained by drawing a smooth curve through data on any hazard paper.

Given next are the different methods for estimating distribution parameters on exponential, Weibull, normal, log normal, and extreme-value hazard papers. Descriptions of these distributions and their parameters are given in Appendix B. The methods are explained with the aid of simulated data from known distributions. Thus, we can judge from the hazard plots how well the hazard-plotting method does.

Exponential Parameter Estimate

Simulated data are shown in Table 7-4 and are a sample from an exponential distribution with a mean 1000 hours. The data are failure censored, which is a special case of multiply-censored data. That is, the data simulate a situation where 50 units were put on test at the same time and, according to a prespecified plan, five unfailed units were removed from test when the fifth failure occurred, five more unfailed units were removed when the tenth failure occurred, . . . and the last five unfailed units were removed when the twenty-fifth failure occurred. Unfailed units would be removed for inspection for deterioration and to hasten the life test. A plot of these data on exponential hazard paper is shown in Figure 7-16. In Figure 7-16 the theoretical cumulative hazard function for the distribution is the straight line through the origin. On exponential hazard paper only, the time scale must start with time zero, and the straight line fitted to the data must pass through the origin.

For the purpose of showing how to obtain from an exponential hazard plot an estimate of the exponential mean time to failure, assume that the straight line on Figure 7-16 is the one fitted to the data. Enter the plot at the 100% point on the horizontal cumulative hazard scale at the bottom of the paper. Go up to the fitted line and then across horizontally to the vertical time scale where the estimate of the mean time to failure is read and is 1000 hours. The corresponding estimate of the failure rate is the reciprocal of the mean time to failure and is $1/100 = 0.001$ failures per hour.

---

*The Poisson distribution is a special form of the normal distribution as described in Appendix B, Figure B-2.

Table 7-4
Simulated Exponential Data and Hazard Calculations

n(K)	Hours	Hazard	Cumulative Hazard	n(K)	Hours	Hazard	Cumulative Hazard
1 (50)	3*	2.00	2.00	26 (25)	385		
2 (49)	52*	2.04	4.04	27 (24)	385		
3 (48)	58*	2.08	6.12	28 (23)	385		
4 (47)	71*	2.13	8.25	29 (22)	385		
5 (46)	77*	2.17	10.42	30 (21)	385		
6 (45)	77			31 (20)	391*	5.00	46.49
7 (44)	77			32 (19)	410*	5.26	51.75
8 (43)	77			33 (18)	562*	5.56	57.31
9 (42)	77			34 (17)	611*	5.88	63.19
10 (41)	77			35 (16)	621*	6.25	69.44
11 (40)	101*	2.50	12.92	36 (15)	621		
12 (39)	130*	2.56	15.48	37 (14)	621		
13 (38)	161*	2.63	18.11	38 (13)	621		
14 (37)	180*	2.70	20.81	39 (12)	621		
15 (36)	235*	2.78	23.59	40 (11)	621		
16 (35)	235			41 (10)	662*	10.00	79.44
17 (34)	235			42 (9)	884*	11.11	90.55
18 (33)	235			43 (8)	1101*	12.50	103.05
19 (32)	235			44 (7)	1110*	14.29	117.34
20 (31)	235			45 (6)	1232*	16.67	134.01
21 (30)	301*	3.33	26.92	46 (5)	1232		
22 (29)	306*	3.45	30.37	47 (4)	1232		
23 (28)	319*	3.57	33.94	48 (3)	1232		
24 (27)	334*	3.70	37.64	49 (2)	1232		
25 (26)	385*	3.85	41.49	50 (1)	1232		

*Denotes failure.

Weibull Parameter Estimates

Table 7-5 contains a randomly censored sample of 90 times from a Weibull distribution with shape parameter equal to 0.8 and scale parameter equal to 1000 hours. The data are randomly censored with independent censoring times from a Weibull distribution with shape parameter equal to 2.0 and scale parameter equal to 300 hours. The data were obtained by generating an observation from the failure distribution and an independent observation from the censoring distribution. If the censoring time was smaller than the failure time, the censoring time was used and vice versa. Such data arise if there are independent, competing modes of failure for a unit and the mode of failure can be determined on a failed unit. The 20 failure times are plotted in Figure 7-17 on Weibull hazard paper. The data follow the cumulative hazard function of the distribution of time to failure but do not fall on a particularly straight line.

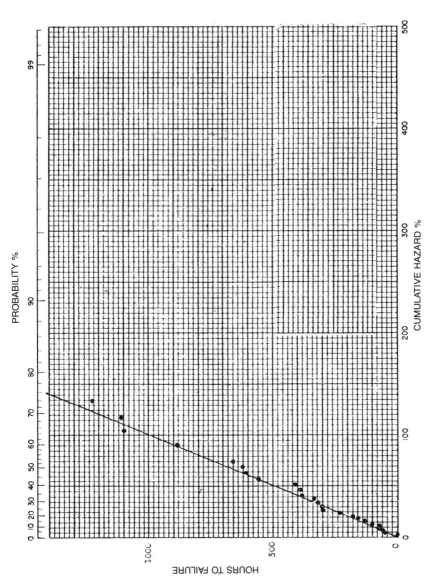

Figure 7-16. Exponential hazard plot of simulated exponential data.

Table 7-5
Randomly Censored Simulated Weibull Data

Number	Time	Cumulative Hazard	Number	Time	Cumulative Hazard	Number	Time	Cumulative Hazard
1	1*	1.11	31	199*	20.26	61	264	
2	5*	2.23	32	199		62	276	
3	7*	3.37	33	200		63	279	
4	21*	4.52	34	201		64	279	
5	37*	5.68	35	205		65	281	
6	38*	6.86	36	206		66	283	
7	49*	8.05	37	207		67	283	
8	78*	9.25	38	212		68	293*	29.59
9	85*	10.47	39	213		69	298	
10	91*	11.70	40	214		70	301*	34.35
11	96		41	217		71	301	
12	98*	12.97	42	217*	22.30	72	309	
13	111*	14.25	43	217		73	315	
14	125*	15.55	44	218		74	318	
15	141		45	220		75	322	
16	145		46	222		76	329	
17	156		47	223		77	330	
18	158		48	224		78	343	
19	160		49	226		79	344	
20	162		50	231		80	345	
21	163		51	232		81	349	
22	165*	17.00	52	239		82	354	
23	168		53	244		83	355	
24	176		54	244		84	363	
25	181		55	245		84	363	
26	182		56	251		86	370	
27	187		57	254*	25.24	87	380	
28	188*	18.59	58	256		88	385	
29	191		59	261		89	407	
30	198		60	262		90	412	

*Denotes failure time.

For the purpose of showing how to obtain from a Weibull hazard plot estimates of distribution parameters, assume that the straight line on Figure 7-17 was fitted to the data points. The estimate of the Weibull scale parameter is given first. Enter the plot at the 100% hazard point on the horizontal cumulative hazard scale on the bottom of the paper. Go up to the fitted line and then horizontally across to the vertical time scale where the scale parameter estimate is read and is 1000 hours. The estimate of the shape parameter is given next. Draw a line that is parallel to the fitted line and passes through the heavy dot in the upper left-hand corner of the grid. This line is shown in Figure 7-17. At the point where the line intersects the shape parameter scale at the top of the paper, read the estimate of the shape parameter, which is 0.80.

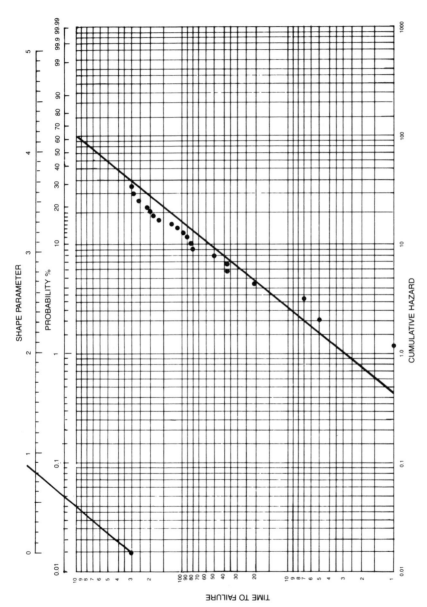

Figure 7-17. Weibull hazard plot of simulated Weibull data.

Similarly, we could investigate the shape parameter of Figure 7-12, the generator fan failures. However, we leave this up to the reader: were those failures the "infant mortality" kind, the random (antifriction bearings?) failure experience, or was it wear-out mode? Also, it would seem logical—after finding an answer to those questions—to start our investigation on the part/component level. Hazard plotting on this level might possibly lead us to new insights.

Some Practical Aspects of Data Plotting

Several aspects of data plotting that a user of hazard and probability plotting should be aware of are discussed here. The discussions cover potential difficulties that may be encountered in using plotting papers. Also, tips are given on how to use plotting papers effectively. Most of the discussions apply both to hazard and probability papers in the sense that good plotting techniques for probability plotting are also appropriate to hazard plotting.

If it is not known through engineering experience which hazard paper is appropriate to use for a set of multiply censored data, it may be useful then to try different papers to determine which does the best job. This may be done most efficiently by first plotting the sample cumulative hazard function on exponential hazard paper, which is just square grid. Then compare the plot with the theoretical cumulative hazard functions in Figures B-1, B-2, B-3, and B-4 of Appendix B and choose the distribution with the cumulative hazard function that agrees most closely with the shape of the sample cumulative hazard function over the range of cumulative hazard values in the sample. Plot the data on the corresponding hazard paper. The cumulative hazard scale on the horizontal axis of exponential hazard paper shown runs from 0% to 500%, which may be too big a range, particularly if the data consist of some early failure times and many censoring times and the sample cumulative hazard values are small. Then the scale should be changed to run from 0% to 50%, say, by adding decimal points. The probability scale then no longer corresponds to the hazard scale and must not be used. This was done in Figure 7-14 where the probability scale has been deleted.

If a data plot on a chosen hazard paper departs significantly from a straight line by being bowed up or down, it is an indication that the data should be replotted on a different hazard paper. Whether a plot of points is curved can best be judged by laying a transparent straight edge along the points. If a plot has a definite double curvature on exponential paper, log normal hazard paper should be tried. If the plot is still not straight, then this may be an indication that the units have two or more failure modes with different distributions of time to failure. Then a mixture of such data from different distributions usually will not plot as a straight line on hazard or probability paper but will be a line with two or more curves to it. Discussions on how to interpret curved probability plots are given elsewhere.[11, 12]

When a set of data does not plot as a straight line on any of the available papers, then one may wish to draw a smooth curve through the data points on one of the plotting papers, and use the curve to obtain estimates of distribution percentiles and probabilites of failure for various given times. With such a nonparametric fit to the data, it is usually unsatisfactory to extrapolate beyond the data because it is difficult to determine how to extend the curve. Nonparametric fitting is best used only if the data contain a reasonably large number of failures.

The behavior of the failure rate as a function of time can be gauged from a hazard plot.[13] If data are plotted on exponential hazard paper, the derivative of the cumulative hazard function at some time is the instantaneous failure rate at that time. Since time to failure is plotted as a function of the cumulative hazard, the instantaneous failure rate is actually the reciprocal of the slope of the plotted data, and the slope of the plotted data corresponds to the instantaneous mean time to failure. For data that are plotted on one of the other hazard papers and that give a curved plot, one can determine from examining the changing slope of the plot whether the true failure rate is increasing or decreasing relative to the failure rate of the theoretical distribution for the paper. Such information on the behavior of the failure rate cannot be obtained from probability plots.

On exponential hazard paper, if a smooth curve through the data points clearly passes through the time axis above the origin instead of through the origin, it is an indication that the true distribution may have no failures before some fixed time. This may arise, for example, if time to failure is figured from date of manufacture, and there is a minimum possible time it takes to get a unit into service. The minimum time is estimated as the time on the axis that the smooth curve passes through. This minimum time is then treated as time zero, and all times to failure are calculated from it for purposes of plotting the data on another hazard paper.

In a data plot on exponential hazard paper, the early failure times are crowded together by the origin in the lower left-hand corner of the plot and the later failure times are spaced farther apart (see Figure 7-14, for example). Thus, on exponential hazard paper, the later failures influence the eye and placement of the best-fit line more than the early failures. If one is interested in the early failure part of the distribution, which is typical of reliability work, then the early failures should be emphasized more in the analysis. This can be done by plotting data on Weibull hazard paper. The data points are then spaced out more equally throughout the range of the data, and the early and later failures are weighted about equally by eye (see Figure 7-12). If an exponential fit to the data is still desired, then a best-fit line with 45° slope should be drawn through the data points. This gives an exponential fit to the data, since the exponential is a special case of the Weibull with the shape parameter equal to one. If a nonparametric fit to the data such as described previously is desired that emphasizes early times to failure, then the data should be plotted on Weibull hazard paper and a smooth curve drawn through the data to provide a nonparametric estimate of the cumulative hazard function of the distribution. Also, percentiles and probabilities of failure for various given times can then be estimated from the smooth curve by the methods given earlier.

If estimates of distribution parameters are desired from data plotted on a hazard paper, then the straight line drawn through the data should be based primarily on a fit to the data points near the center of the distribution the sample is from and not be influenced overly by data points in the tails of the distribution. This is suggested because the smallest and largest times to failure in a sample tend to vary considerably from the true cumulative hazard function, and the middle times tend to lie close to it. Similiar comments apply to probability plotting.

Failure times are typically recorded to the nearest hour, day, month, hundred miles, etc. This is so because the method of measurement has limited accuracy or because the data are rounded. For example, if units are inspected periodically for failure, the exact time of failure is not known but only the period in which the failure occurred. For data plotting, the amount that failure times are rounded off should be considerably smaller than the spread in the distribution of time to failure. A rough rule of thumb for tolerable rounding for data plotting is that the amount of rounding should be no greater than one-fifth of the standard deviation of the distribution. Another way of saying this is that data are coarse if a large number of the failures are recorded for just a few different times. If data are too coarse, a plot on hazard or probability paper will have a staircase appearance with flat spots at those times with a number of failures. Plots of coarse data can be misleading unless the coarseness is taken into account; for example, estimates of failure probabilities and distribution parameters and percentiles can be somewhat in error if coarseness of the data is not taken into account.

Methods for analyzing coarse data can be found in the statistical literature[14, 15] where it is called grouped data. One method of taking into account the coarseness of data in plotting is to plot the failures with a common failure time at equal spaced times over the time interval associated with that time. For example, if failure times are recorded to the nearest 100 hours and 5 failures are at 1000 hours, then the corresponding interval is 950 to 1050, and the corresponding plotting times assigned to the 5 failures are 960, 980, 1000, 1020, and 1040 hours. This smooths the steps out of the plot and makes the plot more accurate.

When a set of data consists of a large number of failure times, one may not wish to plot all failures but just selected ones. This is done to reduce the labor in making the data plot or to avoid a clutter of overlapping data points on the plot. Any selected set of points might be plotted. For example, one might plot every kth point, all points in a tail of the distribution, or only some of the points near the center of the distribution. Care must be taken in plotting just a subset of the data to ensure that the plot of the subset leads to the same conclusions as the plot of the entire set. In practice, the choice of the selected points to be plotted depends on how the plot is going to be used. For example, if one is interested in the lower tail of the distribution, that is, early failures, then all of the early failures should be plotted. In general, one plots the data points that are most relevant to the information wanted from the data. Ideally, no fewer than 20 failure times, if available, should be plotted from a set of data. Often, in engineering practice there are so few failures that all should be plotted. When only a small number of failures is available, it should be kept in mind that conclusions drawn from a plot are based on a limited amount of information. Note that, if only selected failures from a sample are to be plotted on hazard paper, it is necessary to use all of the failures in the sample to calculate the appropriate cumulative hazard values for the plotting positions. Wrong plotting positions will result if some failures in the data are not included in the cumulative hazard calculations. A similar comment applies to the calculation of plotting positions for probability plotting.

For plots of some data, the ranges of the scales of the available hazard and probability papers shown throughout this chapter may not be large enough. To enlarge the time scale on any of the papers, join two or more pages together. To enlarge the hazard scale on Weibull or extreme-value hazard paper to get smaller hazard and probability values, add on extra log cycles for the lower values, and label them in the obvious manner; the cumulative hazard and cumulative probability scales are equivalent for the small probabilities in the added cycles. Weibull hazard papers with two different ranges are available to allow a user to choose the ranges that best suit his data (see Figures 7-12 and 7-17).

Concluding Remarks

The cumulative hazard plotting method and papers presented here provide simple means for statistical analyses of multiply-censored failure data to obtain engineering information. The hazard-plotting method is simpler to use for multiply-censored data than other potting methods given in the literature and directly gives failure-rate information not provided by others.

For further in-depth studies of statistical approaches to machinery failure analysis, please refer to Wayne Nelson's book on applied life data analysis.[11] Dr. Nelson is a consulting statistician with the General Electric Company.

Method to Identify Bad Repairs from Bad Designs*

Earlier we dealt with Weibull analysis and noticed how data-demanding that method can become. This is especially true when one wants to examine a sample of data from a time continuum such as in an established process plant. The sample of data may be a one-year period where the failure pattern can be considered similar to that in the previous year and in the succeeding year. However, this analysis method involves only a knowledge of the total sample population, total number of failures within the time period, the number of units on which these failures occurred, and time intervals between failures on units which failed more than once within the sample period. A good analogy would be a hospital operation with a yearly number of patients, of whom some are returning within that period for repeated treatment. Some of the returning patients only return once, others more than once during that year, and so forth.

In an earlier section, we acquainted ourselves with commonly used reliability terms: mean time between repairs (MTBR), mean time between failures (MTBF), etc. We saw that these terms have similar meanings but often include minor deviations. To avoid confusion, mean time to failure (MTTF) will be used in the following discussion. It can be expressed in terms of any time periods, i.e., days, months, years, etc.:

$$MTTF = \frac{P \times t}{F_\tau} \qquad (7\text{-}2)$$

---

*By permission from "Learn From Equipment Failures," by B. Turner, *Hydrocarbon Processing,* Nov. 1977, pp. 317-320.

where:
  MTTF = mean time to failure.
   P = population.
   t = time periods in sample time.
   F_τ = failures in sample time.

If t is unity, F_τ is failures

$$MTTF = \frac{P}{F_\tau}, \text{years or other period} \qquad (7\text{-}3)$$

While such a number is very useful as an index of performance, it gives no indication of what needs to be done to correct the situation if the index is low.

Inherent defects in either the system or the equipment, or infant mortality cause a low MTTF. Inherent defects will result in units failing more frequently than is desirable. These defects can be corrected by locating and correcting them or by taking action to reduce their effects, thus improving machine life.

Infant mortality results in repaired units failing shortly after their return to service. It can be corrected by simplifying repair techniques, quality control of repairs and repair parts, improved starting techniques, etc. Units which have persistent abnormally short lives may have an inherent defect which cannot be corrected by the previous methods. However, the usual characteristic of infant mortality is its variability. Very good pumps of last year become the very bad ones this year and vice versa. Any pump is a potential bad actor.

Two problems have different solutions.

Consider the beginning of any time period, i.e., this instant. Almost all the units are either running or available to run. Further, the lives of these units will be distributed, from the unit which was repaired yesterday to the unit which has run many years without failure.

If the life at which these units fail is a fixed interval, the failures expected during the period will also occur evenly throughout the year. Alternatively, if the unit life is random, the failures will also be evenly distributed, and so will any combination in between. The result is that the number of units being repaired remains fairly constant on a monthly basis. Such a system can be described as an exponential decay function, i.e., the number of failures occurring in a short period is proportional to the number available to decay (fail) at the start of the period. This is similar to atomic nuclei half life, i.e.,

$$R = Pe^{-Yt} \qquad (7\text{-}4)$$

where:
  R = number of unfailed units at the end of sample time.
  P = original population.
  e = base of natural logarithms.
  Y = decay constant or fraction which will fail in a unit time period.
  t = time periods, in sample time.

Failures can be expressed as $P - R$, or

$$F = P(1 - e^{-Yt}) \qquad (7\text{-}5)$$

where F is the number of units from the original population which fail within R. If a unit fails and is repaired during the sample period, any subsequent failures are not included in F.

Consider a one-year sample period. All of the units which did not fail at all during the period are obviously R. The decay constant Y can be found from

$$-Y = \log_n \frac{R}{P} \qquad (7\text{-}6)$$

since R is 1, but $R = P - F$ where F is the number of units which failed at least once

$$\therefore -Y = \log_n \frac{P - F}{P} = \log_n \left(1 - \frac{F}{P}\right) \qquad (7\text{-}7)$$

Since $-Y$ is based on R, it cannot include any infant mortality effects from the repairs during the sample period.

By definition, F is the total units which fail at least once. However, this is not the total number of failures that can be expected. The total number of failures expected, F_{TE}, is the total of F plus the failures expected on these repaired units, F_1, plus the failures expected on units repaired twice, F_2, and so on. The repaired units have a similar expectation of failure to the original population, i.e.,

$$F_1 = F(1 - e^{-Y}) \qquad (7\text{-}8)$$

These will also have further failures, F_2, where

$$F_2 = F_1(1 - e^{-Y}) \qquad (7\text{-}9)$$

However, there is a difference. All of P are available to fail from the start of the period. If the failures, F are distributed evenly through the sample period, the average length of time that the repaired units are available to fail is only half of the period. Therefore, t is 0.5 for F_1 failures and the exponent to be used to calculate F_1 is $0.5Y$. The use of $t = 0.5$ for F_2, F_3, F_4, etc. is an exaggeration, since the average time available for the third failure will be less than half a year and the average time for the fourth and fifth failures will be even less. However, the assumption that $t = 0.5$ does not significantly distort the total number of failures expected and does simplify the calculations. This simplification is conservative, i.e., it increases the expected failures and slightly reduces the apparent infant mortality.

Total expected failures in a sample period are:

$$\begin{aligned}
\text{Total failures} = F_{TE} &= F + F_1 + F_2 + F_3 + F_4 \qquad (7\text{-}10) \\
&= F + F(1 - e^{-Y/2}) + F_1(1 - e^{-Y/2}) \\
&\quad + F_2(1 - e^{-Y/2}) + F_3(1 - e^{-Y/2}) \qquad (7\text{-}11)
\end{aligned}$$

F_{TE} can be easily calculated using a hand calculator.

If F_{TE} is significantly below the total failures observed, there is a significant infant mortality problem. Further, the effect of the mortality is expressed in numerical terms:

$$F_{TA} - F_{TE} = \text{number of repairs which could be eliminated by solving infant mortality problems.} \qquad (7\text{-}12)$$

where:
F_{TA} = actual failures.
F_{TE} = expected failures.

Also

$$\frac{F_{TA} - F_{TE}}{F_{TA}}\% = \text{percent of actual failures due to infant mortality.} \qquad (7\text{-}13)$$

This conclusion about significant infant mortality can be confirmed in the following manner. Group the repaired machines which had two or more failures within the year into specified intervals of life; we have used 0.1 years and also months, for our graphs. These periods will be some fraction of a year, t. Total the repeated failures in each sample period, i.e., so many failing within 1 month or 0.1 years of being repaired, another number failing between 1 month and 2 months of being repaired, and some failing with a life of more than 0.2 of a year but less than 0.3 years. Now determine sequential failures.

The failures to be expected during the first sample period are

$$\frac{2 - t}{2} (1 - e^{-tY})F_{TE} \qquad (7\text{-}14)$$

The reason for the first multiplier is to correct for the failures which will take place outside the period under study. For example let $t = 0.1$ years. For the sixth period, i.e., units expected to fail in the period between 0.5 and 0.6 of a year, all such units repaired in the first 146 days of the year will have the second failure within the year. All such units repaired later than 183 days or July 2 will not have the second failure within the sample period. For units repaired between 146 and 183 days, there is an equal chance that the second failure will or will not be observed. The multiplier is to correct this effect. The total probability that a unit with life t will fail again within the year is

$$\text{probability} = 1 - \frac{t}{2} = \frac{2 - t}{2} \qquad (7\text{-}15)$$

During the second period, the expected failures will be

$$\frac{2 - 3t}{2}\left[(1 - e^{-2tY}) - (1 - e^{-tY})\right]F_{TE} \qquad (7\text{-}16)$$

For each subsequent period, the multiplier for t in the first expression increases by 2 and, for t in each exponent, it increases by 1.

For example, in the sixth period the expression will be

$$\frac{2 - 11t}{2}\left[(1 - e^{-5tY}) - (1 - e^{-5tY})\right]F_{TE} \qquad (7\text{-}17)$$

While the previous method is rigorous, it is unnecessarily complicated where F_{TE}/P is less than 0.5.

The expression

$$\left[(1 - e^{-ntY}) - (1 - e^{-(n-1)tY})\right] \simeq (1 - e^{-tY}) \qquad (7\text{-}18)$$

will give adequate accuracy. $F_{TE}(1 - e^{-tY})$ can be calculated and then multiplied by the initial multiplier from the rigorous equation for each period. The simplification becomes increasingly inaccurate for the latter periods, but the multiplier becomes increasingly smaller and reduces the significance of the error.

Calculation of the expected failures appears complicated, but it is quite easy with an electronic calculator (Table 7-6).

Significant differences between the actual failures and those expected during the first three periods (and especially in the first) are telltales of infant mortality. However, there are other methods of indicating this. The advantage of the method described is that it indicates what progress is possible if the problem is corrected. This justifies measures to correct the problem.

The expected failure rate F_{TE}/P can be inverted and used in a similar way to MTTF.

P/F_{TE} is expressed in time units and indicates the expected life of the units without infant mortality. Low P/F_{TE} ratio population obviously contain units which are unsuitable for the conditions of service, i.e., have inherent defects. Such cases can only be improved by changes to the machines or the conditions of service. Alternatively, analysis may show that samples with an unacceptable MTTF have an acceptable life when expressed as P/F_{TE}.

This indicates that the problem is due to infant mortality and not to inherent defects. However, units with a high MTTF will probably not justify extensive work to correct a high failure rate due to infant mortality because even the unnecessarily high repair rate is still very low.

Figure 7-18 shows the variation in F_{TA} and F_{TE} by 0.1 years for pumps in a refinery. Figure 7-19 shows the same variation by months for pumps in a chemical process plant. Table 7-7 indicates the range of observed values of MTTF, P/F_{TE} and $(F_{TA} - F_{TE})/F_{TA}$. Table 7-8 shows parameter variations.

Table 7-8 seems to indicate that the in-line pump has a higher MTTF and P/F_{TE} than the horizontal pump. The in-line also appears to be harder to assemble, since its infant mortality rate is higher. It is not necessarily correct that the in-line pump has a better inherent life since, in the sample they may have been installed in less severe services. Such data on the relative effect process services have on pump life is unknown. Therefore, Table 7-8 should not be used as a sole means of selecting a pump for a specified service. However, this approach may offer a way of evaluating pump bids, provided data on the effect of the pumped fluid on service life is established and a larger sample population is collected. *(text continued on page 558)*

Table 7-6
Calculation of F_{TE} Using Electronic Calculator

	Assume P-500 F-200		
	Enter	**Display**	
1	200	200	(F)
2	+	200	
3	500	500	(P)
4	−	0.4	
5	CS	−0.4	
6	+	−0.4	
7	1	0.6	
8	F	0.6	
9	$\log_n X$	−0.510825	(Y)
10	+	−0.510825	
11	2	2	
12	−	−0.2554125	$(Y/2)$
13	F	−0.2554125	
14	e^x	0.774598	
15	CS	−0.774598	
16	+	−0.774598	
17	1	1	
18	×	0.225402	$(1 - e^{-t/y})$
19	MC	0.225402	
20	200	200	(F)
21	M+	200	
22	−	−45	
23	M+	−45	
24	−	10.16	
25	M+	10.16	
26	−	2.29	
27	M+	2.29	
28	−	0.516	
29	M+	0.516	
30	−	0.116	
31	M+	0.116	
32	MR	258.16	(F_{TE})

Note: Most calculators have a single keyboard, and it is necessary to press the F button to activate the functions log, etc.

Other calculators have a double keyboard with separate buttons for the functions. Obviously, there is no need to press the F button on the latter models. The above procedure is based on the single keyboard model.

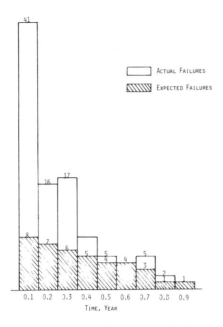

Figure 7-18. Variation between actual and expected failures of pumps (refinery).

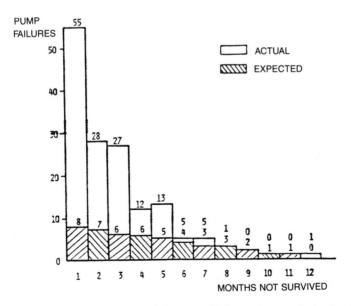

Figure 7-19. Variation between actual and expected failures of pumps (petrochemical process plant).

Table 7-7
Centrifugal Pumps Failure Analysis

Refinery	MTTF Years	$\dfrac{P/F_{TE}}{\text{Years}}$	$\dfrac{F_{TA}\text{-}F_{TE}}{F_{TA}}$ %
A	2.0	2.5	20
B	2.45	3.3	19
C	2.33	2.6	6
D	5.13	5.2	1

Table 7-8
Centrifugal Pumps Failure
Analysis (Type Pump)

Pump Type	MTTF Years	$\dfrac{P/F_{TE}}{\text{Years}}$	$\dfrac{F_{TA}\text{-}F_{TE}}{F_{TA}}$, %
Inlines	2.54	3.45	26.4
Horizontals	2.13	2.54	15.3
Deepwells	2.00	2.09	4.4

Nomenclature

MTTF Mean Time To Failure.
 F Number of units from original population that fail in sample time.
 F_T Total failures in sample time.
 F_1 Expected failures after first repair.
 F_2 Expected failures after second repair.
 F_3 Expected failures after third repair.
 F_{TE} Total failures expected.
 F_{TA} Total failures in time period.
 P Population.
 t Time periods in sample time.
 Y Decay constant or fraction that will fail in a unit time period.

References

1. Grothus, H., "Total Preventive Maintenance," Course Notes by F.K. Geitner, Toronto, Ontario, 1975.

2. Karassik, I.J., "Flow Recirculation in Centrifugal Pumps: From Theory to Practice," *Power and Fluids,* 1982, Vol. 8, No. 2.

3. Summers-Smith, D., "Performance of Mechanical Seals in Centrifugal Process Pumps," *pumps-pompes-pumpen*, 1981-181, pp. 464-467.

4. Czichos, H., "Tribology—A Systems Approach to the Science of Technology of Friction, Lubrication and Wear," Elsevier Scientific Publishing Co., 1978, pp. 237-239.

5. Bergling, G., "The Operational Reliability of Rolling Bearings," *SKF Ball Bearing Journal*, 188, pp. 1-10.

6. Czichos, H., "Tribology—A Systems Approach to the Science of Technology of Friction, Lubrication and Wear," Elsevier Scientific Publishing Co., 1978, pp. 240.

7. TEAM, "TEAM Easy Analysis Methods," Vol. 3, No. 4., Issued by TEAM, Box 25, Tamworth, NH 03886

8. Grothus, H., "Die Total Vorbeugende Instandhaltung", Grothus Verlag, Dorsten, W. Germany, 1974, pp. 63-66.

9. Jardine, A.K.S., *Maintenance, Replacement, and Reliability*, Pitman Publishing, Bath, U.K., 1973, pp. 13-24.

10. Nelson, W., "Hazard Plotting for Incomplete Failure Data," *Journal of Quality Technology*, Vol. 1, No. 1, Jan. 1969, pp. 27-52.

11. Nelson, W., *Applied Life Data Analysis*, J. Wiley and Sons, N.Y., 1982, pp. 634.

12. Hahn, G.J. and Shapiro, S.S., *Statistical Models in Engineering*, Wiley, New York, N.Y., 1967.

13. King, J.R., "Graphical Data Analysis with Probability Papers," TEAM Report, Box 25, Tamworth, NH 03886

14. Buckland, W.R., "Statistical Assessment of the Life Characteristic: A Bibliographic Guide," Hafner, New York, N.Y., 1964.

15. Govindarajulu, Z., "A Supplement to Mendenhall's Bibliography on Life Testing and Related Topics," *Journal of the American Statistical Association*, Vol. 59, No. 308, Dec. 1964, pp. 1231-1291.

Chapter 8
Sneak Analysis*

Major machinery systems are inconceivable without hydraulics, electronic interfaces, control circuitry, shutdown instrumentation, and the like. The increasing complexity of these devices makes them prime initiators of unexpected malfunctions and unscheduled equipment downtime. Advanced troubleshooting methods must be employed to resolve existing problems or, better yet, find and eliminate potential problems before they can lead to expensive outages of plant and equipment. One such method is called sneak analysis, an advanced pre-event troubleshooting and failure prevention tool.

Within the context of overall machinery systems troubleshooting and general reliability reviews, it is deemed appropriate to present a description of the Sneak Analysis (SA) with primary emphasis on its purpose, potential benefits, and limitations, as well as cost factors and tradeoffs. It is not the intent of this section to define the exact methods or techniques for the performance of the analysis. However, it is hoped that our treatment of this topic will serve to introduce the concept of SA to management and engineering personnel responsible for the design and development of systems and equipment for process plants. Being familiar with SA principles will enable them to make appropriate decisions regarding its applicability to a given project. Guidelines are provided for the management and implementation of the SA.

Sneak Analysis Usage

SA is a powerful tool to identify system conditions that could degrade or adversely affect plant safety or basic equipment uptime and reliability. It is a technique that can

---

*Adapted (by permission) from "Sneak Analysis Application Guidelines," Contract No. F30602-81-C-0053, CDRL No. A002, Initiated By Rome Air Development Center, New York, 1982, and "Contracting and Management Guide For Sneak Circuit Analysis," NAVSEA Document TE001-AA-GYD-010/SCA, Washington, D.C., 1980.

be applied to both hardware and software. For hardware, SA is generally based on documentation used in engineering and manufacturing. Its purpose is to identify conditions which cause the occurrence of unwanted functions or inhibit desired functions, assuming all components are functioning properly.

SA examines system operations during normal conditions for design oversights. It consists of two subanalyses: (1) Sneak Circuit Analysis (SCA) for electrical-electronic systems, and (2) Software Sneak Analysis (SSA) for computer programs. SCA represents a mature and useful technique that can be performed on both analog and digital circuitry. The purpose of SSA is to define logic control paths which cause unwanted operations to occur or which bypass desired operations without regard to failures of the hardware system to respond as programmed. After an SCA and an SSA have been performed on a system, the interactions of the hardware with the system software can readily be determined. The effect of a control operation that is initiated by some hardware element can be traced through the hardware until it enters the system software. The logic flow can then be traced through the software to determine its ultimate impact on the system. Similarly, the sequence of action initiated by software can be followed through the software and electrical circuits until its eventual total impact can be assessed. Finally, the analysis should be considered for critical systems and functions where other techniques are not effective but should not be applied to off-the-shelf computer hardware such as memory or data processing equipment.

The SA concept has potential applications in any design or development project for systems, equipment, or software, and its merits should be evaluated as part of the development planning effort. In the past, SA was generally considered applicable only to electrical and electronic circuits, but the concept has evolved to where (in theory at least) it can be used to evaluate other systems, including mechanical, pneumatic, hydraulic, power generation, etc. The complexity and criticality of such systems or devices, the extent of other analyses and test programs, and the associated cost factors are the primary determinants in the decision to require (or not to require) SA.

Definitions

As with any new or emerging technology, there soon develops a jargon which needs to be translated or defined for the uninitiated. Specialized terms unique to the SA technique are defined as follows:

Clue. A known relationship between a historically observed sneak-circuit condition and the underlying causes which created it. Clues are used to evaluate systems for possible sneak circuit conditions.

Glitch. An anomalous circuit response which occurs without apparent reason, generally due to signal overlap or timing problems.

Highlight. The process of isolating a particular design feature which is likely to generate a system malfunction so that it can be subjected to intensive analysis.

Manual SA procedures. SA procedures that are performed without computer assistance.

Network. A group of interconnected elements intended to perform some system function. Elements may be electrical components, electromechanical components, computer program instructions, operator actions, procedures, mechanical or pneumatic functions, or any other portion of a system that can be considered as an independent entity. Although the networks most frequently considered in SA consist of electrical and electromechanical components, the definition is not restricted to networks of this type.

Path search. A process of searching out all possible paths through a network.

Software. A computer program which may be either an independent (stand-alone) program or an embedded (within other system hardware) program.

Tailoring. The process of adapting a specification to fit the constraints of a particular program.

Topological methods. This term simply describes the application of clues at each mode found in the networks when networks are depicted "topologically"; that is, with flow initiating at the top of the diagram and terminating at the bottom.

Trade-off. The process of comparing the relative advantages and disadvantages of different courses of action.

Verification and validation. Terms generally referring to those activities in a software development program that serve to assure the adequacy and the accuracy of the software.

Sneak Circuits and Their Analysis

A sneak circuit is an unexpected path or logic flow within a system which, under certain conditions, can initiate an undesired function or inhibit a desired function. The path may consist of hardware, software, operator actions, or combinations of these elements. Sneak circuits are not the result of hardware failure but are latent conditions, inadvertently designed into the system or coded into the software program, which can cause it to malfunction under certain conditions. Categories of sneak circuits are:

1. *Sneak paths* which cause current, energy, or logical sequence to flow along an unexpected path or in an unintended direction.
2. *Sneak timing* in which events occur in an unexpected or conflicting sequence.
3. *Sneak indications* which cause an ambiguous or false display of system operating conditions and thus may result in an undesired action taken by an operator.
4. *Sneak labels* which incorrectly or imprecisely label system functions, e.g., system inputs, controls, displays, buses, etc., and thus may mislead an operator into applying an incorrect stimulus to the system.

Figure 8-1 depicts a simple sneak circuit example. With the ignition off, the radio turned to the on position, the brake pedal depressed, and the hazard switch engaged, the radio will power on with the flash of the brake lights.

SCA is the term that has been applied to a group of analytical techniques which are intended to identify methodically sneak circuits in systems. SCA techniques may be

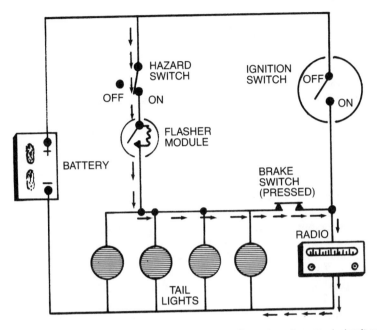

Figure 8-1. The automotive sneak circuit (sneak path). The automotive sneak circuit appears in a few types of automobiles. The radio should be powered by current flowing through the ignition switch; however, the electric current flows along an unintended sneak path (→), causing the radio to "blink" with the tail lights.

either manual or computer-assisted, depending on system complexity. Current SCA techniques which have proven useful in identifying sneak circuits in systems include:

1. *Sneak path analysis.* This method of SCA attempts to discover sneak-circuit conditions by means of methodical analysis of all possible electrical paths through a network. Because of the large volume of data involved, sneak path analysis normally mandates computer data processing. It has been found that sneak-circuit conditions generally have certain common characteristics which are directly related to topological patterns within the network. Sneak path analysis uses these common characteristics as "clues" to look for sneak circuits in the system being analyzed.

2. *Digital sneak circuit analysis.* This SCA method is intended to discover sneak-circuit conditions in digital systems. Digital SCA may involve some features of sneak path analysis, but it may also involve additional computer-assisted techniques such as computerized logic simulation, timing analysis, etc., to handle the multiplicity of system states encountered in modern digital designs. In general, digital SCA will identify the following types of anomalies:

    a. Logic inconsistencies and errors.

    b. Sneak timing, that is, a convergence of signals which causes an erroneous output due to differing time delays along different signal paths through a digital network.

    c. Excessive signal loading or fan-out.

    d. Power supply cross-ties, grounding, or other misconnections of signal pins.

3. *Software sneak path analysis.* Software sneak path analysis was adapted from hardware sneak path analysis. It was found that computer program flow diagrams which contained known sneak paths were most often associated with certain common flow diagram topologies and had other common characteristics. These common characteristics served as a basis for establishing "clues" which could be used to analyze new computer program flow diagrams. Computerized path search programs developed to do SCA on hardware were adapted or rewritten to accept software logical flows, and new clues were developed to analyze them. Software SCA can be done either manually or with computer assistance, depending primarily on the size and complexity of the software. It may be combined with the hardware SCA and is most often used on embedded software in a complete minicomputer or microcomputer-controlled hardware system. It has been used on both assembly language and higher-order language programs.

4. *Other sneak circuit analysis techniques.* Variations of SCA have been developed to analyze particular types or combinations of systems, such as hardware/software interfaces. The application of any new SCA procedure to a particular system or situation must be judged on its demonstrated effectiveness in detecting sneak circuits in similar cases or on its anticipated benefit in the specific situation being considered. It is difficult to predefine a course of action for handling new SCA types which may emerge, but the general ground rules for evaluating applicability hold true also.

All of the SCA types given above, i.e., sneak path analysis, digital SCA, and software sneak path analysis may be performed manually under some limited conditions. Computer-assisted SCA data processing is also possible with each of these types and is absolutely essential for more complex systems. The availability and thoroughness of computer aids strongly influences the selection of a contractor to perform SCA on a complex system.

A few additional words relating to complexity are in order. Sneak conditions result from the following three primary sources: (1) system complexity, (2) system changes, and (3) user operations. Hardware or software system complexity results in numerous subsystem interfaces that may obscure the intended functions or produce unintended functions. The effects of even minor wiring or software changes to specific subsystems may result in undesired system operations. A system that is relatively sneak free can be made to circumvent desired functions or generate undesired functions as a consequence of improper user operations or procedures. As systems become more complex, the number of human interfaces multiplies because of the involvement of more design groups, subcontractors, and suppliers. Hence, the probability of overlooking potentially undesirable conditions is increased proportionately.

Historical Development of SCA

Systems analysis techniques, in one form or another, have been employed throughout the evolution of the various technologies. As systems become more and more complex, and as greater emphasis is placed on safety and reliability, the need for more sophisticated techniques evolves. The development of SCA, as a unique analytical technique under that name, came about when it was recognized that conventional analytical methods did not identify certain subtle anomalies in systems design. The first investigations using SCA techniques were initiated in the 1960s by NASA to identify sneak conditions in missile launch command and control electrical subsystems. The SCA technique can also be applied to hydraulic, pneumatic, and other analogous systems and has been used in the analysis of detailed system operating procedures. As regards industrial applications, electronic governor and two-out-of-three voting logic systems associated with major turbomachinery shutdown systems are prime examples where the recent application of sneak analysis has proved highly valuable.

A major consideration in other analyses is the potential effect of component and/or subsystem failures and methods for preventing or mitigating such effects. On the other hand, SCA considers the potential that an undesired function may occur, or that a desired function may be inhibited, given that the system is performing as designed. Thus, SCA looks at the system from a different perspective and complements these other techniques but does not replace them. While it could be said that SCA will identify many of the same potential problems as other techniques, it would be a serious error to consider this approach totally redundant, or simply a desirable "double-check" for these aspects. In fact SCA *does* address a class of problems that failure-oriented reviews and testing will not cover.

As regards the relation of SCA to test programs, many of these are designed to determine if desired functions occur under given conditions. It is seldom feasible to test all combinations of conditions that might result in undesirable or unexpected functions. Tests frequently can be performed on only a limited number of items, which typically will not represent the range of variables that may be inherently present in the total population of a given item. Tests are expensive, and in many instances, destructive in nature. Thus, test programs are not a viable substitute for analyses of any type. However, the results of analyses (such as SCA) may identify potential problems and allow the test program to be structured such that if these problems are present, they will be detected.

It is appropriate also to highlight computer aids which may be used to assist in performing SCA tasks on large systems. These include:

1. Configuration management programs to handle large volumes of system configuration data.
2. Automated path search, network plotting, and "clue" evaluation programs used in sneak path analysis.
3. Digital logic analzyer programs used on complex digital systems.
4. Circuit analysis and design programs used to analyze subcircuits and functions.
5. Code analyzers, and in some cases, compilers for software programs.
6. Report generation programs.

Purpose and Benefits of SCA

Sneak circuit analysis offers three principal benefits in the analysis of systems which distinguish it in some respects from other analytical techniques:

1. Sneak conditions are distinguished from other types of design concerns primarily by their resistance to conventional forms of analysis. Thus, SCA can identify such problems as unintended current paths, false indications, misleading labels, sneak timing or signal race conditions, "glitch" conditions, and many intermittent problems which occur at only one portion of an operational cycle or are the result of procedural causes rather than faulty components. SCA is a "highlighting" technique which isolates particular areas of the system that are most likely to generate sneak conditions and applies intensive analysis of those areas to flush out possibilities for system malfunctions. Consequently, SCA is much more efficient in finding design flaws which would be classified as "sneak" conditions than other more conventional analyses.

2. SCA inherently results in a reasonably thorough design review if performed by qualified analysts with experience and training in the SCA methods. This is far more important than experience or training with the type of system being analyzed. The analysts generally will possess a list of "clues" which are applied to isolate sneak conditions. Most of these clue lists also contain a variety of other common design oversights to consider when performing the SCA. It has been found that the SCA frequently identifies potential problems in related areas such as:

   a. Overstressed parts.
   b. Single failure points.
   c. Unnecessary or unused circuitry or components.
   d. Lack of relay transient protection.
   e. Excessive fan-out on microcircuits.
   f. Component misapplication.
   g. Drawing errors.
   h. Ambiguous labeling.
   i. Incorrect labeling.
   j. Less-than-optimum indicator/sensor location.

3. SCA complements a number of other analyses commonly performed in a system development program, filling in for "blind spots" which exist in all analytical techniques. For example, it complements failure mode and effect analysis (FMEA) by considering procedural errors rather than part failures. It also generally looks deeper, at individual wire segment levels, than traditional functionally-oriented FMEAs. Moreover, it complements the test program by concentrating on situations or conditions under which the system might not

work rather than on proving that the system does work under particular controlled input conditions. A number of other examples of this complementary nature of SCA when compared to more conventional analyses and test programs are discussed later.

Effect on System Reliability, Safety, and Life-Cycle Cost

SCA benefits the system reliability program by identifying potential system malfunctions which are not the result of part failures. The more conventional methods of reliability prediction and control concentrate on the influence of inherent part failure rates on system reliability. More often than not, inherent or "random" part failures are not the most significant contributor to observed field failures of a system. Design oversights, procedural problems, and interface problems cause a significant number of field failures. By identifying these kinds of malfunctions, SCA attacks the persistent problem that exists of reconciling reliability predictions with observed field failure rates. In fact, SCA may be especially appropriate when a large number of unexplained or non-verifiable field failures have been experienced on a project. By removing operational failures, glitches, intermittent failures, etc., system availability can be improved. The number of non-verifiable failures in the rework cycle should also be reduced, further increasing operational availability of the system.

The US military considers SCA to be a key input to system hazard analysis. It definitely performs this function, although the scope of SCA is considerably broader. Sneak circuits in petrochemical plant or machinery systems can be hazardous, and their removal would improve system safety in these cases.

SCA can act to reduce life cycle cost of a system through early identification and control of system malfunctions. The removal of malfunctions which would occur during the equipment operation and maintenance phase is particularly effective in reducing life cycle costs. Since SCA aims at a class of problems which are often overlooked by other analyses and test programs, the net improvement to life cycle cost through SCA is usually very real.

Just as SCA must be considered for many aerospace system procurement situations, an SCA should be performed on sophisticated machinery systems where the consequences of a sneak circuit would be unusually severe. This decision can often be made by postulating the most severe consequences of possible unwanted operational modes and comparing the resulting cost with that of an SCA. Obviously, some costs cannot be quantified in terms of dollars, such as the cost of human life, or a significant environmental impact, but in these cases, performance of an SCA is clearly indicated. SCA should be performed when one or more of the following conditions is present:

1. The system is critical, to the point that operational malfunctions cannot be tolerated.

2. Improper function of the system could endanger human life or substantially damage expensive equipment.
3. The system complexity is such that sneak circuits may go undetected in normal testing.
4. Correction of events resulting from sneak conditions, if they occurred in operation, would be difficult or impossible.
5. The system involves interfaces of equipment designed or produced by a number of different contractors or vendors.
6. The unique features of an independent SCA would partially compensate for lack of thorough design evaluation by a supplier.
7. Sneak circuits are suspected in an existing system which meets the previous criteria. Before a sneak circuit analysis is performed, a failure analysis of the malfunction(s) should be performed to eliminate causes other than sneaks, such as electromagnetic interference, temperature sensitivity, etc.

Many operational systems develop problems which are intermittent or are not diagnostically repeatable. The application of SCA is a viable technique for identification of potential or actual causes for these anomalies. This is especially true when the system is inaccessible for testing and diagnostic checkout, or unavailable due to fear that the temporary hookup of diagnostic instrumentation might cause costly shutdowns. Large complex systems are difficult to analyze without the aid of an SCA-type technique to determine the possible paths which could cause the observed problem.

Application to electrical and electronic systems. SCA is particularly applicable to electrical and electronic circuits of all types. When schedule, time, and cost are the predominant factors, the analysis may be limited to specific critical subsystems or functions. The hardware and software directly related to specific hazards, safety, reliability, test anomalies, and primary system functions are those most often selected for SCA.

SCA applies to digital circuits of varying complexity, including small-scale integration (SSI), medium-scale integration (MSI), large-scale integration (LSI), and hybrids. Standard integrated circuits (ICs) allow simplification of clue lists and encoding procedures for data processing. Analysis of custom LSI and hybrid devices generally requires the IC to be treated in the same detailed manner as a circuit card. SCA of digital circuits can reveal unexpected cause-and-effect relationships, timing hazards, reliability concerns, and other related problems.

Similarly, SCA of analog circuitry can reveal unexpected circuit configurations (modes), unexpected cause-and-effect relationships, error sources, timing (phase shift and stability) problems, reliability concerns, and other related problems.

Many new systems are software driven or hybrid, that is, software controlled with hardware control backup. This situation has given rise to the development of software SCA techniques which were adapted from SCA of hardware. The SCA technique can be applied to the embedded software program as a separate entity (i.e., the "stand-alone" program), or to the integrated software/hardware as a system. There are definite advantages to applying SCA to the integrated software/hardware system, but this form of embedded code analysis is not widely available at present.

Limitations in Application of SCA

SCA has been successfully applied to a variety of systems and has been responsible for removing many potentially costly sneak conditions from hardware. Nevertheless, there are several limitations of SCA which must be considered by the potential user or purchaser. Some of these limitations are inherent to the SCA technique; others are a result of the present state of development and availability of SCA in the market place.

One typical limitation is the lack of an approved SCA method. There is presently no industry-wide approved technique for performing SCA. As a result, techniques may vary among the contractors performing it. The thoroughness, the degree and quality of computer aids, the level of training of sneak circuit analysts, the validity and completeness of the "clues" used in looking for sneak conditions will differ, and these factors must be resolved when selecting an SCA contractor.

A second limitation relates to our dependence on the capacities of the analyst. The techniques, even when highly automated and subjected to strict quality control, generally depend on an analyst who evaluates each path or network and considers the conditions under which that path may occur. This process is always subject to human error and may permit some sneak conditions to slip through.

Next, we should realize that in all its variations, SCA tends to concentrate on topological similarities of sneak circuits. While most sneaks can be found in this manner, not all possible undesirable conditions will be found because they are not part of the topology of the network as it is defined by the system drawings. For example, the commonly known "sneak circuit" which causes a transistor amplifier to oscillate because of stray capacitance between collector and base circuits would not normally be found in an SCA because the stray elements do not exist in the system drawing definition.

SCA considers that all system components are functioning normally, that is they are failure free. In line with this assumption, the effect of varying environments on component failures is not considered. The analysis does consider the possible "failure" of the system operator to provide correct system inputs, but not component failures. Although the SCA often identifies single-point failures, it cannot be considered comprehensive in this regard.

The SCA technique has found numerous problems in mature systems that were overlooked in other analyses and test programs, but it is virtually impossible to determine whether an SCA has found all or nearly all sneak conditions in a system. In this respect, however, it is difficult to determine absolutely that any analysis technique has completely accomplished its purpose. Despite this, all analyses, including SCA, will identify problems that can be corrected, to the ultimate enhancement of operational availability.

In summary, we can appreciate that other analyses are still necessary. SCA provides a unique way of looking for potential problems in systems and may discover problems which would more properly be found by other analyses, such as single-point failures which are also considered in an FMEA, or excessive fan-out which is considered in a stress analysis. However, SCA must not be used as the sole reason for eliminating these other analyses. Other analyses are generally done to specific standards which define both methods and reporting requirements so that results properly support other activities in the systems acquisition cycle. For example, an FMEA will probably be used to support a maintenance analysis and to establish spares requirements. Although an SCA may consider single-point failures, this is done more incidentally than as a primary objective of the analysis; SCA does not consider all or even most part failures. Similar arguments could be made for other analyses which have some overlap with SCA, such as stress analysis or hazards analysis. SCA necessarily concentrates most heavily on those areas in a system that are most likely to contain sneak circuits or conditions. This leads to differing levels of concentration on different subsystems or their interconnections. Accordingly, the SCA is not necessarily comprehensive in the aspects covered by other analysis techniques, nor is it intended to be.

Trade-offs. The potential benefits of an SCA will always be weighed against other project considerations. Factors such as cost, schedule, the existence of other analyses, the availability of data, the ease with which available system data can be interfaced with SCA programs, and the availability of a qualified SCA contractor, must all be considered. When it is considered that SCA can be tailored to fit the needs of most systems by applying it only to system-level interconnections or only to the more critical subsystems, an argument against SCA based strictly on cost is not really defensible. While there is some overlap between SCA and other analyses, performance of most other analyses and test programs cannot substitute for an SCA (and vice versa).

Software Sneak Analysis

Software sneak analysis (SSA) is used to discover program logic which causes undesired program outputs or inhibits a desired output. The technique involves the reduction of the program source code to topological network tree representations of the program logic.

SSA is envisioned to serve the petrochemical industry in such areas as programmed starting and commissioning of major turbomachinery systems. Data used for SSA should reflect the startup sequencing program as it is actually written. This includes system requirements, system description, coding specifications, detailed and complete source code, a compilation listing, and operating system documentation. All reports are written against these documents.

Direct analysis of program listings is difficult because the system is modular for ease of programming. Also, the code is listed as a file of records, without regard to functional flow. The first task of the software sneak analyst is to convert the program source code into a form usable for analysis. In most cases, this step requires computer conversion. In either case, the program source code is converted with reference to an input language description file into topological network trees, such that program control is considered to flow down the page.

Once the trees have been drawn, the analyst identifies the basic topological patterns that appear in the trees. Five basic patterns exist and are shown in Figure 8-2: (1) the single line, (2) the ground dome, (3) the power dome, (4) the combination dome, and (5) the H-pattern. The topological patterns containing branch or jump

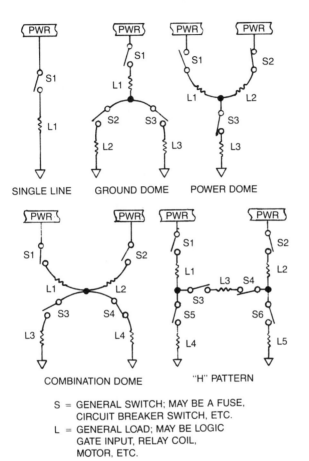

Figure 8-2. Basic topological patterns appearing in network trees.

instructions have the highest incidence of problems. This includes the ground, power, and combination domes. The crossing of module or function interfaces as a result of the branch instruction is a prime problem cause.

Although at first glance a given software tree may appear to be more complex than these basic patterns, closer inspection will reveal that the code is actually composed of these basic structures in combination. As each node in the tree is examined, the analyst must identify which pattern or patterns include that node. The analyst then applies the basic clues that have been found to typify the sneaks involved with that particular structure. These clues are in the form of questions that the analyst must answer about the use and interrelationships of the instructions that are elements of the structure. These questions are designed to aid in the identification of the sneak conditions in the instruction set which could produce undesired program outputs. Software clues are different than the hardware clues referred to earlier and are typically proprietary to the performing contractor. Branch instructions can alter program flow to an incorrect location or address, encounter unitialized variables, and induce timing or sequencing problems.

Software sneaks are classified into four basic types:

1. *Sneak output:* the occurrence of an undesired output.
2. *Sneak inhibit:* the undesired inhibition of an input or output.
3. *Sneak timing:* the occurrence of an undesired output by virtue of its timing or mismatched input timing.
4. *Sneak message:* the program message does not adequately reflect the condition.

When a potential sneak is identified, the analyst must verify that it is valid. The code is checked against the latest applicable listings, and compiler information may be reviewed concerning the language in question. If the sneak is verified, a software sneak report (SSR) is written which includes an explanation, system level impact, and a recommendation for elimination of the sneak.

Cost, Schedule, and Security Factors

Estimating the cost of an SA program can be complicated because of a number of factors. There is a variation in SA techniques among contractors which can lead to significant differences in their cost estimates for the analysis. The process of tailoring or partitioning the system and using different techniques and different levels of scrutiny in the analysis of certain subsystems may also cause a wide variation in cost estimates for what is apparently the same job. Other factors, such as the degree of accessibility to the SA data base, the level of owner involvement, amount of liaison required, frequency of reports and specific data submission requirements, will all have an effect on the price quoted for the SA.

While assessing the technical differences among competing proposals, the owner's review engineer must further assure that the scope of the SA under consideration, including system partitioning and special considerations, had been clearly defined in the request for quotation (RFQ) and that the quotes received reflect the contractors' understanding of these specific tailoring requirements. It is further assumed that the owner's engineer will include contract provisions for monitoring SA performance sufficient to assure that the SA is performed with the same techniques, partitioning, and levels of scrutiny as were quoted. Only after all of these preconditions have been met can the purchaser be sure that the price quotes are truly comparable or, at least, that the price quote really does support a trade-off decision between cost and SA effectiveness.

Cost effectiveness. The application of SA on a project must always be done with a serious consideration for cost effectiveness. There are obvious situations in which SA is clearly mandatory, such as a manned space flight, a nuclear power plant control system, or a nuclear missile system. In such cases the consequences of sneak conditions are potentially disastrous; the costs of a serious error are virtually incalculable. The decision in most cases involving industrial machinery systems or complex offshore systems is not so clear-cut. The owner's engineer must weigh the potential benefits of SA against the cost and decide either to do it or not.

Tailoring the SA to reduce costs while maintaining the essential benefits of the analysis is strongly encouraged. A vigorous assessment of the areas in the design most likely to generate serious sneak conditions can do a great deal to improve the cost effectiveness of SA on a given system. Thus, the built-in test equipment circuitry, or the interconnection to a test set, or the control and indicator subsystems could be singled out as the primary focus for SA. Experience has shown that power distribution circuits are especially susceptible to sneak conditions and are thus potential candidates for SA. Of course, this top-level tailoring of the effort should remain flexible until SA contractor inputs regarding proper tailoring are heard and considered.

Another consideration in assuring that an SA remains cost effective is the efficient use of the system data base. The configuration management costs for maintaining an accurate system data base can be substantial for a large system. An effort should be made to have the SA performed by the same organization that performs other analyses if this is possible, or at least that minimize the number of activities at which the data base must be maintained. The benefits of an SA performed by an independent agency may outweigh the additional data base costs, but this factor should at least be considered. Some cost benefits can be derived by combining analyses, such as, SA with FMEA or SA with stress analysis, especially if computer-aided analysis is anticipated for these other analyses.

In cost of SA on a large system is likely to be high. The cost of preparing accurate and detailed cost estimates for performing SA can likewise be high. The most desirable partitioning of the system for cost effectiveness, or the relative effectiveness of several candidate SA techniques, may not be obvious at the outset of a program. In such cases, the owner should consider separate funding of the SA phases as discussed earlier.

Schedule impact. Both the number and the types of problems found in an SA can have an impact on project schedules. One of the more desirable situations is to run a controlled test to demonstrate the problem and its consequent system effects. Some problems are of this category and are relatively easy to demonstrate. Other problems require repeated cycling in order for the condition to occur. Normally these equipment or software problems are associated with race conditions or equipment sizing and may only occur after some component degradation. These apparently non-repeatable or non-recurring problems give rise to the likelihood of occurrence argument. The reported condition may then be dimissed solely on the basis of a value judgment. Extreme care should be exercised in dismissing the report conditions and the possible system effects.

In some cases, a single SA report may document many conditions, although only one condition is described in the body of the report. For example, multiple points in a design intended to have redundancy may be compromised by a common condition, such as common connectors, power supplies, or electrical isolation. Rather than issue separate reports for each condition, only one report is issued and all conditions are listed. The resolution of the report conditions may be so extensive that only a major program redesign and/or consequent schedule slide is possible.

Once problems identified by SA are determined to have a schedule impact, proposed equipment or software modifications should be sent to the performing SA contractor. The SA contractor should then incorporate the proposed changes into the configuration baseline (network trees), and reanalyze the affected circuitry or software. The intent here is to assure that the modification achieves the desired system operation and that no additional sneak conditions are generated. The analysis can be performed on a proposed modification, thereby saving schedule time in the redesign, implementation, and testing cycle. The later the project development phase, the more the change analysis option should be considered by the procuring group or ultimate equipment owner.

Schedule impacts can also occur in the performance of the SA project itself. The data-acquisition phase of an SA project is critical to successful completion of the task. Timely data acquisition of the correct-type documentation is required so that inspection of the data for completeness and hookup can be performed. Any missing data areas or incompatible configurations, especially at equipment or module interfaces, can be investigated and new drawings or source computer code program listings ordered. In general all data should be received within the first 20% of the SA project, but this may vary depending on the requirements of the particular application. In no case should the data acquisition schedule exceed one-third of the project duration, unless these data are change information. Failure to acquire data in a timely manner can result in schedule slippage and increased analysis cost.

All SA reports should be evaluated by the responsible parties in a timely fashion. Any comments on the reports, including final report dispositioning, should be transmitted to the SA contractor. This is especially important in the initial project phase for the drawing error reports for hardware and software. Discrepancies between various types of documentation must be reported and resolved prior to completing the first one-third of the project schedule.

No evident project schedule impacts occur as a result of the type of equipment or software to be analyzed. Digital-type systems require a slightly longer performance schedule than relay-type logic only because of the amount of circuitry incorporated in the system.

Analysis scope and depth. Scoping factors which should be considered in an SA application include the following:

1. The number and type of components and/or computer instructions.
2. Schedule requirements.
3. Data availability.
4. Change analysis.
5. Estimated project cost.

The most important scoping factor in an SA application is the number and type of components involved in the system or systems to be analyzed. The greater the number of components, the higher the cost and the longer the required project schedule. The type of equipment may be digital, analog, or relay for hardware systems, and high-order or assembly language for software systems. No particularly significant scoping requirements are associated with any of these system types.

Digital systems incorporate much more circuitry in small physical dimensions, but the analysis is basically performed using the same approach, changing only some of the SA clues. Relay and digital systems are the two most prevalent hardware categories selected for SA. Hardware systems currently tend more toward digital-type applications. Assembly language is the most prevalent software category selected for SA. However, this software trend may change as more and more program applications shift to the high-order languages.

A prime consideration in scoping an SA project is the completeness of functions depicted in the drawings and/or source program listings. Much of a system design may be included in the supplied documentation, but there may be a requirement for documentation outside of the scoped area. For example, assume that an electronic governor system interfaces with the electrical power system, compressor surge control system, and process computer system. If the equipment that constitutes only the steam turbine governor control system is selected, undefined interfaces exist beyond which the performing SA contractor cannot see. Assumptions as to function, timing, and configuration must be made in the course of the analysis. Assumptions should be based on the interface documentation as much as possible. While the documentation does not show the actual circuit or code detail, it does specify function and pin or channel number assignments. It would be desirable to include documentation depicting the circuit/software configuration, eliminating the need to make assumptions.

The scoping and scheduling of SA is dependent on the availability of system data. Complete data packages describing an individual system or subsystem should be available for the analysis. If the complete data are not currently available, the analysis can be scoped to those areas that are complete, and the analysis sequenced into the remaining areas as data becomes available. In this way, the complete system or subsystem can be analyzed a section at a time and then the various sections joined in an integration analysis. If many desired functions are missing in the designated system and will not be available by at least one-third of the detailed engineering phase, commencement of SA application should be deferred until the documentation becomes available.

The performance of change analysis is very dependent on project duration. Short-duration projects under three months are not ideal applications for change analysis, unless all baseline data and the complete change package are available at project initiation. Attendant delays in transmittal of data can affect short-duration projects. Projects of six or more months duration are better candidates for change analysis. Formal change analysis is an extra cost option for SA tasks. Approved and released changes can be evaluated under this option. Proposed changes occurring in response to reported sneak conditions can, and should be, evaluated within the scope of the baseline SA effort.

Estimated project cost can be a major consideration in the performance of SA. Cost information relating to aerospace projects was reviewed in efforts to determine applicability to petrochemical industry projects. As of late 1981, there did not appear to be a commonality; consequently, one should approach the SA contractor with requests for cost quotation rather than attempting to estimate on the basis of NASA jobs. If the anticipated scope of the analysis and/or the systems to be analyzed change sufficiently, cost considerations may dictate a rescoping of the task. Also, one should be aware that high-order language software applications typically cost more than an equivalent number of assembly language program applications. Additional tracking and accounting of functions is required in the high-order language application.

Depth of analysis should be at the detailed component/instruction level, instead of at the subsystem or system level. Other analyses are available for considering the higher-level analyses, but very few of the analyses can be implemented on the detailed level. SA is unique in approach and has demonstrated the capability to identify problems not found by these other analyses and extensive testing.

Security considerations. When proprietary data are involved in a SA, it may be necessary to obtain a signed agreement from the SA contractor not to release such data to a third party and to use such data only in the performance of the SA. The SA data base including input data, computer data base, output reports, worksheets, etc. would normally be retained by the performing contractor for a specified period to ensure its availability should the need arise to perform SA on system modifications. Maintenance by the performing contractor will be more cost effective and the risk of loss is less.

Summary

Establishing Need for SA

1. Reliability improvements in the overall project result from the identification and resolution of system problems. SA is very effective in identifying problems which may be missed by other analyses and testing.
2. Independent analysis is currently the only established approach for the analysis. The analysis must be performed by a contractor independent of the design group to preserve the integrity of the effort. It is also an excellent analysis tool which can be used to verify or cross-correlate the results or findings of other analyses.
3. Problem detection to eliminate the need for costly retrofits or redesigns in mass-produced systems and possible loss of critically important one-of-a-kind systems such as offshore platforms are immediate considerations for performing the analysis.
4. High criticality of the systems to be analyzed also warrants the analysis. Critical machinery protection, safety, and shutdown systems are the most likely candidates. Low criticality systems may be elminiated from consideration as long as no active control functions are performed in these systems.
5. Unresolved system problems or "recurring glitches" that have not been found by other analyses or tests are also good candidates for SA. If the analysis is undertaken to identify or isolate these system problems (typically during late-development phases), allow the contractor some additional leeway in cost and interpretation of instructions included in his work scope. Frequently, the unidentified problem causes are located in "unrelated" equipment/software areas. Analysis of only problem-prone functions, or areas where the problem is manifest, such as an instrument panel or test equipment, may be insufficient to locate the cause of the system problems.

6. A high change rate in the baseline design can also be used to justify the analysis. Loss of the design configuration baseline resulting from greater-than-expected change rates can be rectified by the detailed analysis of each change before the change is implemented in the hardware or software system.
7. SA is a cost-effective tool in all phases of project development, but the analysis results exhibit a pronounced effectiveness in early development phases, and particularly in the full-scale engineering development phase.

Determining Application Systems

1. Systems which perform active functions are the primary candidates for SA. Electrical power, distribution, and controls have traditionally been the main areas for hardware analysis. Computer programs which actively control and sequence system functions are good software candidates. In general, those systems which occur in the safety, command, and control areas are the primary candidates. Non-repairable systems are especially good candidates.
2. Passive systems that do not affect the overall plant operation can be omitted from analysis consideration. Systems which do affect plant operation but are highly redundant may also be excluded. Redundancy in control areas, however, is not a ground for eliminating the analysis. There may be design problems which compromise or destroy the redundant design.
3. SA can and has been successfully implemented on complete, unspared turbomachinery applications, as well as limited subsystem or functional applications. The analysis is best performed on configurations involving numerous system interfaces and large-size systems. The high number of interfaces, as well as the complex designs, are primary causes of embedded sneak conditions.
4. The applicable systems should be completely specified by component or instruction-level documentation in the form of schematics, drawings, wire lists, and source computer program code so that the analysis can be conducted at the "as-built" and "as-coded" levels, respectively.
5. Detailed analysis of critical systems can be performed by blending various analysis techniques which bring to bear the best features of each analysis in identifying design and fault-related problems. Favorable project results and costs are obtained in blended analyses. Highly critical functions can be identified by other high-level system analyses such as a preliminary hazard analysis or system hazard analysis.

Calculating Project Cost and Allocation of Budget

1. The cost of SA can be computed on the basis of the number and type of hardware components and the number and type of computer program language instructions. Cost figures relating to NASA-sponsored SA projects do not appear to match those experienced in recent petrochemical industry projects. The user should, therefore, base his cost estimate on formal discussions with, and proposals from, qualified contractors.
2. Limited budgets may force scope reductions and restrict a broad program application of SA. The analysis can and has been scoped to individual systems,

subsystems, and functions. Excessive scoping, however, could limit the analysis effectiveness by eliminating the detailed function tracking which is typically developed across system boundaries. Acceptable project costs are possible by selection of limited program systems.

3. If project funding and/or documentation are major factors restricting performance of SA, then an incremental contracting approach can be undertaken. Perform SA on one or more of the highly critical systems for which documentation is readily available. In a following fiscal period, contract for SA on the remaining systems, with the stipulation that the analysis includes the new systems and interfaces into the previously analyzed systems. This approach is especially desirable when detailed drawing or code instructions are missing for particular equipment or program modules, respectively. Functional diagrams should be made available to the SA contractor for these missing areas.

4. The purchaser can expect annual costs for SA *and* problem resolution to be lowest in the early development phases. There are significant cost and risk penalties associated with late identification and resolution of system problems.

5. Appropriations for the analysis should be allocated in the formulation of the reliability assurance plan and maintained throughout the development cycle for the desired schedule start time. If the project dollars have not been allocated early, they may not be available when required.

6. Since SA can be effectively blended with other analyses, reduced project costs for the combined analyses can be achieved.

Scheduling Requirements

1. SA should be scheduled so that final project results are obtained and can be adequately evaluated by the purchaser and equipment manufacturers prior to the end of the full-scale testing phase. Project costs to implement system changes increase dramatically after this phase.

2. The preferred start time is prior to the full-scale engineering development phase. This is an ideal time to provide a formal input into the design review process.

3. Timely results can be obtained for all scheduled SA projects and also for those projects which are intended to identify a single test or operational problem. For single-problem-oriented SA, limited system scoping and available documentation can provide project results as soon as 1-2 months into the project schedule.

4. Orderly scheduling of SA can be based on an assumed average 4-6-month project duration. Targeting the analysis to a specific project milestone can be performed by moving the start date back the specified number of calendar months. The two most important items affecting successful performance by the designated program milestone date are data availability and contractor performance.

Monitoring Guidelines for the Purchaser

1. Data acquisition has customarily been handled by the ultimate equipment owner. If data acquisition is assigned to the SA contractor, extra cost is

incurred. Proprietary data from vendors and contractors typically requires proprietary data agreements which may require significant time to acquire.

2. SA report evaluation and coordination are important responsibilities of the purchaser. Tracking all reports and their eventual dispositions is an important element in assuring effective program benefits for the project expenditure. The resolution of the identified problems provides a measure of reliability improvement and sneak-finding capability of the performing contractor. Document error reports are typically found early in the project schedule, and once the network tree drawings or diagrams are generated, the primary reports are design and sneak-condition reports.

3. Liaison, contract monitoring, contract modification, and project closeout are the remaining tasks to be handled by the purchaser, his design contractor, or the ultimate equipment owner.

Conclusion

SA is an especially effective potential problem identification tool which is now available to petrochemical plants intending to improve systems and equipment reliability. The analysis tool adds a new dimension to assessing and evaluating reliability of new or mature systems. Most of the reliability tools are fault or failure related. The fault-related analyses are based either on determining eventual system effects produced by specified equipment failures, or alternatively identifying undesirable top-level system events and then determining those functions which can produce the undesired events. Sneak Analysis, however, provides a critical review of systems based on intended modes of operation and assumes no equipment or code failures. The identification of unintended modes of operation and their resultant system effects is the end product of an SA task. As such, SA is complementary to the fault-related techniques. The analysis is best performed by a contractor, agency, or group independent of the equipment or project design group. SA should be based on actual system design produced from "as-built" drawings, schematics, and wire lists for hardware systems, and from "as-coded" computer program source code for software systems.

SA can identify problems before they occur in test or operation so that the cost to modify or redesign should be decreased and the reliability and safety of the system should increase. The analysis method can be specified, funds allocated, and the entire analysis performed early enough in the project development cycle to allow cost-effective system changes to be implemented. Experience shows that regardless of the project development phase, application environment, equipment/software type, criticality ranking, or project cost, SA identifies a significant number of system problems. These taper to lower, but significant, levels in late-development phases. Critical machinery protection systems represent areas of high problem report levels.

Chapter 9

Formalized Failure Reporting as a Teaching Tool

Experienced process plant managers are fully aware of the need to investigate costly equipment failures to prevent, or at least reduce, recurrence of the problem. Failure analysis reports are, of course, a means of presenting the findings in written form for the purpose of informing management. However, the hidden value of formalized reporting should not be overlooked.

A formal failure analysis report is a teaching tool for the investigator. It compels him to think logically, capture all relevant data, consider pertinent background events, and clearly state his conclusions and recommendations. Experienced engineers can testify that the bottom line, or final analysis results contained in their formal written reports, are sometimes quite different from the mentally summarized and verbally enunciated "early" results.

To be useful and effective, formal failure reports do not have to conform to a rigid format. Nevertheless, they must be structured to present the story in a relevant and logical sequence. This sequence typically consists of the following:

1. *Overview.* This is the first paragraph or subsection in a formal failure report. The overview could also be called a "management summary." As this term implies, it contains a synopsis of the entire investigation and answers in a few well-chosen words or sentences the basic questions of what happened, when-where-how, and possibly why.

   Note how these questions are answered in both of the sample failure reports which follow later. In Report No. 1, "Propylene Storage Unit P-2952 HSLF Pump Failure Report," the writers chose to call the lead paragraph an "overview." In Report No. 2, "Primary Fractionator MP-17A Tar Product Pump Failure and Fire Investigation," the first subsection is called "summary" and is comprised of three short paragraphs. However, the lead portions of both reports are about equal in length—roughly 100 words—and manage to convey essential overview information.

2. *Background information and/or summary of events.* A formal failure analysis report will probably be read by management and staff personnel whose familiarity with the equipment, process unit, operation, etc., will vary greatly. It is thus appropriate to include pertinent background information in the failure analysis report so as to make it a "stand-alone" document.

Although many failure analysis reports use the word "background" as the second heading, neither of the two sample failure analysis reports contain this term. Instead, Report No. 1 describes all pertinent background data under "summary of events," while Report No. 2 distributes appropriate background data over two headings, "equipment and process description," and "sequence of events." Recall, again, that even a formal report need not adhere to a rigid, repetitive format. A given topic or writing style lends itself to one or more ways of effective documentation, and both report formats follow a logical thought pattern. Illustrations or graphic data may be included in this section of the report. Very often, these representations are effectively presented directly in the text. Contrary to what we may have learned in college, bunching the representations at the end of a formal report makes very little sense and will, in *all* cases, disrupt the smooth flow of the reader's thoughts and perceptions. Only appendix material or similar supplements should be given last.

3. *Failure analysis.* This is normally the next major heading in a formal report. Detailed data collected and observations made are usually documented under this heading. This is where a competent researcher will put the solid evidence and will prove the soundness of his reasoning. Both of our demonstration reports contain this heading and lay the evidence on the table. Report No. 2 uses such subheadings as "Observations," "Probable Failure Progression," and "Probable Failure Causes" simply because they fit the subject matter and, frankly, the writing style of the committee charged with its issuance.

The bulk of the illustrations or graphic data will probably appear in this section of the failure analysis report.

4. *Conclusions.* The word "conclusion" has several dictionary definitions: the close or termination of something, the last part of a discourse or report, a judgment or decision reached after deliberation. The conclusion of a good failure analysis report should be none of the above. Instead, it should contain the evident and logical outcome or result of the observations, and thought processes documented up to this point. Both of the demonstration reports contain partial restatements of what probably caused the failure, as well as what was judged *not* to have caused the failure.

The investigator should consider it appropriate to list not only the most probable cause but give thought to the perceptions of the uninitiated, or less-experienced readers of his report. Take Report No. 1, for instance. When its conclusions were formulated, the investigating team anticipated questions such as "how do they know the pump didn't explode?" To lay to rest questions of this type, it was decided to let the reader know that this possibility had been looked into and was ruled out. Similarly, the conclusions portion of Report No. 2 lists first the principal conclusion, i.e., statement of failure cause, and follows up by explaining why other scenarios or possibilities must be discounted.

5. *Recommendations*. These are made last. Their ultimate aim is, of course, to prevent recurrence of the problem or failure. However, recommendations can sometimes be separated into near-term and long-term requirements. Also, they can be supplemented with an action plan, or simply a notation of action assignments and action status. Report No. 2 represents this latter format.

Examining the Sample Reports

Much thought was given to the best presentation of practical examples of formal failure analysis reports. The intention is to show actual reports which illustrate a proven approach and have general applicability to the majority of machinery failure situations.

We settled on a format which would first reprint a section of the actual report and then explain the thought processes and rationale which had been employed by the report writer in formalizing his findings.

Writing the report is, of course, the culmination of the investigation and analysis. The writing effort will have been preceded by intuitive processes which chiefly include data collection, examination of the chain of events, gathering and judging of relevant input, fitting the pieces, and defining the root causes of failure.

With this background, we can go directly to the reports. Both reports start with a cover letter, or letter of transmittal. This letter states in a few brief, crisp sentences the principal facts of the failure event, comments on safety-related matters, and reports the status of the investigation (see Figure 9-1). The cover letter is signed by the members of the investigating body and may or may not give their titles or job functions. Both cover letter examples give clues as to the preferred composition of many failure investigating committees in the petrochemical industry: a senior member of the technical staff, line supervisors from operating and support departments, and one or more lower-level operating personnel.

The Case of the High-Speed, Low-Flow Pump Failure

The first of our two example reports deals with a high-speed, low-flow (HSLF) pump which suffered a strange failure in July 1980. The seriousness of the failure in terms of potential injury to personnel and monetary loss to the plant was immediately recognized. As is customary in such cases, plant management requested a formal investigation of the failure event with a view toward establishing the root cause and avoiding a repetition of the failure.

Team Composition. Formal failure investigations start with the formation of the investigative team. The team members are designated by the plant manager or his appointed representative. For the investigation of the HSLF pump failure, management asked for representation by personnel from the operating, maintenance, and technical departments. A senior engineer from one of the mechanical technical services (MTS) groups was assigned to chair the effort. The maintenance department was represented by a mechanical supervisor, and the operations department delegated a supervisor and a technician to participate.

ydrocarbon

August 15, 1980

Propylene Storage Unit P-2952
High-Speed Low-Flow (HSLF) Pump Failure Report
File: 97-4-G

On July 30, 1980, one of the propylene storage pumps (P-2952) blew apart at the casing cover split line. The top half of the pump case, the gearbox, diffuser and motor lifted off and came to rest on the concrete about two feet away. The propane in the suction and discharge lines dispersed rapidly. The incident did not cause any injuries and no fire resulted.

An investigation of this event has been completed. The report of the investigating committee is attached.

John P. Sloper
MTS Engineer, Chairman

George W. Windbreaker
Operating Supervisor

M. E. Person
Mechanical Supervisor

Danny Wright
Operating Technician

HPB:jj
Attachment

cc: ABC   DEF   GHI   JKL   MNO
18B18

Figure 9-1. Letter of transmittal (cover letter) for a formal pump failure report. Note composition of the investigating committee.

Selecting participants or team members from the various segments of the plant organization is considered advantageous for two reasons: (1) it brings together a variety of backgrounds and experiences and leads to a supplementing and complementing of ideas, thoughts, and approaches, (2) it ensures consensus among departments or work groups which are known to have an occasional tendency to pin the blame on the other party.

At the conclusion of their formal failure investigation, the team members summarize their findings in a brief cover letter and affix their signatures. The cover letter, or letter of transmittal, is generally quite brief. In the HSLF pump failure example, the letter states "when-what-where" in the first two sentences. In just two more sentences the reader is given the principal results of the event. The four signatories opted not to state the root cause of the failure in the cover letter, although a different team of investigators may well have decided to highlight the most probable failure cause in one or two additional sentences.

Overview. The body of the investigative report starts with an overview giving details of the when-where-what mentioned earlier in Figure 9-2. Time, location, service details, and equipment particulars are described. The overview paragraph gives the failure cause as well as estimated cost of repair and restoration. It is in the overview paragraph of the report that the reader should be shown how and where the failed machinery fits into the process and interacts with other elements, units, piping circuits, or components. A few well-placed illustrations or schematics will generally help the reader to understand the topic better. These representations belong in the body of the report and should be interspersed in the narrative as appropriate. However, the body of the report is rarely the right place for highly detailed illustrations, calculations, or reproductions of pertinent correspondence. All of these belong in the appendix or attachments to the report, as we will see later.

Summary of events. It is of interest to note that "summary of events" does not refer to failure progression or failure mechanism suspected or identified. Instead, the summary of events section of a failure or incident report relates operating conditions, unit status, process parameters, personnel, and similar matters preceding or accompanying the event. More than any other part of the report, this section documents statements, depositions, conversations, interviews, log-book entries, and chart-readings relating to equipment status before and during the failure event. Furthermore, this section may state the rationale for operating in a certain mode, give observations made immediately after the event, and highlight actions initiated by responsible parties and personnel on the scene (see Figure 9-3).

Accordingly, the example report gives times and identifies shift numbers, personnel, process lineup (fluid routing), switching sequence, and fluid pressures. A restatement of relevant equipment physical details and appropriate pictorial representation follows. The report states in what condition the equipment was found after the event and notes observations made by various people.

Failure analysis. The team member representing the plant technical organization normally takes the lead in pursuing the actual failure analysis. He will contribute to the failure analysis to the maximum extent possible. However, the major

Overview

At 2:15 a.m., Wednesday, July 30, 1980, one of the propylene storage pumps (P-2952 on simplified schematic, Exhibit 1) blew apart at the bolt line. The pump is a Webster Model HSLF, 150-horsepower, 18,300-rpm unit. The top half of the pump case, the gear box, diffuser, and motor were lifted several inches and came to rest several feet away. The bottom half of the case remained on the base. There was no fire and no injuries were caused by this incident. The failure event was attributed to sulfide stress corrosion of the bolts. Estimated cost of repair and restoration: $10,000.

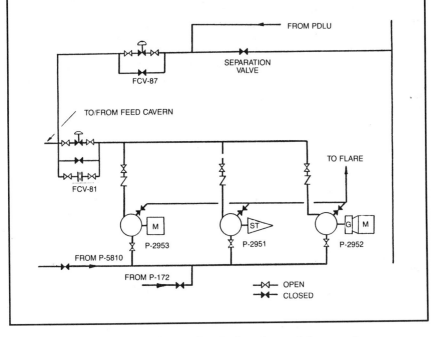

Figure 9-2. Overview section of a formal pump failure report.

contributions of this person are probably best described as perception and resourcefulness. He will have to think about fitting the pieces of the puzzle (sometimes acquiring a thorough understanding of unit operations and processes), separating pertinent observations from extraneous ones, structuring the various findings into a plausible and highly logical sequence, and explaining these matters to the other team members. The person representing the technical organization must show resourcefulness by requesting appropriate input from others in the plant organization, from persons reporting to affiliated plants, colleagues in non-affiliated plants, consultants, outside research organizations, testing laboratories, or perhaps institutions of higher learning.

Summary of Events

The propylene storage unit had been down for a short turnaround. Startup procedures had begun on 11x7 shift Tuesday, July 29. From 11:30 a.m. until 7:30 p.m. on Tuesday. P-2953 had been used to pump the off-test propylene and propane into storage. At 7:30 p.m., the propylene was on-test and routed to our customer.

At that time, P-2953 was shut down and P-2952 was put on line to continue pumping the offtest propane to the feed cavern. The P-2952 was utilized in this service because it is a lower capacity pump. While in this service, P-2952 suction was blocked from the offsite feed line. The depropanizer overhead pump (P-5810) was supplying feed to P-2952. The suction and discharge pressures for P-2952 were 230 psig and 780 psig, respectively. At 1:00 a.m., Wednesday, the propane was on test and routed to the Refinery. P-2952 was shut down at that time.

P-2952 is a high-speed, vertical, Webster pump. The high speed is developed through a gearbox. The pump case is divided into a top half and bottom half which are bolted together with twelve 3/4-inch bolts as indicated on Exhibit 2.

At 2:15 a.m., P-2952 blew apart at the bolt line. The top half of the pump case, the gearbox, diffuser, and motor were lifted up several inches. Evidently, the power cables restrained it from going further. It traveled east approximately two feet. The motor struck and bent a 3/4-inch steam line. The force of the blowout tore the small piping (seal oil, transmission oil, and cooling water) from the case.

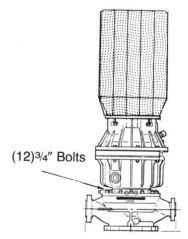

(12)¾″ Bolts

Exhibit 2. Webster HSLF pump.

The propane in the suction and discharge piping formed a vapor cloud which dispersed rapidly. Immediately after the blowout, Messrs. L.E. Hawker, R.D. Smith, R.W. Runner, and J.D. Scranton (Process Technicians) rushed to the unit to investigate. Mr. W.H. Teckel, Mechanical Technical Service, and Mr. L.N. Navasoti, Night Superintendent, were at the West Gate at the time of the blowout. They also proceeded to the unit. The vapor cloud dispersed prior to any personnel reaching the area.

The technicians blocked the suction and discharge valves at the pump and locked the motor out at the Motor Control Building. Mr. Navasoti requested the area be left as it was so pictures could be taken the next morning.

Figure 9-3. Summary section of a formal pump failure report.

In the example report, the subheading "Failure Analysis" (Figure 9-4), introduces a narrative explaining the testing to which the failed components—in this case nuts and bolts—were subjected. The basis for testing is described as visual inspection after showing pre-existing cracks in several bolts. It is further stated that this observation led the reviewer to suspect stress corrosion cracking. The narrative goes on to list the results of laboratory testing and contrasts as-specified physical characteristics with as-found physical characteristics.

Failure Analysis

Failed bolts and stripped nuts were subjected to extensive physical and metallurgical testing. Visual inspection had shown pre-existing cracks in several bolts, and stress corrosion cracking was suspected. Laboratory examination resulted in the following findings:

1. Nuts proved to be ordinary mild steel (AISI 1010) with inferior strength properties. ASTM 194 Grade 2H nuts are required.
2. Bolting was AISI 4140, but lacked ductility because of incorrect heat treatment. Tested versus required properties for B-7 bolting are as follows:

	Test	Required
Yield point at 0.2% offset:	191,200 psi	105,000
Ult. strength :	205,900 psi	125,000 psi
Elongation in 1 inch :	8.2%	16%
Reduction of area :	39.2%	50%

3. Fracture surface analysis with scanning electron microscope showed stress corrosion cracking had initiated bolt failures.

   Chemical analy ʌs of residue in contact with the bolt thread showed sulfur present. Also, the hardness of failed B-7 bolts was quite high, Rockwell C-43 average. The combination of high hardness and sulfur points to hydrogen sulfide stress corrosion cracking.

A brief review of sulfide stress corrosion is deemed appropriate. Numerous investigations have shown that the optimum microstructure for resistance to sulfide cracking is tempered martensite resulting from heat treatment by quenching and tempering. These studies have shown that alloy steels having a maximum yield strength of 90,000 psi and a maximum hardness of Rockwell C-26 are not susceptible to sulfide cracking, even in the most aggressive environments. Our plant standards add an extra margin of safety and list RC-22 as an "out of danger" value. As the strength level increases above 100,000-110,00ſ psi, the threshold stress required to produce sulfide cracking may actually decrease ıı very severe environments. Exhibit 3, below, shows how AISI 4140 bolts hardened to Rockwell C-43 lie in the 100% failure zone if applied stresses exceed 25,000 psi. Most bolts in pump applications are torqued to stresses in the 30,000-40,000-psi region to which a portion of the internal pressurization stresses may have to be added.

(figure continued on next page)

Figure 9-4. Body of a formal pump failure report.

The following conditions must be fulfilled for sulfide cracking to occur in susceptible materials:

1. Hydrogen sulfide must be present.
2. Water must be present in the liquid state.
3. The pH number must indicate acidity.
4. A tensile stress must be present.
5. Material must be in a susceptible metallurgical condition.

· When all of the above conditions are present, sulfide cracking may occur with the passage of time. All of these conditions can be assumed to have been present at P-2952:

1. The automatic transmission fluid in the gearbox of this Webster pump contains 2,600 wppm of sulfur. ATF leakage is a common occurrence on this pump.
2. The bolts are exposed to rainwater.
3. The ATF is slightly acidic.
4. Tensile stresses are present.
5. As shown in Exhibit 3, the material is in a highly susceptible condition.*

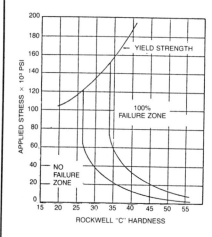

* Source: Cameron, Danowski, Weightman; "Materials for Centrifugal Compressors," Elliott Company, Jeannette PA., Reprint 166.

Exhibit 3. Stress vs. hardness relationships for steels experiencing sulfide stress corrosion cracking.

Figure 9-4. Continued.

At this juncture, there would generally be a temptation to present exact copies of laboratory reports or laboratory correspondence. However, these are generally inappropriate because they lack the narrative ingredient which makes the typical failure report the "readable" document it is meant to be. A failure analysis report is written for an audience not normally familiar with laboratory reports. Report writing should always be done with "audience awareness," and failure reports dealing with machinery problem events in the process industries are certainly not exempt from this rule. Thus, laboratory reports and similar supportive documentation should appropriately be placed at the end of the report.

In the HSLF Pump Failure Report, the results of bolt-strength analysis and material verification for nuts are followed by the findings of fracture surface analysis with a scanning electron microscope. The statement relating to these findings further alludes to the fact that a combination of high hardness, as found in the strength-of-materials investigation, and sulfur, as found in the chemical analysis of residue in contact with the bolt threads, could point to hydrogen sulfide stress corrosion cracking. The writers, for the time being, avoid blaming the failure on this mechanism. Instead, they opt for brief review of sulfide stress corrosion cracking because their "audience awareness" told them that many of the readers would benefit from the explanations.

The explanatory paragraph on sulfide stress corrosion shows that, for the purposes of a failure report, even a relatively complex topic can be adequately explained in less than 10 sentences. Note also that the writers were able to allude to "plant standards." This lets the reader know that responsible personnel are well aware of this failure mechanism. A graphical representation of susceptibility ranges is inserted in the text. It is evident that the particular illustration chosen in this instance facilitates the reader's understanding of explanations made in narrative form. It belongs in the text, and the writer's decision not to relegate the illustration to an appendix location was highly appropriate.

With the statement "the combination of high hardness and sulfur points to hydrogen stress corrosion cracking," the failure analysis section of the report set up a hypothesis. This hypothesis must now be tested for validity by probing whether all contributing factors are, in fact, present. The report therefore lists the various conditions which must co-exist for sulfide cracking to occur in susceptible materials. In this particular case, the researchers identified a total of five such conditions and went on to show that all of these contributory conditions must be assumed to have been present at P-2952. The evidence thus strongly supports the belief that sulfide stress corrosion cracking took place and caused the failure.

Conclusions. In formulating their conclusion, the team members decided not only to say what the failure *was,* but also what the failure was *not* (see Figure 9-5). This was again prompted by "audience awareness," and the anticipation that some readers would be inclined to ask questions which the report did not specifically address. Thus the statements about there not being any evidence of a detonation or explosion, and liquid propane and automatic transmission fluid not causing a chemical reaction violent enough to cause excessive pressure rise in the pump casing.

The remaining statements are firm and definitive: bolt failure was initiated by sulfide stress corrosion of excessively hard B-7 material; leakage of sulfur-containing automatic transmission fluid provided the environment necessary to cause cracking of steel in tension. Also, pump vibration observed prior to failure must have accelerated the propagation of pre-existing cracks in the bolts.

The conclusions also address the question of why the pump came apart at standstill, although one would intuitively expect its component parts to be more

Conclusions

There was no evidence of a detonation or explosion. Available data shows that liquid propane and automatic transmission fluid would not have caused a chemical reaction violent enough to cause excessive pressure rise in the pump casing.

Bolt failure was initiated by sulfide stress corrosion of B-7 material which had been heat treated to excessive hardness values. Occasional leakage of automatic transmission fluid used in the Webster gearbox provided the sulfide environment conducive to this failure mode.

Pump vibration observed prior to failure is thought to have accelerated the propagation of pre-existing cracks in the bolts.

Minor leakage through the discharge check valve caused the pump casing to be pressurized to cavern pressure. While this pressure is, of course, well within the design pressure of the pump, the separating forces (and tensile stresses) on the casing bolts are higher when the pump is so pressurized than when the pump is running. This explains why failure did not occur while the pump was operating. (See Attachment I).

Similarly, the diffuser is exposed to a net upward force with the pump uniformly pressurized at standstill. This caused the diffuser to be ejected from the pump case.

Figure 9-5. Conclusions section of a formal pump failure report.

highly stressed while the pump is running. In the course of the failure investigation the MTS engineer took a close look at the forces acting on pump components of interest. The calculations—appropriately given as Attachment I (Figure 9-7) to the report—established that structurally sound bolts would safely contain the pressure at either standstill or pump operation. The calculations further established that the net separating forces on the cover at standstill of the pump were higher than the forces exerted while the pump was running. This explains why bolt failure occurred at standstill and not during operation of the pump. As a final check, the investigators verified that the bolts would not have been overstressed even if the casing had been exposed to the full design pressure of 1440 psi.

Recommendations. The main purpose of a failure investigation is to identify steps which could be taken to prevent recurrence of the event. These steps are outlined in a subsection given the heading "Recommendations" (see Figure 9-6). Each recommended step or item must be the logical followup to findings, statements, observations, or conclusions given earlier in the report. It would not be appropriate to make recommendations whose rationale or background would be known to the investigator but would not have been disclosed to the reader of the failure report.

For the sake of illustration and emphasis, test each recommendation to verify that it meets the criteria stated previously:

1. *Immediately check all Webster pumps for excessive vibration.* The report had earlier concluded that pump vibration observed prior to failure could have accelerated the propagation of pre-existing cracks in the bolts.

Recommendations

- Immediately check all Webster pumps for excessive vibration.
- Replace all casing cover bolts in Webster pumps equipped with gearboxes. New bolting shall not exceed a hardness of RC-22 or 250 Brinell. Replacement should be initiated immediately and should be completed no later than September 15, 1980.
- Replace all nuts on Webster pumps. ASTM 194 Grade 2H nuts are required.
- Pump repair shops handling Webster repairs for our plant should be notified, in writing, of the above requirements.
- Specifications for bolts on Webster pumps should be amended so as to *disallow* hardness values in excess of 250 Brinell or RC-22.

Figure 9-6. Recommendations section of a formal pump failure report.

2. *Replace all casing cover bolts in Webster pumps equipped with gear boxes.* The report established sulfur stress corrosion to be the prime initiator of the failure event and further submitted evidence that sulfur was contained in the automatic transmission fluid (ATF) used in Webster gear boxes. Replacement of casing cover bolts is clearly a prudent step for pumps equipped with gear speed increasers. This particular recommendation step also restates the bolt hardness limit given in the body of the report.
3. *Replace all nuts on Webster pumps.* Although the hexagon nuts used in conjunction with the cover bolts did not contribute to the overall failure, they were found not to meet the specified strength properties. They should be replaced in an effort to bring all parts of Webster pumps into compliance with purchase specifications.
4. *Repair shops handling Webster pump repairs for the user's plant should be given formal notification of the new requirements.* Obviously, there is a risk of repair facilities replacing parts in kind. Formal notification of new requirements is appropriate for proper emphasis of urgency and to establish the repair shop's legal obligation to provide proper materials. No mention is made of the possibility that incorrect materials were perhaps substituted for proper ones during a recent repair.
5. *Future specifications should clearly disallow bolts with excessive hardness.* The body of the report contained narrative and graphic-illustrative evidence showing highly stressed bolting with high hardness very susceptible to failure. Accordingly, specifications for future procurement of Webster pumps should clearly indicate allowable hardness limits. No clues are given as to whether bolts with excessive hardness were originally furnished by the Webster Company. The reader is left to conclude that the failure investigation effort was either unable to determine the origin of failed bolts or considered it of academic value only to make this determinaton. It can, however, be inferred that the purchase specification for Webster pumps did not contain any specific requirements regarding bolt hardness.

Appendix material. Except for appendix material, the original failure report was issued in exactly the format shown in this text. For the sake of illustration and completeness, we are providing a listing of the original appendixes without actually reprinting all of them here:

1. *Documentation of separating forces acting on pumpover and diffuser.* Shown above as Attachment I (Figure 9-7), these calculations were given in the appendix of the original report.
2. *Tensile test and chemical analysis of bolts and nuts.* This attachment consisted of certification documents from the metallurgical laboratory.
3. *Tabulation of significant prior failure events on LPG pumps in the United States and Europe.* Significant event reports, fire incidents, etc., are often shared by the petrochemical industry. Major industrial insurance carriers or the central engineering offices of large companies are also in a position to assemble and catalog pertinent event reports for future review and statistical reference.

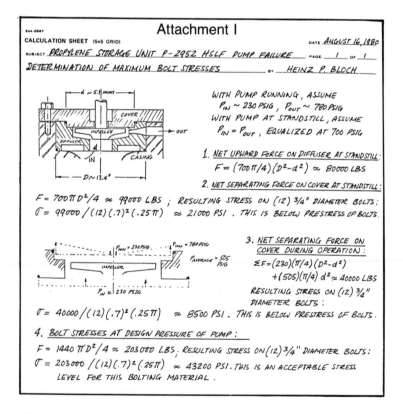

Figure 9-7. Appendix portion of a formal pump failure report would typically include calculations, tabulations, cross-sectional views, etc.

These tabulations were made available in this example case and were appended to the report.

4. *Cross section of Webster HSLF process pump.* It was felt that at least some readers of the original failure analysis report would be interested in a review of the component arrangement found in Webster HSLF pumps. This prompted the investigative team to include a dimensionally accurate cross sectional drawing in the report appendix.

5. *Results of fracture surface analysis (scanning electron microscope examination) made on broken bolt surface.* The findings reported by the research laboratory were appended for future reference and for the purpose of authenticating statements made by the investigating team in this regard.

6. *Specifications for P-2952.* Primarily comprised of the API data sheet, specifications for P-2952 were appended to complete the reference documentation package.

The Case of the Tar Product Pump Fire

The second of two example reports explains a serious failure and subsequent fire event on a single-stage overhung centrifugal pump in tar product service at "Ydrocarbon," a major manufacturing facility. Again, this example is based on actual field experience of a large petrochemical plant in the US Gulf Coast area. The failure investigation and reporting process used to deal with this particular event are explained to show the logic and rationale which led to positive identification of failure mode and guidelines for avoiding similar problems in the future.

Cover Letter. The cover letter, or letter of transmittal, for this failure report (see Figure 9-8) is not very different from that utilized in our earlier example. It is very brief, purely intended to give higher management the main points in a few clear, crisp sentences. Here, the cover letter answers such questions as "when—what—where—why—how" in a four-line paragraph and concludes by stating that the investigation is hereby complete and the most probable sequence of events identified.

From the list of signatories, we can again infer that management has opted to select the members of the investigative team from different departments within the plant. We can identify representatives from the mechanical technical services area, from operating departments, and the plant maintenance organization. There need not be a firm rule to limit representation to only four investigators, nor is there any great merit in restricting team composition to certain departments. If a large plant has safety engineering or loss control engineering functions, there may be additional incentive to assign team membership to these personnel.

In any event, selecting team members from several different segments of the plant organization will bring together a variety of backgrounds and experiences. It is to be expected that the team members contribute complementary ideas to the team effort. Moreover, representation from several different groups will greatly reduce or even eliminate the possibility that others do not accept the findings and recommendations or that one group will blame the other for the failure event.

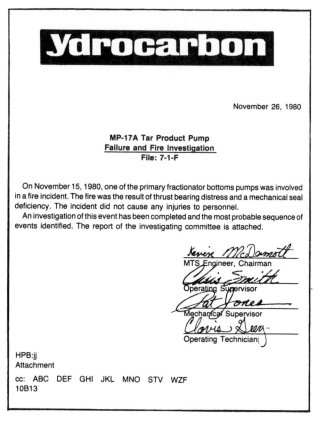

Figure 9-8. Letter of transmittal (cover letter) for a formal pump failure report. Note composition of the investigating committee.

Summary. The reader will recall our earlier statement indicating that failure reports are not necessarily locked into a rigid format. In keeping with this observation, the second example report starts out with the subheading "Summary" (see Figure 9-9) whereas the first sample report used the term "Overview." The summary lists the time at which the subject machinery—MP-17A—experienced catastrophic failure. The lead sentence states that failure of mechanical components resulted in a fire and goes on to report how and when the fire was brought under control. With personnel safety of paramount importance in the petrochemical industry, the investigative team reported "no injuries" in the first paragraph of the summary. Next, the report summary states why the mechanical components failed and concludes with a brief description of extent of damage and cost of restoration.

Equipment and process description. This subsection of the failure report (Figure 9-10) is usually quite straightforward and easy to compile. One or more pictorial representations or schematics such as Figure 9-11 might help the reader to visualize

Summary

At 19:33 hours, Saturday, November 15, 1980, one of the primary fractionator tar bottoms pumps (MP-17A) experienced a thrust bearing and seal failure which resulted in a fire. The fire was extinguished by Team 3 technicians and the area reported secure at 19:45 hours. No injuries were caused by this incident.

The failure event is attributed to inadequate prelubrication of the duplex thrust bearing. When the bearing seized, the mechanical seal opened up.

Physical damage includes a fractured bearing housing, severely bent pump shaft, and burned insulation. Restoration cost is estimated at $7000.

Figure 9-9. Summary section of a formal pump failure report.

Equipment and Process Description

MP-17A is an Atlantic Model BTSG 3 x 13, 250-hp motor-driven, 3560-rpm centrifugal pump. At rated conditions, the pump feeds 320 gpm of tar at temperatures ranging from 520°F to 410°F from the primary fractionator tar boot and coke filter MF-2 through tar coolers ME-13 and ME-15 to tar product tank MTK-31. Pump design inlet and discharge pressure conditions are 10 psig and 300 psig, respectively. Refer to Exhibit 1 for a simplified schematic of the tar system.

Differences between MP-17A and MP-17B/C: The A-pump is made by Atlantic. It did not, at time of failure, incorporate reverse flush. The B and C pumps are made by Unicorn. They incorporate reverse flush (stuffing box to suction).

Figure 9-10. Equipment description portion of a formal failure report.

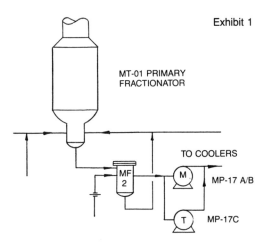

Exhibit 1

MT-01 PRIMARY FRACTIONATOR

TO COOLERS

MF 2

M   MP-17 A/B

T   MP-17C

Figure 9-11. Schematics will always enhance the equipment description portion of a formal failure report.

specifics of process flow or peculiarities of the equipment whose failure is being analyzed. Process conditions such as flows, pressures, temperatures, and equipment details such as manufacturer, model, size, speed, and flow should be listed.

Significant differences between the failed and any existing satisfactory equipment should be explained if both were used under identical process conditions or if the investigative team has reason to expect that the readers would raise this question. In the case of MP-17A, it was noted that seal failure preceded the fire event. It was thus rightly deemed appropriate to state the differences in seal flush arrangements between the failed MP-17A and the well-functioning pumps MP-17B and C. Whether or not these differences had something to do with the failure of MP-17A is, for the time being, unimportant and may not merit further mention until perhaps later.

Sequence of events. The sequence of events leading to equipment failure must be investigated and documented (see Figure 9-12). In this example case it was

---

Sequence of Events

Process Operations Debriefing

   Process operations during the week of November 3, 1980, included tar pump test operations with filters in the circuit. No extreme plugging was experienced. The filters were first bypassed on November 10, 1980, at 13:30 hours and a seal failure is thought to have occurred during the second shift (late evening) on November 12. At that time, the filters were put back into service.

   On November 15, 1980, seal repairs were complete, and the MP-17A motor-driven pump was put in service at 16:30 hours. Team #5 day shift reported the pump in service and operating properly at that time. Team #3 night shift took over duty between 17:30 and 18:00 hours. Satisfactory operation is graphically represented in Exhibit 2, the strip chart obtained from the Console 2 trend recorder for fractionator boot and main vessel level. At 17:30, the chart shows the boot level to be 100%. About 10 minutes later, the boot level is reduced to 35% for approximately 5-10 minutes before returning to 100%. The chart verifies that the boot level never dropped below 35% on the day of the failure incident.

M & C Debriefing

   M & C personnel report commencing MPM-17A motor solo run on Monday, November 10, 1980. The pump was coupled up on Wednesday, November 12, and run with the tar filter bypassed. It is thought that bypassing the filter caused coke fines to enter the seal cavity. The pump was removed from its field location during Thursday afternoon, November 13, to effect replacement of bearings and seals.

   On Friday, November 14, 1980, MP-17A was restarted at 17:00 hours and shut down around 17:10 to replace the discharge pressure gauge. Upon restart, the pump ran for about 10 minutes when it had to be shut down because of seal leakage. The pump was again removed and taken to the shop.

   On Saturday, a new seal arrived at 12:00 hours. This new seal and sleeve combination was installed and measurements taken during assembly. At the impeller end, the seal sleeve was found to be free to displace 0.006" (see Exhibit 3). Dial indicator readings taken on the rotating hard face showed an initial out-of-perpendicularity of 0.027" which, through a series of adjustments, was improved to 0.006" upon final installation. At 14:30 hours, the pump was ready for warm-up and hot-alignment checked at 16:00 hours. MP-17A was started at 16:30 and its operation observed until 17:20. During this 50-minute time span, the discharge pressure was observed to go from initially 280 psi to a low of 230 psi.

   The pump fire was detected around 19:33 hours and extinguished by operations personnel.

Figure 9-12. Failure reports may include a sequence of events section as shown here.

recognized that MP-17A had recently been repaired and that it was thus necessary to describe two such sequences. Process events were obtained in a debriefing of operators who reported that levels, flows, and pressures had been quite normal until the actual failure event. With adequate NPSH critically important to the safe, cavitation-free operation of centrifugal pumps, the availability of a strip-chart recorder tape showing sufficient level in the suction vessel was considered a particular advantage to the investigators. It is included in the report as Figure 9-13.

A debriefing of Mechanical and Construction (M & C) personnel shed some light on maintenance techniques and assembly quality-control procedures employed during repairs which preceded the final failure event. The report states that bypassing the filters probably caused coke particles to enter the seal cavity. It is implied that this could cause seal malfunctions unless a reverse flushing arrangement, i.e., a line from stuffing box to pump suction, was employed. Recall that under the heading "Equipment and Process Description" only MP-17B and C pumps were listed as incorporating reverse flush.

At this stage, the report does not pass judgment on the adequacy of mechanical work such as seal and bearing replacements but continues to list the sequence of repair events: ran for only 10 minutes on November 14; new seal installed on November 15, but dimensional difficulty was encountered when the installation tolerances were checked. Figure 9-14 is inserted in the report to facilitate visualizing the narrative description.

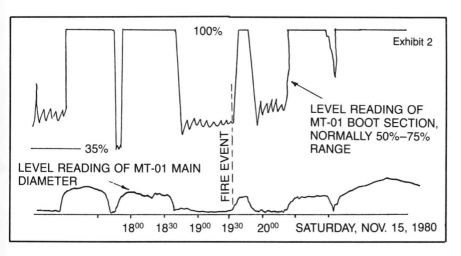

Figure 9-13. Supporting evidence may be in the form of strip charts and should sometimes become part of the formal failure report.

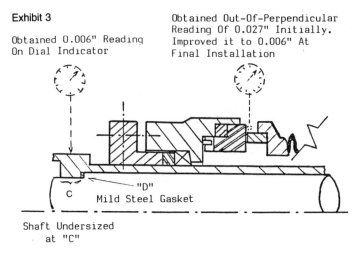

Exhibit 3

Obtained 0.006" Reading
On Dial Indicator

Obtained Out-Of-Perpendicular
Reading Of 0.027" Initially.
Improved it to 0.006" At
Final Installation

"D"
Mild Steel Gasket

C

Shaft Undersized
at "C"

Figure 9-14. Component details are best shown in sketches.

Failure analysis. The observations, conjectures, and analysis results of the investigative team are finally documented under this subheading (see Figure 9-15). The analyst observed the condition of every mechanical component and included Fig-ure 9-16 to again facilitate the reader's understanding of technical jargon.

It can be seen that the massiveness of the bearing failure lent strength to the investigator's belief that failure of the duplex thrust bearing set in motion the chain of events leading to the fire. Explanatory statements follow: excessive vibration resulted and caused the less-than-optimally installed seal to release pumpage. This pumpage could have ignited on the red-hot bearing.

After establishing a probable failure progression, the investigation now turns to probably failure *causes*. Why would a bearing fail so soon after installation? If it were designed for inadequate life, should it not have failed repeatedly in the past? Could it have been incorrectly installed? The report states that the failure was too massive to determine the exact failure cause from an observation of the failed parts and goes on to list a number of possible deviations which could have played a role in the failure event. Certainly, one of these would be inadequate lubrication.

Plant maintenance personnel were asked to describe how the pump bearings were being lubricated. They explained that new bearings were usually taken out of their oil-paper wraps and given a generous bath of "Western Handy Oil." Next, the bearings are pressed on the pump shaft and the entire assembly installed in the pump casing. After the pump is returned to the field, it is connected to an oil-mist lubrication header which supplies "Oilmist-100" in aerosol form to antifriction bearings throughout the plant.

The investigation now shifted to the properties of the "Western Handy Oil." The manufacturer gives its viscosity as 75 SUS at 100°F (~14 cSt at 38°C), which would

---

Failure Analysis

Observations

When the pump was dismantled, the seal area was found in clean and undamaged condition. Some solids were found in the impeller. Impeller wear rings and inboard bearing appeared satisfactory. The duplex thrust (outboard) bearings were totally destroyed. Severe metal loss was noted on virtually every bearing ball. Many balls were deeply imbedded in the inner race; the ball separators had disintegrated. The shaft was severely bent in the region adjacent to the duplex bearing lock nut (see Exhibit 4). Two of the four seal gland nuts had backed off, a third one had fallen off. A pedestal support bracket at point A had not been connected to the pump casing. The malleable iron bearing bracket was fractured at point B.

Probable Failure Progression

It must be assumed that failure of the duplex thrust (outboard) bearing set in motion the chain of events leading to the fire. Excessive friction resulting from severe and near-instantaneous bearing failure caused the shaft to bend. Extreme vibration was generated and caused the less-than-optimally installed seal to release pumpage. By this time, the outboard bearing area is thought to have been red-hot, causing spilled pumpage to ignite.

Probable Failure Causes

Near-instantaneous massive failure of rolling contact bearings is most frequently caused by deficient lubrication and overheating. The failure was too massive and too far progressed to allow determination of the origin of overheating. Installation method, housing bore dimensions, shaft dimensions, driver to pump alignment accuracy, class of bearing (i.e., rolling-element tolerance), and sparking action between auxiliary gland and shaft sleeve could have played a role in the event. However, the primary cause of bearing overheating upon initial operation of new bearings at our plant is probably not related to any of the above. Instead, the most probable cause is our shop's practice of prelubricating with "Western Handy Oil." Western does not market this oil for industrial use. Its extreme low viscosity (75 SUS @ 100°F) makes it suitable only for bicycle and door lock-type of lubrication duties. This oil is very volatile and will evaporate at temperatures well below those anticipated for new antifriction bearings operating at relatively high speeds.

Figure 9-15. Body of formal failure analysis report.

put it into the low-viscosity, high-volatility region. Figure 9-17 was used to determine the minimum lubricant viscosity required for adequate bearing protection of MP-17 and a similar pump, MP-03. The bearing pitch diameter d_m is calculated, and the chart entered at the abscissa. The vertical line intersects a diagonal line representing the shaft speed (3570 rpm), and from there moves to the left where it intersects the ordinate at the required viscosity. For MP-17 pumps with $d_m = 87.5$ mm, the

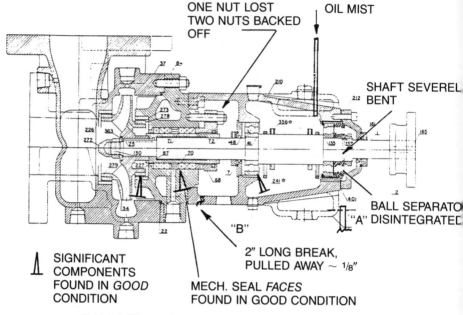

Exhibit 4: MP-17A CROSSECTION AND DAMAGED AREAS.

Figure 9-16. Assembly sketches show the respective locations of damaged parts.

lubricant viscosity should always be 8.3 cSt or higher. Next the investigators had to employ Figure 9-18 to find the maximum allowable bearing rotating-element surface temperature at which "Western Handy Oil" would still meet the 8.3 cSt viscosity requirement.

Figure 9-18 represents an ASTM Viscosity-Temperature Chart on which the investigators plotted the viscosity-temperature relationship of "Western Handy Oil" parallel to that of the more typical industrial lubricants. Entering the chart at 8.3 cSt, they found the maximum allowable temperature for MP-17 to be only about 130°F. At bearing operating temperatures higher than 130°F, this oil would have an even lower viscosity than 8.3 cSt. This would certainly *not* give adequate protection to the bearings.

The report brings out these observations and stresses further that newly installed rolling-element bearings will inevitably experience temperatures in excess of 130°F at the loads and operating speeds encountered in centrifugal pumps. Moreover, connecting the pumps to oil-mist headers supplying the superior "Oilmist-100" could not be expected to result in the immediate establishment of an adequate oil film. By the time a dry bearing has been exposed to oil mist long enough for an adequate film to form, the damage would be done.

Continuing under the heading "probable failure causes," the failure analysis report comments on installation or quality control deficiencies. The seal face was

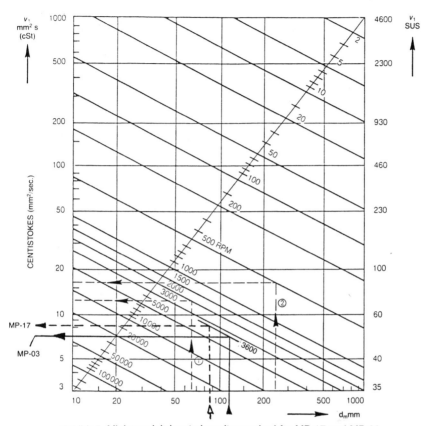

Exhibit 5. Minimum lubricant viscosity required for MP-17 and MP-03.

Minimum Required Lubricant Viscosity
d_m = (bearing bore + bearing OD) ÷ 2
v_1 = required lubricant viscosity for adequate
lubrication at the operating temperature

Figure 9-17. Lubricant-related failure events often can be better explained if viscosity charts are included in the report.

probably not sufficiently square with the shaft centerline. This made the seal inherently more vulnerable, or more likely to leak. Deviations such as an undersized shaft region and an incorrect seal sleeve gasket are pointed out for the record. These deviations were not, however, thought to have caused the seal to fail. Similarly, non-uniform plugging of the impeller had to be ruled out because the impeller wear rings did not show the inevitable contact pattern which would have resulted.

The omission of a bearing housing support bracket at point A of Figure 9-16 is noted, but not considered very significant in terms of the overall failure. But, leaving

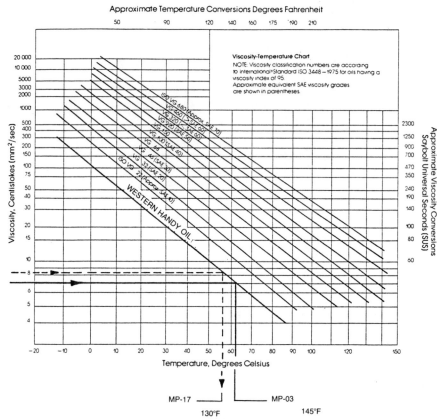

Exhibit 6: MAXIMUM ALLOWABLE BEARING ROTATING ELEMENT SURFACE TEMPERATURE

Figure 9-18. ASTM viscosity-temperature chart and supporting text containing clues to a lubricant-related pump failure event.

out the bracket mounting bolts is an indication of lack of diligent installation followup, to say the least.

Conclusions. There are some noteworthy parallels between the conclusions in the previous failure report and the conclusions written up in this second report (see Figure 9-19). In each of the two reports, the team members decided that the subsection on "conclusions" should not only say what the failure *was*, but also what the failure *was not*. Recall that the former report writers demonstrated "audience awareness" by anticipating that some readers would perhaps ask whether the team had considered this or that possibility before reaching its conclusions. The same thinking went into the conclusions of this second failure analysis report.

Exhibit 5 can be used to determine the viscosity required to lubricate the MP-17A bearings adequately. At a mean diameter of 87.5 mm and a speed of 3570 rpm, a minimum lubricant viscosity of 8.3 cSt is required. As shown in Exhibit 6, maintaining this minimum viscosity is possible only if the operating temperature does not exceed 130°F. Newly installed duplex and double row thrust bearing will, however, experience temperatures well in excess of 130°F. Athough the superior Oilmist-100 lube oil was supplied to the pump bearings via oil mist lube tubing, it could not overcome the diluting effect of the inferior low-viscosity oil which was present in a "trough" formed by the bearing outer race at the 6 o'clock position. The existence of this "trough," or minisump explains why oil mist lubricated bearings in horizontally arranged pumps and drivers survive for periods of 8 or more hours after the oil mist supply has been turned off. Unfortunately, if the minisump is filled with a dilutant, the beneficial effects of applying highly viscous Oilmist-100 cannot come into play until the damage is done.

As to the mechanical seal, the maximum out-of-perpendicular reading of the hard face should not have exceeded 0.001 inch. Excessive displacement of the seal sleeve must be attributed to an undersized shaft dimension near point C, Exhibit 3. The gasket under the seal sleeve (point D, Exhibit 3) was made of mild steel. This is not an effective seal and may have been a leakage path during previous and subsequent pump distress events. Grafoil or soft copper gaskets are required. Shaft undersizing is assumed to have resulted from shaft reconditioning on earlier repair occasions. However, while the seal was thus obviously vulnerable, we do not think that the massive failure event originated with seal problems. Similarly, the failure cannot be attributed to coke plugging of the impeller. A totally plugged impeller would cause pump operation against shutoff head. The resulting thrust and radial loads, while well in excess of normal operating values, could not possibly reach values high enough to cause immediate failure of properly lubricated MP-17A bearings. Uniform plugging would result in no feed forward. This condition was not observed. Non-uniform plugging—the latter associated with impeller unbalance vibration—would primarily affect the radial (inboard) bearing, and impeller distress could be expected to wreck the wear rings. There was, however, no appreciable wear ring contact.

During normal operation, the omission of support-bracket fastening bolts (point A of Exhibit 4) is not considered significant for this style of centrifugal pump. Nevertheless, it is possible that the bearing housing would not have broken had the two bolts been in place.

Figure 9-18. Continued.

The conclusions identify, in the first paragraph, prelubrication with an unsuitable lubricant as the failure initiator. The second paragraph shows why process-initiated impeller unbalance could not have caused this particular problem. Paragraphs three and four anticipate such questions as whether or not a spark could have started the fire, and show why the failure analysis had to rule out this possibility.

In paragraphs five through seven, the writers address component or systems weaknesses which, while in no way responsible for the failure event, could have amplified its severity. Correcting these weaknesses might reduce the possibility of other, not necessarily related, failures in the future.

Conclusions

- It was concluded that failures after a two-hour run length of MP-17A on brand new bearings, and two and eight hour runs of other pumps at our plant on brand new bearings in the same general span of four or five work days follow a pattern pointing to possible commonality of failure causes. The common link in all failures is bearing prelubrication with a lubricant approaching the characteristics of penetrating oil.
- Extreme unbalance vibration originating at the impeller would have been expected to cause wear ring and inboard bearing defects. Seal failure preceding bearing failure is inconsistent with the surprisingly clean appearance of the seal after the fire.
- Sparking action between the non-standard auxiliary gland (made of 316 SS) and seal sleeve (made of 410 SS) is possible but should have resulted in severe galling of the softer of the two materials. No such galling was observed.
- Tar leakage between shaft and sleeve has probably occurred, but is not thought to have started the fire. Experience shows that pump fires brought on by seal distress must reach a very high intensity before outboard thrust bearings disintegrate catastrophically. A low-level fire lasting for 5-10 minutes simply does not fit this scenario.
- Thrust bearings using formed-steel ball separators will work fine as long as the pumps are operated well within intended flow, pressure, NPSH, and lubrication recommendations. However, their margin of safety is inherently lower than that of rolling-element bearings with machined bronze cages.
- Throttle bushings and auxiliary gland plates must be made from non-sparking materials. Both API 610 and Plant Standard P-3002 prohibit the use of ferrous materials.
- Bypassing of filters could have sent higher-than-usual quantities of coke fines into pump and seal cavity. It can be postulated that dead-ending the seal (i.e., the seal is surrounded by pumpage which cannot escape from the seal cavity) is satisfactory for operation with the tar filters in the circuit. However, operation without filters may have, at times, caused excessive amounts of coke fines to accumulate in the seal cavity. It would thus seem appropriate to implement reverse flush on MP-17A to bring it into compliance with the flush method originally specified and actually used on MP-17B and MP-17C.

Figure 9-19. Conclusions portion of a formal pump failure report.

Recommendations. Many of the recommendations made here have already been stated earlier (see Figure 9-20). Others are the logical result of observations made in the report or have, perhaps, been implied in the preceding report text. For instance, recommendations 1, 7, 8, and 9 had been stated either as findings or deviations from good practice. Recommendation 2 represents guidance for the procurement of better components, while recommendations 3 and 10 could be considered steps to improve shop practices. Recommendations 4 and 6 relate to better, more-effective training for maintenance personnel. Finally, recommendation 5 asks for the correction of an installation oversight which, although lubrication-related, did not contribute to the failure being discussed in the report.

Other than the figures described in the main text, the report does not contain any appendix material.

Recommendations

1. Discontinue use of "Western Handy Oil" for prelubrication of pump antifriction bearings. Use Oilmist-100 instead.
   Action: Had discussions with all millwrights on November 25, 1980 (PQR). Follow up by ABC/DEF.

2. To strengthen general quality improvement, implement bearing procurement and incoming inspection procedures aimed at verifying that bearings comply with plant requirements as outlined in PS/P-3002:

   - AFBMA-1 Class C-3 clearance fit for radial bearings.
   - Use machined bronze or phenolic cages for all bearings in hydrocarbon services.
   - Disallow double row antifriction bearings in thrust locations.
   - Disallow thrust bearings with filler notches.
   - Disallow snap rings as a means of axially locating thrust bearings.

   Action: ABC/DEF

3. Develop mechanical seal face and sleeve installation and tolerance criteria (ask Whooper Company to provide, then post in prominent shop location).
   Action: GHI has initiated and will follow up.

4. Arrange for additional training opportunities in bearing installation and maintenance.
   Action: PQR has contacted SKF Company to obtain information on February 1981 training course. SKF reply will be forwarded to JKL.

5. Verify use of directed oil-mist fittings on pumps with inner-race velocities more than 2000 fpm (MP-17 operates at 2600 fpm but did not have directed mist fittings).
   Action: ABC/DEF

6. Conduct a 30-minute information interchange session once every month. This discourse should be between MTS machinery engineers and millwrights.
   Action: First session held on 11/24/80, RSQ. Subsequent sessions with MNO and GHI to be arranged by ABC/DEF.

7. Steel auxiliary gland to be replaced with identical bronze part.
   Action: Initiated by ABC.

8. Sleeve gaskets to be made of soft copper.
   Action: Initiated by ABC.

9. Implement API Flush Plan 13 (reverse flush from stuffing box to suction line) on MP-17A.
   Action: Initiated by DEF.

10. Develop a "pump repair check board" for prominent display in our shop. Include, in large letters, such items as:

    - Auxiliary glands made of non-sparking material?
    - Bearings prelubricated with Oilmist-100?
    - Seal sleeves installed with shaft gasket?
    - Seal faces installed perpendicular?
    - Viton O-rings disallowed for MP-10, MP-18 and MP-38.
    - No riveted steel cages as ball separators?
    - Directed oil-mist fitting if shaft diameter >2.2 inches and speed >3000 rpm?

    Action: ABC/DEF to implement. GHI to assist in completing the above list.

Figure 9-20. Recommendations may be appropriate for inclusion in a formal pump failure report.

Failure Investigation May Require Calculations

Incomplete, inconclusive, or even erroneous identification of failure cause can result if the analysis neglects to perform calculations or, as a minimum, takes into consideration pertinent numerical data from equipment manufacturers. The analysis of thrust bearing failure events on large compressor speed-increaser gears will serve to illustrate the point.

A large North American petrochemical complex, "Petronta," operates a motor-driven refrigeration compressor train as shown in Figure 9-21. After a few years' operation, the train was brought down for general inspection and major preventive maintenance such as cleaning of motor windings. Approximately six months after reassembly and startup, the gear unit suffered a serious axial locating bearing (ALB) failure.

To find out what happened and why, we must resort to both observation and calculation.

The observations can be in the form of a service report written by vendor representatives or maintenance work forces. Here is a pertinent excerpt from a failure analysis report written by the resident machinery engineer and the machinery maintenance section head:

1. . . . the compressor train developed again high axial motor vibrations that started to exceed levels experienced prior to the last shutdown by a factor of (2). Also, serious gearbox vibrations were noted, whereas the compressor itself ran exceptionally quiet without any signs of distress.

Exhibit 1

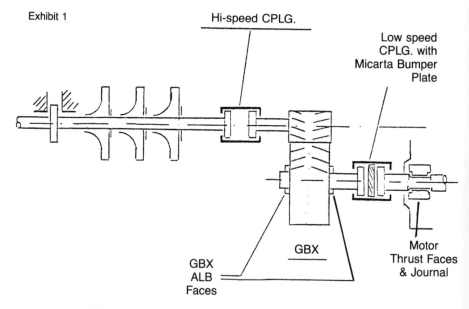

Figure 9-21. Motor-gear-driven compressor train at PETRONTA.

2. A task group looking at the train's past history and the vibration patterns recommended that it be shut-down for an inspection, including the motor, and the correction of possible misalignment between the motor and gearbox.
3. The train was subsequently shut-down. Ensuing inspection work revealed a severely damaged bull gear bump shoulder at the non-drive end, marginal but tolerable alignment between motor and gear, and severely locked up high-speed and low-speed couplings.
4. No signs of distress were observed when compressor thrust and journal bearings were examined.
5. Thorough inspection of the motor internals under direction of OEM personnel revealed no problems.
6. Both couplings were replaced. LS coupling was expedited from OEM, HS coupling was on hand. It had been ordered but not installed during the last turnaround due to good condition of the original coupling.
7. Gearbox bump shoulder arrangement was redesigned and repaired.
8. The unit was started up after resetting and realigning the gearbox. Motor and gear were operating at satisfactory vibration levels.

The gear manufacturer's service representative gave the following account of the incident:

> The gear unit cover was next removed. It was noted that the low-speed gear was fully against the blind end bearing 'bump' face and spalling of babbitt was visible. It was further noted that operating heavy contact was visible on both driving, driven, *and* following flanks of gear and pinion teeth.

> When gear and pinion assemblies were removed it was noted that indeed the 'bump' face babbitt of the low-speed bearing was damaged. As no spare bearings were available, it was decided to modify the existing assembly as follows: . . .

On another occasion the plant maintenance forces discovered a similar axial "bump" shoulder failure on their feed-gas compressor train when they opened its gear unit for routine inspection. Since in this case the machines had not shown any signs of distress during operation, it was concluded that the damage had occurred while the train was allowed to run in reverse after shutdown and during depressurization. The damage also extended to the motor rotor locating shoulders, and it appeared as though thrusting had taken place from the gear into the motor. There was no damage to the compressor thrust bearings. Here, when a coupling defect was suspected also, the coupling manufacturer's local service representative was called in and reported his observations to the factory. The factory responded with an interoffice note which, except for names, is faithfully reproduced in Figure 9-22.

But now for the failure analysis. To find out what really happened, we must:

1. Collect general data.
2. Separate relevant observations from extraneous observations.
3. Perform appropriate calculations.
4. Correlate calculations and relevant observations. This will establish the most probable cause and sequence of failure events.

Superior Coupling Manufacturing Company
Interoffice Correspondence

TO: J. Localrep
FROM: J. F. Manager

Subject: Petronta Ontario
Date: July 7, 1981

The thrust bearing failure in a herringbone gear box was originally attributed to the high-speed gear coupling. Petronta is using our coupling as shown in Exhibit 1 and quoted a technical paper entitled "Why Properly Rated Gears Still Fail" to substantiate their conclusion. After visiting the plant and discussing the failure with plant personnel, we conclude that the failure was caused by improper axial installation of the driving motor. The bearing failure evidenced that the gear was pulled toward the motor. Two possible forces could do that:

1. The compressor shaft expands toward the gear box. If the high-speed coupling has a high friction coefficient, it will transmit to the pinion an axial force. This force will unload one side of the gear mesh and overload the other side. It will be almost totally transmitted to the thrust bearing. Inspection of the unit should reveal some differential wear between the two sides of the herringbone teeth. As we were told, no such uneven wear was visible, and the high-speed coupling teeth were found to be in excellent condition. Hence, we can safely eliminate this possibility.
2. The low-speed coupling between the gear box and the motor is a "limited end float" coupling. The coupling is designed to allow the electric motor rotor to seek its magnetic center, but to prevent it from bumping into the end housings when the power is cut-off. If we assume that the motor was installed too far from the gear box, the coupling will prevent the rotor from going to the magnetic center, and the large forces pulling the rotor will load the thrust bearing. As the low-speed coupling was not available for inspection, we could not confirm this possibility, but as we cannot see a third possibility, we have to assume that improper axial installation caused the bearing to fail.

JF. Manager
J.F. Manager

Figure 9-22. A coupling manufacturer's "quickie" analysis of a serious component failure.

Collecting Data

For the train configuration shown in Figure 9-21, the user organization listed the following data:

Motor output: 8400 hp.
Bull gear speed: 1190 rpm.
Pinion speed: 6198 rpm.
High-speed coupling pitch diameter: 7.38 inches.
Possible axial float of bull gear: 0.005 inches.

Motor bearings thrust face area: 19.4 sq. in., babbitt on steel.
Bull gear bearing thrust face area: 19.6 sq. in., babbitt on steel.
Motor rotor magnetic centering force: 4200 pounds.

Calculating Thrust Forces

Gear couplings are capable of transmitting axial thrusts equal to $(T/r) \mu$, where torque T is (63025) (hp)/rpm, r is the pitch radius of the gear coupling, and μ is the coefficient of sliding friction between the coupling teeth. For a given compressor train, all of these values, except for μ, are assumed to be constant. The coefficient of friction is known to vary from perhaps 0.05 in a well-lubricated, new coupling to as high as 0.25 or more after an appreciable operating time with sometimes marginal lubrication. We conclude from this that any kind or degree of gear-coupling deterioration initially manifests itself in an increase of coefficient of sliding friction. The worst case is usually referred to as friction lock-up, where μ will momentarily become unity, i.e., equals one.

As long as the motor shaft journals are not contacting the motor bearing thrust faces, the bull gear axial bump faces may be exposed to a combined thrust load composed of motor rotor magnetic centering force and axial thrust transmitted by the high-speed gear coupling. The magnetic centering force can be calculated from

$$F_M = 0.0117A \ (60/f)E_o I_{MO}(l_o/l)^2[dl/dx]$$

where:

A = number of phases.
f = frequency, cycles per second.
E_o = phase voltage, with motor in reference position.
I_{mo} = phase current, effective mutual value.
l_o = effective stator (rotor) length, inches (reference position).
l = effective rotor length, inches (displaced position).
dl/dx = rate of change of effective machine lengths when stator (rotor) overhangs rotor (stator) at each end.

Alternatively, we can use a rule of thumb for determining the approximate axial thrust generated by the magnetic centering force in induction motors with two, four or six poles.

---

two-pole motors: 0.1 pounds per hp.
four-pole motors: 0.35 pounds per hp.
six-pole motors: 0.5 pounds per hp.

We suspect this is how the owner obtained the reported 4200 lbs in the data summary for this motor.

If we assume the high-speed coupling to have a coefficient of sliding friction of 0.15, the transmitted axial thrust will be

$$F_{HS} = \left[(63025)(8400) \ (0.15)\right] \big/ \left[(3.69)(6198)\right]$$
$$= 3472 \text{ pounds}$$

The total combined thrust is, therefore,

$$F_t = 4200 + 3472 = 7672 \text{ pounds}$$

and the load on the axial bump faces of the gear 7672/19.6 = 391 pounds per sq. in. This pressure load is substantially in excess of the 250-psi load normally permitted for simple babbitted thrust bumpers.

Does this mean that the thrust bearing failed because of improper axial installation location of the motor rotor? Not necessarily. Let us assume the high-speed coupling did, in fact, have a coefficient of friction of 0.15. This value is not at all unusual in our experience. Then

$$(3472 + F_M) / 19.6 \text{ in}^2 = 250 \text{ pounds per in}^2$$

and

$$F_M = 1428 \text{ pounds}$$

This value may already be reached when the motor rotor is displaced a fraction of an inch from its magnetic center and coupled, via limited end float low-speed coupling, to the bull gear.

What, then, is responsible for causing the thrust bearing failure? Why did the installation in the case of the refrigeration train given reliable service for several years but failed relatively soon after turnaround work was performed on the compressor train? The previous calculations contain all the clues to the answer.

We can see that the gear thrust bump faces are small, 19.6 in². Their maximum allowable pressure loading of 250 pounds per in². is easily exceeded; all it takes is for the total axial thrust to exceed (19.6) (250) = 4900 pounds. The total axial thrust is made up of coupling-generated and motor rotor-generated components. Therefore, the gear thrust bumper faces failed because either the coupling coefficient of friction or the motor rotor magnetic center-seeking force, or perhaps both, were too high. The coupling undoubtedly had a wear pattern which gave a reasonably low coefficient of friction during its earlier, long-term operation. At the time of the turnaround, a new wear pattern had to be established, or the teeth were perhaps driven against a ridge representing the boundary of the original wear pattern. This could have caused the coefficient of friction to go up.

One question in this context will always be difficult to answer, namely: "Were both high- and low-speed coupling assembly match marks observed to avoid random reassembly?" Often this is well understood by mechanics as far as high-speed couplings are concerned because of the impact on coupling balance. In the case of low-speed gear couplings ($n \leq 1800$) less emphasis is encountered. Yet, the previous question should be pertinent in connection with all gear couplings, because as they get older their teeth will no doubt have deformed in plastic flow. This deformation may be minute, but it will be sufficient to cause the coefficient of sliding friction to increase if, after turnaround, the original mesh is not reestablished. In the case of the refrigeration train, the investigators were unable to find any match marks on the failed low-speed coupling. Also, the motor was removed from its pedestal for cleaning work during the turnaround. When it was reinstalled, the rotor was probably no longer in exactly the position where it had been previously. It probably exerted a different axial thrust force on the coupled system.

For installations of this type, it is prudent to insist on larger gearbox thrust bearings. Also, turbomachinery trains require gear couplings whose coefficient of friction is guaranteed to be very low—especially upon reuse after a turnaround. A properly designed non-lubricated disc or diaphragm coupling will probably give more reliable service than a gear-type coupling. As for the motor installation procedure, it should ensure that after coupling up, the rotor will be located directly at the magnetic center of the stator. In the case of the refrigeration train, OEM service representatives were very unclear about the location of the motor's magnetic center. Old plant records showed the magnetic center location of the motor to be at a ⅝-inch nondrive-end shaft protrusion. This was the shaft location during the distress sequence. After repairs, the motor was running without problems with a ⅜-inch protrusion.

In conclusion, we have shown that calculation can be of tremendous help to the engineer investigating equipment failures. Even an educated guess may not be sufficient or may, as the coupling manufacturer's report shows, be only partly correct.

The Short-Form Failure Analysis Report

Every troubleshooter should strive to make himself easily understood. This requires that he be successful in clearly organizing his more or less complex ideas. In the preceding section we offered a few good examples in connection with critical machinery incident reports.

Writing a good technical transmittal is largely a matter of organization. Most people find that when they can arrange their thoughts properly, putting the words on paper is a relatively simple matter. One of the best ways to put your ideas together is by using the typical problem solution process discussed in Chapter 6. In this context steps to organize one's thoughts can be summarized in five points:

1. *Situation analysis.* This first step sets the background for what you want to communicate.
2. *Problem definition.* This arises out of the situation analysis. It may be perfectly clear to you that there was or is a problem and that it has been solved. Make sure that your readers understand exactly what the problem is.
3. *Possible solutions.* This is the possible answer to the problem, and it is stated as short as possible. It introduces Step 4.
4. *Preferred solution.* This details the preferred solution. It also contains all the necessary information to tell the reader why this particular solution was chosen.
5. *Implementation.* This is actually still part of Step 4 but stands on its own. It will show what was done, and what still needs to be done, i.e., it lays out the action plan.

We might want to call this system "SPSI" (Situation-Problem-Solution-Implementation). No doubt, it could well be too cumbersome to be applied in all cases.

There is however, an area where SPSI could be routinely applied—on the smaller, non-critical machinery failure events. Here, a one-page failure analysis report often refreshes memories and keeps the record straight. Figures 9-24 and 9-25 show how so-called "short-form" analysis reports can be easily structured around SPSI.

MEMORANDUM

To: Superintendent, Process

June 4, 1982
CP-11 FAILURE ANALYSIS
File: 4040
From: H. Y.

P-11 was transported to Area 3 for repair on Thursday, May 20, 1982. It ran satisfactorily on Thursday, June 3, 1982. Following is a summary of the failure analysis and work done to repair the pump.

Observation. Pump in as found condition.

- Pump shaft was broken at the thrust bearing.
- Thrust bearing was totally destroyed.
- Radial bearing was severely damaged.
- Very little oil was found in the bearing housing.
- Thrust bearing housing was bored .002" out-of-round.
- Previous bushing for radial bearing was incorrectly machined; No oil return cutouts were made.
- Brazing on OB bearing housing was cracked.

Most Probable Failure Cause. Radial bearing was first damaged by inadequate oil supply. Metal particles were then transmitted to the thrust bearing through oil circulation. Subsequently, thrust bearing was damaged. A lot of heat was generated which also weakened the pump shaft. Eventually the shaft broke off.

Recommendation. The bearing housing has been cracked for some time (exact date not known). The cracks were repaired by brazing. This can only be considered a short-term repair. For long-term reliability, we strongly recommend ordering a new bearing housing for replacement. Replacement cost for a new bearing housing is $1882.

Highlights of Repair Work

- Bearing housing was bored and bushed. Proper cutouts were made for the radial bearing oil return.
- New pump shaft was made out of 4140 material.
- New radial bearing was installed to increase its load-carrying capacity. A SKF 22214C spherical roller bearing was selected for its superior load capacity and also for its ability to adjust for internal misalignment.

/cac

SITUATION ANALYSIS

PROBLEM DEFINITION

POSSIBLE SOLUTION

PREFERRED SOLUTION

IMPLEMENTATION

Figure 9-23. Example of short-form failure analysis report using SPSI concept.

MEMORANDUM

To: Supervisor, Process

April 8, 1981
C-117 FAILURE ANALYSIS
File: 1750
From: J. V.

SITUATION ANALYSIS

[SYMPTOMS:] Motor tripped out and would not restart.
[OBSERVATIONS:] Motor and conduit run contained water. Motor bearings showed signs of rust.

PROBLEM DEFINITION

DIAGNOSIS: Motor burned out due to insulation breakdown caused by excessive water in motor and electrical conduit.

TYPE OF PROBLEM: Operating conditions.

PREFERRED SOLUTION

RECTIFICATION: Motor rewound at Delta, bearings changed.

RECOMMENDATIONS: Water spilling over motor and near it must be stopped. This failure is caused by the same problem referred to in C-117 "Failure Analysis" dated 2 March 1981. Reactor cleaning drains spill into lower level of recovery room as they are not piped to any drain system. Drains should be installed or the affected equipment made "waterproof" by covers similar to those used on LB-110.

IMPLEMENTATION

/cac

c.c. J. J.
N. G.
H. W.

Figure 9-24. Short-form failure analysis report.

Chapter 10

Organizing for Successful Failure Analysis and Troubleshooting

The reader probably need not be reminded that most machinery systems and components in the process industries will sooner or later experience defects, failures, and malfunctions. Unless the root causes of these failures or abnormalities are identified and rapidly remedied, a given plant may risk costly outages or production curtailments. Consequently, the process industries tend to employ specialist personnel to recognize potential failures, determine the failure mode, and perform troubleshooting tasks in efforts to define the many factors causing an unexpected or undesirable event.

Although very costly, dangerous, or environmentally unacceptable failure events are often analyzed by composite teams of investigators whose backgrounds and organizational assignments cross interdepartmental or intra-affiliate boundaries, the more routine analyses are generally handled by one or more members of a so-called mechanical technical services (MTS) group. The typical functions of an MTS group were described in Volume 1 of *Practical Machinery Management for Process Plants* and include two tasks which relate to failure analysis and troubleshooting. These tasks are the investigation of special or recurring maintenance problems, and the review of maintenance costs and service factors. Accordingly, the systematic definition and reduction of costly repeat failures through in-depth failure analysis and troubleshooting is one of their priority tasks.

How the skills of qualified MTS machinery engineers can be utilized is best demonstrated by example. With centrifugal pump failures making up the bulk of machinery problems in the process industries, we will consider the steps necessary to reduce systematically the number of pump failure incidents in a typical petrochemical plant.

Setting Up a Centrifugal Pump Failure Reduction Program

Centrifugal pump repairs account for most maintenance expenditures in many petrochemical companies. Recent statistics show that large refineries with 3000 installed pumps experience approximately 800-1000 failure events per year. A highly conservative estimate assumes average repair costs of $2500 per event, although in 1983 many petrochemical plants reported a more believable average cost of $5000 after burden, shop space, field labor, engineering analysis, and miscellaneous charges have been added.

The incentive to reduce pump failures is quite evident and hardly merits further discussion. An effective program of reducing pump failures is always desirable and should have the following objectives:

1. To improve centrifugal pump reliability through more accurate determination and subsequent elimination of failure cause.
2. To reduce process debits resulting from pump outages.
3. To reduce centrifugal pump maintenance costs by effective analysis of component condition, replacing only those parts which actually require change-out.

However, the success of a pump failure reduction program is influenced by a number of factors. These range from administrative-managerial to technical-clerical and could include such items as management commitment, willingness of technical staff to follow-up on tasks previously handled by maintenance work forces, and clerical support for better documentation and record keeping.

Definition of Approach and Goals

There are two basic implementation methods for a program of this type:

1. *The team approach.* This method requires the formation of a team of troubleshooting specialists who, over a short period of time, will make a concentrated effort to define and eliminate sources of frequent pump failures.
2. *The engineering support approach.* Here the effort is centered around the field maintenance organization, supplemented by support and guidance from a staff engineer trained in pump selection and failure analysis. The staff engineer belongs to the MTS group.

Although still popular with some petrochemical companies, the team approach was found to have a number of significant shortcomings:

1. It lacks continuity. Team members are often unavailable if follow-up needs should arise.
2. The team is sometimes considered as "outsiders" effort. Operating personnel and field maintenance forces fail to develop the necessary rapport with the team.
3. A lack of commitment and reluctance to communicate have sometimes been observed. The field forces may tend to stand back and let the "brains" struggle with the problem. Field personnel see the team as a vote of non-confidence in their own capabilities or past efforts.
4. Extra manpower is generally required with this approach.

In view of these drawbacks, petrochemical plants would do well to consider the "engineering support" approach for a pump failure reduction program. Let us examine how this works.

Based on an analysis of frequency and cost and severity of pump failures, the plant mechanical technical service group should designate a certain number of pumps as "problem pumps." These could be pumps which failed more than twice in a 12-month time period, or pumps which have a history of one failure per year but cost more than $10,000 to repair. Of course, these guidelines are for illustrative purposes only and can be modified to suit any given need or situation.

A plant or operating unit should set itself an improvement goal. One typical goal is to strive for reducing the number of "problem pumps" by 50% in one year's time, or to reduce pump maintenance expenditures by a certain amount in a designated time period. In the spirit of participative management, the initiative for goal setting and program execution should originate with field personnel rather than office or headquarter staff personnel. While being guided by staff MTS personnel, field forces would make the program work and be credited with its successful implementation.

Action Steps Outlined

A number of action steps are required to get the pump failure reduction program going, and management must designate responsibilities for the execution of these steps. Responsible MTS personnel will have to:

1. Designate "problem pumps" after reviewing past failure history, cost of repairs, cost of product losses, etc.
2. Identify these pumps for record purposes and by actually tagging them in the field. Tagging will alert shop and field foremen to the need to arrange for a designated technician or engineer to be present whenever work is performed.
3. Update pump records to include failure history for past 18 months. Develop and include step-by-step instructions for assembly, disassembly, tolerance checks, etc., as necessary.
4. Follow the disassembly and reassembly of all problem pumps, record measurements, and revise detailed instructions, as required.
5. Identify where, within the plant or elsewhere, pumps identical or similar to the "problem pumps" are operating. Compare their operating histories and examine significant deviations for clues to the source of the problem.
6. Provide checklists and tabulations of critical mechanical and hydraulic parameters, e.g., bearing fits, piping strains, excessive suction specific speeds, operation away from BEP (best efficiency point), inadequate clearance between impeller periphery and cutwater, etc.
7. Provide checklists and guidance on critical process parameters (solids or polymers in pumpage, non-optimized seal flush arrangements, equipment start-up and shutdown, preheating, cooldown, venting, etc.).
8. Assist mechanical work force in finding the true failure causes and train mechanical personnel in identifying ways of eliminating these causes.

9. Monitor progress of program and communicate results to management. To fulfill this responsibility, the MTS engineer should:

- Issue monthly activity summaries.
- Tabulate interim results after six months.
- Report accomplishments after one year.
- Develop plans for next phase of program.

Documentation and Reference Data Requirements

It is hoped that pump maintenance data folders already exist at the plant which is interested in implementing a pump failure reduction program. If data folders are not available, this would be a good time to at first put them together for problem pumps and then expand the data bank to cover all pumps.[1]

Ideally, the data folders should contain the following documents:

1. API data sheets.
2. Performance curves.
3. Supplemental specifications invoked at time of purchase.
4. Dimensionally accurate crossection drawing.
5. Assembly instructions and drawings.
6. Seal drawings with installation dimensions and specific instructions.
7. Manufacturing drawings of shaft and seal gland.
8. Small-bore piping isometric sketch.
9. Spare-parts list and storehouse retrieval data.
10. Maintenance instructions, including critical dimensions and tolerances, running clearances, shop test procedures for seals and bearings, and impeller balancing specifications.
11. Repair history, including parts replaced, analysis of parts condition, and cost of repair.
12. Design change documentation.
13. Initial startup and check-out data.
14. Lubrication-related data.
15. Special installation procedures.
16. Alignment data.
17. Vibration history.

Needless to say, the data files should be as complete as possible, kept current, and readily available to every person involved in the implementation of the pump failure reduction program.

Development of Checklists and Procedures

Checklists, procedures, or similar guidelines must be used if the objectives outlined earlier are to be reached in an expeditious and well-defined fashion. These checklists or procedures should cover topics which fall into roughly three classification categories:

1. Fieldwork-related reviews.
2. Shopwork-related reviews.
3. Technical service reviews (failure prevention, procurement guidelines, and pre-purchase reliability assurance topics) for MTS use.

In the first category, fieldwork-related checklists must be developed and applied to accomplish a number of necessary steps which will aid in problem identification and thus will lead to a reduction of problem incidents in the future. These checklists must address problem pump:

1. Lubrication instructions.
2. Routine surveillance instructions.
3. Data-taking requirements preparatory to shutdown for subsequent shop repair.
4. Installation precautions and data requirements prior to restart after shop repair.
5. Special commissioning instructions such as cool-down of cold service pumps, air-freeing of seal cavities, etc.

Checklist development can be a joint effort involving plant operating, technical service, and pump vendor personnel. Table 10-1 shows one such joint-effort procedure; it gives guidance on the commissioning of a vulnerable cold-service pump[2] and serves as a typical example.

In the second category, shopwork instructions and procedures must be developed and conscientiously used to ascertain both quality workmanship and uniformity of product inprovement efforts. Typical instructions and checklists should make maximum use of technical information which is routinely available from pump, bearing, and mechanical seal manufacturers. Included in this second category are such guidelines as:

1. Wear ring and throat bushing clearances.
2. Antifriction bearing housing fit tolerance to be used for pumps, motors and turbines.[3]
3. Antifriction bearing shaft fit tolerance to be used for pumps, motors and turbines.[3]
4. Dimensional checking of mechanical seal components.
5. Maximum allowable rotor unbalance for pumps and small steam turbines.
6. Dimensions for steel bushings converting standard bearing housings to accept angular contact bearings.
7. Others (see Table 10-2 and 10-3).

Material in the third category is intended for failure analysis use by plant engineering or technical service groups. It is a "memory jogger" and should

Table 10-1
Field-Posted Commissioning Instructions*

Barrier fluid system dry out procedure (mechanical technicians)
1. Open flanges on barrier fluid return line.
2. Verify reservoir isolation valves in barrier fluid system piping are open.
3. Pump methanol into barrier fluid reservoir drain until clean methanol flows from open flanges.
4. Block reservoir drain valve and remove barrel pump hose.
5. Follow Pump casing/process piping dry out procedure below.
6. Prime barrel pump with fresh barrier fluid until clean fluid flows from open flanges.
7. Reconnect hose to reservoir drain valve and unblock valve.
8. Pump barrier fluid into system until clean fluid flows from open flanges.
9. Tighten open flanges using new gasket.
10. Continue pumping barrier fluid into system until reservoir is filled.
11. Block reservoir drain valve and remove pump hose.
12. Repeat procedure for remaining barrier fluid system.

Pump casing/process piping dry out procedure (process technicians)
1. Block all valves into pump casing, suction, and discharge process piping.
2. Open atmospheric vent valves A1 on casing drain, A2 on seal flush piping, A3 on suction piping, and A5 on discharge check valve bypass. Verify vent valve A4 on check valve bypass closed.
3. Open casing drain valve D1, verify drain valve D2 closed.
4. Open seal flush piping vent valve V1, verify vent valve V2 to reflux drum firmly closed.
5. Open methanol injection valve M1 into pump casing and begin filling with methanol. CAUTION: DO NOT PRESSURIZE PUMP CASING WITH METHANOL!
6. Block individual vent valves as methanol appears. Continue filling.
7. Block methanol injection valve M1 when methanol appears at vent valve A5 on discharge check valve bypass.
8. Block vent valve A5 on discharge check valve bypass.
9. Slow roll turbine at 50-60 rpm for about 5 minutes.
10. Open atmospheric vent A2 on seal flush piping briefly to clear collected air.

Pressure out procedure (process technician)
1. Open seal flush piping vent V2 to reflux drum to pressurize pump casing.
2. Open pump casing drain D2 to cold blowdown.
3. Allow pump casing to blowdown until casing drain piping begins to frost.
4. Verify methanol completely purged by cracking vent valve A1 on casing drain.
5. Block pump casing drain D1, D2 out of cold blowdown when light frost begins to form on pump casing.

Cold soak procedure (process technician)
1. Crack suction block valve B1 2-3 turns.
2. Open suction block valve B1 completely when casing is well frosted.
3. Open discharge block valve B2 completely.
4. Allow pump to cold soak for 2 hours.
5. Verify pump shaft turns freely by hand.

Startup/shutdown (process technicians—follow normal operating procedures)

*For vulnerable, cold-service pump. Condition: pump is completely blocked in, cleared, and depressurized.

Table 10-2
Typical Shop Repair Instructions*

Brg. bore, mm	Shaft sizes Min	Max	Hsg. ID, mm	Housing fits Min	Mean	Max
35	1.3781	1.3785	40	1.5744	1.5749	1.5754
40	1.5749	1.5753	42	1.6531	1.6536	1.6541
45	1.7718	1.7722	47	1.8500	1.8505	1.8510
50	1.9686	1.9690	52	2.0467	2.0474	2.0480
55	2.1655	2.1660	55	2.1649	2.1656	2.1662
60	2.3623	2.3628	62	2.4404	2.4411	2.4417
65	2.5592	2.5597	68	2.6767	2.6774	2.6780
70	2.7560	2.7565	72	2.8341	2.8348	2.8354
75	2.9529	2.9534	75	2.9523	2.9529	2.9535
80	3.1497	3.1502	80	3.1491	3.1498	3.1504
85	3.3466	3.3472	85	3.3460	3.3467	3.3474
90	3.5434	3.5440	90	3.5428	3.5435	3.5442
95	3.7403	3.7409	95	3.7397	3.7404	3.7411
100	3.9371	3.9377	100	3.9365	3.9372	3.9379
105	4.1340	4.1346	110	4.3302	4.3309	4.3316
110	4.3308	4.3314	115	4.5271	4.5278	4.5285
115	4.5277	4.5283	120	4.7239	4.7246	4.7253
120	4.7245	4.7251	125	4.9207	4.9215	4.9223
125	4.9214	4.9221	130	5.1175	5.1183	5.1191
130	5.1182	5.1189	140	5.5112	5.5120	5.5128
140	5.5119	5.5126	145	5.7081	5.7089	5.7097
150	5.9056	5.9063	150	5.9049	5.9057	5.9065
160	6.2993	6.3000	160	6.2986	6.2994	6.3002
170	6.6930	6.6937	170	6.6923	6.6931	6.6939
180	7.0867	7.0874	180	7.0860	7.0868	7.0876

*For pumps with rolling element bearings.

incorporate helpful hints from the numerous publications dealing with pump reliability improvements.[4-18] The plant MTS engineer is encouraged to structure his review in the following manner.

General Review

1. Use the Generalized Machinery Problem Solving Sequence which was described in Chapter 6 to assist in defining the most probable failure cause.
2. Identify identical pumps in same or similar service within affiliated plants. Analyze maintenance history for these pumps.
3. If no identical pumps exist, or if identical pumps are applied differently, compare construction and design details. Do not neglect installation details such as piping, etc.
4. Verify that rotating elements are dynamically balanced per applicable checklist or procedure.
5. Review applicable field and shop checklists for accuracy and supervisor's signature at conclusion of work.

The results of the failure analysis should be documented on a short form similar to the one shown in Figure 10-1.[4] Far more detailed forms have been devised and are appropriate for plants with computerized failure records.

Table 10-3
Detailed Pump Assembly and Disassembly Instructions
for a Specific Pump

A. Field disassembly
 1. Remove coupling spacer.
 2. Drain pump completely.
 3. Disconnect all external piping (flush, steam tracing, etc.) and interfering insulation.
 4. Remove casing and foot support bolts.
 5. Separate casing from distance piece by means of pusher screws. Leave bearing housing connected to distance piece unless bearing replacement is contemplated.

B. Shop disassembly
 1. Unscrew impeller nut and remove impeller.
 2. Remove stuffing box housing by carefully applying pressure from two crowbars contacting seal plate.
 3. Remove seal plate from stuffing box housing and also remove stationary and rotating seal elements.
 4. Remove shaft sleeve without undue force. Apply heat, if necessary.
 5. Examine seal parts for warpage, fractures, and solids build-up on faces.

C. Shop reassembly
 1. Verify flatness of seal faces to be within two light bands.
 2. Measure inside diameter of throat bushing and outside diameter of shaft sleeve.
 3. Machine inside diameter of throat bushing to achieve 0.0200-0.030-in. clearance with shaft sleeve O.D.
 4. Assemble stationary seal ring and seal plate, taking care to insert required gaskets.
 5. Carefully tighten plate hold-down nuts. Use depth micrometer to verify that stationary seal ring is installed parallel to product side of stuffing box cover within .0005 in.
 6. Place stuffing box cover on distance piece.
 7. Place rotating seal assembly on shaft sleeve and slide over shaft. Make sure all dimensions are per CHEMEX Dwg. B-541599.
 8. Insert keys.
 9. Assemble impeller, taking care to place gaskets between hub and shaft sleeve, and between shaft and impeller nut.

D. Measurements required for record
Before and after replacing worn parts with new parts, please measure and record the following dimensions:

	Manufacturer's recommended value, mils	With old parts, mils	With new parts, mils
Shaft runout (impeller region)	1 mil or less		
Shaft runout (seal region)	Not to exceed 2 mils		
Seal face compression	Per CHEMEX DWG. B-541599		
Depth micrometer Check to ensure stationary seal is parallel to seal gland.	0.0005 in. or less		
Clearance between impeller disc vanes and stuffing box housing	0.032-0.047 in.		
Clearance between shaft sleeve and throat bushing	0.020 in. diametral. (max 0.032")		

E. Field reassembly
 1. Back out pusher screws and bolt distance piece to casing. Make sure proper gasket is placed in between.
 2. Replace foot support bolts.
 3. Reconnect all auxiliary piping.
 4. Refill bearing housing with Synesstic-32 lubricant.
 5. Hot-align coupling to zero-zero setting (max. allowable deviation: place motor 0.002 in. higher than pump shaft) while observing dial indicators on casing.
 6. Replace coupling spacer. Ascertain coupling disc stretch or compression does not exceed 4 mil.

Date __OCTOBER 17, 1981__

Anytown Chemical
Winddrift, Texas

FAILURE ANALYSIS DATA FORM
(BREAKDOWNS OTHER THAN PLANNED PREVENTIVE MAINTENANCE)

° IDENTIFICATION: UNIT __HEPA-2__   COST CENTER __48217__

Equip. Type __CENT. PUMP__   Ident. No. __P-2257__   Yard No. __225-17101__

Serial No. __GOOFER-DURR 832121__   System (If Not Number Equip.) __ACID PREWASH__

° MECHANICAL AVAILABILITY DATA: Total Hours Unavailable ____48____

	Date	Time
Equipment Down	10/12/81	9:00 AM
Ready for Service	10/14/81	8:20 AM

° BRIEF DESCRIPTION OF FAILURE: __SEAL FLUSH PIPING DEVELOPED CRACK, BROKE OFF.__
__SEAL FAILURE CAUSED BEARING WASHOUT. BEARING SEIZED, CAUSING__
__SHAFT TO BREAK__

° COMPONENT FAILURE ANALYSIS ATTACHED

° SERIOUSNESS OF FAILURE   YES   NO

	YES	NO
Safety		
Production Loss	☒	☒
Air or Water Quality	☒	
Problem		
Conservation Loss	☒	

° REPAIR ANALYSIS
☒ Routine
☐ Delay For Parts
☐ Shop Quality Control Problem
☐ Other Delay

° CORRECTIVE ACTION   Complete
Design Change   ☒
Revise Operating Instr. ☐
Additional Training   ☐
Revise P.M. Schedule   ☐

Follow Up Action Required: __NONE. HAVE__
__CHANGED TO SCH. 160 PIPING AND HAVE__
__SEAL WELDED__

° DIRECT COST DATA:
Estimate ☒
Actual ☐

	Field	Central Shop	Contract	Subtotals
Labor	2000	4000		6000
Material	200	2000		2200

Product Loss _____ x _____ = _____
                 Hours $/Hour

Total = __8200__

° BASIC CAUSE OF FAILURE:

1. °MAINTENANCE
   A. _____ Improper Repair
   B. _____ Inadequate P.M.
   C. _____ Faulty analysis of trouble
   D. _____ Other
   E. _____ Design and/or selection
   F. _____ Improper field installation

2. °OPERATING
   A. _____ Overload
   B. _____ Improper start-up, shutdown, or oper. tech.
   C. _____ Lubrication lacking, wrong, contamination
   D. _____ Failure to release for scheduled P.M.
   E. _____ Other

3. °NOT SUITED FOR SERVICE DUE TO
   A. ✓ Faulty design assumptions
   B. _____ Incorrect material
   C. _____ Change in oper. cond. since design
   D. _____ Unexpected corrosion
   E. _____ Other

4. °MANUFACTURING DEFECTS
   A. _____ Design
   B. _____ Workmanship
   C. _____ Material
   D. _____ Other

5. °END OF DESIGN LIFE
   A. _____ Normal Wear
   B. _____ Reached retiring thickness
   C. _____ Other

Distribution List:

__D. P. SONDERBERG__
__S. A. YONDREWICZ__
__F. T. ARBENZ__

Analysis By __Elmer Breckhorn Jones__
Operating Supervisor __George Thublixuwon III__

Figure 10-1. Failure analysis form used in pump failure reduction program.

Review of Hydraulic Parameters

1. Examine hydraulic unbalance potential of single volute pumps at low flow.
2. Verify that low flow operation is not a problem (e.g., excessive heat build-up, internal recirculation, thermal distortion, vaporization of fluid, etc., have all been ruled out).

Metallurgy Review

1. Examine component condition. Initiate comparison study, as required.
2. Verify that wear ring materials and clearances reflect best applicable experience. High-temperature pump clearances may have to be as large as [(API clearance) + (0.02 x °F)] mils, diametral.

Installation Review

1. Evidence of pipe stresses (due to routing, non-sliding supports, incorrect hangers, etc.)
2. Coupling misalignment and lockup problems.
3. Baseplate resonance, out-of-parallelism, grout defects.
4. Preferential flow, if suction elbows are too close to double-suction pump.

Bearing and Lubrication Review

1. Ascertain bearing fits comply with applicable checklist.
2. Verify filler notch bearings have been replaced by angular contact bearings per applicable checklist. No spacers between duplex rows!
3. Check to see that thermal expansion of bearing outer race is not restricted by coolant surrounding it.
4. Calculate bearing B-10 life under normal operating conditions.
5. Verify that oil-mist lubrication reaches every row of bearing balls. Directed-mist fittings may be required instead of plain-mist application fittings. Consult Reference 19 for specific guidance.
6. Verify that motor windage does not impair effective venting of oil mist.
7. Ensure that approved grade oils only are used in oil-mist lubrication systems. Recall that paraffinic mineral oils and oils with high pour point can cause plugging of applicator fittings. Consider approved synthetic lube at temperatures below 25°F.

Seal and Quench Gland Design and Troubleshooting Review

1. Verify API flush plan conforms to applicable experience. Reverse flush may not be possible if balance holes are used (see earlier comment).
2. Ascertain flush fluid has adequate lubricity, vapor pressure at seal face temperature, supply quantity and pressure, and is free from solid particles. How about during abnormal operation? At startup?

3. Tandem seal lubrication and installation details should conform to latest experience. Consult pages 257-264 of Reference 20.
4. Seal installation tolerances should be verified.
5. Verify that only balanced seals are used at pressures above 75 psi.
6. Ascertain that quench steam or water is applied only if required to keep solid particles from forming external to seal components.

Tackle your mechanical seal troubleshooting tasks by:

1. Examining the entire seal.
2. Examining the wear track.
3. Examining the faces.
4. Inspecting the seal drive.
5. Checking spring condition.
6. Checking the elastomer.
7. Checking for rubbing.

Vertical Pump Considerations

1. Verify that critical speed is not a problem.
2. Check for proper lubrication of column bearings.
3. Some vertical pumps may need vortex baffles in suction bell.

Safety Considerations Must Not Be Overlooked

1. If minimum flow bypasses are used, verify each pump has its own bypass arrangement.
2. Do not allow excessively oversized pumps.
3. Verify isolation valves are accessible in case of fire. Are weight-closing valves possible?
4. Piping must be braced and gusseted. Flexible instrument connections may be advisable—but only if well engineered.
5. Is onstream condition monitoring instrumentation feasible? Should high- or low-frequency vibration monitoring be used? Consult Reference 20, pages 292-343.
6. Verify that substantial coupling guards are used.

Armed with this checklist and after reading the reference material cited, plant engineers and mechanical technical service personnel should feel confident in their ability to diagnose recurring pump problems.

Program Results

The results of centrifugal pump failure reduction programs are generally very significant. One large plant reported a 29% reduction in failures after the first year. Although the failure reduction rates and percentages leveled off at 37% after three years, the ensuing savings exceeded this percentage because the reductions were concentrated on many maintenance-intensive problem pumps.

Failure rate comparisons are given in Figure 10-2 for a medium-size chemical plant. They are again quite representative of typical first-year comparisons.

Conclusion

Experience shows that significant reductions in centrifugal pump failure events are possible without, in most cases, adding to the work force of typical petrochemical plants. The key to the successful implementation of a failure analysis and reduction program can be found in management commitment, thorough involvement of plant technical personnel, and good documentation.

Management commitment manifests itself in many ways. Briefing sessions and reporting requirements can be arranged so as to lend high visibility to the program. The thoroughness of plant engineering and technical personnel is exemplified by checklist and procedure development and by their ability to convince mechanical or field work forces of the need to follow these guidelines conscientiously. Finally, the importance of good documentation cannot be overemphasized. Computerized failure records are ideally suited to keep track of a pump failure reduction program,

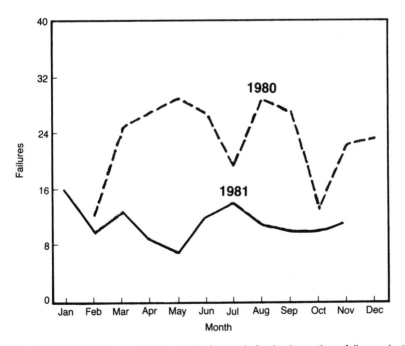

Figure 10-2. Centrifugal pump failure rates before and after implementing a failure reduction program in a medium-size chemical plant.

but the computer output can only be as good as the input. Manual data keeping is perfectly acceptable. Close cooperation between technical and clerical staff have gone a long way toward ensuring the overall success which, happily, can be reported for centrifugal pump failure reduction programs using the approach outlined here.

References

1. Bloch, H. P., "Machinery Documentation Requirements For Process Plants," ASME Paper 81-WA/Mgt-2, presented in Washington D.C., Nov. 1981.

2. Sangerhausen, C. R., "Mechanical Seals In Light Hydrocarbon Service: Design and Commissioning of Installations For Reliability," paper presented at ASME Petroleum Mechanical Engineering Conference, Denver, Colorado, September 14-16, 1980.

3. SKF Bearing Company, *Engineering Data Manual,* King of Prussia, Pennsylvania.

4. James, R., "Pump Maintenance," *Chemical Engineering Progress,* Feb. 1976, pp. 35-40.

5. Taylor, I., "The Most Persistent Pump—Application Problem For Petroleum and Power Engineers," ASME Paper No. 77-Pet 5, presented in Houston, September 1977.

6. Nelson, W. E., "Pump Curves Can Be Deceptive," *Proceedings of NPRA Refinery and Petrochemical Plant Engineering Conference,* 1980, Pages 141-150.

7. Eeds, J. M., Ingram, J. H., and Moses, S. T., "Mechanical Seal Applications—A User's Viewpoint," *Proceedings of the Sixth Turbomachinery Symposium,* Texas A&M University, Pages 171-185, 1977.

8. Bush, Fraser, Karassik, "Coping With Pump Progress: The Sources and Solutions of Centrifugal Pump Pulsations, Surges and Vibrations." *Pump World,* 1976, Volume 2, Number 1, pp. 13-19 (A publication of Worthington Group, McGraw-Edison Company).

9. McQueen, R., "Minimum Flow Requirements for Centrifugal Pumps." *Pump World,* 1980, Volume 6, Number 2, pp. 10-15 (Ibid).

10. Bloch, H. P., "Improve Safety and Reliability of Pumps and Drivers," *Hydrocarbon Processing,* January through May, 1977.

11. Bloch, H. P., "Optimized Lubrication of Antifriction Bearings for Centrifugal Pumps," ASLE Paper No. 78-AM-1D-1, presented in Dearborn, Michigan, April 17, 1978.

12. Sprinker, E. K. and Patterson, F. M., "Experimental Investigation of Critical Submergence For Vortexing in a Vertical Cylindrical Tank," ASME Paper 69-FE-49, June, 1969.

13. Bloch, H. P., "Mechanical Reliability Review of Centrifugal Pumps For Petrochemical Services," Presented at the 1981 ASME Failure Prevention and Reliability Conference, Hartford, Connecticut, September 20-23, 1981.

14. Bussemaker, E. J., "Design Aspects of Baseplates For Oil and Petrochemical Industry Pumps," IMechE (UK) Paper C45/81, Pages 135-141.

15. Karassik, Igor J., "So, You Are Short on NPSH?" Presented at Pacific Energy Association Pump Workshop, Ventura, California, March, 1979.

16. Makay, Elemer and Szamody, Olaf, "Survey of Feed Pump Outages," FP-754 Research Project 641 for Electric Power Research Institute, Palo Alto, Ca. 1978.

17. Fraser, W. H., "Flow Recirculation in Centrifugal Pumps," *Proceedings of the Tenth Turbomachinery Symposium,* Texas A&M University, College Station, Texas, Dec. 1981.

18. Ingram, J. H., "Pump Reliability—Where Do You Start?" Presented at ASME Petroleum Mechanical Engineering Workshop and Conference, Dallas, Texas, September 13-15, 1981.

19. Bloch, H. P., "Large Scale Application of Pure Oil Mist Lubrication In Petrochemical Plants," ASME Paper 80-C2/Lub 25, Presented at Century 2 ASME/ASLE International Lubrication Conference, San Francisco, California, August 18-21, 1980.

20. Bloch, H. P., *Improving Machinery Reliability,* Gulf Publishing Company, Houston, Texas, 1982, Pages 257-264.

Appendix A

Machinery Equipment Life Data[1,2]

Table A-1
Life Spans of Selected Machinery Components and Equipment

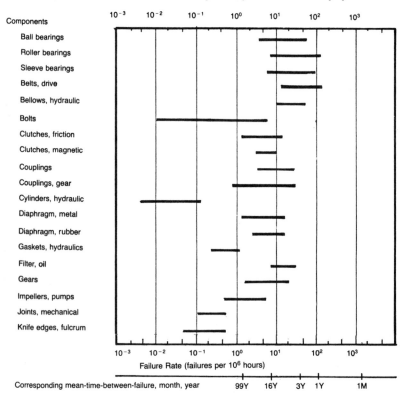

Components

	10^{-3}	10^{-2}	10^{-1}	10^0	10^1	10^2	10^3

Ball bearings
Roller bearings
Sleeve bearings
Belts, drive
Bellows, hydraulic

Bolts
Clutches, friction
Clutches, magnetic
Couplings
Couplings, gear
Cylinders, hydraulic
Diaphragm, metal
Diaphragm, rubber
Gaskets, hydraulics
Filter, oil
Gears
Impellers, pumps
Joints, mechanical
Knife edges, fulcrum

10^{-3} 10^{-2} 10^{-1} 10^0 10^1 10^2 10^3

Failure Rate (failures per 10^6 hours)

Corresponding mean-time-between-failure, month, year   99Y   16Y   3Y   1Y   1M

628

Table A-1 (cont.)

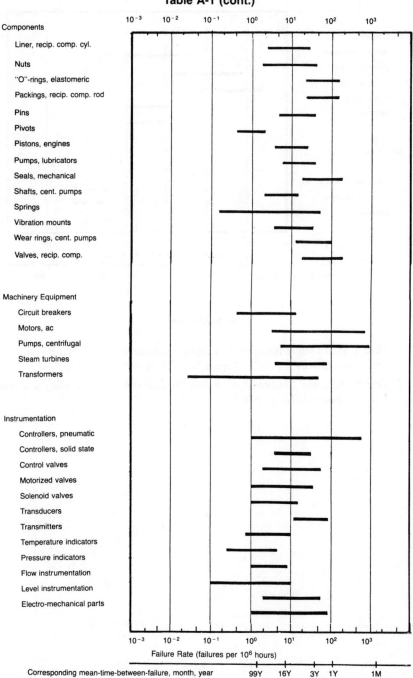

Failure Rate (failures per 10⁶ hours)

Corresponding mean-time-between-failure, month, year

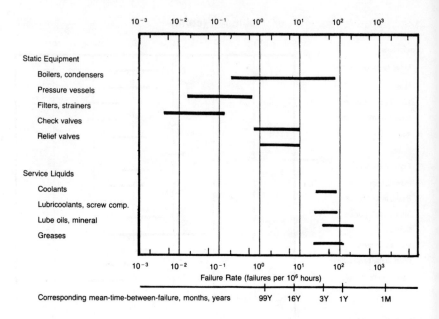

References

1. Hauck, D., "Failure Rates of Mechanical components for Nuclear Reactors: A Literature Survey," AECL Report, No. CRNL-739 (SP-R-10).

2. Green, A.E. and Bourne, A.J., *Reliability Technology,* John Wiley and Sons Ltd., London, 1972, p. 538.

Appendix B

Theory of Hazard Plotting*

This appendix contains a presentation of theory underlying hazard plotting. Also, theoretical properties of the exponential, Weibull, normal, log normal, and extreme-value distributions are presented briefly to aid us in deciding which hazard paper to use for a set of data. In the formulas given in the appendix, ln x denotes the natural logarithm (base e) of the number x, and log x denotes the common logarithm (base 10).

General Theory and Application

General theory underlying hazard plotting is presented here and then applied to the exponential, Weibull, normal, log normal and extreme-value distributions.

A probability model for the distribution of time to failure is given by a cumulative distribution function $F(x)$, which is the probability of failure by time x. $F(x)$ is defined for $x \geq 0$, where time zero is the starting time, and increases from zero to one as x goes from zero to ∞. The derivative of a cumulative distribution function is called a probability density function and denoted by $f(x)$. That is,

$$f(x) = \frac{d}{dx} F(x) \qquad \text{(B1)}$$

or, equivalently,

$$F(x) = \int_0^x f(x) \, dx \qquad \text{(B2)}$$

---

*See Chapter 7, Reference 10. By permission.

The hazard function $h(x)$ of a distribution is defined by $x \geq 0$ by

$$h(x) = \frac{f(x)}{1 - F(x)} \tag{B3}$$

and is the conditional probability of failure at time x given that failure has not occurred before then. In other words, $h(x)$** is the instantaneous failure rate at time x; that is, in a short time Δ from x to $x + \Delta$, a proportion $\Delta \times h(x)$ of the units at age x can be expected to fail. The cumulative hazard function of a distribution is defined for $x \geq 0$ by

$$H(x) = \int_0^x h(x)\ dx \tag{B4}$$

From (B3) and (B4), one gets

$$H(x) = -\ln\ [1 - F(x)] \tag{B5}$$

and, equivalently,

$$F(x) = 1 - e^{-H(x)} \tag{B6}$$

This relationship allows one to convert probability paper for a distribution into hazard paper. A cumulative hazard value H is located on probability paper at the cumulative probability value F given by (B6). For example, the cumulative hazard value 100% is located on probability paper at the cumulative probability value $F = 1 - e^{-1.00} = 63.2\%$. The relationship (B6) between the probability and hazard scales can be seen on all hazard papers. In actuality, the hazard and probability scales on the hazard papers were obtained from numerical calculations, rather than probability paper, to ensure accuracy.

For small values of the cumulative hazard function $H(x)$, (B6) simplifies to the approximation

$$F(x) \approx H(x) \tag{B7}$$

That is, the cumulative probability value and the cumulative hazard value coincide for small probabilities, say, less than a few percent. This can be seen on the various hazard papers.

The principle underlying plotting on hazard paper is essentially that underlying plotting on probability paper. For a complete sample of n failures plotted on probability paper, the increase in the sample cumulative probability distribution function at a failure time is equal to the probability $1/n$ for the failure. Then the sample cumulative distribution function, based on the sum of the probabilities of failure, approximates the theoretical cumulative distribution function, which is the

---

**h(x)$ is also referred to as $r(t)$; see Figure 7-1.

integral of the probability density. Similarly, for a sample plotted on hazard paper, the increase in the sample cumulative hazard function at a failure time is equal to the conditional probability $1/K$ for the failure, which is equal to one divided by the number K of units in operation at the time of the failure, including the failed unit. Then the sample cumulative hazard function, based on the sum of the conditional probabilities of failure, approximates the theoretical cumulative hazard function, which is the integral of the conditional probability of failure.

It must be emphasized that the hazard plotting method is based on the assumption that the censoring time associated with a censored unit is statistically independent of the time when the unit would fail. In other words, the distribution of potential failure times of censored units is the same as the distribution of failure times of uncensored units.

The theoretical distributions and their hazard plotting papers are presented next. These results are obtained from this general theory.

Exponential Distribution

The exponential cumulative distribution function is

$$F(x) = 1 - e^{-x/\theta} \quad x \geq 0 \tag{B8}$$

where $\theta > 0$ is the value of the mean time to failure. The probability density function is

$$f(x) = \frac{1}{\theta} e^{-x/\theta} \qquad x \geq 0 \tag{B9}$$

The hazard function is

$$h(x) = 1/\theta \qquad x \geq 0 \tag{B10}$$

which is constant over time. $\lambda = 1/\theta$ is called the failure rate. Because the failure rate (B10) for the exponential distribution is constant over time, the distribution is suitable for "random failures" rather than for wearout or infant mortality failures. The cumulative hazard function is

$$H(x) = x/\theta \qquad x \geq 0 \tag{B11}$$

Thus, from (B11), time to failure x as a function of the cumulative hazard H is

$$x(H) = \theta H \tag{B12}$$

Since time to failure is a linear function of the cumulative hazard, x plots as a straight-line function of H on square-grid graph paper. Thus, exponential hazard paper is merely square-grid graph paper. The slope of the line is the mean time to failure θ of the distribution. More simply, θ is the value of the time for which $H = 1$; this fact is used to estimate θ from data plotted on exponential hazard paper.

For more detailed discussions of the theoretical properties of the exponential distribution, refer to Reference 1. For sketches of the previous functions for the exponential distribution, see Figure B1 for the functions labeled $\beta = 1$, substituting θ for α.

Weibull Distribution

The Weibull cumulative distribution function is

$$F(x) = 1 - e^{-(x/\alpha)^\beta} \qquad x > 0 \tag{B13}$$

where $\beta > 0$ is the value of the shape parameter of the distribution and $\alpha > 0$ is the value of the scale parameter. α is a characteristic time to failure, since it is the $100 \times (1 - e^{-1}) = 63.2$ percentile of the distribution regardless what the value of the shape parameter is. As a special case with $\beta = 1$, the Weibull reduces to the exponential distribution. The probability density function is

$$f(x) = \frac{\beta}{\alpha^\beta} x^{\beta-1} e^{-(x/\alpha)^\beta} \qquad x > 0 \tag{B14}$$

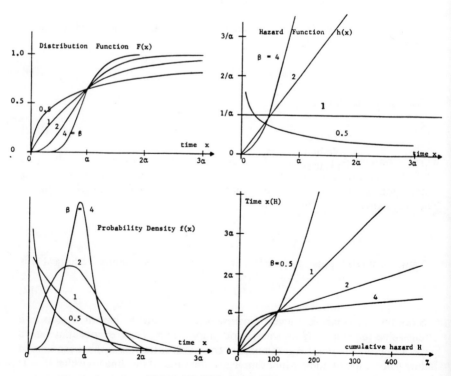

Figure B1. Weibull distributions.

The hazard function is

$$h(x) = \frac{\beta}{\alpha^\beta} x^{\beta-1} \qquad x \geq 0 \qquad \text{(B15)}$$

which is just a power of the time x. The failure rate $h(x)$ increases with time for values of the shape parameter β greater than one and decreases with time for values less than one. Thus, the Weibull distribution is flexible enough to describe either an increasing or a decreasing failure rate according to the value of the shape parameter. For $\beta = 1$, the exponential distribution results and the failure rate is constant. The Weibull distribution is an extreme-value distribution for the smallest time to failure from a large sample of times from some failure distribution. In other words, if a unit can be regarded as a system of many parts (each with a failure time from the same distribution) and if the unit fails with the first part failure, then the Weibull distribution may be appropriate for time to failure of such units. The cumulative hazard function is

$$H(x) = (x/\alpha)^\beta \qquad x \geq 0 \qquad \text{(B16)}$$

which is a power function of time x. Then, from (B16), time to failure x as a function of the cumulative hazard H is

$$x(H) = \alpha H^{1/\beta} \qquad \text{(B17)}$$

or, taking logarithms (base 10),

$$\log x = (1/\beta) \log H + \log \alpha \qquad \text{(B18)}$$

By (B18), x plots as a straight-line function of H on log-log graph paper, since $\log x$ is a linear function of $\log H$. Thus, Weibull hazard paper is merely log-log graph paper. The slope of the straight line equals $1/\beta$; this fact is used to estimate β from data plotted on Weibull hazard paper. Also, for $H = 1$, the corresponding time x equals α; this fact is used to estimate α from data plotted on Weibull hazard paper. For sketches of the functions for Weibull distributions, see Figure B1.

Normal Distribution

The normal cumulative distribution function is

$$F(x) = \frac{1}{\sigma \sqrt{2\pi}} \int_{-\infty}^{x} e^{-(x-\mu)^2/2\sigma^2} \, dx \quad -\infty < x < \infty \qquad \text{(B19)}$$

where μ is the value of the mean time to failure and is the center of the distribution, and $\sigma > 0$ is the value of the standard deviation of the distribution. This distribution is defined for all values of time to failure x both positive and negative. If μ is at least two or three times as great as σ, then the distribution gives a probability of failure at a negative time that is small enough to be neglected, and the distribution may serve as a satisfactory model for time to failure. The probability density function is

$$f(x) = \frac{1}{\sigma \sqrt{2\pi}} \, e^{-(x-\mu)^2/2\sigma^2}$$

(B20)

The hazard function is

$$h(x) = \frac{f(x)}{1 - F(x)}^*$$

(B21)

where $F(x)$ and $f(x)$ are given by (B19) and (B20). The failure rate is an increasing function of time. Because the failure rate for the normal distribution is strictly increasing, this distribution may be appropriate to wearout types of failure. The normal distribution is symmetric about the mean value μ; this symmetry should be considered in deciding on the appropriateness of the distribution for a set of data.

The functions

$$\Phi(u) = \frac{1}{\sqrt{2\pi}} \int_{-\infty}^{u} e^{-u^2/2} \, du$$

(B22)

and

$$\varphi(u) = \frac{1}{\sqrt{2\pi}} \, e^{-u^2/2}$$

(B23)

are called the standard normal distribution function and probability density function and are widely tabulated. They are used to give

$$F(x) = \Phi\left(\frac{x-\mu}{\sigma}\right)$$

(B24)

$$f(x) = \frac{1}{\sigma} \varphi\left(\frac{x-\mu}{\sigma}\right)$$

(B25)

and

$$h(x) = \frac{\frac{1}{\sigma}\varphi\left(\frac{x-\mu}{\sigma}\right)}{1 - \Phi\left(\frac{x-\mu}{\sigma}\right)}$$

(B26)

---

*Also referred to as r(t).

The cumulative hazard function is

$$H(x) = -\ln\left(1 - \Phi\left(\frac{x - \mu}{\sigma}\right)\right) \tag{B27}$$

From (B27), time to failure x as a function of the cumulative hazard H is

$$x(H) = \mu + \sigma\Phi^{-1}(1 - e^{-H}) \tag{B28}$$

Φ^{-1} is the inverse of the standard normal distribution function Φ defined in (B22) and gives the values of the percentage points of the standard normal distribution. By (B28), time x plots as a straight-line function of $\Phi^{-1}(1 - e^{-H})$. Thus, on normal hazard paper, a cumulative hazard value H is located on the horizontal axis at the position $\Phi^{-1}(1 - e^{-H})$, and the vertical time scale is linear. The slope of the straight line equals the value of the standard deviation; this can be used to estimate σ from data plotted on normal hazard paper. Also, for $H = \ln 2$ or, equivalently, for the 50% point on the probability scale, the corresponding time x equals μ; this fact is used to estimate μ from data plotted on normal hazard paper.

For more detailed discussions of the theoretical properties of the normal distribution, refer to Reference 1. For sketches of the functions for the normal distributions, see Figure B2.

Log Normal Distribution

The log normal cumulative distribution function is

$$F(x) = \Phi\left(\frac{\log x - \mu}{\sigma}\right) \qquad x > 0 \tag{B29}$$

where the standard normal distribution function Φ is defined in (B22), μ is the value of the mean log time to failure, and $\sigma > 0$ is the value of the standard deviation of log time to failure. Note that $\log x$ is a base 10 logarithm. The probability density function is

$$f(x) = \frac{0.4343}{x\sigma} \varphi\left(\frac{\log x - \mu}{\sigma}\right) \qquad x > 0 \tag{B30}$$

where the standard normal density φ is defined in (B23). The hazard function is

$$h(x) = \frac{\dfrac{0.4343}{x\sigma} \varphi\left(\dfrac{\log x - \mu}{\sigma}\right)}{1 - \Phi\left(\dfrac{\log x - \mu}{\sigma}\right)} \qquad x > 0 \tag{B31}$$

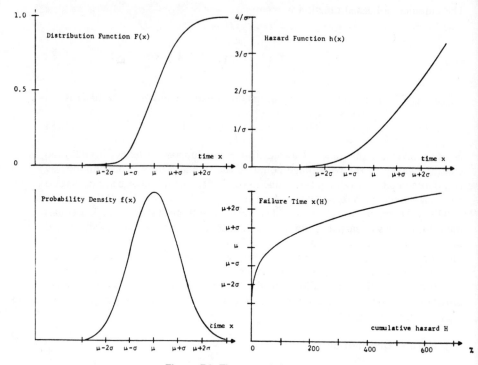

Figure B2. The normal distribution.

The failure rate (B31) is zero at time zero, increases with time to a maximum, and then decreases back down to zero with increasing time. The failure rate decreasing to zero after some time is a property of the log normal distribution that may make it an unsuitable model for wearout failure. Weibull cumulative distribution functions with a shape parameter value less than two are close to certain log normal distribution functions in the lower tail. Thus, it may be difficult to choose between a Weibull plot and a log normal plot for a set of data, since both may fit the data equally well in the lower tail. This may be a problem, since the two distributions can lead to different conclusions about the failure rate in the upper tail of the distribution of time to failure. The cumulative hazard function is

$$H(x) = -\ln\left(1 - \Phi\left(\frac{\log x - \mu}{\sigma}\right)\right) \qquad x > 0 \qquad (B32)$$

Note that base e and base 10 logarithms are used in (B32). From (B32), the log of the failure time x, as a function of the cumulative hazard H, is

$$\log x(H) = \mu + \sigma \Phi^{-1}(1 - e^{-H}) \tag{B33}$$

where Φ^{-1} is the inverse of the standard normal distribution function Φ defined by (B22). By (B33), $\log x(H)$ is a straight-line function of $\Phi^{-1}(1 - e^{-H})$. Thus, normal hazard paper also serves as log normal hazard paper, if the vertical time scale is made logarithmic rather than linear. The slope of the straight line is the value of the parameter σ; this fact can be used to estimate σ from data plotted on log normal hazard paper. Also, for $H = \ln 2$ or, equivalently, for the 50% point on the probability scale, the corresponding $\log x$ equals the value μ which can be read directly from the logarithm scale; this fact is used to estimate μ from data plotted on log normal hazard paper.

For more detailed discussions of the theoretical properties of the log normal distribution, refer to Reference 1. For sketches of the functions for log normal distributions, see Figure B3.

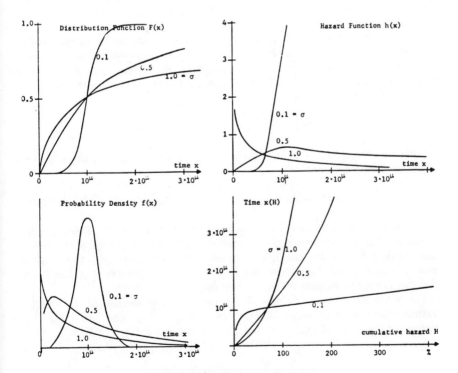

Figure B3. Log normal distributions.

Smallest Extreme-Value Distribution

The cumulative distribution function of the smallest extreme-value distribution is

$$F(x) = 1 - \exp\left[-e^{(x-b)/a}\right] \quad -\infty < x < \infty \tag{B34}$$

where b is the value of location parameter of the distribution, and $a > 0$ is the value of the scale parameter. b is a characteristic time to failure, since it is the $100 \times (1 - e^{-1})$ = 63.2 percentile of the distribution regardless what the value of the scale parameter is. The probability density function is

$$f(x) = \frac{1}{a} \, e^{(x-b)/a} \exp\left[-e^{(x-b)/a}\right] \quad -\infty < x < \infty \tag{B35}$$

The hazard function is

$$h(x) = \frac{1}{a} e^{(x-b)/a} \quad -\infty < x < \infty \tag{B36}$$

By (B36), the failure rate for the smallest extreme-value distribution increases exponentially with time. Like the Weibull, the smallest extreme-value distribution is an extreme-value distribution for the smallest time to failure for a large sample of times from some failure distribution which is far from the time origin. In other words, if a unit can be regarded as a system of identical parts with times to failure from a distribution and if the unit fails with the first part failure, then the smallest extreme-value distribution may be appropriate for time to failure of the units. The cumulative hazard function is

$$H(x) = e^{(x-b)/a} \quad -\infty < x < \infty \tag{B37}$$

which is an exponential function of time. From (B37), time to failure x as a function of the cumulative hazard H is

$$x(H) = b + a \ln H = b + 2.3026 \, a \log H \tag{B38}$$

By (B38), time x plots as a straight-line function of $\log H$. Thus, smallest extreme-value hazard paper is merely semilog graph paper. The slope of the straight line equals $2.3026 \, a$; this fact is used to estimate a from data plotted on smallest extreme-value hazard paper. Also, for $H = 1$, the corresponding time x equals b; this fact is used to estimate b from plotted data.

For more detailed discussions of the theoretical properties of the smallest extreme-value distribution, see References 1 and 2. For sketches of the functions above for the smallest extreme-value distribution, see Figure B4.

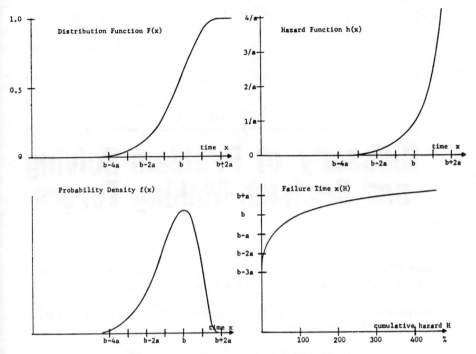

Figure B4. Smallest extreme-value distribution.

References

1. Hahn, G. J. and Shapiro, S. S., *Statistical Models in Engineering,* Wiley, New York, 1967, Chapter 3.
2. Gumbel, E. J., *Statistics of Extremes,* Columbia University Press, New York, 1958.

Appendix C

Glossary of Problem-Solving and Decision-Making Terms

Action Plan   This is what you intend to do. It should include your planned action, against what, how much, and when started and completed.

Adaptive Action   Steps taken to allow achievement of goals in spite of a trouble. You live with it, rather than fix it.

Ex.: A man with a bad leg may decide to buy a cane rather than get the leg fixed.

Cause   Usually refers to a past event or events which resulted in a problem. It is often a compound series of things which created a specific non-standard effect.

Cause Analysis   The systematic search for the cause of a problem. It is a questioning technique to gather the relevant facts and analyze them.

Contingent Action   Your fall-back plan. A plan set up ahead of time to be implemented if trouble occurs.

Ex.: A sprinkler system is set up ahead of time. It does not operate unless there is a fire.

Decision Analysis   A systematic technique to select the action which best fits your criteria or your objectives.

Fix   What you do to really take care of a trouble. It ought to be *safe*, be *legal*, and *remove the cause*. A good fix is not a patch and does not cause additional problems.

Interim Action   Temporary action to meet goals while the problem is worked on. A patch.

Ex.: Hiring temporary workers while cleaning up a spill.

Planning for Change   A systematic approach to identifying future troubles or opportunities and making an Action Plan to avoid them, reduce their effects, capitalize on them, or to reinforce their benefits.

Preventive Action   A plan to remove causes of future problems. (This is different from contingent action—here we want to avoid fire by removing its cause—for instance, piles of oily rags, etc.)

Situation Analysis   The breaking apart of a general situation into more specific problems, picking the most important ones and working on them. It generally includes selecting appropriate analysis techniques, if any.

Index

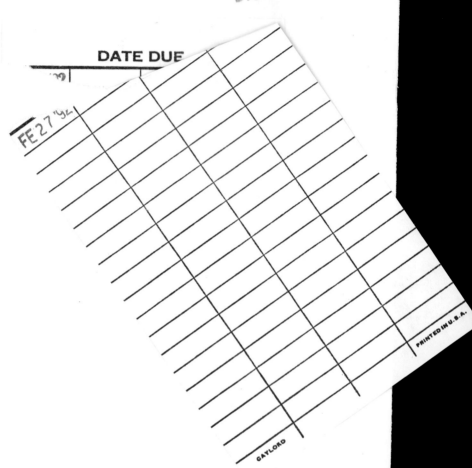

DATE DUE

FE 27 '92

PRINTED IN U.S.A.

GAYLORD